Advances in Applied Mechanics

Volume 27

Contributors to Volume 27

John P. Boyd

Ru Ling Chou

C. K. Chu

V. I. Fabrikant

Jorg Imberger

John C. Patterson

Chang-lin Tien

Viggo Tvergaard

Kambiz Vafai

ADVANCES IN
APPLIED MECHANICS

Edited by

John W. Hutchinson

DIVISION OF APPLIED SCIENCES
HARVARD UNIVERSITY
CAMBRIDGE, MASSACHUSETTS

Theodore Y. Wu

DIVISION OF ENGINEERING AND APPLIED SCIENCE
CALIFORNIA INSTITUTE OF TECHNOLOGY
PASADENA, CALIFORNIA

VOLUME 27

ACADEMIC PRESS, INC.
Harcourt Brace Jovanovich, Publishers
Boston San Diego New York Berkeley
London Sydney Tokyo Toronto

THIS BOOK IS PRINTED ON ACID-FREE PAPER. ♾

ACADEMIC PRESS, INC.
1250 Sixth Avenue, San Diego, CA 92101

United Kingdom Edition published by
ACADEMIC PRESS LTD.
24–28 Oval Road, London NW1 7DX

LIBRARY OF CONGRESS CATALOG CARD NUMBER: 48-8503

ISBN 0-12-002027-0

PRINTED IN THE UNITED STATES OF AMERICA

89 90 91 92 9 8 7 6 5 4 3 2 1

Contents

List of Contributors

Numbers in parentheses indicate the pages on which the authors' contributions begin.

JOHN P. BOYD (1), Department of Atmospheric, Oceanic and Space Sciences and Laboratory for Scientific Computation, University of Michigan, 2455 Hayward Avenue, Ann Arbor, Michigan 48109

RU LING CHOU (283), Lamont Doherty Geological Observatory, Columbia University, New York, New York 10017 and NASA Goddard Institute for Space Studies, Greenbelt, Maryland 20771

C. K. CHU (283), Department of Applied Physics and Nuclear Engineering, Columbia University, New York, New York 10017

V. I. FABRIKANT (153), Department of Mechanical Engineering, Concordia University, Montreal, H3G 1M8, Canada

JÖRG IMBERGER (303), Centre for Water Research, The University of Western Australia, Nedlands, Western Australia 6009

JOHN C. PATTERSON (303), Centre for Water Research, The University of Western Australia, Nedlands, Western Australia 6009

CHANG-LIN TIEN (225), Department of Mechanical Engineering, University of California at Berkeley, Berkeley, California 94720

VIGGO TVERGAARD (83), Department of Solid Mechanics, The Technical University of Denmark, Lyngby, Denmark

KAMBIZ VAFAI (225), Department of Mechanical Engineering, 206 West 18th Avenue, Ohio State University, Columbus, Ohio 43210

Preface

This volume of *Advances in Applied Mechanics* contains six expository articles that are surveys of active development in several major fields of applied mechanics. John Boyd elucidates the limitations of the celebrated advances in mathematical methodology made in the modern theory of solitons and nonlinear waves as they apply to the determination of solutions to more general problems, such as those involving boundaries at finite distances. New directions in research are delineated for this field of great and increasing importance, with refreshing ideas furnished on how to cope with challenging problems that may arise from applications to physical oceanography and other soliton-bearing systems.

Viggo Tvergaard presents a micromechanical study on the fracture mechanism of material failure due to coalescence of microscopic voids in ductile and polycrystalline metals. This is a problem of significant basic value to the rapidly developing field of micromechanics as well as one of practical importance.

V. I. Fabrikant's article describes how a general solution can be derived for various punch and crack problems involving an isotropic elastic half-space, with or without axisymmetry, here exemplified with results available for some immediate and future applications. This method is of value not only to the study of elasticity, but to other branches of engineering science where potential theory is being used.

Chang-Lin Tien and Kambiz Vafai bring forth an authoritative essay on convective and radiative heat transfer and multiphase transport processes in porous media, with or without phase change. This is a subject that is known for its complexities and for its wide range of applicability in high technology.

In discussing the solitons induced by boundary conditions, C. K. Chu and Ru Ling Chou bring to our attention the urgent need for analytical methods for resolving boundary-value problems associated with the Korteweg-de Vries and other nonlinear evolution equations. This field is undoubtedly emerging as an area of great importance.

The article on physical limnology by Jörg Imberger and John C. Patterson is an in-depth study of the hydrodynamics of lakes, which involves physical processes no less complicated than those pertaining to the ocean. Yet as a discipline, physical limnology has received far less attention until only the last few years. In this field, we may find problems of significance to environmental issues of today.

We are indebted to these authors for their interest and enthusiasm in preparing these articles with great care to share with our readers their expertise and scholarly views.

T.Y.W. and J.W.H.

ADVANCES IN APPLIED MECHANICS, VOLUME 27

New Directions in Solitons and Nonlinear Periodic Waves: Polycnoidal Waves, Imbricated Solitons, Weakly Nonlocal Solitary Waves, and Numerical Boundary Value Algorithms

JOHN P. BOYD

Department of Atmospheric, Oceanic, and Space Sciences
and
Laboratory for Scientific Computation
University of Michigan
Ann Arbor, Michigan

ISBN 0-12-002027-0

I. Introduction

Until recently, the study of solitary waves has been dominated by three methodologies: (i) the inverse scattering transform, (ii) multiple scales perturbation theory, and (iii) numerical initial value codes. For a time, inverse scattering was widely regarded as the magic bullet of applied mathematics, the philosopher's stone that would turn all mystery into truth. Even after it became clear that this method was limited to a very special class of (mostly one-dimensional) wave equations, the enchantment continued. Almost all books on solitons, even now, are treatises on inverse scattering.

The method of multiple scales is important because it is the justification for inverse scattering. The perturbation theory shows that when waves are only weakly nonlinear and satisfy certain structural assumptions, such as a narrowly peaked Fourier spectrum or a very long length scale, their dynamics is *approximately* described by one of those very special wave equations that can be solved by inverse scattering. The combination of inverse scattering with multiple scales perturbation theory has been a high-powered microscope for viewing the hidden structure of nonlinear waves. Unfortunately, the real world is three-dimensional, and the collapse of three coordinates into one is legitimate only for weak nonlinearity. For overcoming the limits of perturbation theory and inverse scattering, the main tool has been numerical integration of the initial value problem.

Time-marching is a flexible and very general strategy because it requires no *a priori* assumptions about amplitudes, structure, or anything else. The emergence of solitons from very nonsolitonic initial conditions in one-, two-, and three-dimensional systems has been a compelling motive for the study of solitary waves.

Unfortunately, time-marching is also clumsy and inelegant because the numerical solution is always a potpourri of transients, dispersing wave-trains, and solitons. Time-marching is a kind of numerical zoology, trying to understand the beast by observing it in life. This is both useful and necessary, but there is still a compelling need for a kind of computational biochemistry to examine solitons in isolated, purified form.

In this review, we will describe four new strategies for exploring solitary and nonlinear periodic waves. All break away from the limitations of the "Big Three" methodologies to open new horizons for what is still a very lively field.

The first topic is the theory of polycnoidal waves. These are the spatially periodic generalizations of multiple solitons. Their theory explores what happens when solitary waves are not solitary, but instead form periodic wave trains.

In this review, we will emphasize four different methodologies for polycnoidal waves. The Nonlinear Fourier Transform and theta function formalism are both extensions of inverse scattering. In contrast, the Stokes series and the Newton/Fourier/continuation polyalgorithm are direct methods that bypass inverse scattering completely.

The second topic is superimposing solitons to generate spatially periodic waves, the method of "imbricate series." This is an offshoot of polycnoidal theory that has acquired a life of its own. Remarkably, inverse scattering-soluble equations often allow an *exact* nonlinear superposition principle. Imbricate series show that each soliton may behave as an independent dynamical system even when very close to other solitary waves.

The third topic is the direct numerical computation of multidimensional solitary waves. New methods—continuation in physical and artificial parameters, rational Chebyshev pseudospectral schemes, and preconditioned iterations—have enormously improved our capabilities. Instead of marching for thousands of time steps, waiting for the transients to disperse, the new iterative schemes, which can be interpreted as integrations in an artificial time variable, compute the soliton directly in only ten or twenty steps. Crude finite difference discretizations have been replaced by exponentially accurate pseudospectral algorithms. Truncated domains have been rendered obsolete by new spectral basis sets, such as rational Chebyshev functions, which compute on the infinite domain.

The fourth topic is an important generalization of the traditional concept of a solitary wave: the "weakly nonlocal" soliton. These are coherent structures that are stable and long-lived, but violate the classical definition of a soliton because they leak radiation to infinity. For the class known as "nanopterons," the leakage is exponentially small in the amplitude, so the lifetime of the nonlocal soliton may be longer than the age of the universe! Because of this exponential smallness, special analytical and numerical tools are needed. Section V shows that the necessary tool-making is well-advanced and gives seven examples of nonlocal solitons.

II. Polycnoidal Waves

A. Definition and Overview

The N-polycnoidal wave is an exact, spatially periodic solution to an integrable nonlinear wave equation. The special case $N=1$ is the ordinary cnoidal wave. This is defined to be a nonlinear, spatially periodic wave that translates at a constant phase speed c. The N-polycnoidal wave for $N>1$ is similar except that it is characterized by N independent phase speeds. The polycnoidal wave is the spatially periodic generalization of the cnoidal wave in the same way that the N-soliton solution generalizes the single, isolated solitary wave.

In this review, we shall concentrate on the polycnoidal waves of the Korteweg–deVries (KdV) equation

$$u_t + uu_x + u_{xxx} = 0. \tag{2.1}$$

This is the simplest, most studied, and best understood case. However, it is known that polycnoidal waves exist for other wave equations including the cubic Schroedinger (Ma and Ablowitz, 1981), the Toda lattice (Ferguson *et al.*, 1982), the sinh–Gordon equation (Forest and McLaughlin, 1983), and the sine–Gordon equation. The latter has been heavily studied: Forest and McLaughlin (1982), Zagrodzinski (1982, 1983), Zagrodzinski and Jaworski (1982), and Jaworski and Zagrodzinski (1982) are but a partial list.

The simplest nontrivial polycnoidal wave is the case $N=2$, which was dubbed the "double cnoidal" wave by Hyman (1978). Figure 1 illustrates its four regimes. In the limit of very small amplitudes,

$$u(x, t) \approx a \cos[k_1(x - c_1 t) + \phi_1] + b \cos[k_2(x - c_2 t) + \phi_2], \tag{2.2}$$

where a and b are amplitudes, k_1 and k_2 are wavenumbers, and ϕ_1 and ϕ_2 are phase constants. To $O(a^2, b^2)$, the two cosine terms are independently propagating linear waves. The shape of $u(x, t)$ changes with time as the crests and troughs of the two components, each traveling at its own speed, reinforce and cancel in turn.

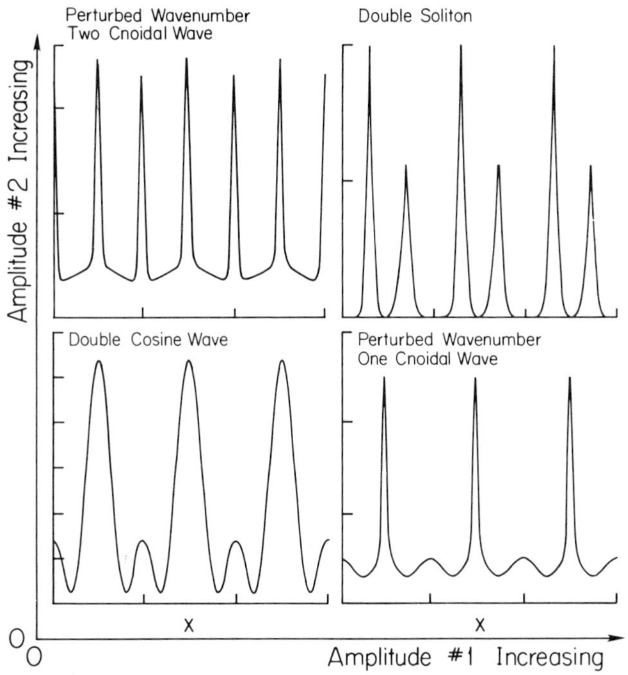

FIG. 1. Regimes of the double cnoidal wave. All 2-polycnoidal waves have these same four regimes, but the schematics show the shapes for the "principal" branch. Each diagram illustrates an interval that is three times the width of the spatial period.

In the opposite limit of very large amplitudes, the double cnoidal wave may be approximated by two solitary waves of different sizes:

$$u(x, t) \approx 12s_1^2 \operatorname{sech}^2[s_1(x - c_1 t) + \phi_1] + 12s_2^2 \operatorname{sech}^2[s_2(x - c_2 t) + \phi_2], \quad (2.3)$$

where s_1 and s_2 are the pseudowavenumbers and where (2.3) applies only when the solitons are separated. The peaks are very narrow in comparison to the spatial period ($s_1, s_2 \gg 1$), so the pair of solitons on one period interacts very weakly with the pair on the spatial period to the left or to the right. However, when the peaks overlap, one should replace (2.3) by the well–known KdV exact solution for the double solitary wave.

When $s_1 = s_2$, the double cnoidal wave degenerates into an ordinary cnoidal wave with half the period; near equality between the two pseudowavenumbers gives the "perturbed wavenumber two cnoidal wave" shown in Figure 1. The fourth regime is a cnoidal wave that is weakly perturbed by a sine wave.

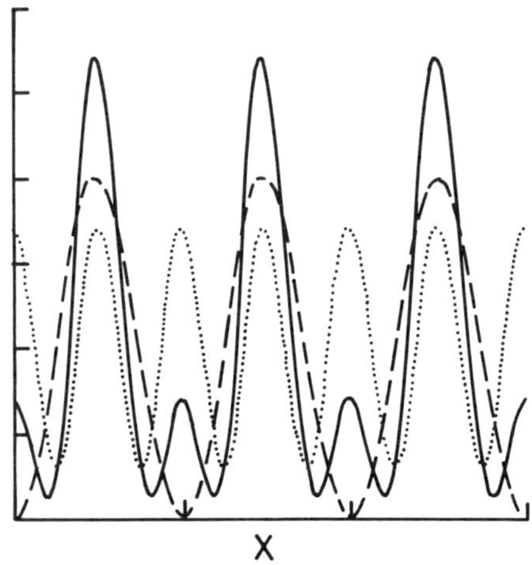

FIG. 2. The superposition of wavenumber one and two components to generate the small amplitude double cnoidal wave. The dashed and dotted curves are the individual components; the solid curve is the 2-polycnoidal wave at a time when the two cosine waves are in phase. Because each wave is travelling at its own phase speed (wavenumber two is roughly four times faster than wavenumber one for the KdV equation), the shape of the double cnoidal alters. However, the in-phase pattern shown here will periodically recur forever.

Figure 2 illustrates how the two Fourier components reinforce and interfere to create two peaks for small amplitudes. The relationship between the nonlinear normal modes and the two crests changes dramatically with amplitude. At large amplitudes, the nonlinear normal modes are the crests themselves. When the amplitude is small, however, the tall peak is the *sum* of the two normal modes (cosine waves) while the smaller crest is the *difference* between the two independently propagating waves.

It is striking, however, that the shape of $u(x, t)$ does *not* change with amplitude (along the diagonal in Figure 1) in the sense that there is always one tall peak and one short crest on each spatial period during at least part of each recurrence cycle. Because of this, the small amplitude and large amplitude approximations (Equations (2.2) and (2.3), respectively) strongly overlap. For intermediate amplitude, one may describe the double cnoidal wave as *either* the superposition of sine waves or as the sum of two solitons. The error is small for both because both approximations predict two unequal crests.

Strictly speaking, Figure 1 and Equation (2.3) describe only the "principal" branch of the double cnoidal wave. This branch has $k_2 = 2k_1$ in (2.2) and has precisely two solitons, one tall and one short, on each spatial period. Mathematically, one may have spatially periodic solutions with m tall solitons and n short solitons on each period, where m and n are arbitrary integers. Haupt and Boyd (1989) and Hyman (1978) have computed explicit examples where m is 1 and $n = 2$, 3, or 4. If the ratio k_2/k_1 is irrational, then the double cnoidal wave is almost-periodic in space. Forest and McLaughlin (1982) discuss possible applications of almost-periodic polycnoidal waves. Whenever we discuss a specific solution branch, however, it will be the principal branch as shown in Figure 1.

Polycnoidal waves have a bewildering variety of names. Mathematicians have used "finite band," "finite gap," "waves on a circle," and "N-phase wavetrains" as synonyms for "polycnoidal." The last of these is very descriptive because the N-polycnoidal wave can always be written in terms of N independent phase variables

$$\zeta_i \equiv k_i(x - c_i t) + \phi_i, \qquad i = 1, \ldots, N, \qquad (\text{"phase variables"}) \quad (2.4)$$

each with its own independent wavenumber and phase speed. However, we shall always refer to these solutions as "N-polycnoidal" waves.

One major issue is that polycnoidal waves are obviously very special. How relevant are these special waves to the *general* spatially periodic solution? The answer is that any $u(x, t)$ may be approximated to arbitrarily high accuracy for an arbitrary, finite time interval by an N-polycnoidal wave of sufficiently large N and appropriately chosen parameters. Thus, to understand the polycnoidal special solutions is to understand the general solution, too.

Polycnoidal methodologies include the following:

(i) Variational Principle
(ii) Nonlinear Fourier Transform
(iii) Theta Function Residual Equations
  (a) Numerical solutions
  (b) Perturbative solutions
(iv) Phase Variable Boundary Value Problem
  (a) Stokes (Poincaré–Lindstedt) series
  (b) Newton/Fourier/continuation computations.

In what follows, we shall discuss each in turn.

B. Variational Principle

The Korteweg–deVries equation has an infinite number of independent, conserved integrals (Ablowitz and Segur, 1981). Lax (1976) showed that these provide a variational characterization of the N-polycnoidal wave. If the first $N-1$ of these invariants are fixed, then any solution which minimizes the N-th invariant is an N-polycnoidal wave. The minimizing solution is not unique because the polycnoidal wave is a function of N phase variables; as t increases, each phase variable increases linearly, too, and the shape of $u(x, t)$ alters. Thus, the minimizing solution is unique except for N arbitrary phase parameters.

Hyman (1978) translated Lax's principle into FORTRAN. He discretized the functionals and performed a constrained minimization of the N-th functional. This crude solution was refined by solving an ordinary differential equation, the usual Euler equation of variational calculus.

Unfortunately, all his solutions had either four or five solitons on each spatial period—typically one large solitary wave and three or four identical shorter, broader peaks. He missed the "principal" branch and did not thoroughly explore the parameter space.

Nevertheless, this approach has interesting possibilities. Its strength is that one may compute N-polycnoidal waves for large N by solving an ordinary differential equation (albeit one of order $2N+1$). However, Hyman's work has never been continued.

C. Nonlinear Fourier Transform ("Hill's Spectrum" or "Generalized Inverse Scattering")

The early work on polycnoidal waves was the exclusive province of pure mathematics. The result was many beautiful and intricate formulas, existence proofs, and absolutely no attempt to actually calculate anything. Very recently, however, Osborne and Bergamasco have developed a fast algorithm for both the inverse scattering transform on the infinite interval and its generalization for spatial periodicity. They call the latter the "Nonlinear Fourier Transform."

Unfortunately, a full explanation of their algorithm requires a heavy dose of spectral theory, far beyond what is feasible to include in this review. The computer program, however, is not particularly long or complex. Osborne and Bergamasco's two numerical articles have not yet been published, but see Osborne and Bergamasco (1985, 1986, 1987) and Tracy, Larson, Osborne, and Bergamasco (1988, 1989).

In brief, their algorithm computes a set of amplitudes from an arbitrary initial condition. When $|u(x, t=0)| \ll 1$, these amplitudes are simply the coefficients of the Fourier series of the initial condition. For larger solutions, however, some or all of the amplitudes become (in effect) the soliton pseudowavenumbers (the s_i in (2.3)) that determine the width and amplitude of each solitary wave. Thus, the Osborne–Bergamasco algorithm is a procedure for computing the amplitudes of normal modes—but it is a generalization of ordinary Fourier analysis because the normal modes are nonlinear. When a mode is sufficiently strong, it is not a single, steadily propagating sine wave but rather a single, steadily propagating soliton.

Of course, the word "amplitude" must be used with care because *all* the amplitudes must be *simultaneously* fed to the algorithm to compute the polycnoidal wave when two or more solitons are strongly interacting. Nonetheless, the Nonlinear Fourier Transform is still very powerful because one may *predict* the solitary waves (and smaller, quasi-sinusoidal components) that will emerge from an arbitrary, spatially periodic initial condition.

An amusing illustration is Osborne and Bergamasco's (1986) reanalysis of Zabusky and Kruskal (1965), the now-legendary note that first described the elastic collisions of Korteweg–deVries solitary waves and coined the word "soliton." Zabusky and Kruskal's initial sine wave broke up into nine solitons. By carefully tracing each solitary wave before, during, and after each soliton–soliton collision in the numerical output, they were able to plot the x–t trajectory of each solitary wave. In contrast, Osborne and Bergamasco (1986) were able to predict the number of solitions and the whole time-evolution directly from the initial condition. No Runge–Kutta or leapfrog for them!

Osborne and Bergamasco are currently applying their algorithm to deduce the spectrum of ocean long waves from surface height measurements. Another extension (Tracy *et al.*, 1988) is to examine the relationship between the Nonlinear Fourier Transforms of the Korteweg–deVries and Cubic Schroedinger equations: one equation passes into the other and vice-versa in the appropriate asymptotic limit (Johnson, 1976).

Thus, there is no doubt that this line of attack will continue. The studies by Osborne, Bergamasco, and their collaborators are the direct and logical extension of the vast amount of research on the inverse scattering transform for the infinite interval.

Unfortunately, the algorithm does have some drawbacks. First, it is not a good tool for exploring the parameter space of the double cnoidal or triple cnoidal waves. An arbitrary initial condition is an infinite genus polycnoidal wave (that is, an N-polycnoidal wave with $N \rightarrow \infty$). The Nonlinear Fourier Transform does not provide any obvious way to specialize the initial condition to a polycnoidal wave of finite N.

A second limitation is that the method is restricted to those very special wave equations which can be solved via the inverse scattering transform: the Kortweg–deVries, the sine–Gordon, and so on. It seems probable that there are quasi-polycnoidal waves for nonintegrable wave equations, too, but the Nonlinear Fourier Transform is useless for these.

The third and most serious limitation is that the theory behind the generalized inverse scattering method is very complicated. Few applied mathematicians or fluid dynamicists will have the stamina to master dozens of theorems and new terms to understand a technology whose range of applicability is limited—limited to the narrow parameteric range where perturbation theory can justify the Korteweg–deVries equation.

D. Theta Functions

Another remarkable property of N-polycnoidal waves is that they can be expressed in terms of the N-dimensional Riemann theta function (Its and Matveev, 1975). Although the theta series are infinite, all the Fourier coefficients may be expressed in terms of just $N(N+1)/2$ independent parameters. These are the elements of the *symmetric* N-dimensional "theta matrix." Because an infinite number of coefficients has been replaced by a small *finite* set of unknowns, theta functions permit some remarkable numerical and perturbative algorithms.

Theta functions do not themselves satisfy the KdV equation, but rather give the KdV solution through the transformation

$$u = 12(\ln[\theta])_{xx}. \tag{2.5}$$

This transformation was introduced by Hirota (1974) because it provides a remarkably compact way of writing the N-soliton solution to the KdV equation. He subsequently extended this formalism to most of the other inverse scattering-soluble wave equations as reviewed by Matsuno (1984).

Nakamura (1979) and Boyd (1982b) independently showed that the theta function satisfies Hirota's transformed bilinear version of the KdV equation. When the theta function is substituted into the Hirota–KdV equation, however, one is confronted by a paradox: an infinite number of Fourier terms must vanish, but the $N = 2$ theta solution contains only four parameters! The Boyd–Nakamura Theorem shows that when four independent terms in the residual function vanish, this implies that all the rest are zero, too: the theta formalism reduces a *partial* differential equation to a *small* set of *algebraic* equations ("theta residual" set).

For the double cnoidal wave, for example, the unknowns are the two phase speeds, c_1 and c_2, plus the off-diagonal element of theta matrix, T_{12}, and a physically irrelevant constant of integration. The diagonal elements of the theta matrix are parameters that are equivalent to the amplitudes of the two independent wave components or solitons.

One technical complication is that the Boyd–Nakamura Theorem reduces the Hirota–KdV equation to 2^N residual equations whereas the N-polycnoidal wave has only $[1 + (N + 1)N/2]$ parameters—more equations than parameters when $N > 2$. Hirota and Ito (1981) tried the bold expedient of solving 7 of the 8 residual equations for $N = 3$ and then verifying that the remaining equation was also satisfied to machine precision. The numerical evidence is that the "theta residual" equations are always soluble, but a rigorous proof is lacking.

A second technical complication is that for large amplitudes, the theta-Fourier series converges very slowly. Via the Poisson summation formula, however, the theta function may be alternatively expressed as a "Gaussian" or "imbricate" series. The lowest term is the KdV double soliton and the series converges more rapidly as the polycnoidal wave becomes larger. Boyd (1982b) shows that the Boyd–Nakamura Theorem applies to the Gaussian series, too. In the seven years since the review of Dubrovin (1981), polycnoidal wave theory has been extended in three directions. First, Nakamura (1980) and his collaborators (Nakamura and Matsuno, 1980) have generalized the theta formalism to many other wave equations besides the KdV equation (Matsuno, 1984).

Second, several groups of mathematicians have filled in the theoretical gaps, especially for equations other than the KdV. Tracy *et al.* (1989), Ferguson, Flaschka, and McLaughlin (1982), and Forest and McLaughlin (1982, 1983) are especially interesting. Zagrodzinski (1982, 1983), Zagrodzinski and Jaworski (1982), and Jaworski and Zagrodzinski (1982) apply the theta formalism to the sine–Gordon equation. Other interesting discussions are given in the reviews by Date and Tanaka (1976), Dubrovin (1981), and Zagrodzinski (1984).

Third, Boyd (1982b, 1984b, 1984c, 1984d) has computed perturbative solutions to the KdV theta-residual equations for both the small amplitude (Fourier) and large amplitude ("imbricate-Gaussian") theta series. These articles discuss a number of subtle details that must be omitted here. The beauty of the theta formalism is that one may express the complete answer via just three perturbation series, one for each of the following: c_1, c_2, and the off-diagonal theta matrix element. Each expansion is a double series in the two amplitudes (or equivalently, in the diagonal theta matrix elements).

This simplicity in the final answer is one of the great strengths of the theta formalism. The other is beauty; the multi-dimensional theta functions, conserved integrals, trace formulas, logarithmic transformations, and hyperelliptic functions fit together as grandly and as astonishingly as a painting of Jan Vermeer.

The great weakness of the theta formalism is that the simple answer is obtained only at the price of great complexity in the derivation, a weakness shared with the inverse scattering transform. This is not unusual in applied mathematics; the method of multiple scales, for example, is notorious for generating simple answers from long and tedious calculations.

The difference is that multiple scales is a very general tool, applicable to boundary layers and limit cycles as well as inviscid waves. The theta formalism, alas, is only applicable to a handful of differential equations which, on an infinite interval, can all be solved via inverse scattering. If one likened these tools to languages, the method of multiple scales would be a major tongue like French or English; the theta function theory would be Basque or the dialect of a bush tribe in New Guinea.

Because of its intricacy and narrow scope, the theta method and the Osborne–Bergamasco algorithm described in the previous section are likely to remain the province of specialists. In the next subsection, we shall discuss a much less elegant but more general approach to polycnoidal waves.

E. The Phase Variable Boundary Value Problem and the Stokes Series

The double cnoidal wave is a function only of the phase variables $X \equiv k_1(x - c_1 t) + \phi_1$ and $Y \equiv k_2(x - c_2 t) + \phi_2$. By substituting $u(X, Y)$ into the KdV equation and using the definitions of the phase variables, one finds that the double cnoidal wave must solve the Phase Variable Boundary Value Problem (PVBVP):

$$(u - c_1)u_X + 2(u - c_2)u_Y + u_{XXX} + 6u_{XXY} + 12u_{XYY} + 8u_{YYY} = 0. \quad (2.6)$$

The numerical coefficients in (2.6) are those for the "principal branch" with $k_1 = 1$, $k_2 = 2$, and a spatial period of 2π, but Haupt and Boyd (1989) also discuss the branch with $k_2 = 3$.

Equation (2.6) is rather unusual in that it is a nonlinear problem with *two* eigenvalues c_1 and c_2. Nonetheless, the problem is well-posed on both the periodic and infinite plane.

Haupt and Boyd (1989) discuss two ways of solving (2.6). The numerical Newton/Fourier/continuation polyalgorithm will be explained in the next section. The second method is the Stokes expansion, which is known in celestial mechanics as the Poincaré–Lindstedt series.

Stokes observed that in nonlinear wave problems, one must expand not only the wave height but also the phase speed—for the double cnoidal wave, *both* phase speeds—as functions of the amplitude a:

$$u = au_1 + a^2u_2 + a^3u_3 + \cdots, \quad (2.7a)$$

$$c_1 = c_{10} + ac_{11} + a^2c_{12} + \cdots, \quad (2.7b)$$

$$c_2 = c_{20} + ac_{21} + a^2c_{22} + \cdots, \quad (2.7c)$$

The lowest order solution is the sum of two sine waves of different wavenumbers ($k_1 \neq k_2$):

$$u_1 = \cos X + b \cos Y. \quad (2.8)$$

The amplitude parameters a and b are independent so that the double cnoidal wave is a two-parameter family of solutions. However, the most interesting case is when the two amplitudes are the same order of magnitude; otherwise, the double cnoidal wave is merely an ordinary cnoidal wave with a small perturbation. Consequently, we assume that b is $O(1)$ and use the amplitude of $\cos X$ as the order of magnitude for the amplitude of $\cos Y$, too. The perturbation series may be computed by substituting (2.7) into (2.6) and matching powers of a.

At each order, we solve

$$-c_{10}u_{i,X} - 2c_{20}u_{i,Y} + u_{i,XXX} + 6u_{i,XXY} + 12u_{i,XYY} + 8u_{i,YYY} = F_i(X, Y), \quad (2.9)$$

where

$$F_i(X, Y) = -\sum_{j=1}^{i-1} [(u_j - c_{1j})u_{i-j,X} + 2(u_j - c_{2j})u_{i-j,Y}]. \quad (2.10)$$

To solve (2.9), we expand both the solution and $F_i(X, Y)$ as Fourier series:

$$u_i(X, Y) = \sum_{m=0}^{i} \sum_{n=-i}^{i} u_{i,mn} \cos(mX + nY), \tag{2.11}$$

$$F_i(X, Y) = \sum_{m=0}^{i} \sum_{n=-i}^{i} f_{i,mn} \sin(mX + nY). \tag{2.12}$$

One may prove recursively that each Fourier series in (2.11) and (2.12) is finite as expressed by the upper limits on the sums. Matching trigonometric factors gives

$$u_{i,mn} = f_{i,mn}/(c_{10}m + c_{20}2n + m^3 + 6m^2n + 12mn^2 + 8n^3). \tag{2.13}$$

When $m = 0$ and $n = 1$ or $m = 1$ and $n = 0$, the denominator of (2.13) vanishes. The phase speed corrections are chosen so that $f_{i,10}$ and $f_{i,01}$ are 0. In the language of the method of multiple scales, the phase speeds are found by imposing the "nonsecularity" conditions. Because these conditions set $u_{i,10} = u_{i,01} = 0$, the amplitudes of $\cos X$ and $\cos Y$ are a and ab to all orders in perturbation theory.

At ninth order, the denominator of the ($m = 5$, $n = -4$) term is also zero. Numerical solutions show that this trigonometric term, which would is expected to be $O(a^9)$, actually varies proportionally to a^7. Haupt and Boyd (1987) show, using a simple one-dimensional wave equation, how to modify the Stokes series to cope with such a resonance.

The basic idea is to assume that the resonant term occurs at lower order (in this case, seventh order instead of ninth order) with an *undetermined* coefficient. At ninth order, this extra degree of freedom is chosen so as to make $f_{9,5,-4}$ zero. This strategy of carrying undetermined parameters to higher order and then choosing them to eliminate resonance is precisely what is done with the phase speed corrections.

Table 1 gives the first few terms in the Stokes series for the principal branch of the double cnoidal wave. It is less compact than the theta-Fourier series, but in compensation, one does not need to take the logarithm and differentiate twice: the terms in Table 1 give $u(x, t)$ directly.

Like all small amplitude expansions, the Stokes series fails when the soliton peaks are too narrow in comparison to the spatial period. Unless one employs the theta formalism, there is no large amplitude perturbation theory to switch to. However, Padé approximants, as shown for a one-dimensional Stokes series in Boyd (1986a), can double the range of usefulness of a power series.

TABLE 1

PHASE SPEEDS AND FOURIER COEFFICIENTS FOR THE KdV DOUBLE CNOIDAL WAVE AS COMPUTED VIA STOKES' EXPANSIONS[a]

$$c_1 \approx -1 + \frac{a^2}{24}$$

$$c_2 \approx -4 + \frac{a^2 b^2}{96}$$

Fourier term	Coefficient
$\cos X$	a
$\cos Y$	ab
$\cos 2X$	$\frac{a^2}{12}$
$\cos 3X$	$\frac{a^3}{192}$
$\cos 2Y$	$\frac{a^2b^2}{48}$
$\cos 3Y$	$\frac{a^3b^3}{3072}$
$\cos X + Y$	$\frac{a^2b}{12}$
$\cos X - Y$	$-\frac{a^2b}{12}$
$\cos X + 2Y$	$\frac{25a^3b^2}{10368}$
$\cos X - 2Y$	$-\frac{a^3b^2}{128}$
$\cos 2X + Y$	$\frac{a^3b}{162}$
$\cos 2X - Y$	0

[a] $X \equiv x - c_1 t + \phi_1$ and $Y \equiv 2(x - c_2 t + \phi_2)$, where ϕ_1 and ϕ_2 are phase constants. These definitions imply that the series are for the "principal branch" of the double cnoidal wave ($k_2 = 2k_1$). Taken from Haupt and Boyd (1989), which also gives the expansion for the $k_2 = 3k_1$ branch.

Stokes expansion is very *general* because its derivation is purely mechanical. It is unnecessary that the evolution equation be integrable; Stokes' algorithm merely manipulates finite sums of trigonometric functions. The algebra is easily delegated to a computer, especially using an algebraic manipulation language such as REDUCE, MACSYMA, or Mathematica.

The weakness of Stokes' method is that precisely because it is so mechanical, it will always generate an answer even if none exists. Generation of a few terms, or even many terms, is not an existence proof. Vivid examples are given by the series for nonlocal solitons, which are discussed in Section V.

As the amplitude increases, the accuracy of the Stokes series deteriorates to uselessness, even with Padé improvement. The remedy is to solve the Phase Variable Boundary Value Problem *numerically* via the Newton–Kantorovich/Fourier/continuation polyalgorithm (Haupt and Boyd, 1987, 1989). Such numerical methods are the theme of Section IV.

Haupt (1987, unpublished Ph.D. thesis) has applied the Stokes series to the wave equation of Peregrine (1966) and Benjamin *et al.* (1972). This equation is known to be *nonintegrable.* Nevertheless, when the Stokes series is used to initialize a time integration, the numerical solution remains close to the Stokes series for several recurrence periods at least in some parameter ranges. The applicability of the concept of polycnoidal waves to the Benjamin–Bona–Mahony equation and other nonintegrable wave equations remains an intriguing open question.

F. Polycnoidal Waves in Two Space Dimensions

The Kadomtsev–Petviashvili equation

$$(u_t + 6uu_x + u_{xxx})_x + 3u_{yy} = 0 \qquad (\text{"KP" equation}) \tag{2.14}$$

is a two-dimensional generalization of the Korteweg–deVries equation. (Kadomtsev and Petviashvili, 1970). It, too, has polycnoidal wave solutions. At small amplitudes, the double cnoidal is again the sum of two noninteracting sine waves. At large amplitudes, the KP double cnoidal wave is the superposition of solitons of two different sizes.

The difference from the Korteweg–deVries equation is that because the KP equation has two space dimensions, the two independent components of the double cnoidal wave are free to propagate at an *angle* to each other. In the special case when this angle is zero, the KP reduces (after a rotation of coordinates, if needed) to the Korteweg–deVries equation. Thus, the KP double cnoidal wave includes the KdV double cnoidal wave as a special case.

Once again, however, the double cnoidal wave is a "genus 2" hyperelliptic function, which may be expressed in terms of two-dimensional Riemann theta functions. Once again, there are two independent phase speeds and amplitudes.

Segur and Finkel (1985) give a good treatment of the KP double cnoidal wave. In particular, they show that the theory predicts patterns that qualitatively resemble the "X-shaped" patterns often seen in real water waves. Hammack (1989) reviews laboratory experiments that have confirmed the theta function theory for shallow water waves. Pierini (1986) has performed initial value experiments.

III. Periodic ("Cnoidal") Waves as Exact Imbricate Series of Solitons

A. Background: Constructing Periodic Solutions from Soliton Trains

As noted earlier, one obvious approximation to a large amplitude spatially periodic wave is a train of evenly spaced, identical solitons. If the solitary wave is denoted by $U(x)$, then the function

$$u(x) \equiv \sum_{n=-\infty}^{\infty} U(x - 2n\pi) \tag{3.1}$$

is manifestly periodic with a period of 2π. To show that this is a very good approximate solution whenever the solitons are tall and narrow in comparison to the periodicity interval, an example is helpful.

The solitary wave of the Korteweg–deVries equation is

$$U(x; s) \equiv 12s^2 \operatorname{sech}^2(sx), \tag{3.2}$$

where s is the "pseudowavenumber" and x is the spatial coordinate in a frame of reference that is traveling with the wave. Denote the individual terms in (3.1) by

$$U_n(x) \equiv U(x - 2n\pi). \tag{3.3}$$

If the soliton series is substituted into the KdV equation, one obtains

$$\sum_{n=-\infty}^{\infty} \left[\{U_{n,xxx} + (U_n - c)U_{n,x}\} + \sum_{\substack{m=-\infty \\ m \neq n}}^{\infty} (U_n U_m)_x \right] \equiv r(x), \tag{3.4}$$

where the residual function $r(x)$ would be identically zero if $u(x)$ were an exact solution. Since each $U_n(x)$ is a solution of the KdV equation, the terms in braces { } in (3.4) are identically 0. The cross-terms in (3.4) alas, are not zero—but they are exponentially small. The maximum overlap of the soliton centered at $x = 0$ with its nearest neighbors

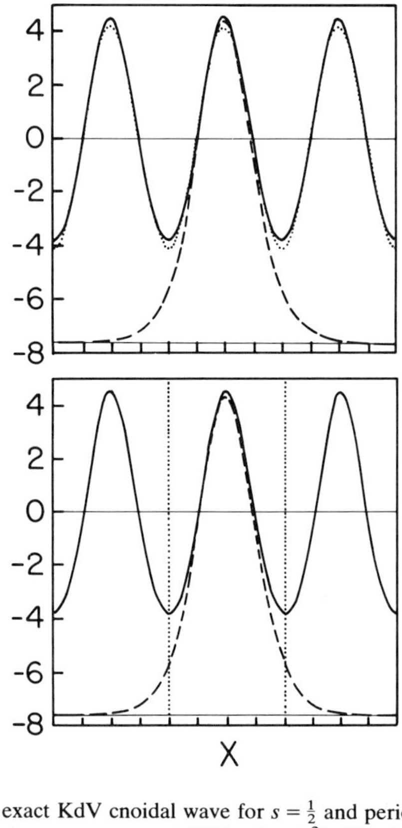

FIG. 3. (a) Solid curve: exact KdV cnoidal wave for $s = \frac{1}{2}$ and period 2π. Dotted curve: $u_{\cos} = 1.03908 \cos X$. Dashed curve: $u_{\text{sol}} = -1.9099 + 3\,\text{sech}^2(x/2)$. (b) Solid curve: exact cnoidal wave, same as in (a). Dashed: constant-plus-soliton approximation u_{sol} as in (a). The horizontal dividing line is at $u = -6/\pi = -1.9099$ to show that the soliton asymptotes to this negative number rather than to zero. Although the soliton is not a spatially *uniform* approximation to the cnoidal wave, it is accurate on $x \in [-\pi, \pi]$.

centered on $x = \pm 2\pi$ is

$$\Omega \equiv \max_{x\in[-\pi,\,\pi]} |u(x;s) - U(x;s)| \approx 48s^2 e^{-2s\pi}, \qquad s \gg 1, \tag{3.5}$$

which is attained at $x = \pm\pi$. Thus, the residual $r(x) \sim 0(\exp[-2s\pi])$ so that the error of the soliton series is *exponentially* small in the amplitude-and-width parameter s.

All this is very straightforward and obvious. It is also irrelevant! Toda (1975) showed that the cnoidal wave of the Korteweg–deVries equation is given *exactly* by

$$u(x;s) = -\frac{12s}{\pi} + 12s^2 \sum_{n=-\infty}^{\infty} \operatorname{sech}^2[s(x - 2n\pi)]. \tag{3.6}$$

The constant $-12s/\pi$ has been added to the sum of solitons so that the spatial mean of $u(x;s)$ is zero, which is the conventional normalization. The *phase speed c* of the cnoidal wave is different from that of its constituent solitons, but only by an exponentially small amount when $s \gg 1$. Remarkably, however, (3.6) is error-free for all s, even when s is very small and $u(x;s) \approx 24\exp(-\pi/s)\cos x$. Figure 3 illustrates how the individual terms in (3.6) superpose to give the cnoidal wave.

TABLE 2

SUMMARY OF ARTICLES ON IMBRICATE SERIES OF SOLITONS

Reference	Wave equation
Toda (1975)	Toda Lattice, Korteweg–deVries
Korpel and Banerjee (1981)	Korteweg–deVries
Zaitsev (1983)	Benjamin–Ono, Joseph, Cubic Schroedinger, Modified Korteweg–deVries Kadomtsev–Petviashvili[a]
Whitham (1984)	Korteweg–deVries, Modified Korteweg–deVries, Boussinesq and Burger's[b]
Boyd (1984a)	Korteweg–deVries, Cubic Schroedinger, Modified Korteweg–deVries
Miloh and Tulin (1989)	Benjamin–Ono
Boyd (1988e)	Quartic Korteweg–deVries[c]
Boyd (1989b)	Resonant Triad

[a] Zaitsev alone has succeeded for an equation in more than one space dimension, but his "pattern function" is restricted to the rational solitons of the KP equation.

[b] Burger's equation does not have soliton solutions, but Whitham shows that it does have periodic solutions that are the exact superposition of shocks.

[c] Boyd shows that the superposition is *not* exact even though the solitons and periodic waves are hyperelliptic functions. All other articles prove the superposition principle is exact for the wave equations treated.

Toda's Theorem—published eighty years after Korteweg and deVries had discovered and named the steadily translating, spatially periodic solution to the differential equation—has spawned a kind of cottage industry. Table 2 summarizes extensions of Toda's work. The conclusion is that the superposition of solitons is exact for all periodic waves and for all solitons that are hyperbolic or rational functions.

In the next subsection, we shall give a general methodology for proving and understanding such soliton-superposition principles.

B. Imbricate Series and the Poisson Summation Formula

Theorem 3.1. (Imbricate Series). *Any periodic function $f(x)$ with period L has two series representations. If the usual Fourier series is*

$$f(x) = \alpha(2\pi)^{-1/2} \sum_{n=-\infty}^{\infty} g(\alpha n) e^{i2\pi x n/L}, \tag{3.7a}$$

then the alternative expansion is the imbricate series

$$f(x) = \sum_{m=-\infty}^{\infty} G\left(\frac{2\pi}{L\alpha}[x - mL]\right), \tag{3.7b}$$

where $G(k)$ is the Fourier transform of $g(n)$, that is,

$$G(k) \equiv (2\pi)^{-1/2} \int_{-\infty}^{\infty} g(x) e^{ikx}\, dx, \tag{3.8}$$

and where α is an arbitrary positive constant.

Definition 3.1. The expansion (3.7b) is said to be an *imbricate Fourier series* or simply *imbricate series.*

Definition 3.2. $G(k)$ is said to be the *pattern function* for the imbricate series.

Definition 3.3. The Fourier series and the function defined by it are both said to be the *imbrication* of the pattern function $G(k)$.

These rather odd-sounding names come from a Latin word, still used by geologists, which means "to overlap like tiles" (Boyd, 1986a). The term is apt because various copies of the pattern function do overlap.

Many functions also have a second species of imbricate series that is equally important for solitons and elliptic functions.

Theorem 3.2 (Alternating Imbricate). *If $f(x)$ is the sum of an odd Fourier series of period $2L$, that is, a series whose terms are restricted to only the sines and cosines of* odd *degree, i.e.,*

$$f(x) = \alpha(2\pi)^{-1/2} \sum_{n=-\infty}^{\infty} g(\alpha[n + \tfrac{1}{2}])e^{i(2n+1)\pi x/L}, \tag{3.9a}$$

where α is an arbitrary scaling constant, then $f(x)$ also has the alternating imbricate *series*

$$f(x) = \sum_{m=-\infty}^{\infty} (-1)^m G\left(\frac{2\pi}{L\alpha}[x - mL]\right), \tag{3.9b}$$

where $G(k)$ is again the Fourier transform of $g(x)$.

Theorem 3.3. *The pattern function $G(x)$ is not unique.*

The reason is that the Fourier coefficients are defined only by the values of $g(n)$ for *integer* n (or half-integer n for (3.9)). Consequently, there are an infinite number of different functions $g(n)$ that yield the same Fourier coefficients. For example, one may add $w(n)\sin(\pi n)$ to $g(n)$, where $w(n)$ is abritrary. Different choices for $g(n)$ will yield different pattern functions; however, the differences between various pattern functions cancel out when the imbricate series is summed.

Theorem 3.4. *The imbrication of a given $G(x)$ is always a unique function $f(x)$.*

It is only the reverse, function-to-pattern relationship, $f(x) \rightarrow G(x)$, which is ambiguous.

The proofs of all these theorems are given by the Poisson summation formula (Morse and Feshbach, 1953) together with other elementary Fourier transform theorems. (Some Fourier monographs actually use (3.7) as the definition of the Poisson summation formula.) The scaling constant α has been included in the theorem to show that the narrower the pattern function, the more slowly the Fourier coefficients decrease as $|n| \rightarrow \infty$ and vice-versa.

TABLE 3

EXAMPLES OF IMBRICATE SERIES

Function[a]	Fourier term[b]	Imbricate term[b]
$u_{\text{KDV}} + \dfrac{12s}{\pi}$	$6n \operatorname{cosech}\left(\dfrac{n\pi}{2s}\right) e^{inx}$	$12s^2 \operatorname{sech}^2[s(x - 2\pi n)]$
$u_{\text{Benjamin–Ono}}$	$\pi s e^{-s\|n\|} e^{inx}$	$\dfrac{s^2}{s^2 + (x - 2n\pi)^2}$
$\alpha_1^{(\text{triad})}$ (dn)	$\operatorname{sech}\left(\dfrac{n\pi}{s}\right) e^{2\pi int}$	$s \operatorname{sech}(s[t - n\pi])$
$\alpha_2^{(\text{triad})}$ (sn)	$\dfrac{1}{2i} \operatorname{cosech}\left(\dfrac{(n+\frac{1}{2})\pi}{s}\right) e^{2\pi i(n+1/2)t}$	$s(-1)^n \tanh(s[t - n\pi])$
$\alpha_3^{(\text{triad})}$ (cn)	$\dfrac{1}{2} \operatorname{sech}\left(\dfrac{(n+\frac{1}{2})\pi}{s}\right) e^{2\pi i(n+1/2)t}$	$s(-1)^n \operatorname{sech}(s[t - n\pi])$
θ_3	$e^{-(\pi/s)n^2} e^{2inx}$	$s^{1/2} e^{-(s/\pi)(x-\pi n)^2}$

[a] The Korteweg–deVries (KdV) cnoidal wave u_{KdV} is normalized to have a zero mean. The Benjamin–Ono cnoidal wave has the closed form expression

$$u_{\text{Benjamin–Ono}} = \frac{s\delta/2}{1 - (1 - \delta^2)^{1/2} \cos x}, \qquad \text{where } \delta = \tanh s.$$

The triad expressions are those for unit frequency, but Boyd (1989b) shows that it is trivial to construct the general triad solution from the functions displayed here. Each triad component is proportional to the elliptic function which is shown in parentheses; the more complicated formulas for the dn, sn, and cn functions themselves are given in Boyd (1984a). Note that α_1 has a period of π while the other components have periods of 2π so that the triad as a whole repeats only every 2π units of time.

[b] The summation is over $n = -\infty$ to ∞.

Happily, the Fourier coefficients of the elliptic functions and of simple rational functions of $\cos x$ are known in analytical, closed form—and so are their transforms. Table 3 is a collection of typical imbricate series and the corresponding Fourier sums. The expansions for the second and third triad components (elliptic functions sn and cn) are alternating imbricate series.

The resonant triad solutions have been included to illustrate that imbricate expansions are useful not only in space, but also in time. These unit frequency triad solutions are special, but Boyd (1989b) shows that the *general* triad solution can be obtained from these special solutions merely by rescaling and a phase translation.

Each triad component is proportional to a different elliptic function; the dn, sn, and cn functions (in the space coordinate) are also the spatially periodic solutions of various wave equations such as the Cubic

Schroedinger, Modified Korteweg–deVries, and so on. Lack of time precludes cataloging all these relationships. The crucial point is that all the hyperbolic and rational solitons can be "imbricated" to form exact, spatially periodic solutions.

No exceptions are known. However, the phase speed of the periodic wave is almost always different from that of the constituent solitons.

Imbricate series like those in Table 2 have been useful in many branches of physics and applied mathematics. Besides the soliton/cnoidal wave theory, other applications include the following: (i) constructing adaptive coordinate grids for periodic problems (Boyd, 1989e), (ii) comparing Fourier and Chebyshev domain truncation for spatially decaying solutions (Boyd, 1988f), (iii) crystal lattice sums (Wimp, 1981), (iv) periodic solutions of the heat equation (via theta functions like θ_3 in the table), (v) asymptotic approximations to Mathieu, spherical harmonic, and prolate spheroidal functions (Morse and Feshbach (1953) and Boyd (1989d)), and (vi) periodic solutions of Burger's equation as the imbrication of shocks (Whitham, 1974).

C. Generalizations and Open Problems

An immediate question is, Does the exact soliton superposition principle extend from ordinary cnoidal waves to N-polycnoidal waves for $N \geq 2$? The same reasoning given earlier shows that, at least *approximately,* the answer is yes. The double soliton formula, for example, approximates the 2-polycnoidal wave to within an error that is exponentially small in the inverse of the widths of the two solitary waves.

Boyd (1982b, 1984b, 1984c, 1984d) shows that the superposition is *exact* for the N-dimensional Riemann theta functions whose second logarithmic derivative (for the KdV equation) is the N-polycnoidal wave. However, it is still an open problem whether the imbricate-soliton series is exact or merely approximate for the hyperelliptic functions that describe $u(x, t)$ itself.

Boyd (1988e) shows that the superposition is *not* exact (although it is a good approximation even for very small amplitudes) for the steadily translating, spatially periodic solutions of the quartically nonlinear KdV equation. The cnoidal waves of this one-dimensional wave equation are hyperelliptic functions. Clearly, merely being a special function—and hyperelliptic functions are *very* special—is insufficient for the imbricate series of solitons to solve a wave equation exactly.

The theme developed in this section—that the crests of a periodic wave behave as solitary waves, dynamically independent of one another—has been known to water wave specialists for a long time (Peregrine, 1983). Munk (1949) noted, simply by observing the shapes of waves running up a beach, that the waves become steeper and narrower as they approach the shore until they *look* like solitons. Stiassnie and Peregrine (1980) gave a more rigorous demonstration that such shoaling waves interact very weakly. Ocean wave crests are not identical in either shape or spacing, but it is the narrowness of the crests in comparison to their spacing that is important. Yasuda and Tsuchiya (1982)—with good success—fit observed water wave spectra by a sequence of solitons with variable spacing and amplitude.

It is plausible that similar ideas should work in more than one dimension. For example, McWilliams (1984, 1989) has shown that two-dimensional turbulence and three-dimensional, quasi-geostrophic turbulence decay into a sea of small, intense vortices, which, apart from occasional collisions, interact very weakly.

Lastly, the resonant triad solutions show that imbricate series may be useful for time-dependence. Lorenz (1960) showed that the triad equations are the simplest model for instability and nonlinear vacillation. The triad series are not the imbrication of solitary waves, but rather are the superposition of isolated growth/equilibration/decay events. It remains to be seen whether similar imbricate expansions will be useful for other examples of time-periodic or time-repetitive vacillating instability.

IV. Numerical Boundary Value Algorithms for Direct Computation of Solitons

A. Introduction

Numerical solutions to initial value problems have been an important tool for understanding nonlinear waves. However, IVP codes are crude instruments for investigating solitary waves themselves. Because the solitons are unknown, the initial conditions will necessarily be crude approximations or guesses to the solitary wave. The modeller is then forced to try to distill the pure soliton from a sea of transients.

A new direction is to compute solitons directly by solving a nonlinear boundary value problem. The soliton is assumed to be steadily translating, which reduces the time-dependent wave equation to a nonlinear boundary value problem in the spatial coordinates with the phase speed c as the nonlinear eigenparameter.

In this section, we shall describe and catalog a rather diverse collection of direct, solve-the-boundary-value-problem attacks on solitons. Each investigation is numerically characterized by the three parts of its computer code: (i) initialization (generating the first guess), (ii) iteration (usually Newton's), and (iii) discretization (finite difference, spectral, or a special scheme).

In the next subsection, we give an overview of fifteen representative studies. Subsection C is an in-depth analysis of one particular numerical polyalgorithm: the continuation/Newton–Kantorovich/pseudospectral strategy of Boyd (1986a, 1986b, 1988b, 1988d). The final subsection is focused on iteration methods, which are crucial to efficiency.

B. A Catalog of Direct Computations of Solitons

1. *Initialization* (*Generating the First Guess for the Iteration*)

One special curse of nonlinear problems is that it is almost always necessary to iterate. This in turn requires a first guess to initialize the iteration. If the first guess differs too greatly from the true solution, the iteration will diverge, so a good initialization is essential. Initialization strategies include the following four: (i) an analytical solution, (ii) continuation ("homotopy"), (iii) residual inhomogeneity and other artificial parameter techniques, and (iv) a one degree-of-freedom numerical solution.

Small amplitude perturbation theory is a fertile source of analytical solutions. Boyd (1988d), for example, applies the method of multiple scales to approximate equatorial waves as Korteweg–deVries solitons. For nonlinear spatially periodic waves, an even simpler approximation—the infinitesimal cosine wave—is sufficient. For small amplitude, the perturbative approximation will be sufficiently close to the exact answer so that the iteration will converge.

Unfortunately, numerical solutions are most desirable precisely for those regions in parameter space where the analytic approximations are terrible. The need to have a crude answer in order to compute an accurate answer is very restrictive. The second initialization scheme, the "continuation" or "homotopy" method, transcends this limitation. Continuation is so important that it will be given "star billing" in the next subsection. The basic idea is very simple, however.

If one solves the wave equation for small steps in the amplitude a (or any other parameter), then $u(a+\delta a) \approx u(a) + O(\delta a)$. This implies that one may use the solution for $u(a)$ as a good first guess for $u(a+\delta a)$, which in turn furnishes the initialization for $u(a+2\delta a)$, and so on. The parameter march begins at a small amplitude where the perturbative approximation is the initialization for the first computed amplitude. For all larger amplitudes, the converged result of the iteration for each a is the first guess for the next amplitude. One may thus march from very small amplitude to very large amplitude in small, smooth steps.

Unfortunately, there are some interesting solitons that do not have small amplitude limits! The geophysical and plasma vortices known as "modons" (Flierl, 1987) and the solitons of the Kuramoto–Sivashinsky equation (Chang, 1986) and the Flierl–Petviashvili monopole (Boyd, 1988d) exist only for finite amplitude. There are two popular strategies for generating a first guess when analytical solutions are unavailable.

The first is to solve the wave equation using a very *low order* finite difference or spectral scheme. Boyd (1988d), for example, faced the problem

$$W_{rr} + \frac{1}{r} W_r - W - W^2 = 0. \tag{4.1}$$

Neglecting the nonlinear term gives an equation whose only solution is the trivial one, so perturbation theory fails. Instead, Boyd (1988d) truncated the rational Chebyshev pseudospectral series to the lowest basis function:

$$W(r) \approx a_1 \phi_1(r) \equiv a_1\left[\mathrm{TB}_2\left(\frac{r}{2}\right) - 1\right] = a_1\left(\frac{-2}{1 + r^2/4}\right) \tag{4.2}$$

Substituting (4.2) into (4.1) and demanding that the residual function should vanish at the single interpolation point $r = 2$ gives a *linear* equation for a_1 whose solution is $a_1 = -1$. This one-degree-of-freedom approximation has a maximum error of 16% of the maximum in $|u|$, but it is nonetheless a successful initialization for more accurate calculations. Boyd (1986b) is another example.

The mysteries of choosing basis functions and collocation points will be explained in the next subsection. The important thing is that the analytical solution was obtained merely by specializing the numerical procedure to a small number of unknowns. (In this case, one!)

It is not necessary that the low order system be analytically soluble. For example, if we attacked (4.1) with two basis functions, the pseudospectral procedure would yield two quadratic equations in two unknowns. Each equation is the equation of a conic section, so one can determine all four solutions of the algebraic system by graphing both hyperbolas (or ellipses) and identifying the points of intersection. Polynomial systems of higher degree can be solved through library software ("black boxes") described by Watson and Scott (1987) and Morgan (1987). Most of the roots of the algebraic system do not correspond to differential equation solutions, but one can test all low order roots by inserting them into the high resolution code. The "good" solutions to the low order system are those that initialize convergent iterations when the numerical resolution is increased.

The fourth strategy for initialization is to modify the homogeneous boundary value problem by adding an artificial inhomogeneity. Usually, one knows the general shape of the solitary wave; most one-dimensional solitons qualitatively resemble the hyperbolic secant function, for example. (The functions $\text{sech}^2(x)$ and $\exp(-x^2)$ are alternatives of similar shape.) Let us denote such a qualitative guess by $u_0(x)$. If Newton's method diverges when initialized with u_0, we can artificially modify the original problem.

Suppose that the goal is to solve $L(u)=0$, where L is a nonlinear operator. The "residual inhomogeneity" method modifies this wave equation to

$$L(u)=(1-\tau)L(u_0), \tag{4.3}$$

where τ is an artificial parameter. When $\tau=0$, the initialization u_0 is, *by construction,* the *exact* solution of the *modified* equation (4.3). Obviously, Newton's iteration will converge! After solving (4.3) for $\tau=0$, we can then apply the *continuation* method to march in small steps in τ until we reach $\tau=1$. This technique of introducing an artificial parameter τ so that an arbitrary first guess is the exact solution for $\tau=0$ is very popular for solving nonlinear systems of equations. It is often called the "homotopy method" with (4.3) as the "homotopy mapping".

Boyd (1988d) also solved (4.1) using this residual inhomogeneity method. The first guess $W \approx \exp(-r)$ is a poor approximation—in polar coordinates, $W(r)$ has a cusp at the origin! This does not matter, however, because the continuation march in the dummy parameter τ will continuously deform the solution from the initial guess to the true solution $W(r)$.

2. *Iterations*

The basic algorithm for any nonlinear problem is Newton's iteration. If the discretization is applied first, then the wave equation is transformed into N nonlinear algebraic equations for the values of the soliton at each of the N grid points. The iteration for a system of algebraic equations is usually called the Newton–Ralphson method. Alternatively, as discussed at greater length in the next subsection, one may apply the iteration directly to the nonlinear differential equation. This tactic is the Newton–Kantorovich method; one must solve a *linear* boundary value problem at each iteration. Either way, the convergence is second order in the sense that, close to a root, the number of correct digits doubles at each iteration.

Unfortunately, the Newton iteration is rather expensive because one must solve a system of N linear equations in N unknowns at each step. For this reason, quasi-Newton variations are very common. The simplest tactic is to compute and factorize the Jacobian matrix only once and merely backsolve to compute all the later iterates. The second order rate of convergence is reduced to *geometric* convergence, that is, $e_{i+1} \approx e_i/\lambda$, where e_i is the error of the i-th iterate and λ is a constant > 1. The cost per iteration is greatly reduced, however.

Many other quasi-Newton schemes are explained in texts on optimization. However, there have been several successful applications of non-Newton methods to solitons.

Eydeland and Turkington (1987) recast the quasi-geostrophic equation for Rossby solitons in an infinitely long channel into a variational problem. Their iteration has only a geometric rate of convergence, but there are two major rewards for this (relatively) slow convergence rate. First, they are able to rigorously *prove* that their iteration will converge from an arbitrary first guess, eliminating the initialization problem and simultaneously placing their work on a firm mathematical footing. Second, Poisson's equation is solved at each iteration instead of a more complex partial differential equation. Since this is separable, the solution requires only $O(N^2 \log_2 N)$ operations, where N is the number of grids points in each coordinate. In contrast, a standard Newton–Kantorovich/finite difference solution of the quasi-geostrophic equation would require solving a *nonseparable* partial differential equation at each step, which is $O(N)$ more costly. Ingersoll and Cuong (1981) and Verkley (1989) used similar iterations. In Subsection D, we show how these schemes can be interperted as time-marching solutions to artificial diffusion problems.

Finally, a third strategy is to employ a nonlinear generalization of Richardson's iteration ("preconditioned Newton flow"). This, too, has only a geometric rate of convergence. However, the nonlinear Richardson iteration combines the high accuracy of spectral algorithms with the low cost of finite difference methods. This very promising tool will receive "star billing" in Subsection D.

3. *Discretizations*: *Finite Difference, Spectral, and Special*

Second-order finite differences have been popular with many authors including Tung, Chang, and Kubota (1982), Ingersoll and Cuong (1981), and Eydeland and Turkington (1987, 1988). Since these are discussed in every numerical analysis text, we merely note that these algorithms are easy to program but give only moderate accuracy unless one uses a prohibitive number of grid points.

Special numerical discretizations are quite common because most nonlinear wave specialists are good (and ingenious) mathematicians. Deem and Zabusky (1978) and later papers reviewed by Dritschel (1988) employ "contour dynamics". The Euler equations for a fluid are simplified by assuming that the flow consists entirely of *patches* of *uniform vorticity* surrounded by irrotational motion. With this simplification, the two-dimensional dynamics is reduced to computing the shapes of the one-dimensional contours that bound the patches of constant vorticity. Most of this work has focused on vortex merging, stability, and other time-dependent processes. However, Deem and Zabusky (1978) show some interesting vortex pairs that are independent of time except for a uniform translation. Pierrehumbert (1980), Wu *et al.* (1984), and Tanveer (1986) also give solutions for pairs of uniform vorticity patches.

Free surface water waves are commonly attacked via complex variable techniques, which reduce the dimensionality by (effectively) eliminating the vertical coordinate. Chen and Saffman (1980) solved an integro-differential equation via sixth-order differences, then checked their code by solving a different but equivalent problem via pseudospectral methods.

Vanden-Broeck (1986) solved a complex free surface condition by expanding the complex velocity W in terms of the modified Fourier expansion

$$W(f) = (1 - \beta e^{-2\pi i f})^{1/3}\left\{1 + \sum_{n=0}^{\infty} d_n e^{-2\pi i n f}\right\}, \tag{4.4}$$

where β is, along with the phase speed c and a finite number of the coefficients d_n, one of the unknowns. For the highest wave, Stokes showed that the crest has a corner with an angle of 120°, which implies $\beta = 1$ and $W(f) \sim f^{1/3}$. The variable β allows the expansion to mimic the near-singularity for very steep waves that are just short of maximum steepness.

The last special method discussed here is the Stokes perturbation series in powers of the wave amplitude. This technique was illustrated for the double cnoidal wave at the end of Section II. The order-by-order calculation of terms is usually regarded as a sequence of symbolic computations rather than a numerical iteration. Nevertheless, there is an initialization: the linear cosine wave is the lowest order approximation. Like the quasi-Newton iterations, the Stokes series converges geometrically: an error that is proportional to a^n for some a is the definition of geometric convergence. The perturbation theory closely resembles the quasi-Newton scheme which computes and factorizes the Jacobian just once. At each order in perturbation theory, one solves a linear differential equation with the *same* linear operator; only the inhomogeneous terms vary with n. The perturbation theory also increases the accuracy by one additional power of the amplitude a at each order. In contrast, each Newton iterate is a rational function of the amplitude. If the denominators are expanded in powers of the amplitude a to convert this into a polynomial in a like the perturbative approximation, one finds that the n-th iterate is accurate to the 2^n-th power of a. The degree (in amplitude) doubles at each Newton iteration. Thus, perturbation theory is a quasi-Newton iteration.

Since the Stokes series has a finite radius of convergence (or perhaps is asymptotic), it has a smaller range than other algorithms. However, the series often converges well into the soliton regime, and Padé approximants constructed from the Stokes terms usually double the amplitude range where a given order of approximation is accurate. Schwartz (1974), Boyd (1986a), Andersen and Geer (1982), and especially the reviews by Van Dyke (1975, 1984) illustrate Padé and other series-extension methods.

The final discretization strategy is to expand the unknown in a series of global expansion functions. The choice of basis functions is determined partly by the geometry and partly by personal preference.

When the geometry is periodic in a given coordinate, the natural basis functions are the sines and cosines of a Fourier series. Examples include Haupt and Boyd (1987, 1989), Boyd (1986a), and most applications of

Stokes series to water waves including Schwartz (1974). Meiron, Saffman, and Yuen (1982) computed free surface waves by expanding the velocity potential in a two-dimensional Fourier series with coefficients that vary with height so that each term satisfies the three-dimensional Laplace equation.

When the domain is a channel, i.e., finite in one coordinate with nonperiodic boundary conditions at the sidewalls, the best basis functions for the cross-channel coordinate are Chebyshev polynomials. Numerical computations of the geophysical solitary waves known as "modons" also use Chebyshev polynomials even though the domain is unbounded. The reason is that modons have nonzero vorticity only within a finite region bounded by a closed curve—an ellipse in Boyd and Ma (1989). Because the flow outside this ellipse is analytically known, the computational task is confined to the interior of the closed curve. Boyd and Ma solved their interior problem in elliptical coordinates by combining Fourier functions in the quasi-angular variable with Chebyshev polynomials in the quasi-

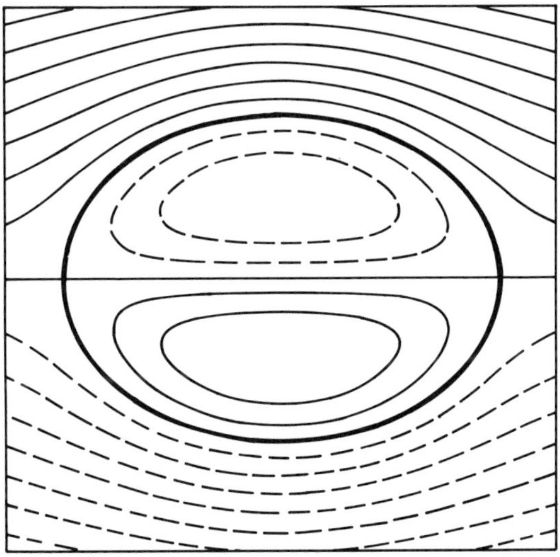

FIG. 4. Streakfunction for an elliptical modon from Boyd and Ma (1989). (The streakfunction is the streamfunction as seen in a coordinate system moving with the wave.) Positive contours are solid, while negative values of Ψ are dashed. The boundary between the interior and exterior flow is the contour $\Psi = 0$, an ellipse that is shown with a bold line. By exploiting the four-fold symmetry (symmetric in x and antisymmetric in y), one may compute this case to six decimal place accuracy with only 200 basis functions.

radial coordinate. Figure 4 shows the streamfunction (including the analytically evaluated exterior flow) for a typical case.

On an infinite interval, one has a variety of good choices: rational Chebyshev functions, sinc series, and domain truncation with either Chebyshev or Fourier functions. All give spectral accuracy, but the rational Chebyshev functions (Boyd, 1982, 1987a, 1987b) are the most flexible and will be discussed at greater length in Subsection C. Boyd (1988b, 1988d) gives examples.

For all these two-dimensional problems, these one-dimensional basis functions are combined as tensor products. For example, a good choice for the infinite channel is to define $\phi_{mn}(x, y) \equiv \mathrm{TB}_n(x) T_n(y)$, where the TB_n are the rational Chebyshev functions (infinite interval in x) and the T_n are the Chebyshev polynomials (for the cross-channel coordinate y).

C. The Newton/Pseudospectral/Continuation Polyalgorithm: A Closer Look

To avoid the review-as-a-list-of-references disease, this section will give a detailed description of how to compute a solitary wave using particular choices for the initialization, iteration, and discretization. The range of options is so great that one cannot possibly cover all the possibilities in depth. The particular choices we shall discuss here, however, represent the mainstream of solving nonlinear boundary value problems.

The Newton–Ralphson and Newton–Kantorovich algorithms are equivalent and require the same work. We will discuss the latter—linearizing first and then discretizing the partial derivatives—because this makes it easier to understand important techniques such as matrix preconditioning and the nonlinear Richardson iteration explained below.

To illustrate these ideas, we shall solve the particular wave equation

$$u_{xx} + u_{yy} - y^2 u - \frac{u}{c} - \frac{yu^2}{c} = 0 \qquad \text{(EMR equation)} \tag{4.5}$$

on the doubly infinite interval, $x \in [-\infty, \infty]$, $y \in [-\infty, \infty]$. The physical background and the reason for the name "EMR equation" are given in Boyd (1988d). Here, it will suffice to note that the EMR equation is an approximate model for planetary-scale Rossby waves which are trapped by Coriolis forces in a narrow band around the equator, $y = 0$.

The Newton–Kantorovich iteration begins by writing

$$u^{(i+1)} \equiv u^{(i)}(x, y) + \Delta(x, y), \qquad \max|\Delta| \ll 1, \tag{4.6}$$

where the superscript is the iteration number. Strictly speaking, the correction $\Delta(x, y)$ should have a superscript (i) also, but this has been omitted for notational simplicity. The key step is to substitute (4.6) into (4.5) and then linearize by neglecting terms of $O(\Delta^2)$. This gives the *linear,* inhomogeneous boundary value problem that must be solved at each step:

$$\Delta_{xx} + \Delta_{yy} + \left[-\frac{1}{c} - y^2 - \frac{2yu^{(i)}}{c}\right]\Delta = -\left\{u_{xx}^{(i)} + u_{yy}^{(i)} + \left[-\frac{1}{c} - y^2\right]u^{(i)} - \frac{y[u^{(i)}]^2}{c}\right\}. \tag{4.7}$$

One then iterates (4.6) and (4.7) until $\Delta(x, y)$ is negligibly small.

To solve (4.7), we employ the rational Chebyshev functions defined by

$$\mathrm{TB}_n(x) \equiv \cos(nt), \tag{4.8}$$

where $t(x)$ is defined by

$$x = L \cot t, \tag{4.9}$$

where L is a user-choosable constant ("map parameter"), which allows one to adjust the scale of the basis functions to that of $u(x, y)$. The x-derivatives of the basis functions can be calculated by exploiting the fact that the $\mathrm{TB}_n(x; L)$ are merely the terms of a Fourier cosine series in disguise. The first six derivatives are given in Boyd (1987a). For the present application, it suffices to note that

$$\begin{aligned} \mathrm{TB}_{n,x} &= -\frac{\sin^2 t}{L}[-n\sin(nt)], \\ \mathrm{TB}_{n,xx} &= \frac{\sin^4 t}{L^2}[-n^2\cos(nt)] + \frac{2\cos t \sin^3 t}{L^2}[-n\sin(nt)]. \end{aligned} \tag{4.10}$$

When the basis functions are substituted into the differential equation, $H\Delta = f$, one obtains the *residual* function

$$R(x; d_1, \ldots, d_N) \equiv f(x) - \sum_{n=1}^{\infty} d_n[H\phi_n(x)], \tag{4.11}$$

where the d_n are the spectral coefficients of $\Delta(x)$ and the $\phi_n(x)$ are the basis functions. The goal of all spectral methods is to determine the a_n so as to make the residual function as small as possible in some sense. $R(x) \equiv 0$ for the exact solution.

Galerkin's method obtains N conditions by multiplying (4.11) by each of the first N basis functions in turn and integrating. It is usually simpler and almost equally accurate to apply the "pseudospectral" or "collocation" method, which imposes the condition

$$R(x_i; d_1, \ldots, d_N) = 0, \qquad i = 1, \ldots, N, \tag{4.12}$$

where the "interpolation" or "collocation" points in x are the mapped images of an evenly spaced grid in the trigonometric coordinate t:

$$x_i = L \cot\left(\frac{\pi(2i-1)}{2N}\right), \qquad i = 1, \ldots, N. \tag{4.13}$$

The grid (4.11) does not include the endpoints ($t = 0, \pi \leftrightarrow x = \pm\infty$), so (4.9) and (4.10) are well-defined and nonsingular at all x_i.

Both Galerkin's method and the pseudospectral algorithm convert the differential equation into an $N \times N$ matrix problem whose solution is the vector of basis coefficients. For the latter, the linear algebra equation is

$$\mathbf{J}\mathbf{d} = \mathbf{r}, \tag{4.14}$$

where $\mathbf{d}$ is the column vector of the spectal coefficients of the correction $\Delta(x)$; $\mathbf{r}$ is the column vector of residuals of $u^{(i)}(x)$ at the i-th interpolation point

$$r_i \equiv R(x_i; a_1, \ldots, a_N), \qquad i = 1, \ldots, N, \tag{4.15}$$

where the a_i are the coefficients of $u^{(i)}(x)$; and $\mathbf{J}$ is the Jacobian matrix whose elements are

$$J_{ij} \equiv (H\phi_j)|_{x=x_i}, \tag{4.16}$$

where H is the linear operator acting on Δ in the linearized differential equation. For (4.5),

$$H \equiv \frac{\partial^2}{\partial x^2} + \frac{\partial^2}{\partial y^2} + \left(-\frac{1}{c} - y^2 - \frac{2yu^{(i)}}{c}\right).$$

Unlike finite difference matrices, the pseudospectral Jacobian matrix is dense rather than sparse, but the reward is that to obtain moderate accuracy (say 5%), the number of degrees of freedom is smaller—typically by at least a factor of two in each coordinate. The rational Chebyshev functions are so efficient that Boyd (1988d) computed the solitons of (4.5) to three correct digits using only a total of 120 basis functions in *two* dimensions. Although there are exceptions such as the solitons of the Kuramoto–Sivashinsky equation, solitary waves are usually rather smooth functions. Good two-dimensional solitons from 100 basis functions are common.

Because the pseudospectral matrix is small, one can afford to employ a simple, direct method to solve the linear algebra problem: Gaussian elimination (Crout reduction). Factoring a 100×100 matrix requires only about 60 seconds on an IBM AT and less than 10 seconds on a work station. For difficult problems ($N \gg 100$), one can drastically reduce both operations and storage by using the preconditioned nonlinear Richardson iteration described in the next subsection.

Two other tricks also greatly improve efficiency. First, solitons often have symmetries. The most interesting solitons of (4.5), for example, are symmetric in x and antisymmetric in y. One may halve the size of the basis set in each coordinate by keeping only those basis functions that match the symmetry of the solution. For the rational Chebyshev functions, it is easy to show that all the even degree functions are symmetric about the origin, while TB_1, TB_3, and all the other odd functions are antisymmetric. When the basis is halved, one must simultaneously halve the interpolation grid by replacing the $1/(2N)$ in (4.13) by $1/(4N)$.

The second trick is that since solitons vanish at infinity, one may modify the basis by defining combinations of the TB_n that vanish at infinity. (This is strictly *optional,* but does make the computation more efficient and is especially helpful when computing a low order spectral solution to initialize the Newton iteration.) We may take

$$\begin{aligned}\phi_{2n}(x) &\equiv \mathrm{TB}_{2n}(x) - 1,\\ \phi_{2n-1}(x) &\equiv \mathrm{TB}_{2n+1}(x) - \mathrm{TB}_1(x).\end{aligned} \tag{4.17}$$

The new basis functions retain the even or odd symmetry of the $\mathrm{TB}_n(x)$ from which they are constructed.

For simplicity, we have explained the key ideas in one dimension, but the extension to two dimensions is trivial. Normally, one employs a tensor product basis. That is,

$$\phi_{ij}(x, y) \equiv \phi_i(x)\phi_j(y), \qquad i = 1, \ldots, M, \quad j = 1, \ldots, N, \tag{4.18}$$

for a total of MN basis functions. The collocation points are of the form (x_i, y_j), where x_i and y_j are both given by (4.13). The number of collocation points and the map parameter L may be chosen independently for each coordinate.

The third major component of the polyalgorithm is the continuation method. This, as briefly described earlier, is a strategy for tracing an entire solution branch from a single starting point. It is still necessary to use the other initializations described above—analytical solutions, perturbative approximations, or low order spectral solutions—to obtain a decent first guess for *one* point on the solution branch. After $u(\varepsilon_0)$ has been computed, however, one may apply continuation to compute the solution for all ε.

Zeroth-order continuation means using the solution for $u(\varepsilon_{n-1})$ as the first guess for $u(\varepsilon_n)$. A better procedure—after one has computed u for a second value of ε via zeroth-order continuation—is to fit a straight line through the two previous values:

$$u(\varepsilon_n) \approx u(\varepsilon_{n-1}) + (\varepsilon_n - \varepsilon_{n-1}) \frac{u(\varepsilon_{n-1}) - u(\varepsilon_{n-2})}{\varepsilon_{n-1} - \varepsilon_{n-2}} \tag{4.19}$$

Note that one may use a variable step size in the parameter ε. This "first-order" continuation may be generalized by fitting polynomials of higher degree. Boyd (1988d), for example, used a library routine that fit a polynomial of up to fourth degree. It was called twice for each value of ε: once at the top of the loop to generate a first guess and once at the bottom of the loop with a different flag to store $u(\varepsilon_{n-1})$ (along with previously computed solutions) in preparation for the next extrapolation.

The only difficulty is that continuation fails whenever the Jacobian matrix is singular. This occurs wherever the discretized, linearized differential equation has a zero eigenvalue. One such breakdown occurs at a bifurcation point where two independent solution branches intersect. However, Keller (1977) showed that it is possible to "shoot the bifurcation point" by trying various parameter steps in ε until the Newton iteration converges on the far side of the bifurcation point. Seydel (1988) and Chen and Saffman (1980) describe how to switch branches at a bifurcation point to trace the second branch, too.

The other class of breakdown is a "limit point" or "fold" where the solution branch curves back upon itself so that no solutions exist for $\varepsilon > \varepsilon_{\text{limit}}$, while there are two solutions for $\varepsilon > \varepsilon_{\text{limit}}$. The classic example is a small amplitude periodic solution. For the Korteweg–deVries cnoidal wave of period 2π, for example, two solutions exist for all $c > -1$, while there are no solutions with smaller phase speed. The two solution branches that join at $c = -1$ are approximately given by

$$u(x; a) = \pm |a| \cos(x - ct), \qquad c \approx -1 + \frac{a^2}{24}, \quad |a| \ll 1. \tag{4.20}$$

For all amplitudes, these two branches differ only through a phase shift of π.

Parenthetically, we note that the Stokes expansion differs from other quasi-Newton iterations because of the nonsecularity condition that gives the corrections to $c(a)$. This arises because the Stokes series is an expansion about a limit point ($a = 0$) rather than one where the Jacobian matrix $\mathbf{J}$ is nonsingular.

The standard remedy for a "fold" is to use the arclength of the curve as the marching variable instead of ε (Seydel, 1988). Arclength continuation will smoothly trace the solution around the limit point from the lower branch to the upper branch; the price is that the linear algebra problem is increased in size from N to $N + 1$ and more programming is needed.

For many wave problems, however, a simpler strategy is to use the amplitude rather than the phase speed as the marching parameter.

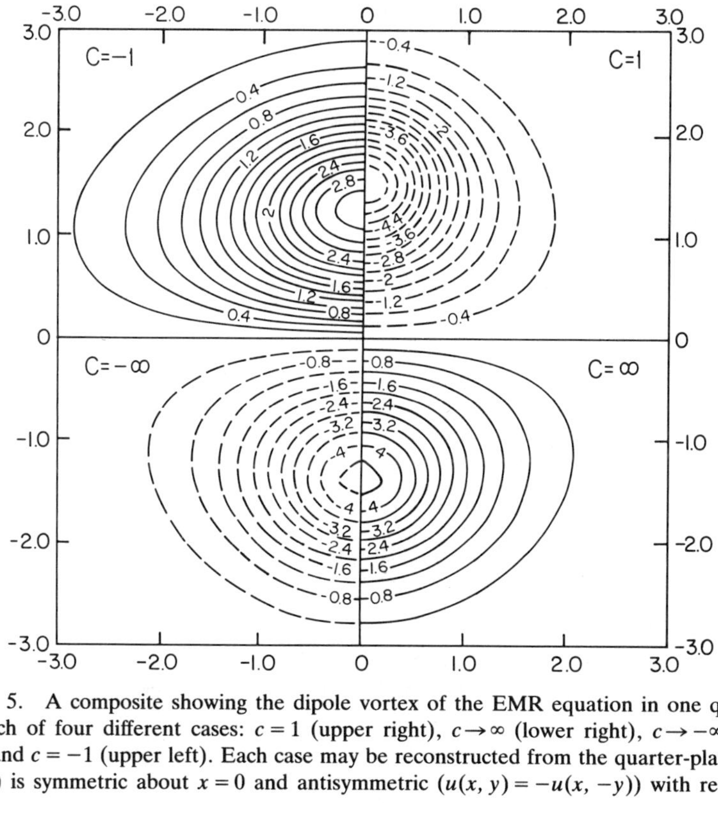

FIG. 5. A composite showing the dipole vortex of the EMR equation in one quadrant for each of four different cases: $c = 1$ (upper right), $c \to \infty$ (lower right), $c \to -\infty$ (lower left), and $c = -1$ (upper left). Each case may be reconstructed from the quarter-plane since $u(x, y)$ is symmetric about $x = 0$ and antisymmetric ($u(x, y) = -u(x, -y)$) with respect to $y = 0$.

Equation (4.20) and the double cnoidal Stokes series (Section II) are examples. Both define the amplitude parameter to be the lowest Fourier coefficient, that of $\cos x$. This choice is easy to program, but other definitions of amplitude are mathematically and computationally acceptable. For some examples such as the fifth-degree Korteweg–deVries equation (Boyd, 1986a), this change-of-parameter is sufficient to avoid limit points except for the trivial one at zero amplitude. (Because the lowest order approximation (4.20) gives a good initialization for the branches that meet at zero amplitude, arclength continuation is unnecessary at that fold.) Another virtue of the fixed-amplitude numerical solutions is that they are easier to compare with Stokes series, which also fix a. However, if there are no limit points, then one may use the phase speed c as the marching parameter over all of parameter space. The author uses whichever of c and a is most convenient for a given problem.

As noted earlier in discussing the residual inhomogeneity initialization, one is always free to march in an artificial parameter when a physical parameter is inconvenient. The "four-eyed" contour plot in Figure 5 is a good illustration. The lower two cases are $c \to \pm\infty$. The obvious way to compute them is to simply march in small steps from small negative c (where a perturbative solution is accurate) to large $|c|$. However, this is very costly because the limit of infinite c is approached very slowly, as $O(1/c)$.

A much better strategy is to think. First, a simple scale analysis shows that the $-u/c$ term is negligible in comparison to the others as $|c| \to \infty$, which reduces (4.5) to

$$u_{xx} + u_{yy} - y^2 u - \frac{yu^2}{c} = 0 \qquad (\text{"Large-}|c|\text{ EMR equation"}). \qquad (4.21)$$

Second, it is trivial to show that the solution of (4.21) for all c is given exactly by

$$u(x, y) = cv(x, y), \qquad (4.22)$$

where $v(x, y)$ is a single, universal function that solves (4.21) for the special case $c = 1$. Equation (4.22) further shows that the soliton for $c = -\infty$ is merely the negative of that for $c = \infty$: the two branches connect through $|c| = \infty$, and it is necessary to compute only a single function $v(x, y)$ to obtain both cases in the lower half of Figure 5.

This parameter-free function was computed by marching in c to $c = -1$ and then solving

$$u_{xx} + u_{yy} - y^2 u + yu^2 + (1 - \tau)u = 0. \qquad (4.23)$$

When $\tau = 0$, (4.23) is simply the EMR equation for $c = -1$. Marching to $\tau = 1$ then gives $u = -v(x, y)$. Instead of calculating the limit $|c| \to \infty$ by marching hundreds of steps in c, we compute the limiting solution through (4.23) in just four or five steps in the artificial parameter τ.

The four-fold symmetry of the solutions to this wave equation is very helpful. This symmetry not only reduces the number of basis functions by a factor of four, as described above, but also allows equal economy in displaying the answer: four cases are combined in a single graph.

Clearly, the Newton/pseudospectral/continuation polyalgorithm is very powerful and general. However, Werner Rheinboldt's quote is still relevant: "Applied mathematics is mathematics-plus." In the next subsection, we discuss some methods for reducing the cost of what is almost always the rate-determining step: solving the matrix problem (4.14).

D. The Nonlinear Richardson ("Newton Flow") Iteration and Other Artificial Time Methods

Newton's iteration for the nonlinear system $\mathbf{r}(\mathbf{u}) = \mathbf{0}$ is equivalent to applying the Euler forward marching scheme with unit time step to the system of differential equations

$$\frac{d\mathbf{u}}{dT} = -\left(\frac{1}{\mathbf{J}}\right)\mathbf{r} \qquad \text{("Newton flow" equation)}, \tag{4.24}$$

whre T is a fictitious time variable and $1/\mathbf{J}$ is the inverse of the Jacobian matrix. Converting the problem from an algebraic to a differential system seems crazy at first, but it is a very powerful conceptual and computational tool. All the theorems and all the algorithms of differential equations can now be brought to bear to solve (4.24), which is known as the 'Newton flow" equation. Chu (1988) is a good review of such "continuous realizations of iterative processes," as his paper is titled. See also Peitgen and Richter (1986), who show that the attractor basin boundaries for the Newton flow are smooth while those for the Newton iteration are usually fractal.

If we expand $\mathbf{r}(\mathbf{u})$ in a multi-dimensional Taylor series centered on a root $\mathbf{u}_0$, (4.24) becomes

$$\frac{d\mathbf{u}}{dT} = -\left(\frac{1}{\mathbf{J}}\right)\{\mathbf{r}(\mathbf{u}_0) + \mathbf{J}(\mathbf{u} - \mathbf{u}_0) + O([\mathbf{u} - \mathbf{u}_0]^2)\} \tag{4.25}$$

$$= -(\mathbf{u} - \mathbf{u}_0) + O([\mathbf{u} - \mathbf{u}_0]^2) \tag{4.26}$$

Thus, *every root* of $\mathbf{r}(\mathbf{u}) = \mathbf{0}$ is a *local attractor*. However, the rate of convergence is geometric rather than quadratic except for Euler's method with unit time step. The residual decays with the artificial time T as

$$\mathbf{r}(\mathbf{u}(T)) = e^{-T}\mathbf{r}(\mathbf{u}(0)) \tag{4.27}$$

for all initial conditions, even those far from a root. However, (4.27) is not sufficient to guarantee success because the solution could diverge to infinity or converge to the trivial solution $\mathbf{u} = \mathbf{0}$.

The Newton flow concept opens up intriguing computational possibilities because one may *replace* the inverse Jacobian matrix by an *arbitrary* matrix without altering the fact that each root of the residual is a critical point of the differential equation. This freedom is enticing because the pseudospectral Jacobian matrix is dense. Factorizing the Jacobian to compute the Newton correction may be prohibitively expensive for multi-dimensional problems. If we use N basis functions in each coordinate, then the cost is $O(2N^6/3)$ operations for each iteration in two dimensions. Ouch!

The simplest replacement for the inverse Jacobian is the identity matrix, but unfortunately there is no guarantee (or even likelihood) of convergence. If, however, the inverse Jacobian is replaced by a *preconditioning* matrix, that is, a matrix whose inverse *approximates* the inverse Jacobian, then it may be possible to prove fast convergence.

Orszag (1980) showed that an excellent pseudospectral preconditioning matrix is that of the second order finite difference approximation to the same problem on the same grid. Let the finite difference matrix be denoted by $\mathbf{H}$. Orszag demonstrated that for a second-order differential equation, the matrix $(1/\mathbf{H})\ \mathbf{J}$ has eigenvalues that all lie on the interval $\alpha_n \in [1, 2.47]$. If the Newton flow equation is modified to

$$\frac{d\mathbf{u}}{dT} = -\left(\frac{1}{\mathbf{H}}\right)\mathbf{r}(\mathbf{u}) \qquad \text{("Preconditioned Newton flow")}, \tag{4.28}$$

then Taylor expansion near a root gives

$$\frac{d\mathbf{u}}{dT} = -\left(\frac{1}{\mathbf{H}}\right)\mathbf{J}(\mathbf{u} - \mathbf{u}_0) + O([\mathbf{u} - \mathbf{u}_0]^2). \tag{4.29}$$

Orszag's theory implies that each root is still a local attractor. If the error $\mathbf{u} - \mathbf{u}_0$ is decomposed into the eigenvectors of $(1/\mathbf{H})\mathbf{J}$, then the error in different components will decay with T at various rates ranging between $\exp(-T)$ and $\exp(-T/2.47)$.

Because of these varying rates of decay, it is not possible to find a magic time step that will retain the quadratic convergence of Newton's method. However, the Euler forward scheme does give very rapid geometric convergence: with a time step of $\frac{4}{7}$, one can guarantee that each component of the error will decay by at least a factor of 2.33 at each iteration. Since the factorization of the finite difference matrix is cheaper than that of the pseudospectral Jacobian by a factor of $O(N^2)$ in both one and two dimensions, the thirty or so preconditioned iterations that are needed to achieve spectral accuracy are orders of magnitude cheaper than the corresponding four or five Newton iterations. When $N = 100$ per coordinate, the preconditioned iteration may be cheaper by a factor of 2000!

Orszag developed this preconditioning for *linear* equations; in this context, the forward Euler scheme is known as the Richardson iteration. Thus, the "Preconditioned Newton Flow Iteration", which follows from applying the Euler method to (4.28), may alternatively be dubbed the "Nonlinear Richardson Iteration":

$$\mathbf{H}\mathbf{u}^{n+1} = \mathbf{H}\mathbf{u}^{n} - \tau\mathbf{r}(\mathbf{u}^{n}) \qquad \text{("Preconditioned Newton Flow Iteration")}, \tag{4.30}$$

where τ is the time step.

Although the author devised (4.30), it is such an obvious idea that others have independently invented it, too. This iteration is briefly mentioned in the book by Canuto *et al.* (1987, p. 390) and also in a conference paper/technical report by Streett and Zang (1984), but the author has not yet found it in a regular journal article. Nonetheless, it is a very powerful technique.

The finite difference preconditioning requires that the unknowns be the values of $u(x)$ on the collocation grid, rather than the Chebyshev or Fourier coefficients. However, this requires only the trivial modification of replacing the Chebyshev polynomials or functions by the corresponding "cardinal functions" (Boyd, 1989d).

For expository purposes, we shall use the simplest of cardinal functions: the sinc or "Whittaker cardinal" functions (Stenger, 1981) functions: the sinc or "Whittaker cardinal" functions (Stenger, 1981) defined by

$$\operatorname{sinc}(x) \equiv \frac{\sin(\pi x)}{\pi x}, \tag{4.31}$$

which have the property that

$$\operatorname{sinc}(0) = 1 \quad \text{and} \quad \operatorname{sinc}(j) = 0 \qquad \text{for all integers } j \neq 0. \tag{4.32}$$

It follows that if we write

$$u(x) = \sum_{j=-\infty}^{\infty} u_j \phi_j(x; h), \tag{4.33}$$

where h is the spacing of the grid, a constant, and

$$\phi_j(x; h) = \operatorname{sinc}\left(\frac{x - jh}{h}\right), \qquad j = 0, \pm, \pm 2, \ldots, \tag{4.34}$$

then the values of $u(x)$ on the evenly spaced grid

$$x_j = jh \tag{4.35}$$

are simply

$$u(x_j) = u_j, \tag{4.36}$$

so that the spectral coefficients are the grid point values.

The mechanics of differentiating the sinc functions, computing second-order centered differences on a uniform grid, and solving (4.30) with a band-solver are described in Boyd (1989d). The infinite interval is truncated by assuming that all u_j with $|j| > N$ are negligibly small and can be replaced by zero. Figures 6 and 7 show the computation of the solitary wave of the second order form of the Korteweg–deVries equation.

The accuracy—a maximum pointwise error which is 200,000,000 times smaller than the maximum of the soliton—clearly illustrates the power of pseudospectral methods. Although the Fast Fourier Transform cannot be combined with sinc functions so that the cost of computing $\mathbf{r}(\mathbf{u})$ is $O(N^2)$ (versus $O(N \log_2 N)$ for Chebyshev and Fourier pseudospectral algorithms), each preconditioned Newton flow iteration required only about 4 seconds on a 6 Mhz IBM AT, even with 201 grid points. In contrast, the LU factorization of a 201×201 dense matrix would cost about 400 seconds. Although we need about 20 iterations to achieve spectral accuracy instead of the 4 to 5 iterations of Newton's method, the

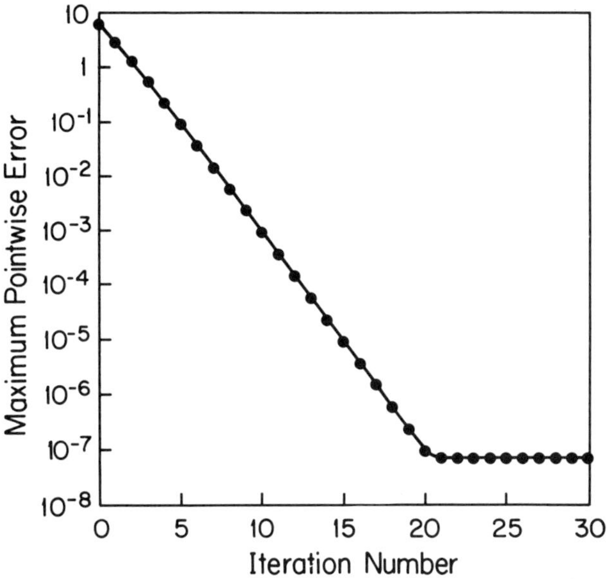

FIG. 6. The maximum pointwise error as a function of iteration number for the preconditioned Newton flow iteration. The differential equation is the once-integrated Korteweg–deVries equation, $u_{xx} + (u/2 - c)u = 0$ for a phase speed of 4, subject to the boundary conditions $u(\pm\infty) = 0$. The time step is 0.6 and the error decreases by a factor of a little over 2.5 at each iteration until plateauing at about 5×10^{-8}, which is the maximum pointwise error of the spectral calculation with 201 grid points and a uniform grid spacing of $\frac{1}{10}$.

preconditioned Newton flow is about 20 times faster overall. Better yet, memory storage is also reduced by a factor of $N/3$. (Note that one may compute $\mathbf{r}(\mathbf{u})$ *without* calculating and storing the Jacobian matrix.)

It is possible to solve the third derivative form of the Korteweg–deVries equation, too. The crucial trick is to use a staggered grid in which the finite differences and pseudospectral derivatives are evaluated at

$$x_j = h(j + \tfrac{1}{2}), \qquad j = -N, \ldots, N, \tag{4.37}$$

while the sinc coefficients, i.e., the unknowns, are the $2N + 1$ values of $u(x)$ on the unstaggered symmetric grid (4.35). The nonlinear Richardson iteration converges more slowly than that for the second order equation—Orszag (1980) explains that the condition number of the preconditioned matrix is poorer as the order of derivatives increases—but it is still fast and efficient.

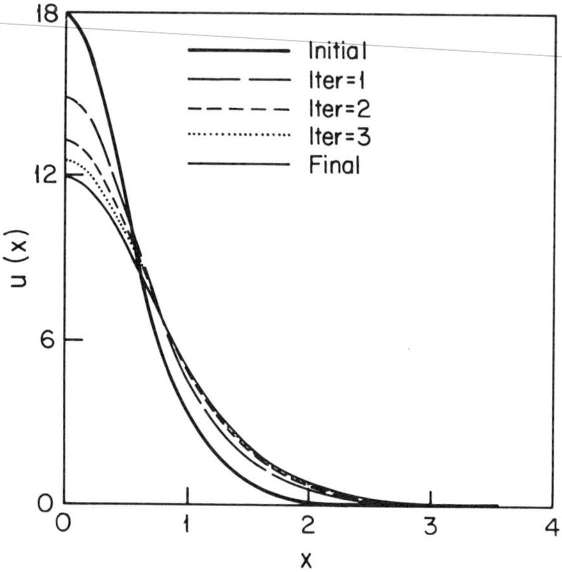

FIG. 7. Same case as the previous figure: $u(x)$ for the first guess, the first three iterates, and the final converged result, which is indistinguishable from the exact soliton. The computational interval is $[-10, 10]$, but only a subinterval is shown because the soliton is symmetric about $x = 0$ and exponentially small for large $|x|$.

The book by Boyd (1989d) explains the use of other cardinal functions with extensive differentiation tables, etc. However, the Fourier and Chebyshev cardinal functions are very similar in shape to the sinc function and are also given by simple analytical expressions.

Canuto *et al.* (1987) thoroughly discuss preconditioning and iterations. Incomplete *LU* factorizations are a popular alternative to strict finite difference preconditioning for multi-dimensional equations because even finite difference matrices are expensive to invert in more than one dimension (unless the equation is separable). Pseudospectral multigrid may be the ultimate weapon for direct computation of solitons. This, too, however, fits into the framework of the "Newton flow."

The iterations of Eydeland and Turkington (1987, 1988) and Petviashvili (1976) and Petviashvili and Tsvelodub (1978) may also be justified by applying Euler's forward marching method to a differential equation in artificial time. (These authors justified their schemes via variational principles, however.) For example, Eydeland and Turkington solve

$$\psi_{xx} + \psi_{yy} + F(\psi + cy - \mu) = 0, \tag{4.38}$$

where c and μ are eigenparameters and $F(z)$ is a known function. If write this as the generalized diffusion equation

$$\psi_T = \psi_{xx} + \psi_{yy} + F(\psi + cy - \mu), \tag{4.39}$$

where T is the artificial time coordinate, the Euler one-step method is equivalent to the "power" method for solving eigenvalue problems.

If F were linear (as for analytically soluble modons (Flierl, 1987)), then one could solve (4.39) via separation of variables. The high order modes are (approximately) proportional to $\exp(ikx + imy)$ and decay with time as $\exp(-[k^2 + m^2]T)$. The gravest mode, that is, the eigenmode with the simplest spatial structure, may either grow or decay, depending upon $F(z)$. In any event, this mode will decay (perhaps at a negative rate!) more slowly than any of the higher modes. Consequently, for sufficiently large T, the solution of (4.39) is dominated by the lowest eigenmode. Thus, the structure of ψ converges geometrically to that of the gravest eigenfunction of the operator on the right hand side of (4.39). The eigenvalue (if there is only one) may be determined by monitoring the growth or decay rate of $|\psi|$ for $T \gg 1$.

When the problem is nonlinear, the qualitative behavior is exactly the same. The power method is a very simple and effective way to determine the gravest mode for both linear and nonlinear problems. The only complication is that when $F(z)$ is nonlinear, the changes in the amplitude of ψ will alter the shape of F, and this in turn will change both the eigenvalue and eigenfunction. To specify a unique solution, one must demand

$$\|\psi\| = a, \tag{4.40}$$

where a is some amplitude parameter and the $\|\quad\|$ denotes some norm or measure of amplitude, and then enforce this constraint at each iteration. Actually, most texts recommend applying (4.40) even for linear problems (where it is not absolutely necessary) to avoid roundoff problems.

To impose this constancy-of-amplitude condition, note that the Euler forward scheme gives

$$\psi^{j+1} = \psi^j + \tau\{\psi^j_{xx} + \psi^j_{yy} + F(\psi^j + c^{j+1}y - \mu^{j+1})\}, \tag{4.41}$$

where τ is the artificial time step. If ψ^j is known, then taking the norm of both sides of (4.41) gives a nonlinear equation that can be solved for c^{j+1}. For Eydeland and Turkington's problem, it is necessary to impose *two* constraints like (4.40) to specify a unique solution. Applying these (independent) conditions to (4.41) gives two nonlinear equations in the two unknowns c^{j+1}, μ^{j+1}. The principle is the same, however, independent of the number of constraints.

One simplification is that it is not necessary that the "norm" be a norm in a strict mathematical sense. For example, Eydeland and Turkington prove that the vorticity is positive definite for their problem, so they use the area integral of the vorticity as one of their constraints. This definition of amplitude does not satisfy the mathematical definition of a norm, but has the virtue of being easy to impose on the right hand side of (4.41).

There is one remaining problem: the rapid decay of the high modes in (4.39) demands that the time step τ in (4.41) should be very, very small. Eydeland and Turkington eliminate this by replacing ψ_T in (4.39) by $-\psi_{xxT} - \psi_{yyT}$. If we define the vorticity ω (strictly speaking, the negative of the vorticity) and also the Green's function for the Poisson equation G via

$$-(\psi_{xx} + \psi_{yy}) \equiv \omega \qquad \leftrightarrow \qquad \psi = G\omega, \tag{4.42}$$

then the pseudotime equation becomes

$$\omega_T = -\omega + F(G\omega + cy - \mu). \tag{4.43}$$

The high order modes, which are approximate eigenfunctions of the Laplacian operator, now all decay like $\exp(-T)$ due to the first term on the right hand side in (4.43)). Consequently, a unit time step is stable and the iteration becomes simply

$$\omega^{j+1} = F(G\omega^j - c^{j+1}y - \mu^{j+1}), \tag{4.44}$$

which is the form Eydeland and Turkington actually use. Verkley (1989) independently discovered and used the same iteration to compute vortex pairs on the sphere.

The philosophy of eliminating the huge spread in decay rates is similar to the preconditioning of the Newton flow algorithm. However, the "preconditioning" here is not intended to approximate the full Jacobian matrix but only the part that controls the high order modes. (Here, this part is the discretization of the Laplacian operator.) The reward for this cruder preconditioning is that at each iteration, one only needs to invert the Laplace operator—not a general, nonseparable partial differential equation. The cost of solving the Poisson equation on an $N \times M$ grid is only $O(10\,NM)$ operations via multigrid (Briggs, 1987).

The danger of "imperfect" preconditioning is that (4.44) may have modes that might diverge *if* the iteration were applied with fixed c and μ. The beauty of the nonlinear power method is that c and μ are *not fixed,* but rather are adjusted at each iteration so that the norm of ω always stays the same.

Petviashvili (1976) used an iteration very similar to (4.44) except that he chose to fix the phase speed rather than the amplitude with the result that (4.44) *is* unstable and divergent. However, he devised a simple fix.

In his case, the right hand side of (4.44) is quadratic in the unknown. This implies that if the first guess is too large by a factor of two, the next iterate will be four times too large and the iteration diverges. The behaviour is the same as that for the simple iteration/ordinary differential equation

$$u_T = u^2 - u \qquad \leftrightarrow \qquad u^{j+1} = (u^j)^2. \tag{4.45}$$

Equation (4.45) has a fixed point at $u = 1$, but it is unstable.

Petviashvili's remedy is to multiply the nonlinear term in (4.45) by the α-th power of the norm of the unknown divided by the α-th power of the nonlinear term where α is a user-choosable parameter. For (4.45), this gives

$$u_T = \left(\frac{u}{u^2}\right)^{\alpha} u^2 - u \qquad \leftrightarrow \qquad u^{j+1} = (u^j)^{2-\alpha}. \tag{4.46}$$

If we write $u = 1 + \Delta$ where $\Delta \ll 1$, then near the fixed point $u = 1$, (4.46) becomes

$$\Delta_T = (1 + \Delta)^{2-\alpha} - (1 + \Delta) \approx (1 - \alpha)\Delta + O(\Delta^2). \tag{4.47}$$

This shows clearly that the fixed point is unstable for $\alpha < 1$, but when $\alpha > 1$, Δ decreases exponentially fast with the pseudotime T. The Euler one-step marching scheme converts (4.47) into a stable, geometrically converging iteration. The best choice is $\alpha = 2$, whcih gives the exact answer in one iteration.

When u is a vector of spectral coefficients, $\alpha = 2$ is still the best choice for quadratic nonlinearity. Equation (4.44) is replaced by

$$\omega^{j+1} = \left\{\frac{||u||}{||F(G\omega^j - cy - \mu||}\right\}^{\alpha} F(G\omega^j - cy - \mu). \tag{4.48}$$

We have omitted the superscripts on c and μ because one can specify these quantities with Petviashvili's iteration. In his scheme, the two amplitude constraints are variable and (c, μ) are fixed; Eydeland and Turkington's method fixes the amplitude constraints and computes c and μ during the iteration. These two alternatives of fixed amplitude/variable phase speed versus fixed c/variable amplitude are almost always available in soliton studies. The choice is strictly a matter of convenience.

The "nonlinear power" method of Eydeland and Turkington (and Verkley), the "iteration-with-stabilizing-factor" of Petviashvili, and the "nonlinear Richardson iteration" of the author and Streett and Zang (1984) all have geometric rates of convergence in contrast to the quadratic convergence of Newton's method. The reward for slower convergence is much lower cost per iteration. Furthermore, all three of these algorithms require only $O(NM)$ storage where NM is the total number of grid points, and all can be combined with any type of discretization.

E. Suggestions and Guidelines

The pseudospectral method with a Fourier, Chebyshev, or rational Chebyshev basis is the recommended discretization. Finite differences and finite elements work just fine, but are limited to rather modest accuracy. In contrast, the pseudospectral error decreases exponentially fast with N so that one can often obtain six decimal place accuracy with only twenty or thirty grid points in each coordinate. For computing the very small radiation coefficients of weakly nonlocal nanopterons (next section), pseudospectral methods are almost essential.

For solving the linear algebra problem, we have always endorsed this maxim: Never use anything but Gaussian elimination (Crout reduction) except when absolutely necessary. The iterative schemes are much faster and use less storage, but a few seconds' savings is not worth the bother. Furthermore, all iterations, even the robust schemes described above, have limits. Eydeland and Turkington, for example, prove their scheme converges for vortex pairs, but not necessarily for vortex quadrapoles.

Nonetheless, it is very comforting to know that there are fast, low storage alternatives to Gaussian elimination. There is no longer any good excuse for crude, low resolution calculations of multi-dimensional solitons. A mainframe is not needed. A microcomputer and a couple of hundred lines of code will do just fine.

V. Weakly Nonlocal Solitary Waves

A. Introduction

The classical definition of a soliton is: A "solitary wave" is a *localized, finite amplitude, steadily translating* disturbance of *permanent shape* and *form* and constant phase speed c. Unfortunately, this definition is too narrow. Weakly nonlocal solitons are nonlinear waves that violate one or more of the essential conditions in this definition, and yet closely resemble classical solitary waves by being long-lived, stable, and coherent.

The adjective "nonlocal" means that the quasi-soliton fails to meet the test of "localization" because it asymptotes not to zero, but to a constant amplitude oscillation as $|x|$ increases. The adjective "weakly" means that the wave is similar to a solitary wave, in spite of its nonlocal character, because the amplitude of the "far field" radiation is very *small* in comparison to the maximum amplitude of the solitary wave. It is implicitly assumed that the "weakly nonlocal" wave *cannot be altered into a strict soliton through any small changes or tuning.*

It is useful to subdivide "weakly nonlocal" waves into two subcategories, depending on precisely how the strict definition of a soliton is violated. A "radiatively decaying soliton" is a nonlinear solution that has the following characteristics:

(i) It is strictly *localized* in the sense that $u(x, y, t) \to 0$ as $|x| \to \infty$.

(ii) It satisfies all other requirements of a solitary wave except that it *decays* very *slowly* with time through radiation of energy to large $|x|$.

(iii) The radiative decay is very *slow* because there is an *almost perfect* cancellation between nonlinearity and dispersion.

(iv) It violates the strict definition of a soliton because the wave is not steadily translating and permanent, but slowly disperses as it propagates.

When a localized initial condition is given to a wave equation that allows only nonlocal solitons, radiatively decaying solitons will dominate the wave field for $t \gg 1$. The nonsolitonic transients will disperse quickly, leaving behind a coherent structure that also disperses, but at a much slower rate. The nonlocal solitary wave is a structure in which *most* but *not all* of the dispersion is balanced by nonlinearity.

This subcategory can be subdivided into "algebraically slowly" and "exponentially slowly" decaying solitons, depending on whether the rate of decay is an *algebraic* or *exponential* function of the appropriate small parameter. We shall elaborate on these distinctions below.

The second subcategory, that of "steadily translating, nonlocal" solitons may be similarly divided into two subclasses. A "nanopteron" or micropteron" is a nonlinear solution on $x \in [-\infty, \infty]$ that has the following characteristics:

(i) It is permanent and steadily translating at a constant phase speed c.
(ii) It satisfies all other requirements of a solitary wave except that it is not strictly *localized* because the wave asymptotes to a small amplitude oscillation when $|x| \gg 1$.
(iii) The amplitude of the far field oscillation is an *exponentially* small function of the maximum amplitude of the wave for the nanopteron. The far field oscillation of the micropteron is an algebraic function of some small parameter $\delta \ll 1$, that is, the oscillation is proportional to δ^n for some n.

Figure 8 (top pair) compares these two species of weakly nonlocal solitary waves. The radiatively decaying soliton is localized but unsteady; the nanopteron (or micropteron) is steadily translating but nonlocal.

The theme of Section III was the close relationship between solitons and spatially periodic (cnoidal) waves. Similarly, one may superimpose nonlocal solitons with even spacing to generate a periodic "nanopteroidal" wave. This has an infinite number of evenly spaced peaks or troughs, each differing only slightly from the corresponding features of the nanopteron. The space between the large amplitude peaks is filled with very weak oscillations whose amplitude and wavenumber match the far field of the nanopteron. The bottom half of Figure 8 compares the nanopteron and nanopteroidal wave. Such periodic generalizations of nonlocal solitons are discussed at length in Boyd (1988b, 1989a, 1989d), but the need for brevity prohibits a longer discussion of nanopteroidal waves here.

In this rest of this article, we concentrate on nanopterons. Because these have trivial time-dependence—steady propagation at a constant phase speed—one may write $u(x, y, t) = u(x - ct, y)$ and compute the waves by solving a boundary value problem (BVP). This is the same BVP strategy as we applied numerically to strict solitons in Section IV. The

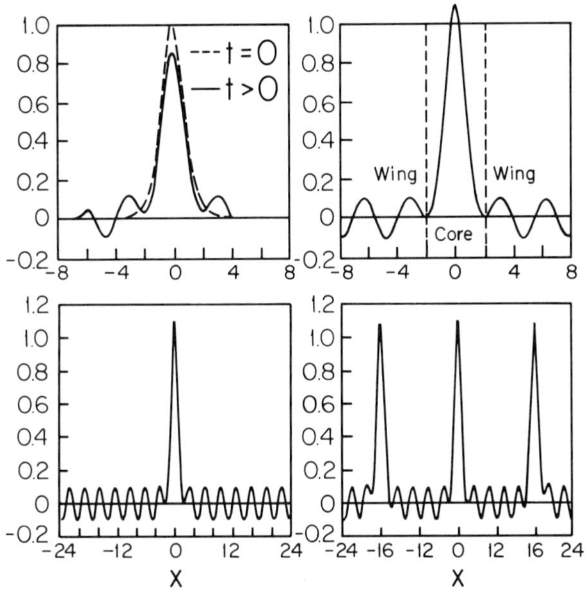

FIG. 8. Top left: schematic of a "radiatively decaying soliton" for $t = 0$ (dashed) and some later time (solid). The dispersion need not be symmetrical. Top right: schematic of a nanopteron. This is nonlocal, but is steady in a coordinate system travelling with the wave. Bottom left: schematic of a nanopteron. Bottom right: the corresponding nanopteroidal wave with a spatial period of 16.

only complication is that one must modify the discretization to cope with the far field oscillation as described in Subsection C below.

The "radiatively decaying solitons" are much harder to analyze because of their complex variations in both time and space. Boyd (1988c) shows that the ordinary differential equation $d\varepsilon/dt = \exp(-1/\varepsilon)$, which is the simplest model for the decay of a nanopteron, has multiple logarithmic time scales. The time evolution of an actual nonlocal wave is likely to be even more complex. Nonetheless, it is necessary to define them because radiatively decaying solitons are what will be observed in a numerical model with a localized initial condition. Williams and Wilson (1988) is a good example.

Unfortunately, it is difficult to quantitatively measure the far field radiation while the disturbance, initially confined to small $|x|$, spreads wider and wider through the emission of small amplitude waves. Consequently, we shall concentrate on nanopterons instead. The two species of weakly nonlocal solitons are closely related, however. One may in principle use energy arguments to infer the decay rate of the radiatively decaying soliton from the magnitude of the far field ripples of the nanopteron (Segur and Kruskal, 1987).

The unfamiliar terminology is borrowed from zoology (Boyd, 1989a). In classical Greek, "micropteron" means "small-wing" while "nanopteron" is "dwarf-wing". The "wings" of the nanopteron, that is, the very small ripples surrounding the large amplitude core, are marked in Figure 8 (top right).

It is unpleasant to wade through so many definitions, but this is the price of assaulting a scientific frontier. Nonlocal solitons are still a very new concept even though we shall discuss eight examples below. The terms and the schematic, Figure 8, are needed to give a language to describe these new ideas. The reward is the excitement of the pioneer, splashing ashore onto undiscovered land.

In the next subsection, we shall show through example how to analyze the far field oscillations. The following two subsections describe strategies for coping with such oscillations: perturbation theory (which fails) and a special numerical method (which spectacularly succeeds). Subsection E is a catalogue of physical examples of both micropterons and nanopterons.

B. Far Field Analysis

The "far field" is defined as the region $|x| \gg 1$ (with the soliton peak at $x = 0$) where the solitary wave has decayed to such small amplitude that its dynamics becomes *linear*. Because of the linearity, it is easy to prove the following crucial result.

Theorem 5.1 (Far Field Analysis). *If the phase speed of a nonlinear coherent structure lies* outside *the range of the* linear *phase speeds of* all *waves in the system, then it is possible that the soliton decays exponentially as* $|x| \to \infty$. *If* c *is equal to the speed of some linear wave mode for some real wavenumber* k, *then the nonlinear coherent structure will be* oscillatory *rather than* decaying *in the far field, and cannot be a classical, localized soliton.*

Proof-by-Example. Pomeau, Ramani, and Grammaticos (1988) consider the generalized Korteweg–deVries equation which, after one integration, is

$$u_{xxxx} + u_{xx} + \left(\frac{u}{2} - c\right)u = 0, \tag{5.1}$$

where c is the phase speed and x is the spatial coordinate in a frame of reference traveling with the wave. For conformity with other examples, we have made some slight changes in notation. For brevity, we shall refer to (5.1) as the "PRG" equation.

In the absence of the fourth derivative, Equation (5.1) reduces to the KdV equation, which has the exact soliton

$$u(x) = 12\varepsilon^2 \operatorname{sech}^2(\varepsilon[x - ct]) \qquad \text{and} \qquad c = 4\varepsilon^2. \tag{5.2}$$

When $\varepsilon \ll 1$, it is trivial to show that (5.2) is also a good approximate solution of the PRG equation (5.1) because the fourth derivative is $O(\varepsilon^2)$ smaller than the other terms in the equation.

In the far field where $|x| \gg 1$, the nonlinear term is exponentially small (at least to the extent that (5.2) can be believed), so the PRG equation reduces to

$$u_{xxxx} + u_{xx} - cu = 0 \qquad (\text{"Far field PRG"}). \tag{5.3}$$

The solutions of this constant-coefficient ordinary differential equation are exponentials, each proportional to $\exp(ikx)$; the wavenumbers are given by the *linear* dispersion relation

$$k^2 = \tfrac{1}{2} \pm \tfrac{1}{2}(1 + 4c)^{1/2}. \tag{5.4}$$

When the phase speed $c > 0$, as it is for the soliton, then one root for k^2 is negative, implying that $k \approx \pm ic^{1/2}$. This root is approximately the same as for the KdV equation and corresponds to the exponential decay of the $\operatorname{sech}^2(\varepsilon x)$ in (5.2). The other root, however, implies that in the far field

$$u(x) \sim \alpha_{\pm} \sin(k_f x) + \beta_{\pm} \cos(k_f x), \qquad |x| \gg 1, \tag{5.5}$$

where

$$k_f = 1 + O(c) \qquad (\text{"Far field wavenumber"}) \tag{5.6}$$

and where the subscripts on the constants α, β indicate that these may be different for $x \to \infty$ and $x \to -\infty$. Unless all the α's and β's are zero, the PRG wave flunks the localization test because it asymptotes to an oscillation.

The PRG soliton is nonlocal because its positive phase speed matches that of a free, traveling oscillation with wavenumber $k = k_f(c)$. In contrast, the Kortweg–deVries soliton is strictly localized because all the linear sine waves of the KdV equation have negative phase speeds while the soliton has a positive phase speed.

Strictly speaking, this proof of the Far Field Theorem is only half a proof. (The other half is to prove that $\alpha_\pm \neq 0$ and/or $\beta_\pm \neq 0$.) Asymptotic analysis as $|x| \to \infty$ is nonetheless a very simple and powerful tool for identifying candidates for nonlocal solitons.

The concept of the solitary wave as a robust, long-lived, and concentrated disturbance may still apply even when the soliton is nonlocal provided that the amplitude of the far field oscillations is sufficiently small. For some micropterons, such as the long-lived but radiating two-layer modons of Flierl (1984), the smallness is directly proportional to a small parameter with a small proportionality constant. For other classes of nonlocal waves, a much stronger statement is possible.

Theorem 5.2 (Exponential Smallness). (i) (*General*) *If the solitary wave is derived via the method of multiple scales with the assumption that the pseudowavenumber* ε *is small and if the zonal scale of the far field oscillations is* $O(1)$, *then the* amplitude *of the far field radiation is* exponentially small *as a function of* $1/\varepsilon$.

(ii) (*Restricted*) *If the soliton is proportional to some positive integral power of* $\operatorname{sech}(\varepsilon x)$, *then the amplitude* α *of far field radiation of zonal wavenumber* k_f *is*

$$\alpha \sim \nu \exp\left(\frac{-k_f \pi}{2\varepsilon}\right), \qquad \varepsilon \ll 1, \tag{5.7}$$

where ν *is either a constant or an algebraic function of* ε.

Proof. See Boyd (1988b, 1989d). The key is the separation of length scales. The core of the solitary wave generates the small amplitude waves that propagate to infinity. This is clear for the radiatively decaying solitary wave: if $u(x, t=0)$ decays exponentially with $|x|$, then this initial crest is the source of whatever waves appear at large $|x|$ for large times. However, the same is true for the nanopteron: insofar as the oscillation for large $|x|$ is concerned, the core is merely a stationary antenna.

In turn, the Fourier spectrum of the nanopteron core determines how much energy is available in waves with $k = k_f$ to be radiated to infinity. The Fourier transform of any smooth function, such as the core of a nanopteron, decays exponentially fast with wavenumber k. The broader the soliton, the more rapidly its tranform falls off. The Fourier transform of $\text{sech}^2(\varepsilon x)$, for example, is proportional to $\exp(-k\pi/[2\varepsilon])$. This implies that the excitation of the waves with $k = k_f$ will be exponentially small in ε—small proportional to $\exp(-k_f\pi/[2\varepsilon])$.

The separation-of-length scales between the nanopteron core and the nanopteron far field is what makes the large $|x|$ oscillation so small. The slowly varying core and the rapidly varying far field wings are coupled exponentially weakly because of their differing scales, $O(1/\varepsilon)$ versus $O(1)$.

The arguments in Boyd (1988b, 1989c) are more systematic than this brief verbal explanation but still short of a rigorous proof. Nevertheless, for each of the physical applications in Subsection E including the PRG soliton, (5.7) has been checked by explicit calculation.

The nanopteron is long-lived only when the radiation coefficient $\alpha \ll 1$. The theorem shows that this condition is met, perhaps to an almost ridiculous degree, when ε is only moderately small. For the PRG nanopteron, for example, $\varepsilon = \frac{1}{10}$ implies $\alpha = O(10^{-7})$!

C. Perturbation Theory

It is easy to calculate a perturbation series in powers of the amplitude and scale parameter ε. For the PRG equation, for example, the coefficients are polynomials in $\text{sech}^2 x$ and the lowest order is the Korteweg–deVries soliton (5.2). Unfortunately, the perturbation series diverges.

One reason for the divergence is that a function like $\exp(-\pi k_f/[2\varepsilon])$, which is smaller than any finite power of ε as $\varepsilon \to 0$, cannot possibly be represented by a power series in ε. A second reason is that a polynomial in $\text{sech}^2(\varepsilon x)$ cannot approximate an oscillation for $|x| \gg 1$ because each term in the polynomial is exponentially small in the far field. Boyd (1988a) shows, however, that even if the sign of the fourth derivative in the PRG equation were changed so that its solutions are localized, the ε power series still diverges. The real villain is the assumption of multiple length scales, which almost always yields divergent but asymptotic series.

Often, divergent power series can be summed via Padé approximants. Boyd (1988a) shows that for nonlocal solitons, even Padé methods fail.

However, it is possible to calculate exponentially small quantities like the radiation coefficients by applying the method of matched asymptotic expansions in the *complex* plane. The pioneers were Pokrovskii and Khalatnikov (1961) and Meyer (1980), who solved the linear problems of WKB reflection without a turning point and quasi-trapped oscillations around islands. Segur and Kruskal (1987) and Pomeau, Ramani, and Grammaticos (1988) have extended this method to nonlocal solitons.

The key idea is to deform the contour of integration into the complex x-plane. For the PRG, for example, the zeroth-order nanopteron has simple poles at $x = \pm i\pi/(2\varepsilon)$. The neighborhood of these poles is the "inner region"; in this disc-shaped boundary layer, the far field oscillation is *not* transcendentally small in comparison to the rest of $u(x)$. Matching this inner solution to the outer solution, which vanishes as $|x| \rightarrow \infty$ along the real axis, gives the radiation coefficients (the α's and β's in (5.5)) in the limit $\varepsilon \rightarrow 0$.

This improved perturbation theory is creative and deepens our understanding of nonlocal solitons. However, it does have drawbacks. First, as illustrated below, the method seems to generate rather inaccurate approximations unless ε is so small that the radiation coefficients are below the limits of machine precision. In contrast, the numerical strategies below can calculate the far field oscillation for *all* ε, not just in the limit $\varepsilon \rightarrow 0$.

A more serious failing is that contour deformation implicitly assumes that no singularity lies between the new and old integration paths. For the linear, second order differential equation of Pokrovskii and Khalatnikov (1961), the singularities of $u(x)$ are the known poles and branch points of the coefficients of the differential equation. Unfortunately, no such theory is available for nonlinear problems. Segur and Kruskal (1987) and Pomeau, Ramani, and Grammaticos (1980) are forced to assume—without proof—that the singularities of the perturbed nanopteron are those of the unperturbed, zeroth order soliton. The numerical calculations support their conjectures, but it is not clear that future workers will be so lucky!

Thus, their theories, which predict that the radiation coefficients are nonzero, strengthen the argument that the ϕ^4 breather and PRG soliton are weakly nonlocal. However, the matched asymptotics method is not rigorous. A pure mathematician can still find some challenging problems in nonlocal soliton theory!

D. Numerical Methods: Ignorance, the Radiation Basis, and Cnoidal Matching

Nanopterons can be calculated by solving a nonlinear boundary value problem on $x \in [-\infty, \infty]$. The pseudospectral rational Chebyshev method of Section IV is very effective when $u(x)$ decays to 0 as $|x| \to \infty$, but nonlocal solitons do not. There are three strategies for coping with this problem.

The first is to simply ignore it! Boyd (1988a) shows that the coefficients of the rational Chebyshev functions, $TB_n(x)$, decrease exponentially until they are somewhat smaller than α—and then level off. Thus, this strategy of "ignorance" is as effective as taking the multiple scales series to optimum order: the error is $O(\alpha)$ but no better.

The second strategy is to augment the rational Chebyshev basis by adding one or more special "radiation basis functions." These are chosen to mimic the desired far field behavior. For example, if

$$u(x) \sim \alpha(\varepsilon)\, \mathrm{sgn}(x) \sin(kx), \tag{5.8}$$

which is appropriate for computing symmetric (with respect to $x = 0$) nanopterons of all the examples discussed in the next subsection, then a good radiation basis function is

$$\phi_{\mathrm{rad}}(x; \varepsilon) \equiv \tanh(\varepsilon x) \sin(kx). \tag{5.9}$$

In the far field ($|x| \gg 1$), $\tanh(\varepsilon x) \sim \mathrm{sgn}(x)$, so this clearly reproduces the asymptotic behavior of $u(x)$. However, this choice is not unique and $\tanh(\varepsilon x)$ could be replaced by any smooth, antisymmetric function.

This special basis function replaces the N-th Chebyshev basis function in the series for $u(x)$. Otherwise, the pseudospectral method is applied exactly as in Section IV.

Boyd (1988a) shows that the mixed basis is a great improvement. Because the Chebyshev functions are only required to do what they do best—represent the exponentially decaying core of the nanopteron—the coefficients do not level off at $O(\alpha)$. Figure 9 shows a PRG soliton and its two constituent parts; one part is the sum of the rational Chebyshev functions and the other is $\alpha\phi_{\mathrm{rad}}$.

For the *linear* problem of Boyd (1988a), a single radiation function is sufficient. For nonlinear problems, however, the self-interaction of the far field wave will give an $O(\alpha^2)$ correction to the far field $u(x)$. (This

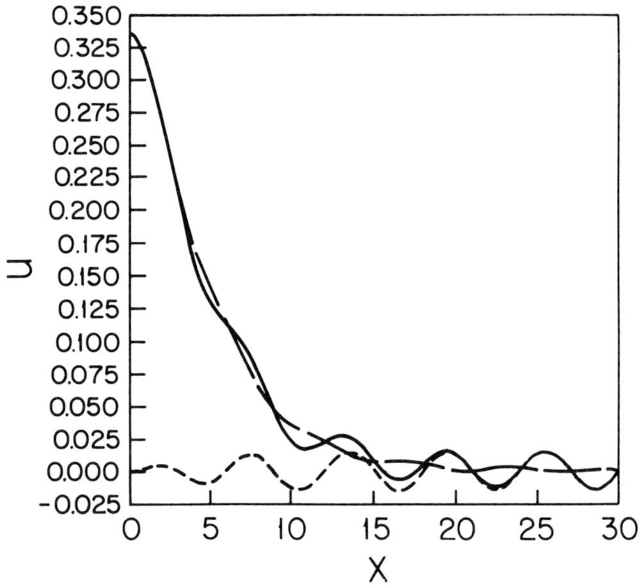

FIG. 9. Solid: the PRG nanopteron for $\varepsilon = 0.16$. The core ($|x| < 10$) is approximately equal to $12\varepsilon^2 \operatorname{sech}^2(\varepsilon x)$. Long dashes: the sum of the rational Chebyshev functions in the numerical solution. Short dashes: $\alpha\, \phi_{\text{rad}}(x; \varepsilon)$, where α is the "radiation coefficient" and ϕ_{rad} is the radiation basis function.

correction is faintly visible in Figure 9 as the oscillations in the sum of the Chebyshev functions.)

The third strategy is to correct for this far field self-interaction by matching the core to the appropriate cnoidal wave, that is, to a *nonlinear, spatially periodic* oscillation. Because $\alpha \ll 1$ if the nanopteron is long-lived and only weakly non-local, Stokes' perturbation method can calculate the far field to as great an accuracy as one pleases. The twist on the procedure described in Subsection II.E is that it is the far field wavenumber k, rather than the phase speed c, which is expanded:

$$k \sim k_0 + \alpha^2 k_2 + \cdots. \tag{5.10}$$

Turning the crank for the PRG equation gives

$$u_{\text{cn}}(x) = \alpha \sin(k\bar{x}) + \alpha^2[u_{10} + u_{12}\cos(2k\bar{x})] - \alpha^3\left[\frac{u_{12}}{162k_0^4 - 18k_0^2 - 2c}\right]\sin(3k\bar{x}), \tag{5.11a}$$

$$k_0 = \{\tfrac{1}{2}[1 + (1 + 4c)^{1/2}]\}^{1/2}, \qquad k_2 = \frac{u_{12}/2 - u_{10}}{4k_0^3 - 2k_0}, \tag{5.11b}$$

$$u_{10} = \frac{1}{4c}, \qquad u_{12} = \frac{1}{64k_0^4 - 16k_0^2 - 4c}, \tag{5.11c}$$

$$\bar{x} \equiv x + \text{phase constant}. \tag{5.11d}$$

These Stokes expansions must be separately matched to the core as $x \to -\infty$ and as $x \to \infty$. However, if the nanopteron is symmetric about the origin (see below), we can combine the cnoidal waves for large positive and large negative x into a single basis function by multiplying $\sin(kx)$ (and at higher order than (5.7), all antisymmetric terms in the cnoidal wave) by $\tanh(\varepsilon x)$.

This gives the basis function

$$\phi_{\text{cn}}(x; \alpha) = \alpha^2[u_{10} + u_{12}\cos(2kx)] + \alpha \tanh(\varepsilon x)\left[\sin(kx) - \frac{\alpha^2 u_{12}\sin(3kx)}{162k_0^4 - 18k_0^2 - 2c}\right]. \tag{5.12}$$

Then, assume the expansion

$$u(x) = \sum_{n=1}^{N-1} a_n[\text{TB}_{2n}(x) - 1] + \phi_{\text{cn}}(x; \alpha). \tag{5.13}$$

Although $\phi_{\text{cn}}(x; \alpha)$ is a *nonlinear* function of its coefficient α, this does not cause any real complication. The PRG equation is nonlinear, and we must iterate no matter what the basis.

This third strategy is the most general because the Stokes series for the far field cnoidal wave may be extended to any order. This series converges rapidly if $\alpha \ll 1$ and gives an explicit analytic representation of the basis function.

There is one subtlety: (5.8) assumes that $u(x)$ is symmetric with respect to $x = 0$. In reality, the symmetric, "minimal radiation" nanopteron is only a special case of a much larger family. To the extent that linear theory is accurate in the far field, i.e., to $O(\alpha)$, we have the freedom to add $\sin(k_f x)$ to the far field solution. Boyd (1988b, 1989c) shows that this same freedom exists for the nonlinear solution, too. For example, it is possible to compute a "one-sided" nanopteron that asymptotes to

$$u(x) \sim \begin{cases} 2\alpha \sin(k_f x) & \text{as} \quad x \to \infty, \\ 0 & \text{as} \quad x \to -\infty. \end{cases} \tag{5.14}$$

In other words, we may force the nanopteron to vanish at one endpoint at the price of twice as much oscillation as $x \to \infty$. It is quite trivial, however, to modify the radiation basis functions to compute whatever member of the nanopteron family is desired.

Boyd (1988b, 1989c) gives a much fuller discussion of the far field. Here, we choose for simplicity to compute only nanopterons that are symmetric about the origin and have "minimal" far field radiation. Then the β's in (5.5) are zero and $\alpha_{\pm} = \pm\alpha$, so there is only a single radiation coefficient α.

E. Physical Illustrations: Rossby Waves, Higgs Bosons, and the Slow Manifold

1. *Rossby Waves: Equatorial and Otherwise*

Rossby waves are the planetary-scale free oscillations of the atmosphere and ocean. Pedlosky (1979) and Holton (1979) are good reviews of the importance of these modes in geophysical fluid dynamics.

When the Rossby waves are confined to a channel, one may apply the method of multiple scales to derive the Korteweg–deVries equation. However, the KdV equation describes only the (lowest order) dependence on x and t. The structure in the cross-channel coordinate y (latitude) is the solution of a linear eigenvalue problem. For each latitudinal mode, there is a distinct branch of solitons that have most of their energy in that mode.

Several different mechanisms may confine Rossby waves to a latitudinal channel including ocean topography, variations in the mean zonal flow, and Coriolis refraction. The latter, which creates so-called "equatorial" waves that are confined to a narrow band of latitude centered on the equator, gives the simplest class. However, the analysis of topographically trapped and wind shear-trapped Rossby nanopterons is very similar.

The soliton phase speed for the n-th equatorial Rossby is, letting ε denote the wave amplitude,

$$c = -\frac{1}{2n+1} - O(\varepsilon^2), \tag{5.15}$$

where n must be an odd integer (Boyd, 1980). The linear phase speed for a Rossby wave of x-wavenumber k is

$$c_{\text{linear}} = -\frac{1}{2n+1+k^2}. \tag{5.16}$$

For the gravest, $n = 1$ mode, the soliton phase speed lies outside the range of all linear equatorial waves. Theorem 5.1 then implies that this solitary wave should be a localized, long-lived soliton. This has been confirmed by the initial value experiments of Kindle (1983), Greatbatch (1985), and Williams and Wilson (1988).

In contrast, the higher mode Rossby solitons have nonlinear phase speeds that match the speeds of linear waves of lower mode number. For example, the $n = 3$ soliton has $c \approx -\frac{1}{7}$, which matches the phase speed of $n = 1$ Rossby waves with $k = 2$. The consequence is that the $n = 3$ Rossby soliton is weakly nonlocal, leaking energy along the equator in the form of very small amplitude $n = 1$ Rossby waves with the far field wavenumber $k_f = 2$.

Although quantitative comparisons between nanopteron theory and numerical experiments are not yet possible, the qualitative agreement is excellent. Williams and Wilson (1988) find that the $n = 1$ mode is long-lived and stable even for very large amplitude. However, the higher modes are long-lived only for small amplitude (and then only because the far field radiation is exponentially small in the amplitude). For moderate amplitudes, the solitons radiatively decay through emission of Rossby waves of lower latitudinal mode number, just as predicted. See Boyd (1989a) for more details. Figure 10 shows the amplitudes of the $n = 1$ and $n = 3$ Rossby modes for a simple two-mode model of equatorial waves.

2. *The ϕ^4 Breather (Higgs Boson)*

The so-called "ϕ^4" model has been intensely studied by elementary particle physicists (Boyd, 1988b). The partial differential equation is

$$u_{xx} - u_{tt} - 2u - 3u^2 - u^3 = 0. \tag{5.17}$$

The method of multiple scales gives the perturbative solution (Dashen, Hasslacher, and Neveu, 1974)

$$u(x, t) \sim 0.816\varepsilon \operatorname{sech}(\varepsilon x) \cos(\omega t), \qquad \varepsilon \ll 1, \tag{5.18}$$

$$\omega = 2^{1/2}\left[1 - \frac{\varepsilon^2}{4} + O(\varepsilon^4)\right]. \tag{5.19}$$

Because this nonlinear solution is a standing oscillation that rises and falls without propagation like the chest of a person, this species of soliton is called a "breather".

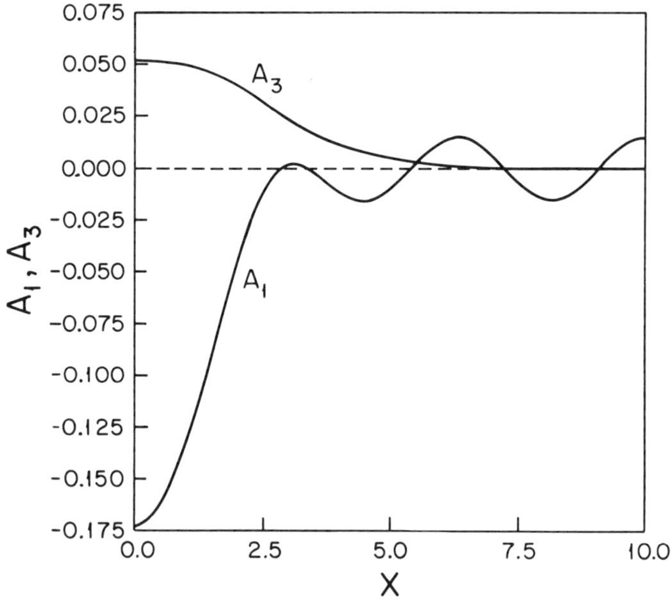

FIG. 10. The amplitudes of the $n = 1$ and $n = 3$ Rossby waves in a two-mode model of the $n = 3$ equatorial nanopteron. The East–West dependence of the $n = 3$ mode, $A_3(x)$, is approximately proportional to $\mathrm{sech}^2(\varepsilon x)$, where $\varepsilon = \frac{1}{3}$. This is the usual form for a Korteweg–deVries soliton; the complete contribution of this term to $u(x, y)$ is $A_3(x)\psi_3(y)$, where the latitudinal structure function is the sum of two Hermite functions.

The longitudinal structure function for the other latitudinal mode has a core proportional to $\varepsilon^4 \mathrm{sech}^4(\varepsilon x)$, which is forced by the self-interaction of the $n = 3$ mode. The nanopteron is non-local because the $n = 1$ mode has oscillatory wings for large $|x|$. For clarity, the amplitude of the $n = 1$ mode has been exaggerated.

Segur and Kruskal (1987) showed that the ε power series is only asymptotic because the breather is weakly nonlocal. Their argument employs the *ansatz*

$$u(x, t) = \sum_{n=0}^{\infty} A_n(x) \cos(n\omega t), \tag{5.20}$$

which is equivalent to the assumption—very likely, but not rigorously proved—that the breather is periodic in time. Substituting (5.20) into (5.17) gives a coupled set of nonlinear, ordinary differential equations in x for the x-dependent coefficients of the Fourier series.

In the far field, $|x| \gg 1$, the breather has decayed to such small amplitude that the nonlinear terms may be neglected. The coupled, nonlinear ODE's reduce to the *uncoupled, linear* set

$$A_{n,xx} + (n^2\omega^2 - 2)A_n = 0, \qquad n = 0, 1, \ldots . \tag{5.21}$$

Because $\omega^2(\varepsilon) \leq 2$ (Segur and Kruskal, 1987), one finds

(i) $A_0(x)$ decays as

$$\exp(-2^{1/2}|x|), \qquad |x| \gg 1; \tag{5.22a}$$

(ii) $A_1(x)$ decays as

$$\exp(-\varepsilon |x|), \qquad |x| \gg 1; \tag{5.22b}$$

(iii) $A_2(x)$ oscillates as

$$\alpha(\varepsilon)\,\mathrm{sgn}(x)\sin(kx), \qquad |x| \gg 1, \tag{5.22c}$$

where the far field wavenumber is

$$k_f \equiv (4\omega^2 - 2)^{1/2} \approx 6^{1/2} + O(\varepsilon^2), \qquad |x| \gg 1; \tag{5.23}$$

(iv) All higher harmonics oscillate in the far field, too, but with larger wavenumbers.

This situation is similar to that for Rossby waves. The dominant component $A_1(x)$ is localized, but the wave is nonlocal because of higher modes. These higher modes are forced by the self-interaction of the dominant mode. $A_0(x)$ and $A_2(x)$ enter at $O(\varepsilon^2)$, A_3 at $O(\varepsilon^3)$, and so on. For the breather, "n-th mode" means "coefficient of $\cos(n\omega t)$"; for equatorial waves, "n-th mode" is the component whose latitudinal shape is that of the n-th Rossby wave. Nevertheless, even though the nonlocal modes are only weakly excited, the breather and the $n = 3$ Rossby wave are both nanopterons.

Segur and Kruskal show

$$\alpha(\varepsilon) \sim \nu \exp(-[3/2]^{1/2}\pi/\varepsilon), \qquad \varepsilon \ll 1, \tag{5.24}$$

exactly as predicted by Theorem 5.2(ii). The far field contributions of all the higher harmonics, $A_3(x), A_4(x), \ldots$, can be neglected because the larger wavenumbers imply, according to the theorem, that their amplitudes will be exponentially small in comparison to that of $A_2(x)$.

The constant ν, alas, can only be computed numerically. Segur and Kruskal combined shooting and boundary value methods (in the complex plane) to find (in our notation)

$$\nu(0) = -0.018 \pm 0.004. \tag{5.25}$$

Note the rather large uncertainty of 22%. They calculated only in the limit $\varepsilon \to 0$.

Boyd (1988b) computed $\nu(\varepsilon)$ for the symmetric, minimally nonlocal nanopteron using the mixed Chebyshev/radiation basis algorithm de-

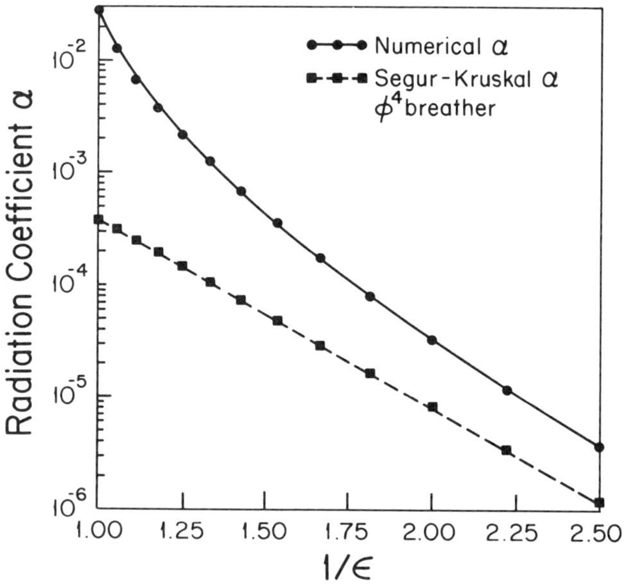

FIG. 11. Solid curve: numerically computed radiation coefficient α as a function of $1/\varepsilon$ for the ϕ^4 breather from Boyd (1988b). Theory asserts that for small ε, $\alpha \sim \nu(\varepsilon)\exp(-[3/2]^{1/2}\pi/\varepsilon)$, where ν varies algebraically (i.e., slowly) with ε. Dashed curve: the radiation coefficient as predicted by Segur and Kruskal (1987). The prediction is linear on this log-reciprocal graph because these authors calculated only $\nu(0)$.

scribed earlier. Figure 11 compares his $\alpha(\varepsilon)$ with the extrapolation of (5.24) and (5.25). Boyd's curve is clearly asymptoting to a line with the same slope as that of Segur and Kruskal, corresponding to decay proportional to $\exp(-3.84/\varepsilon)$ as predicted.

However, Boyd's $\nu(\varepsilon)$ for $\varepsilon = 0.4$ is roughly three times as large as Segur and Kruskal's. Possibly the two curves are tangent in the limit $\varepsilon \to 0$; the graph ends at the relatively large value of $\varepsilon = 0.4$. What is certain is that the extrapolation of the Segur–Kruskal formula is accurate only in order-of-magnitude for $\varepsilon \geq 0.4$.

Figure 12 illustrates $u(x, t)$ for $\varepsilon = 1$. The maximum and minimum are of equal amplitude in the perturbative approximation (5.18), but for $\varepsilon = 1$, the minimum is twice as deep as the maximum. Nonetheless, the far field oscillation is still small ($\alpha = 0.0274$ versus a maximum of $|u(x, t)| > 1$) so that the concept of the breather is still meaningful. If we chopped off the far field oscillations and used the core of the nanopteron for $\varepsilon = 1$ as the initial condition, the solution (after the radiation of initial

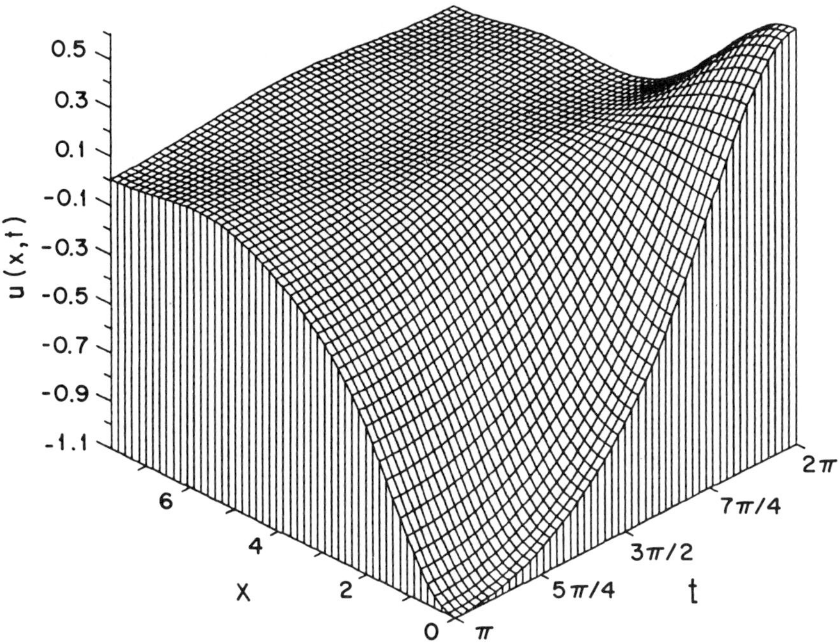

FIG. 12. A surface plot of the amplitude of the ϕ^4 breather as a function of x and t. Only a quarter of the x–t plane is shown because the breather is symmetric about the origin in both space and time. Although the perturbation parameter ε is not small ($\varepsilon = 1$), the amplitude of the wings is only 0.027, almost invisible on the scale of the graph.

transients) would lose only a small fraction of its energy during each cycle of the breather.

Indeed, numerical experiments done in the 1970s (see the references in Boyd, 1988b) show that for small to moderate amplitude, the ϕ^4 breather is a stable and long-lived species. The mathematical proof that the soliton does not exist is irrelevant to the longevity and coherence of ϕ^4 solutions.

Physicists often criticize mathematical existence proofs as being unnecessary demonstrations of what physicists already know. Here, however, the nonexistence proof is not merely irrelevant but extremely misleading, almost the complete opposite of reality.

3. *The "Slow Manifold" of Numerical Weather Prediction*

The "slow manifold" is a hypothetical state of atmospheric or oceanic motion that is completely devoid of high frequency gravity waves. If a numerical weather prediction model contains $3N$ degrees of freedom, then the "slow manifold" is the N-dimensional manifold within the $3N$-dimensional phase space in which the motion consists entirely of slow, low-frequency Rossby modes.

As a description of the observed atmosphere, the slow manifold is a fiction: the amplitude of high frequency gravity waves is very small but nonzero. More than forty years ago, however, it was recognized that the slow manifold is an essential fiction for numerical weather prediction. The reason is that the gravity waves oscillate too rapidly to be measured by existing observational networks. If raw data is used to initialize a prediction model, the amplitudes of the gravity waves will be mostly noise. Unfortunately, the very high frequency of these waves means that this noise can dominate predicted changes even though the amplitude of the noise is very small.

The earliest remedy was to use a filtered set of equations, the so-called quasi-geostrophic system, which eliminated the gravity waves completely. This is too drastic. The nonlinear interaction of the Rossby waves will drive *low-frequency, forced* oscillations in the gravitational modes, too. A more realistic strategy is to modify the raw observations so that the numerical forecast is initialized on the slow manifold and remains there throughout the forecast.

The current state of the art is the so-called "nonlinear normal mode" initialization (Baer and Tribbia, 1977). The method of multiple scales is used to exploit the order-of-magnitude separation between gravity wave and Rossby wave time scales. Nonlinear normal mode initialization has evolved into a whole family of closely related algorithms; operational forecast models now use the first or second order schemes with great success.

The only difficulties are that (i) the slow manifold does not exist and (ii) the perturbation series is divergent. (At least for some models, this is true; the question of whether these problems are shared by *all* feasible weather models is still unknown.) The mathematics of both the success and failure of nonlinear normal mode initialization is very closely related to that of our previous examples. The main difference is that the localization requirement breaks down as $t \to \infty$ instead of as $|x| \to \infty$.

To show this connection, we shall use the simplest possible model, the set of five ordinary differential equations we shall dub the "L–K Quintet" (Lorenz and Krishnamurthy, 1987):

$$\left.\begin{aligned} V_t &= UW - bUz, \\ U_t &= -VW + bVz, \\ W_t &= -VU, \end{aligned}\right\} \text{Rossby Triad} \qquad (5.26a\text{–}c)$$

$$\left.\begin{aligned} X_t &= -z, \\ z_t &= x + bVU. \end{aligned}\right\} \text{Gravity Dyad} \qquad (5.27a,b)$$

(Rossby Triad and Gravity Dyad together: *L–K Quintet*)

The five unknowns are the time-dependent coefficients of five spatial basis functions in this drastically truncated forecasting model. To simplify normal mode initialization, the basis functions are not spherical harmonics or Chebyshev functions but rather are the spatial structure factors of the linear free oscillations of the model. The first three equations describe the interaction of three Rossby waves. The second pair of equations predicts the changes in the amplitude of the two gravity waves. Our notation is that of Lorenz and Krishnamurthy (1987) except that we capitalize the Rossby coefficients and write the gravity wave amplitudes as lower case variables to emphasize that the Rossby modes are of much larger amplitude.

Because the gravity wave dyad is linear in x and z, these two equations may always be reduced to the single equation

$$z_{tt} + z = b(VU)_t. \qquad (5.28)$$

This is of the same form as that solved in Boyd (1988a) except that the coordinate is time instead of space. The variables have been nondimensionalized so that the frequency of the homogeneous solutions of (5.28), i.e., that of the gravity wave free oscillations, is unity. Since Rossby modes have frequencies small in comparison to those of the gravity waves, it follows that $V(t)$ and $U(t)$ should vary with t only on a *slow* time scale, equivalent to nondimensional frequencies much smaller than one. Thus, (5.28) is a differential equation with rapidly varying homogeneous solutions and slowly varying forcing.

The resemblance becomes even closer if we exploit the fact that the nondimensional parameter $b \ll 1$. Neglecting the b-dependent terms in (5.26) gives a closed set of three equations in three unknowns. Lorenz and Krishnamurthy chose to consider the special triad solution

$$U = \varepsilon\,\mathrm{sech}(\varepsilon t), \qquad V = -\varepsilon\tanh(\varepsilon t), \qquad W = -\varepsilon\,\mathrm{sech}(\varepsilon t). \qquad (5.29)$$

As noted in Section III, the *general* triad solution is merely the imbrication of this solution.

For this special case, one is once again in the realm of hyperbolic functions, just as in our previous examples! Lorenz and Krishnamurthy solve (5.28) exactly to show that while it is always possible to impose the condition that $z(t)$ and $x(t)\to 0$ as $t\to -\infty$, the opposite limit gives

$$z(t) \sim -2b\pi e^{-\pi/(2\varepsilon)} \sin t \quad \text{as } t\to\infty \quad \text{for} \quad \varepsilon \ll 1. \tag{5.30}$$

The high frequency tail (5.30) implies that the slow manifold does not exist. Even though the initial state (for large negative t) contains no gravity waves and even though the gravity wave dyad is forced only by the low-frequency interaction of the Rossby waves among themselves, a high-frequency component persists even after the Rossby wave interaction has faded into history.

T. Warn apparently was the first to suspect the nonexistence of the slow manifold and the divergence of the nonlinear normal modes initialization. However, the equivalence relation

$$\text{nonexistence} \leftrightarrow \text{divergence} \tag{5.31}$$

is not true. Modifying (5.28) to

$$z_{tt} - z = b(VU)_t, \tag{5.32}$$

solely to make a mathematical point, gives a differential equation that does have a slow manifold in the sense that its solution tends to 0 as $|t|\to\infty$. Nevertheless, the method of multiple scales still diverges. The reason is that the Fourier transform of the hyperbolic functions in (5.29) and (5.32) does not have a sharp cutoff frequency ω_c. Instead, the transform decays smoothly with $|\omega|$ as $|\omega|\to\infty$. This implies that the "low-frequency" forcing terms in (5.29) and (5.32) have nonzero high-frequency components, contrary to the separation of time scales, which is the key assumption of multiple scales perturbation theory. Fortunately, the exponential smallness of the Fourier transform for $|\omega|\approx 1$ does permit the series to be *asymptotic*, and extremely successful in operational forecasting.

Two questions remain. First, do these conclusions for the L–K Quintet apply to more complicated models? J. Curry (private communication via J. McWilliams) has shown that if one makes the L–K system more consistent by including four additional gravity modes, the slow manifold exists. This suggests that the multiple scales series may converge. What happens for more complex models, with and without dissipation, is still a mystery.

Second, may one modify the L–K Quintet so that it does have a true slow manifold? Boyd (1989c) shows that by applying the "band-limiting" operator in frequency to the nonlinear terms in the gravity wave dyad, one may suppress the exponentially small wing (5.30). It is feasible to solve the band-limited L–K Quintet, but is is not yet clear how this strategy could be applied to a more realistic forecasting model.

One thing is clear. Nonlocalization in *time* may have as much physical significance as nonlocalization in *space*.

4. *The Pomeau–Ramani–Grammaticos (PRG) Nanopteron*

Since we have used the PRG equation (5.1) as a prototype throughout most of this section, our discussion will be brief. Although this equation is a generalization of the important Korteweg–deVries equation, Pomeau, Ramani, and Grammaticos (1988) observe correctly that the PRG model is more interesting as a mathematical example than as a physical illustration. The reason is that extending the method of multiple scales to higher order (beyond the Korteweg–deVries equation) usually adds in additional cross-channel modes rather than merely tacking on an extra term to the differential equation. However, at least in special parameter ranges, the PRG equation does have applications to water waves, electrical circuits, and other phenomena (Boyd, 1989). In particular, Hunter and Scheurle (1988) have shown that (5.1) is a consistent model for capillary-gravity waves for Bond number $\approx 1/3$.

Pomeau, Ramani, and Grammaticos (1988) solve their model using an improved version of the matched-asymptotics-in-the-complex-plane method of Segur and Kruskal (1987). One striking conclusion is that when they compute a series that is accurate near the poles of the nanopteron at $x = \pm i\pi/(2\varepsilon)$, the nonlinearity drops out of the recursion for large n. One finds that the high order coefficients are

$$a_n \sim K(-1)^n(2n-1)! \qquad n \gg 1. \tag{5.33}$$

The *form* of the coefficients is determined by the linear terms; only the proportionality constant K is influenced by the nonlinear part of the PRG equation.

If one applies their method to the *linear, forced* problem

$$u_{xxxx} + u_{xx} = f(\varepsilon x), \tag{5.34}$$

where $f(\varepsilon x)$ has poles of the same order and location as the nonlinear term in the PRG equation, one finds that the coefficients are given by (5.33) exactly. It does not matter that (5.1) is nonlinear while (5.34) is linear. For both the PRG equation and the linear model (5.34), it is these coefficients a_n that determine the radiation coefficient α.

Thus, insofar as the far field is concerned, the PRG equation behaves like (5.34). The nonlinear core of the nanopteron is merely an effective forcing that excites the sinusoidal oscillations in the wings. The dynamics of the far field is linear (at least to within $O(\alpha^2)$); the nonlinear interactions of the core merely determine the effective forcing function, $f(\varepsilon x)$ in (5.34).

We have applied the mixed Chebyshev/radiation basis pseudospectral algorithm to numerically compute the PRG nanopteron, and found an error in Pomeau *et al.* (1988). Their Equation (18) is missing a factor of $\frac{1}{2}$. The correct radiation coefficient is given by (in our notation and for the form of Equation (5.1))

$$\alpha(\varepsilon) \sim \nu(\varepsilon) e^{-\pi/(2\varepsilon)}, \qquad \varepsilon \ll 1, \tag{5.35}$$

where

$$\nu(0) = 376.41. \tag{5.36}$$

Fortunately, this typographical error has no effect on the main conclusions of their article. Fitting a straight line though our computed radiation coefficients for $\varepsilon = 0.05$ and $\varepsilon = 0.075$ gives

$$\nu(\varepsilon) \approx \frac{\nu(0)}{1 + 6.2\varepsilon}, \qquad \varepsilon \ll 1, \tag{5.37}$$

and also confirms Pomeau *et al.*'s (corrected) value for $\nu(0)$ to within 0.7%. However, it is not possible to check (5.36) to higher accuracy or to compute the linear term in (5.37) to more than two decimal places without heroic measures. The problem is that for $\varepsilon = \frac{1}{20}$, $\alpha = 6.5 \times 10^{-12}$! This is sufficiently small so that roundoff error is significant even in double precision. Yet even when the radiation coefficient is this tiny, $\nu(\varepsilon)$ differs from $\nu(0)$ by more than 30%.

As is also suggested by Figure 11, the matched asymptotic radiation coefficients are quantitatively accurate only when so small that they cannot be accurately confirmed even by double precision computations. With the special pseudospectral method, however, one may compute the wings of the nanopteron even when ε is not small, as shown in Figures 9 and 12.

This algorithm could be extended to very, very small ε by deforming the contour of integration into the complex plane as described by Boyd (1985). Segur and Kruskal (1987) employed numerical integration in the complex plane to complete their matched asymptotics calculation of $\nu(0)$ for the ϕ^4 breather.

5. *Plasma Modons in Magnetic Shear*

Modons are dipolar vortices that solve the quasi-geostrophic equation of oceanography and the Hasegawa–Mima equation of plasma physics. Meiss and Horton (1983) studied plasma modons in a magnetic shear. In the absence of shear, modons are solitons in the classical sense, decaying exponentially for large $|x|$.

Meiss and Horton show that in the spatially varying environment created by magnetic shear, the index of refraction changes sign at large distances from the modon center so that the flow is oscillatory in the far field. Because the modon has exponentially decayed to very small amplitude at the turning radius, however, the amplitude of the far field radiation is exponentially small in the shear strength—a nanopteron. This mechanism of nonlocalization via spatial inhomogeneity is probably important in many other physical applications.

Because the decay-to-oscillation transition occurs in the far field, one may compute the radiation by performing a standard WKB turning point analysis. It is not necessary to deform the integration contour into the complex plane in order to apply matched asymptotics at complex WKB critical points, as was true for the PRG nanopteron and ϕ^4 breather.

6. *Quasi-Geostrophic Monopoles*

When β, the y-derivative of the Coriolis parameter, is zero, any radially symmetric vortex is an exact nonlinear solution of the quasi-geostrophic equation. McWilliams and Flierl (1979) show that when the (weak) β-effect is added, all one-signed (monopole) vortices slowly decay with time by radiating Rossby waves. Flierl, Stern, and Whitehead (1983) prove a theorem that shows that this fate is inevitable. Only when the area integral of the vorticity is zero, as it can be for two-signed (dipole) vortices, is a steadily translating soliton possible.

Nevertheless, the rate of dispersion of a strongly nonlinear monopole is quite weak. At least in the initial value experiments, monopoles are "algebraically decaying" because the rate of decay is directly proportional to β instead of being exponentially small.

Gulf Stream rings are the observational motive for the study of monopoles. These quasi-solitons are strong ocean eddies as much as 200 km in diameter. When the observed shapes are used to initialize quasi-geostrophic models that have artificially suppressed the nonlinear terms, the vortex disperses in less than two months (Mied and Lindemann, 1982). When the nonlinear terms are retained, the same initial condition gives a vortex that radiates slowly, but has a lifetime of several years—just like *observed* Gulf Stream rings.

McWilliams (1989) reviews monopoles in more detail. Steadily translating but nonlocal monopoles, i.e., micropterons, may well exist. However, there have been no explicit calculations of micropterons for quasi-geostrophic monopoles or for either of the next two examples.

7. *The Flierl–Petviashvili Monopole*

Flierl (1979) and Petviashvili (1981) independently derived Equation (4.1) as a model for radially symmetric Rossby solitary waves. Swenson (1986) pointed out that although Flierl's analysis is correct to lowest order, the first order perturbation equation is insoluble. Boyd (1988d) shows that the Flierl–Petviashvili equation (4.1) is also a consistent lowest order model for the EMR equation, but again fails at first order.

In both cases, the mathematical reason for the perturbative breakdown is that the lowest order model is translationally invariant. This implies that the y-derivative of the monopole is an eigenfunction of the first (and higher) order perturbation problems. Unless the perturbation is orthogonal to this eigenfunction, the perturbation theory fails.

The interesting difference is that Swenson's perturbation is weak mean shear whereas Boyd's is the variation of the Coriolis parameter with latitude. However, Petviashvili (1981) successfully solved his model to first order including weak y-variations; his first order terms are different from those of Swenson and Boyd. It is clear that higher order effects can be very subtle, and that small inconsistencies in deriving models from the hydrodynamic equations can have a profound impact on the existence of true solitons.

Nevertheless, Swenson notes, "If no appropriate stationary solutions exist, it is still possible that the Flierl solution is close to stationary. That is, it might be more long-lived that a similarly strong vortex which does not nearly meet the nonlinear dispersion relation."

8. *Two-Layer Modons with a Weakly Radiating Lower Layer*

Flierl (1984) studied modons in a two-layer model with a very deep lower layer (as is true of the ocean). He showed that to lowest order in δ, the relative thickness of the two layers, the solution is a modon in the upper layer. However, the modon will force untrapped Rossby wave radiation in the lower layer. Since there is no separation of scales between the modon and the far field radiation, this nonlocal solution is "algebraically decaying": the amplitude of the far field oscillation is proportional to δ, not to the exponential of the inverse of δ.

Nevertheless, Flierl shows that this quasi-soliton is long-lived: $\delta \ll 1$ over most of the ocean and the proportionality constant is also small. As he writes in Flierl (1987): ". . . even nonisolated structures can be long lived if the nonlinear advection is sufficiently strong and properly organized."

F. Summary: The Gereralization of the Concept of a "Solitary" Wave

One major theme is that for a soliton, existence is overrated. An analogue is the property of convergence for an infinite series. Abel's remark that "divergent series are the invention of the devil" is now regarded as nonsense; with a careful definition of asymptoticity, divergent series may be very useful indeed. It was, however, necessary to generalize "converges to" to "asymptotic to" in order to exploit divergent series. In the same way, the concept of "soliton" is still useful for nonlocal waves provided that we generalize "soliton" to allow exponentially small far field oscillations.

The list of open problems is extremely long. One goal is to investigate more systems with weakly nonlocal solitary waves; the probability is that nanopterons are just as ubiquitous as true solitons. Another is to explore the evolution of general initial conditions into radiatively decaying solitons or nanopterons. A third is to find examples of micropterons; the effects of perturbations on geophysical monopoles (Examples 6 and 7) are poorly understood. A fourth goal is to replace the plausibility arguments by rigorous proofs. Hunter and Scheurle (1988) prove the existence of nonlocal solitons to the PRG equation, but even Segur and Kruskal (1987) and Pomeau *et al.* (1988) have mathematical loose ends.

Weakly nonlocal solitons have now become respectable, but a vast amount of polishing and refinement is still needed.

VI. Summary: *Quo Vadis*?

The history of solitons over the last two decades has been a tale of three heroes: inverse scattering, the method of multiple scales, and numerical initial value codes. This review has described four up-and-coming newcomers.

The theory of polycnoidal waves is a mathematical poem with deep roots in the past. The Nonlinear Fourier Transform algorithm of Osborne and Bergamasco is the numerical realization and generalization of inverse scattering. The Nakamura and Boyd theta function methods are the periodic extension of Hirota's bilinear transform for computing N-solitons.

Landau once dreamed that turbulence was merely an N-polycnoidal wave with $N \gg 1$, but this now seems as illusory as the Seven Cities of Gold. In the era of dynamical systems and chaotic attractors, polycnoidal theory looks backward to nineteenth century studies of hyperelliptic functions.

And yet, there is indisputable evidence that turbulence is not merely a sea of randomness, but rather has strong, coherent, long-lived structures, as stable and as conspicuous as icebergs. Understanding this interplay of randomness and coherence requires not only a deep understanding of solitons, but also of the multiply periodic coherent flows, the polycnoidal waves. As Landau recognized, an N-polycnoidal wave of sufficiently large N can mimic many features of true chaotic motion. Polycnoidal waves and chaotically perturbed polycnoidal waves will still be useful models.

To reinforce this assertion, Hammack (1989) reviews laboratory experiments that have observed two-dimensional Kadomtsev–Petviashvili double cnoidal waves in surface water waves. The Torino group of Osborne, Bergamasco and their collaborators are actively applying the Nonlinear Fourier Transform to ocean time series.

Polycnoidal waves will probably never be as fashionable as dynamical systems theory because their range of applicability—integrable differential equations—is relatively narrow. Nevertheless, polycnoidal wave theory will continue as a small but active area. More than 150 years after John Scott Russell first observed solitary waves on a Scottish canal, we still have much to learn about water waves, and their counterparts in optics, plasma physics, and many other fields.

Imbricate series theory is also a concerto rather than a symphony, but it, too, will continue as a minor but steady line of research. It is quite amazing that one may imbricate the soliton to generate the *exact* periodic solution to many important wave equations. However, this nonlinear superposition principle is more than a mathematical curiosity. The concept of the periodic disturbance as a repetition of independent, distinct pulses is often much closer to intuition—and sometimes observation—than the model of the disturbance-as-sine-wave.

For example, it is very difficult to teach hydrodynamic stability theory to meteorology undergraduates because the thought-model of a single, growing Fourier component is irreconcilably different from what they have already learned from weather maps. The naive stability theory predicts a ring of six equally strong storms with equidistant spacing around a circle of latitude. (Wavenumber six is the most rapidly growing linear mode.) On the maps, however, each storm is as unique and individual as a fingerprint, and the spacing between the storms is irregular.

Unfortunately, this growing sine wave is the starting point for the most sophisticated nonlinear theories now available. Is it possible to develop an instability model that is pattern-function-based instead of Fourier-based? The students see pattern functions on the maps, not sine waves. A similar question may be asked in the time domain, too, as posed by the imbricate series for the resonant triad, which is the simplest model for a vacillating (time-periodic) instability.

On a long time scale, the dynamics of the atmosphere is chaotic and unpredictable. On a time scale of several days, however, it is possible to make a good forecast because each individual storm is predictable. It is mostly the amplitude and phase of each storm which is irregular and chaotic. Is it possible to generalize imbricate series to explicitly display this pattern of order within disorder, and to handle phenomena that are not simply periodic in time?

Direct numerical calculations of solitary waves, especially in two and three dimensions, will be a major activity of the 1990s because we finally have the algorithmic and electronic technology to solve multi-dimensional nonlinear boundary value problems with comparative ease. Currently, this capability is badly under-exploited in spite of the fifteen studies reviewed in Section IV. The reason is that the community of nonlinear wave theorists has almost no overlap with the tribe of supercomputers; the few exceptions tend to be scorned by both groups. (I have the bruises to prove it!) Nevertheless, as number-crunching has migrated from the university computer center to the desktop, even theorists have learned a little bit of programming. When the right algorithms are used, multi-dimensional soliton calculations are simple and fast.

Lastly, as the study of solitons matures, there will be an increasing need to generalize the concept. "Nonlocal" solitary waves will be only one of these generalizations. Breathers, excited states, solitons-with-dissipation, and other extensions will also be important. Some work has already been done on most of these generalizations, but only to scratch the surface.

At the moment, solitons are the "old" theory of nonlinearity, and chaos/dynamical systems is the new and the trendy. Scientific fashions aside, the reality is that we will not be able to truly understand either until we know how to blend solitons and chaos, short-range structure and long-range unpredictability. This is turn will require a much deeper understanding of solitons and their generalizations. An intellectual alloy is only as sound as the metals that make it.

Acknowledgments

This work was supported by grants OCE8509923, OCE8812300, and DMS8716766 from the National Science Foundation. I thank C.-S. Yih and the editor, T. Y. Wu, for suggesting this review. I thank Elsevier for permission to reprint Figs. 8 and 10 from Boyd (1989a).

References

Ablowitz, M. J., and Segur, H. (1981). "Solitons and the Inverse Scattering Transform." SIAM, Philadelphia.

Andersen, C. M., and Geer, J. F. (1982). Power series expansions for the frequency and period of the limit cycle of the van der Pol equation. Frequency and period of the limit cycle of the van der Pol equation. *SIAM J. Appl. Math.* **42**, 678–693.

Baer, F., and Tribbia, J. J. (1977). On complete filtering of gravity waves through nonlinear initialization. *Mon. Wea. Rev.* **105**, 1536–1541.

Benjamin, T. B., Bona, J. L., and Mahony, J. J. (1972). Model equations for long waves in nonlinear dispersive systems. *Philos. Trans. Roy. Soc. London Ser. A* **272**, 47–73.

Boyd, J. P. (1980). Equatorial solitary waves, Part I: Rossby solitons. *J. Phys. Oceangr.* **10**, 1699–1717.

Boyd, J. P. (1982a). The optimization of convergence for Chebyshev polynomial methods in an unbounded domain. *J. Comput. Phys.* **45**, 43–79.

Boyd, J. P. (1982b). Theta functions, Gaussian series, and the spatially periodic solutions of the Korteweg–deVries equation. *J. Math. Phys.* **23**, 375–387.

Boyd, J. P. (1984a). Cnoidal waves as exact sums of repeated solitary waves: New series for elliptic functions. *SIAM J. Appl. Math.* **44**, 952–955.

Boyd, J. P. (1984b). The double cnoidal wave of the Korteweg–deVries equation: An overview. *J. Math. Phys.* **25**, 3390–3401.

Boyd, J. P. (1984c). Perturbation theory for the double cnoidal wave of the Korteweg–deVries equation. *J. Math. Phys.* **25**, 3402–3414.

Boyd, J. P. (1984d). The special modular transformation for polycnoidal waves of the Korteweg–deVries equation. *J. Math. Phys.* **25**, 3415–3423.

Boyd, J. P. (1985). Complex coordinate methods for hydrodynamic instabilities and Sturm–Liouville eigenproblems with an interior singularity. *J. Comput. Phys.* **57**, 454–471.

Boyd, J. P. (1986a). Solitons from sine waves: Analytical and numerical methods for non-integrable solitary and cnoidal waves. *Phys. D* **21**, 227–246.

Boyd, J. P. (1986b). An analytical and numerical study of the two-dimensional Bratu equation. *J. Sci. Comput.* **1**, 183–206.

Boyd, J. P. (1987a). Spectral methods using rational basis functions on an infinite interval. *J. Comput. Phys.* **69**, 112–142.

Boyd, J. P. (1987b). Orthogonal rational functions on a semi-infinite interval. *J. Comput. Phys.* **70**, 63–88.

Boyd, J. P. (1988a). A comparison of numerical and analytical methods for the reduced wave equation with multiple spatial scales. *J. Comput. Phys.*, submitted.

Boyd, J. P. (1988b). Numerical calculation of a weakly nonlocal soliton: The ϕ^4 breather. *Nonlinearity*, submitted.

Boyd, J. P. (1988c). An analytical solution for a nonlinear differential equation with logarithmic singularity. *Adv. in Appl. Math.* **9**, 358–363.

Boyd, J. P. (1988d). Monopolar and dipolar vortex solitons in two dimensions. *Wave Motion*, submitted.

Boyd, J. P. (1988e). Exact and approximate nonlinear waves generated by the periodic superposition of solitons. *Z. Angew. Math. Phys.*, in press.

Boyd, J. P. (1988f). Chebyshev domain truncation is inferior to Fourier domain truncation for solving problems on an infinite interval. *J. Sci. Comput.* **3**, 109–120.

Boyd, J. P. (1989a). Weakly non-local solitary waves. *In* "Mesoscale/Synoptic Coherent Structures in Geophysical Turbulence" (J. C. J. Nihoul and B. M. Janart ed.) Elsevier, Amsterdam, 103–112.

Boyd, J. P. (1989b). Nonlinear equatorial waves. *In* "Nonlinear Topics in Ocean Physics: Proceedings of the Enrico Fermi School" (A. R. Osborne and L. Bergamasco eds.). North-Holland, Amsterdam, in press.

Boyd, J. P. (1989c). Weakly non-local solitary waves. *In* "Nonlinear Topics in Ocean Physics: Proceedings of the Enrico Fermi School" (A. R. Osborne and L. Bergamasco eds.). North-Holland, Amsterdam, in press.

Boyd, J. P. (1989d). "Chebyshev and Fourier Spectral Methods." Springer-Verlag, New York, in press.

Boyd, J. P. (1989e). The arctan/tan mapping for periodic solutions with a region of rapid variation. *J. Comput. Phys.*, submitted.

Boyd, J. P. and Haupt, S. E. (1989). Polycnoidal waves: spatially periodic generalizations of multiple solitons. *In* "Nonlinear Topics in Ocean Physics: Proceedings of the Enrico Fermi School" (A. R. Osborne and L. Bergamasco eds.). North-Holland, Amsterdam, in press.

Boyd, J. P., and Ma, H. (1989). Lamb–Batchelor vortex pairs (modons) bounded by an ellipse. *J. Fluid Mech.*, submitted.

Briggs, W. L. (1987). "A Multigrid Tutorial." SIAM, Philadelphia.

Canuto, C., Hussaini, M. Y., Quarteroni, A., and Zang, T. A. (1987). "Spectral Methods in Fluid Dynamics." Springer-Verlag, New York.

Chang, H.-C. (1986). Traveling waves on fluid interfaces: Normal form analysis of the Kuramoto–Sivashinsky equation. *Phys. Fluids* **29**, 3142–3147.

Chen, B., and Saffman, P. G. (1980). Numerical evidence for the existence of new types of gravity waves of permanent form on deep water. *Stud. Appl. Math.* **62**, 1–21.

Chu, M. T. (1988). On the continuous realization of iterative processes. *SIAM Rev.* **30**, 375–387.

Dashen, R. F., Hasslacher, B., and Neveu, A. (1974). Particle spectrum in model field theories from semiclassical functional integral techniques. *Phys. Rev. D* **10**, 4130–4150.

Date, E., and Tanaka, S. (1976). Periodic multi-soliton solutions of Korteweg–deVries equation and Toda lattice. *Progr. Theoret. Phys.* **59**, 107–125.

Deem, G. S., and Zabusky, N. J. (1978). Vortex waves: Stationary "V states," interactions, recurrence, and breaking. *Phys. Rev. Lett.* **40**, 859–862.

Dritschel, D. C. (1988). Contour dynamics. *J. Comput. Phys.* **77**, 240–266.

Dodd, R. K., Eilbeck, J. C., Gibbon, J. D., and Morris, H. C. (1982). "Nonlinear Wave Equations." Academic Press, New York.

Dubrovin, B. A. (1981). Theta functions and non-linear equations. *Uspekhi Mat. Nauk* **36**, 11–80.

Dubrovin, B. A., Matveev, V. B., and Novikov, S. P. (1976). Nonlinear equations of Korteweg–deVries type, finite zoned linear operators, and Abelian varieties. *Russian Math. Surveys* **31**, 59–146.

Eydeland, A., and Turkington, B. (1987). A computational study of Rossby solitary waves. *Stud. Appl. Math.* **76**, 37–67.

Eydeland, A., and Turkington, B. (1988). A computational method of solving free-boundary problems in vortex dynamics. *J. Comput. Phys.* **78**, 194–214.

Ferguson, W. E., Flaschka, H., and McLaughlin, D. W. (1982). Nonlinear normal modes for the Toda chain. *J. Comput. Phys.* **45**, 157–209.

Flierl, G. R. (1979). A simple model for the structure of warm and cold core rings. *J. Geophys. Res.* **84** (C2), 781–785.

Flierl, G. R. (1984). Rossby wave radiation from a strongly nonlinear warm eddy. *J. Phys. Oceangr.* **14**, 47–58.

Flierl, G. R. (1987). Isolated eddy models in geophysics. *Ann. Rev. Fluid Mech.* **19**, 493–530.

Flierl, G. R., Stern, M. E., and Whitehead, J. A., Jr. (1983). The physical significance of modons: Laboratory experiments and general physical constraints. *Dyn. Atmos. Oceans* **7**, 233–263.

Forest, M. G., and McLaughlin, D. W. (1982). Spectral theory for the periodic sine–Gordon equation: A concrete viewpoint. *J. Math. Phys.* **23**, 1248–1277.

Forest, M. G., and McLaughlin, D. W. (1983). Modulations of sinh–Gordon and sine–Gordon wavetrains. *Stud. Appl. Math.* **68**, 11–59.

Greatbatch, R. J. (1985). The response of the equatorial ocean to impulsive wind stress. *J. Geophys. Res.* **90**, 9097–9109.

Hammack, J. L. (1989). Laboratory experiments with nonlinear water waves. *In* "Nonlinear Topics in Ocean Physics: Proceedings of the Enrico Fermi School" (A. R. Osborne and L. Bergamasco eds.). North-Holland, Amsterdam, in press.

Haupt, S. E., and Boyd, J. P. (1987). Modeling nonlinear resonance: A modification to the Stokes' perturbation expansion. *Wave Motion* **10**, 83–98.

Haupt, S. E., and Boyd, J. P. (1989). Double cnoidal waves of the Korteweg–deVries equation: A boundary value approach. *Physica D*, submitted.

Hirota, R. (1974). A new form of Backlund transformations and its relation to the inverse scattering transform. *Progr. Theoret. Phys.* **52**, 1498–1520.

Hirota, R., and Ito, M. (1981). A direct approach to multi-periodic wave solutions to nonlinear evolution equations. *J. Phys. Soc. Japan* **50**, 338–342.

Holton, J. R. (1979). "An Introduction to Dynamic Meteorology." Academic Press, New York.

Hunter, J. K., and Scheurle, J. (1988). Existence of perturbed solitary wave solutions to a model equation for water waves. *Phys. D* **32**, 253–268.

Hyman, J. M. (1978). Time evolution of almost periodic solutions of the KdV equation. *Rocky Mountain J. Math.* **8**, 95–104.

Ingersoll, A. P., and Cuong, P. G. (1981). Numerical model of long-lived Jovian vortices. *J. Atmospheric Sci.* **38**, 2067–2076.

Its, A. R., and Matveev, V. B. (1975). Hill's operator with finitely many gaps. *Functional Anal. Appl.* **9**, 65–66.

Jaworski, M., and Zagrodzinski, J. (1982). Quasi-periodic solutions of the sine–Gordon equation. *Phys. Lett. A* **92**, 427–430.

Johnson, R. S. (1976). On the modulation of water waves on shear flows. *Proc. Roy. Soc. London Ser. A* **338**, 101–114.

Kadomtsev, B. B., and Petviashvili, V. I. (1970). On the stability of solitary waves in weakly dispersing media. *Soviet Phys. Dokl.* **15**, 539–543.

Keller, H. B. (1977). Numerical solution of bifurcation and nonlinear eigenvalue problems. *In* "Applications of Bifurcation Theory" (P. Rabinowitz ed.), pp. 359–384. Academic Press, New York.

Kindle, J. C. (1983). On the generation of Rossby solitons during El Nino. *In* "Hydrodynamics of the Equatorial Ocean" (J. C. J. Nihoul ed.), pp. 353–371. Elsevier, Amsterdam.

Korpel, A., and Banerjee, P. P. (1981). Exact decomposition of cnoidal waves into associated solitons. *Phys. Lett. A* **82**, 113–115.

Lax, P. (1976). Almost periodic solutions of the KdV equation. *SIAM Rev.* **18**, 351–365.

Lorenz, E. N. (1960). Maximum simplification of the dynamic equations. *Tellus* **12,** 243–254.

Lorenz, E. N., and Krishnamurthy, V. (1987). On the nonexistence of a slow manifold. *J. Atmos. Sci.* **44**, 2940–2950.

Ma, Y.-C., and Ablowitz, M. J. (1981). The periodic cubic Schroedinger equation. *Stud. Appl. Math.* **65**, 113–158.

Matsuno, Y. (1984). "The Bilinear Transformation Method." Academic Press, New York.

McWilliams, J. C. (1984). The emergence of isolated, coherent vortices in turbulent flow. *J. Fluid Mech.* **146**, 21–43.

McWilliams, J. C. (1989). Geostrophic vortices. *In* "Nonlinear Topics in Ocean Physics: Proceedings of the Enrico Fermi Summer School" (A. R. Osborne and L. Bergamasco eds.). North-Holland, Amsterdam, in press.

McWilliams, J. C., and Flierl, G. R. (1979). On the evolution of isolated, nonlinear vortices. *J. Phys. Oceangr.* **9,** 1155–1182.

Meiron, D. I., Saffman, P. G., and Yuen, H. C. (1982). Calculation of steady three-dimensional deep-water waves. *J. Fluid Mech.* **124**, 109–121.

Meiss, J. D., and Horton, W. (1983). Solitary drift waves in the presence of magnetic shear. *Phys. Fluids* **26**, 990–997.

Meyer, R. E. (1980). Exponential asymptotics. *SIAM Rev.* **22**, 213–224.

Mied, R. P., and Lindemann, G. J. (1982). The birth and evolution of eastward propagating modons. *J. Phys. Oceangr.* **12**, 213–230.

Miloh, T., and Tulin, M. P. (1989). Periodic solutions of the DABO equation as sum of repeated solitons. *J. Phys. A* **22,** 921–923.

Morgan, A. P. (1987). "Solving Polynomial Systems Using Continuation for Engineering and Scientific Problems." Prentice–Hall, Englewood Cliffs, New Jersey.

Morse, P. M., and Feshbach, H. (1953). "Methods of Theoretical Physics," Vol. 1, p. 395. McGraw–Hill, New York.

Munk, W. H. (1949). The solitary wave and its application to surf problems. *Ann. New York Acad. Sci.* **51**, 376–424.

Nakamura, A. (1979). A direct method of calculating periodic wave solutions to nonlinear evolution equations. I. Exact two-periodic wave solution. *J. Phys. Soc. Japan* **47**, 1701–1710.

Nakamura, A. (1980). A direct method of calculating periodic wave solutions to nonlinear evolution equations. II. Exact one- and two-periodic wave solution of the coupled bilinear equations. *J. Phys. Soc. Japan.* **48**, 1365–1371.

Nakamura, A., and Matsuno, Y. (1980) Exact one- and two-periodic wave solutions of fluids of finite depth. *J. Phys. Soc. Japan* **48**, 653–657.

Osborne, A. R., and Bergmasco, L. (1985). The small amplitude limit of the spectral transform for the periodic Korteweg–deVries equation. *Nuovo Cimento B* **85**, 229–242.

Osborne, A. R., and Bergamasco, L. (1986). The solitons of Zabusky and Kruskal revisted: Perspective in terms of the periodic spectral transform. *Phys. D* **21**, 26–46.

Osborne, A. R., and Bergamasco, L. (1987). Numerical implementation of the Nonlinear Fourier Transform for the Korteweg–deVries equation. *J. Comput. Phys.,* submitted.

Orszag, S. A. (1980). Spectral methods for problems in complex geometries. *J. Comput. Phys.* **37**, 70–92.

Pedlosky, J. (1979). "Geophysical Fluid Dynamics." Springer-Verlag, New York.

Peitgen, H.-O. and Richter, P. H. (1986). "The Beauty of Fractals." Springer-Verlag, New York.

Peregrine, D. H. (1966). Calculations of the development of an undular bore. *J. Fluid Mech.* **25**, 281–300.

Peregrine, D. H. (1983). Breaking waves on beaches. *Ann. Rev. Fluid Mech.* **15**, 149–178.

Petviashvili, V. I., and Tsvelodub, O. Yu. (1978). Horseshoe-shaped solitons on a flowing viscous film of fluid. *Soviet Phys. Dokl.* **23**, 117–118.

Petviashvili, V. I. (1976). Equation of an extraordinary soliton. *Soviet J. Plasma Phys.* **2**, 257–260.

Petriashvili, V. I. (1981). Red spot of Jupiter and the drift soliton in a plasma. *JETP Lett.* **32**, 619–622.

Pierini, S. (1986). Solitons in a channel emerging from a three-dimensional initial wave. *Nuovo Cimento C* **9**, 1045–1061.

Pierrehumbert, R. T. (1980). A family of steady, translating vortex pairs with distributed vorticity. *J. Fluid Mech.* **99**, 129–144.

Pokrovskii, V. L., and Khalatnikov, J. M. (1961). On the problem of above-barrier reflection of high-energy particles. *Soviet Phys. JETP* **13**, 1207–1210.

Pomeau, Y., Ramani, A., and Grammaticos, B. (1988). Structural stability of the Korteweg–deVries solitons under a singular perturbation. *Phys. D* **31**, 127–134.

Schwartz, L. W. (1974). Computer extension and analytic continuation of Stokes' expansion for gravity waves. *J. Fluid Mech.* **62,** 553–578.

Segur, H., and Finkel, A. (1985). An analytical model of periodic waves in shallow water. *Stud. Appl. Math.* **73**, 183–200.

Segur, H., and Kruskal, M. D. (1987). On the nonexistence of small-amplitude breather solutions in ϕ^4 theory. *Phys. Rev. Lett.* **58**, 747–750.

Seydel, R. (1988). "From Equilibrium to Chaos: Practical Bifurcation and Stability Analysis." Elsevier, Amsterdam.

Stenger, F. (1981). Numerical methods based on Whittaker cardinal, or sinc functions. *SIAM Rev.* **23**, 165–224.

Stiassnie, M., and D. H. Peregrine (1980). Shoaling of finite amplitude surface waves on water of slowly varying depth. *J. Fluid Mech.* **97**, 783–805.

Streett, C. L., and Zang, T. A. (1984). Spectral methods for solution of the boundary-layer equations. AIAA Paper 84-0170, 9 pp.

Swenson, M. (1986). A note on baroclinic solitary waves with radial symmetry. *Dyn. Atmos. Oceans* **10**, 243–252.

Tanveer, S. (1986). A steadily translating pair of equal and opposite vortices with vortex sheets on their boundaries. *Stud. Appl. Math.* **74**, 139–154.

Toda, M. (1975). Studies of a nonlinear lattice. *Phys. Rep.* **18**, 1–124.

Tracy, E. R., Larson, J., Osborne, A. R., and Bergamasco, L. (1988). The relationship between the spectral theories for the periodic Korteweg-deVries and nonlinear Schroedinger equations. *Phys. D* **21**, 83–106.

Tracy, E. R., Larson, J., Osborne, A. R., and Bergamasco, L. (1989). The relationship between the spectral theories for the periodic Korteweg–deVries and nonlinear Schroedinger equations. *In* "Nonlinear Topics in Ocean Physics, Proceedings of the Enrico Fermi Summer School" (A. R. Osborne and L. Bergamasco eds.). North-Holland, Amsterdam, in press.

Tung, K.-K., Chan, T. F., and Kubota, T. (1982). Large amplitude internal waves of "permanent form". *Stud. Appl. Math.* **66**, 1–44.

Vanden-Broeck, J.-M. (1986). Steep gravity waves: Havelock's method revisted. *Phys. Fluids* **29**, 3084–3085.

Van Dyke, M. (1975). Computer extension of perturbation series in fluid mechanics. *SIAM J. Appl. Math.* **28**, 720–734.

Van Dyke, M. (1984). Computer extension of perturbation series in fluid mechanics. *Ann. Rev. Fluid Mech.* **16**, 287–309.

Verkley, W. T. M. (1989). A variational principle for modons. *J. Atmospheric Sci.*, submitted.

Watson, L. T., and Scott, L. R. (1987). Solving Galerkin approximations to nonlinear two-point boundary value problems by a globally convergent homotopy method. *SIAM J. Sci. Statist. Comput.* **8**, 768–789.

Whitham, G. B. (1974). "Linear and Nonlinear Waves," pp. 109–110. John Wiley, New York.

Whitham, G. B. (1984). Comments on periodic waves and solitons. *IMA J. Appl. Math.* **32**, 353–366.

Williams, G. P., and Wilson, R. J. (1988). The stability and genesis of Rossby vortices. *J. Atmospheric Sci.* **49**, 207–240.

Wimp, J. (1981). "Sequence Transformations and Their Applications." Academic Press, New York.

Wu, H. M., Overman, E. A., and Zabusky, N. J. (1984). Steady-state solutions of the Euler equations in two dimensions: Rotating and translating V-states with limiting cases. I. Numerical algorithms and results. *J. Comput. Phys.* **53**, 42–71.

Yasuda, T., and Tsuchiya, Y. (1982). Theory of soliton spectra and its application to waves in shallow water. *Trans. JSCE* **14**, 428–431.

Zabusky, N. J., and Kruskal, M. D. (1965). Interaction of solitons in collision-less plasma and the recurrence of initial states. *Phys. Rev. Lett.* **15**, 240–243.

Zagrodzinski, J. (1982). Dispersion relations and a comparison of different quasi-periodic solutions of the sine–Gordon equation. *J. Phys. A* **15**, 3109–3118.

Zagrodzinski, J. (1983). Direct approach to the periodic solutions of the multidimensional sine–Gordon equation. *J. Math. Phys.* **24**, 46–52.

Zagrodzinski, J. (1984). Solitons and wavetrains: Unified approach. *J. Phys. A* **17**, 3315–3320.

Zagrodzinski, J., and Jaworski, M. (1982). Mixed solutions of the sine–Gordon equation. *Z. Phys. B* **49**, 75–77.

Zaitsev, A. A. (1983). Formation of stationary nonlinear waves by superposition of solitons. *Soviet Phys. Dokl.* **28**, 720–722.

Material Failure by Void Growth to Coalescence

VIGGO TVERGAARD

Department of Solid Mechanics
The Technical University of Denmark
Lyngby, Denmark

I. Introduction

Failure by coalescence of microscopic voids is an important fracture mechanism in ductile metals. The voids nucleate mainly at second phase

ISBN 0-12-002027-0

particles, by decohesion of the particle–matrix interface or by particle fracture, and subsequently the voids grow due to plastic straining of the surrounding material. The first micromechanical studies of these phenomena focussed on the growth of a single void in an infinite elastic-plastic solid (McClintock, 1968; Rice and Tracey, 1969), and the results were used to estimate critical strains for coalescence. Most of the subsequent work has focussed on incorporating the effect of the interaction between neighbouring voids, and there has also been some interest in micromechanical modelling of nucleation and of void deformations in the final stages prior to failure by coalescence.

The most widely known porous ductile material model is that developed by Gurson (1977a) based on averaging techniques similar to those used by Bishop and Hill (1951). The characteristic volume element considered by Gurson (1977a) is an aggregate of voids and rigid plastic matrix material rather than a polycrystalline aggregate, and approximate upper-bound solutions on the micro-level have been used to derive a macroscopic yield condition for the material. Alternative approximate yield conditions have been proposed by a number of other authors, e.g., Shima and Oyane (1976), who considered powder metallurgical materials on a purely empirical basis.

A number of micromechanical models, including that employed by Gurson (1977a), make the simplifying assumption that voids are spherical (or cylindrical). This gives a reasonable approximation in some cases, dependent on the shapes of particles from which voids nucleate; but near final failure this is usually a very poor approximation, as is illustrated by the micrograph in Figure 1. This micrograph shows a local failure region in the center of a notched tensile test specimen machined from a sintered iron powder compact, which was initially fully dense (Becker *et al.*, 1987a). It is seen that some voids have already coalesced to form larger voids and that voids can link up by rather localized bands of smaller voids as well as by necking of the ligament.

The porous ductile material models accounting for the nucleation and growth of voids are discussed in some detail in Section III, while cell model studies that are a major tool in understanding material behaviour are considered in Section V. Such model studies have been used to investigate the nucleation of voids and void coalescence, as well as the interaction between different size-scales of voids.

Applications of the material model to predict fracture mechanisms in structural components or test specimens is one of the main purposes of

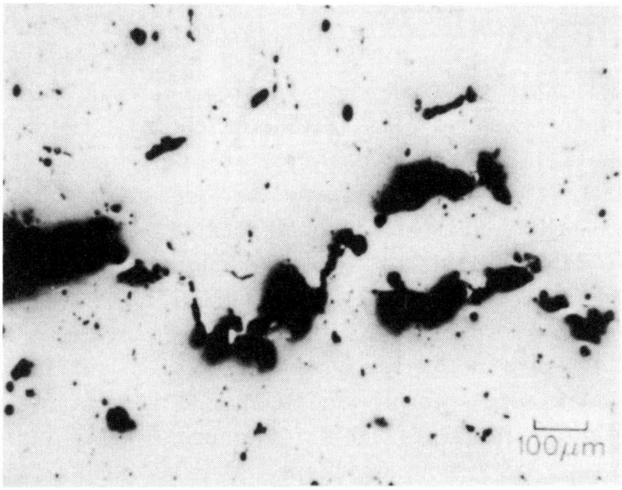

FIG. 1. Local failure region in sintered iron powder compact, with no initial porosity (Becker *et al.*, 1987a).

developing these constitutive relations. Localization of plastic flow in shear bands leading to final shear fracture by a void-sheet mechanism is one such application discussed in Section IV, and ductile crack growth by void coalescence is another discussed in Section VI. Extensions of porous ductile material models to account for the effect of a rounded vertex on the yield surface or material strain rate sensitivity are presented in Sections VII and VIII, respectively. The viscoplastic material model used to represent rate sensitivity is illustrated by analyses of the fracture mode transition in the Charpy V-notch test.

Void growth and coalescence is also the dominant cause of failure in polycrystalline metals subject to creep at elevated temperatures. However, at high temperatures the voids appear mainly in the grain boundaries, and diffusion gives a significant contribution to the growth rate. Due to the interaction of diffusion and dislocation creep, the rate of void growth can here be constrained by creep. This rather different mode of material failure by void coalescence is described in Section IX in terms of a material model proposed by Tvergaard (1984b). Applications of the model to analyses of creep rupture in structural components are included in the discussion.

II. Basic Equations

The growth of voids to coalescence by plastic yielding of the surrounding material involves so large geometry changes that finite strain formulations of the field equations are a necessary tool. This applies to micromechanical studies that focus on the local stress and strain fields around each single void as well as to studies based on continuum models that describe the average macroscopic effect of porosities in ductile solids.

The finite strain formulations in this paper are based on a Lagrangian framework, in which the reference configuration is usually identified with the initial undeformed configuration. Relative to a fixed Cartesian frame, the position of a material point in the reference configuration is denoted by the vector $\mathbf{r}$, and the position of the same point in the current configuration is $\bar{\mathbf{r}}$. The displacement vector $\mathbf{u}$ and the deformation gradient $\mathbf{F}$ are given by

$$\mathbf{u} = \bar{\mathbf{r}} - \mathbf{r}, \qquad \mathbf{F} = \frac{\partial \bar{\mathbf{r}}}{\partial \mathbf{r}}. \tag{2.1}$$

A convected coordinate formulation of the governing equations is used (see, for example, Green and Zerna (1968) and Budiansky (1969)). Convected coordinates ξ^i are introduced, which serve as particle labels. The convected coordinate net can be visualized as being inscribed on the body in the reference state and deforming with the material. The displacement vector $\mathbf{u}$ is considered as a function of the coordinates ξ^i and a monotonically increasing time-like parameter t. Latin indices range from 1 to 3, and the summation convention is adopted for repeated indices.

Covariant base vectors $\mathbf{e}_i$ and $\bar{\mathbf{e}}_i$ of the material net in the reference configuration and the current configuration, respectively, are given by

$$\mathbf{e}_i = \frac{\partial \mathbf{r}}{\partial \xi^i}, \qquad \bar{\mathbf{e}}_i = \frac{\partial \bar{\mathbf{r}}}{\partial \xi^i}. \tag{2.2}$$

The metric tensors in the reference and current configurations, g_{ij} and G_{ij}, are given by the dot products of the base vectors

$$g_{ij} = \mathbf{e}_i \cdot \mathbf{e}_j, \qquad G_{ij} = \bar{\mathbf{e}}_i \cdot \bar{\mathbf{e}}_j. \tag{2.3}$$

The determinants of g_{ij} and G_{ij} are denoted g and G, respectively, and the inverse of the two metric tensors are denoted by g^{ij} and G^{ij}.

Components of vectors and tensors on the embedded coordinates are obtained by dot products with the appropriate base vectors. Thus, the displacement components on the reference base vectors satisfy

$$u_i = \mathbf{e}_i \cdot \mathbf{u}, \qquad u^i = \mathbf{e}^i \cdot \mathbf{u}, \qquad \mathbf{u} = u^i \mathbf{e}_i. \tag{2.4}$$

Substituting (2.4c) into (2.1) and using this in (2.2) gives

$$\bar{\mathbf{e}}_i = \mathbf{e}_i + u^k_{,i}\mathbf{e}_k, \tag{2.5}$$

where $(\ )_{,i}$ denotes the covariant derivative in the reference frame. The Lagrangian strain tensor $\eta_{ij} = \frac{1}{2}(G_{ij} - g_{ij})$, expressed in terms of displacement components, is then found using (2.3) and (2.5)

$$\eta_{ij} = \tfrac{1}{2}(u_{i,j} + u_{j,i} + u^k_{,i}u_{k,j}). \tag{2.6}$$

The true stress tensor $\boldsymbol{\sigma}$ in the current configuration (the Cauchy stress tensor) has the contravariant components σ^{ij} on the current base vectors, where

$$\boldsymbol{\sigma} = \sigma^{ij}\bar{\mathbf{e}}_i\bar{\mathbf{e}}_j, \qquad \sigma^{ij} = \bar{\mathbf{e}}^i \cdot \boldsymbol{\sigma} \cdot \bar{\mathbf{e}}^j. \tag{2.7}$$

The contravariant components τ^{ij} of the Kirchhoff stress tensor on the current base vectors are defined by

$$\tau^{ij} = \sqrt{\frac{G}{g}}\,\sigma^{ij}, \tag{2.8}$$

where $\sqrt{G/g} = d\bar{V}/dV = \rho/\bar{\rho}$, expressed in terms of the volume element dV and the density ρ.

The requirement of equilibrium can be specified in terms of the principle of virtual work

$$\int_V \tau^{ij}\,\delta\eta_{ij}\,dV = \int_S T^i\,\delta u_i\,dS, \tag{2.9}$$

where V and S are the volume and surface, respectively, of the body in the reference configuration, and the specified surface tractions $\mathbf{T} = T^i\mathbf{e}_i$ per unit area in the reference configuration have the components on reference base vectors

$$T^i = (\tau^{ij} + \tau^{kj}u^i_{,k})n_j. \tag{2.10}$$

Here, $\mathbf{n} = n_j\mathbf{e}^j$ is the surface normal in the reference state.

For the solution of boundary value problems, incremental equilibrium equations are needed due to the material path dependence. When the current values of all field quantities, e.g., stresses τ^{ij}, displacements u_i, and tractions T^i, are assumed known, an expansion of (2.9) about the known state gives to lowest order

$$\Delta t \int_V (\dot{\tau}^{ij}\,\delta\eta_{ij} + \tau^{ij}\dot{u}^k_{,i}\,\delta u_{k,j})\,dV$$

$$= \Delta t \int_S \dot{T}^i\,\delta u_i\,dS - \left[\int_V \tau^{ij}\,\delta\eta_{ij}\,dV - \int_S T^i\,\delta u_i\,dS\right]. \tag{2.11}$$

Here, $(\dot{\ }) \equiv \partial(\ )/\partial t$ at fixed ξ^i, and Δt is the prescribed increment of the "time" t, so that $\Delta T^i = \Delta t \dot{T}^i$ are the components of the prescribed traction increments. The terms bracketed in (2.11) vanish according to (2.9) if the current state satisfies equilibrium. However, in linear incremental analyses the solution tends to drift away from the true equilibrium path, due to incrementation errors, and here the bracketed terms in (2.11) can be included to avoid such drifting.

In the constitutive relations to be discussed here, the total strain rate is taken to be the sum of the elastic and plastic parts, $\dot{\eta}_{ij} = \dot{\eta}^{\mathrm{E}}_{ij} + \dot{\eta}^{\mathrm{P}}_{ij}$. Thus, with an elastic relationship of the form $\overset{\nabla}{\sigma}{}^{ij} = \mathcal{R}^{ijkl}\dot{\eta}^{\mathrm{E}}_{kl}$, the constitutive relations can be written as

$$\overset{\nabla}{\sigma}{}^{ij} = \mathcal{R}^{ijkl}(\dot{\eta}_{kl} - \dot{\eta}^{\mathrm{P}}_{kl}), \tag{2.12}$$

where the Jaumann (co-rotational) rate of the Cauchy stress tensor $\overset{\nabla}{\sigma}{}^{ij}$ is related to the convected rate by

$$\overset{\nabla}{\sigma}{}^{ij} = \dot{\sigma}^{ij} + (G^{ik}\sigma^{jl} + G^{jk}\sigma^{il})\dot{\eta}_{kl}. \tag{2.13}$$

For the elastic moduli, Budiansky (unpublished work, see also Hutchinson (1973)) has suggested using the finite strain generalization

$$\mathcal{R}^{ijkl} = \frac{E}{1+\nu}\left\{\tfrac{1}{2}(G^{ik}G^{jl} + G^{il}G^{jk}) + \frac{\nu}{1-2\nu}G^{ij}G^{kl}\right\}, \tag{2.14}$$

where E is Young's modulus and ν is Poisson's ratio. The elastic incremental stress–strain relationship with moduli (2.14) is actually hypo-elastic, since it cannot be derived from a work potential. However, it is noted that in the limit of small stresses relative to Young's modulus the relationship reduces to Hooke's law on the current base vectors. In elastic-plastic analyses the elastic contribution to the total straining is usually very small so that the use of this hypo-elastic relationship rather than a truly elastic one is a reasonable approximation.

An incremental stress–strain relationship in terms of $\dot{\tau}^{ij}$ is needed in

(2.11). This expression is obtained from (2.12) by substituting (2.13) and the incremental form of (2.8)

$$\dot{\tau}^{ij} = \sqrt{\frac{G}{g}}\,\dot{\sigma}^{ij} + \tau^{ij}G^{kl}\dot{\eta}_{kl}. \tag{2.15}$$

For time-independent plasticity, where t is a loading parameter since time does not enter the formulation, the resulting constitutive relations are of the form

$$\dot{\tau}^{ij} = L^{ijkl}\dot{\eta}_{kl}. \tag{2.16}$$

On the other hand, viscous material models are time-dependent, so that here t denotes the time. In visco-plasticity, where $\dot{\eta}_{ij}^{\mathrm{P}}$ represents non-linearly viscous behaviour, the resulting constitutive relations are of the form

$$\dot{\tau}^{ij} = L_*^{ijkl}\dot{\eta}_{kl} + \dot{\tau}_*^{ij}, \tag{2.17}$$

where $\dot{\tau}_*^{ij}$ represents part of the viscous terms.

III. Continuum Models of Porous Ductile Solids

Early micromechanical studies of void growth in ductile materials focussed on the behaviour of a single void in an infinite block of plastic material (McClintock, 1968; Rice and Tracey, 1969). It was found both for a cylindrical void and a spherical void that the rate of growth is strongly dependent on the level of hydrostatic tension, so that fracture by coalescence of voids would be promoted by a high level of triaxial tension. This influence of the triaxial tension was also found in a series of tests on notched tensile specimens, carried out to study the stress state dependence of ductile failure in high strength steels (Hancock and Mackenzie, 1976; Hancock and Brown, 1983).

A. Approximate Yield Conditions

For a ductile material containing a certain volume fraction f of voids, Gurson (1977a,b) has suggested the use of an approximate yield condition of the form $\Phi(\sigma^{ij}, \sigma_{\mathrm{M}}, f) = 0$. Here, σ^{ij} is the average macroscopic Cauchy stress tensor, and σ_{M} is an equivalent tensile flow stress representing the actual microscopic stress-state in the matrix material.

Gurson (1977a) carried out a number of rigid-plastic upper bound analyses to evaluate the yield function $\Phi = 0$. Some of these analyses focussed on a cylindrical volume element, containing a concentric cylindrical void, some focussed on the analogous spherical problem, and in some of the cases the effect of elastic unloading regions around the void was investigated as well. The approximate yield condition derived on the basis of a spherical model problem with no unloading is of the form

$$\Phi = \frac{\sigma_e^2}{\sigma_M^2} + 2q_1 f^* \cosh\left(\frac{q_2}{2}\frac{\sigma_k^k}{\sigma_M}\right) - [1 + (q_1 f^*)^2] = 0 \tag{3.1}$$

for $q_1 = q_2 = 1$ and $f^* = f$. Here, $\sigma_e = (3s_{ij}s^{ij}/2)^{1/2}$ is the macroscopic effective Mises stress, and $s^{ij} = \sigma^{ij} - G^{ij}\sigma_k^k/3$ is the stress deviator. The implication of this analysis is that the voids are assumed to be randomly distributed, so that the macroscopic response is isotropic. Furthermore, it is noted that (3.1) reduces to the standard Mises yield condition when $f = 0$.

Micromechanical studies for materials containing periodic distributions of cylindrical voids or spherical voids have been carried out by Tvergaard (1981, 1982a) and compared with predictions based on the yield condition (3.1). This comparison led to the conclusion that the approximation obtained by (3.1) is improved by taking $q_1 = 1.5$ and $q_2 = 1$.

Final failure of a porous ductile solid occurs by void coalescence. According to the Gurson model, loss of material stress carrying capacity occurs when the voids have grown so large that (3.1) has shrunk to a point (see Figure 2), i.e., at the ultimate value $f = 1/q_1$. However, this value is unrealistically large. Based on a number of experimental studies

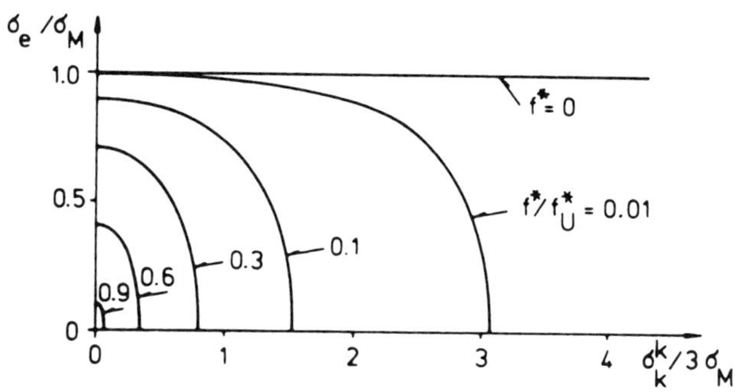

FIG. 2. Yield surface dependence on the hydrostatic tension for various values of the function $f^*(f)$ in (3.1).

Brown and Embury (1973) have found that after nucleation cavities elongate along the major tensile axis and that two neighbouring cavities coalesce when their length has grown to the order of magnitude of their spacing. This local failure occurs by the development of slip planes between the cavities or simply necking of the ligament. Using a simple model, they estimated a critical void volume fraction for coalescence of about 0.15, much smaller than $1/q_1$. A micromechanical model study by Andersson (1977) for the growth of a spherical void in the stress field in front of a crack tip gave coalescence at $f \simeq 0.25$.

Some ways of incorporating the effect of coalescence into the Gurson model have been discussed by Tvergaard (1982c). The function $f^*(f)$ in (3.1) was introduced by Tvergaard and Needleman (1984) to model the complete loss of material stress carrying capacity at a realistic void volume fraction. The function was chosen as

$$f^*(f) = \begin{cases} f, & \text{for } f \leq f_C, \\ f_C - \dfrac{f_U^* - f_C}{f_F - f_C}(f - f_C), & \text{for } f > f_C, \end{cases} \tag{3.2}$$

where the void volume fraction at final fracture is denoted by f_F, so that $f^*(f_F) = f_U^* = 1/q_1$ (see Figure 2). According to (3.2) the modification of the yield condition due to the effect of coalescence starts when the void volume fraction reaches a certain critical value f_C. Based on experimental results (Brown and Embury, 1973) and numerical model analyses (Andersson, 1977) the values $f_C = 0.15$ and $f_F = 0.25$ were chosen.

Alternative yield conditions for porous ductile solids have been developed for the purpose of studying powder compacted metals. In most cases, a quadratic expression in terms of the Mises stress and the mean stress has been assumed, and the coefficients of the different terms have been chosen to fit experimental results for uniaxial, or uniaxial plane strain, tension or compression. Shima and Oyane (1976) have used this type of procedure to fit experiments for specimens with known void volume fractions, varying from 0.02 to 0.38. Written in the same notation as (3.1), the yield function of Shima and Oyane (1976) is

$$\Phi_S = \frac{\sigma_e^2}{\sigma_M^2} + 6.20 f^{1.028} \frac{(\sigma_k^k/3)^2}{\sigma_M^2} - (1 - f)^5 = 0. \tag{3.3}$$

The expression (3.3) was fitted with experiments for sintered copper, but reasonable agreement was also found with some results for sintered iron and sintered aluminium.

Experiments with specimens of known porosity, produced by powder metallurgy, have been used to test the material model based on (3.1), developed with the main focus on ductile fracture. Hancock (1982) has made comparisons with uniaxial compression tests for compacted steels with a wide range of porosities, $0<f<0.18$, and has found good agreement with (3.1) for $q_1=1.5$. Uniaxial compressive tests for a powder compacted aluminium alloy have been carried out by Doraivelu *et al.* (1984), who have compared the results with a number of different yield criteria for ductile porous solids, finding rather good agreement with (3.3). They find poor agreement with the Gurson model, but, in fact, their comparison only relates to a yield function that approximates (3.1). A reexamination shows that the agreement is good if $q_1=2$ is chosen, at least for f up to 0.15. Becker and Needleman (1985) mention that experimental uniaxial tension data obtained by Spitzig and Richmond for porous iron specimens are well represented by taking $q_1=2.38$ and $q_2=0.748$. Quite recently, a comparison with a series of experiments on notched tensile specimens made of sintered iron has been carried out by Becker *et al.* (1987a), and good agreement with predictions based on (3.1) has been found, using $q_1=1.25$ and $q_2=0.95$.

On the basis of all the experiments mentioned above, for sintered materials with known porosity, it appears that using a value of q_1 in the range from 1.25 to 2 and $q_2 \simeq 1$ gives a reasonably accurate representation of ductile porous solids. This agrees with the micromechanical studies of Tvergaard (1981, 1982a) for periodically distributed voids, where the main emphasis was on localization predictions.

The rigid-plastic upper bound analyses of Gurson (1977a) were based on expanding the displacement fields in terms of a few smooth functions. More accurate predictions would require a larger number of functions, or a finite element solution with a reasonably fine mesh. The latter approach has been used by Guennouni and Francois (1987) for a planar array of cylindrical voids. For some stress states, these numerical plane strain solutions give rather localized yield zones, and the resulting yield function has a slightly different shape than that found by Gurson (1977a). The numerically found yield points have been fitted with a quadratic expression in terms of σ_e and $\sigma_k^k/3$, analogous to (3.3), but with different coefficients.

B. Nucleation and Growth of Voids

During plastic yielding, the void volume fraction changes, partly due to void growth and partly due to the nucleation of new voids. Experimental investigations show that the nucleation of new voids occurs mainly at second phase particles, by decohesion of the particle–matrix interface or by particle fracture (see, for example, Puttick (1960), and Goods and Brown (1979)). In some cases, nucleation appears to be controlled by the plastic strain, whereas in other cases nucleation occurs when the maximum normal stress on the particle–matrix interface reaches some critical value.

The rate of growth of the void volume fraction is written on the form

$$\dot{f} = (\dot{f})_{\text{growth}} + (\dot{f})_{\text{nucleation}}. \tag{3.4}$$

Since the matrix material is plastically incompressible, the increment due to growth is given by

$$(\dot{f})_{\text{growth}} = (1-f)G^{ij}\dot{\eta}_{ij}^{\text{P}}, \tag{3.5}$$

where $\dot{\eta}_{ij}^{\text{P}}$ is the plastic part of the macroscopic strain increment. Various nucleation models have been discussed by Gurson (1977b). The model to be used here is of the form

$$(\dot{f})_{\text{nucleation}} = \mathscr{A}\dot{\sigma}_{\text{M}} + \mathscr{B}(\sigma_k^k)^{\cdot}/3 \tag{3.6}$$

proposed by Needleman and Rice (1978).

The effective plastic strain $\varepsilon_{\text{M}}^{\text{P}}$, representing the microscopic strain-state in the matrix material, is related to σ_{M} by the incremental equation

$$\dot{\varepsilon}_{\text{M}}^{\text{P}} = \left(\frac{1}{E_{\text{t}}} - \frac{1}{E}\right)\dot{\sigma}_{\text{M}}, \tag{3.7}$$

where E_{t} is the slope of the uniaxial true stress–natural strain curve for the matrix material at the current stress level σ_{M}. Thus, a material for which void nucleation is controlled by the plastic strain can be modelled in terms of (3.6) by taking $\mathscr{A} > 0$ and $\mathscr{B} = 0$. Chu and Needleman (1980) have suggested a normal distribution for nucleation, so that $\mathscr{A}$ and $\mathscr{B}$ are given by

$$\begin{aligned} \mathscr{A} &= \left(\frac{1}{E_{\text{t}}} - \frac{1}{E}\right)\frac{f_{\text{N}}}{s\sqrt{2\pi}}\exp\left[-\frac{1}{2}\left(\frac{\varepsilon_{\text{M}}^{\text{P}} - \varepsilon_{\text{N}}}{s}\right)^2\right], \\ \mathscr{B} &= 0 \qquad \text{for} \quad \varepsilon_{\text{M}}^{\text{P}} = (\varepsilon_{\text{M}}^{\text{P}})_{\max} \quad \text{and} \quad \dot{\varepsilon}_{\text{M}}^{\text{P}} > 0. \end{aligned} \tag{3.8}$$

Here, ε_{N} is the mean strain for nucleation, s is the corresponding standard deviation, and f_{N} is the volume fraction of void nucleating particles.

For materials, in which nucleation is controlled by the maximum normal stress on the particle–matrix interface, Needleman and Rice

(1978) have suggested using the sum $\sigma_{\mathrm{M}} + \sigma_k^k/3$ as an approximate measure of this maximum stress, thus taking $\mathscr{A} = \mathscr{B}$. Then, by analogy with (3.8), $\mathscr{A}$ and $\mathscr{B}$ are given by

$$\mathscr{A} = \mathscr{B} = \frac{f_{\mathrm{N}}}{s\sqrt{2\pi}} \exp\left\{ -\frac{1}{2}\left[\frac{(\sigma_{\mathrm{M}} + \sigma_k^k/3) - \sigma_{\mathrm{N}}}{s}\right]^2 \right\}$$
$$\text{for} \quad \sigma_{\mathrm{M}} + \sigma_k^k/3 = (\sigma_{\mathrm{M}} + \sigma_k^k/3)_{\max} \quad \text{and} \quad (\sigma_{\mathrm{M}} + \sigma_k^k/3)^{\cdot} > 0, \tag{3.9}$$

where σ_{N} is the mean stress for nucleation. If both strain-controlled and stress-controlled nucleation takes place in the material, then $\mathscr{A}$ is the sum of the values given by (3.8) and (3.9). Furthermore, the non-zero values of $\mathscr{A}$ and $\mathscr{B}$ are only used if the conditions (3.8b) or (3.9b) are satisfied.

C. Stress–Strain Relations

Assuming that the total strain rate is the sum of the elastic and plastic parts leads to an incremental stress–strain relationship of the form (2.12) or (2.16).

In the model of Gurson (1977a), the plastic part of the macroscopic strain–rate for the porous ductile material is taken to be given by

$$\dot{\eta}_{ij}^{\mathrm{P}} = \Lambda \frac{\partial \Phi}{\partial \sigma^{ij}} \tag{3.10}$$

since normality of the plastic flow rule for the matrix material implies macroscopic normality (Berg, 1970; Bishop and Hill, 1951). It will be assumed here that (3.10) also applies during the nucleation of new voids.

The microscopic equivalent tensile flow stress σ_{M} and equivalent plastic strain $\varepsilon_{\mathrm{M}}^{\mathrm{P}}$ are averages that do not specify the actual microscopic fields in the matrix material surrounding the voids. However, it is assumed that the rate of equivalent plastic work in the matrix material equals the macroscopic rate of plastic work

$$\sigma^{ij}\dot{\eta}_{ij}^{\mathrm{P}} = (1-f)\sigma_{\mathrm{M}}\dot{\varepsilon}_{\mathrm{M}}^{\mathrm{P}}. \tag{3.11}$$

Thus, using (3.7), the following expression is found for the rate of the equivalent tensile flow stress in the matrix material

$$\dot{\sigma}_{\mathrm{M}} = \frac{EE_{\mathrm{t}}}{E - E_{\mathrm{t}}} \frac{\sigma^{ij}\dot{\eta}_{ij}^{\mathrm{P}}}{(1-f)\sigma_{\mathrm{M}}}. \tag{3.12}$$

Now, the value of the parameter Λ in (3.10) is determined from the consistency condition $\dot{\Phi} = 0$, required during plastic loading. Substituting (3.4)–(3.7) and (3.12) into this condition gives

$$\dot{\eta}_{ij}^{\mathrm{P}} = \frac{1}{H} m_{ij}^{\mathrm{G}} m_{kl}^{\mathrm{F}} \overset{\nabla}{\sigma}{}^{kl}, \tag{3.13}$$

where $\overset{\nabla}{\sigma}{}^{kl}$ is defined by (2.13), and

$$m_{ij}^{\mathrm{G}} = \frac{3}{2} \frac{s_{ij}}{\sigma_{\mathrm{M}}} + \alpha G_{ij}, \qquad m_{ij}^{\mathrm{F}} = \frac{3}{2} \frac{s_{ij}}{\sigma_{\mathrm{M}}} + \beta G_{ij}, \tag{3.14}$$

$$\alpha = \frac{f^*}{2} q_1 q_2 \sinh\left(\frac{q_2}{2} \frac{\sigma_k^k}{\sigma_{\mathrm{M}}}\right), \qquad \beta = \alpha + \frac{\mathscr{B} \sigma_{\mathrm{M}}}{6} \frac{\partial \Phi}{\partial f}, \tag{3.15}$$

$$H = \frac{\sigma_{\mathrm{M}}}{2} \left[-3\alpha(1-f) \frac{\partial \Phi}{\partial f} - \left(\frac{\partial \Phi}{\partial f} \mathscr{A} + \frac{\partial \Phi}{\partial \sigma_{\mathrm{M}}} \right) \frac{E E_{\mathrm{t}}}{E - E_{\mathrm{t}}} \frac{1}{1-f} \left(\frac{\sigma_{\mathrm{e}}^2}{\sigma_{\mathrm{M}}^2} + \alpha \frac{\sigma_k^k}{\sigma_{\mathrm{M}}} \right) \right]. \tag{3.16}$$

Plastic yielding initiaties when $\Phi = 0$ and $\dot{\Phi} > 0$ during elastic deformation, and continued plastic loading requires $\Phi = 0$ and $m_{kl}^{\mathrm{F}} \overset{\nabla}{\sigma}{}^{kl}/H \geq 0$.

After substituting (3.13) into (2.12), a few standard transformations (Tvergaard, 1982d) lead to an incremental stress–strain relationship of the form (2.16), with the instantaneous moduli specified by

$$L^{ijkl} = \mathscr{L}^{ijkl} - \mu M_{\mathrm{G}}^{ij} M_{\mathrm{F}}^{kl}. \tag{3.17}$$

Here,

$$\mathscr{L}^{ijkl} = \sqrt{\frac{G}{g}} [\mathscr{R}^{ijkl} - \tfrac{1}{2}(\sigma^{ik} G^{jl} + \sigma^{jk} G^{il} + \sigma^{il} G^{jk} + \sigma^{jl} G^{ik}) + \sigma^{ij} G^{kl}], \tag{3.18}$$

$$M_{G}^{ij} = \mathscr{R}^{ijkl} m_{kl}^{\mathrm{G}}, \qquad M_{\mathrm{F}}^{kl} = m_{rs}^{\mathrm{F}} \mathscr{R}^{rskl}, \tag{3.19}$$

$$\mu = \begin{cases} 0 & \text{for elastic unloading,} \\ \sqrt{\dfrac{\mathrm{G}}{\mathrm{g}}} [H + m_{rs}^{\mathrm{F}} \mathscr{R}^{rskl} m_{kl}^{\mathrm{G}}]^{-1} & \text{for plastic loading.} \end{cases} \tag{3.20}$$

It is noted that the symmetry $\mathscr{R}^{ijkl} = \mathscr{R}^{klij}$, satisfied by the tensor of elastic moduli (2.14), is not satisfied by the present instantaneous moduli L^{ijkl}. This is partly due to the last term in (3.18), which has a noticeable effect for the dilatant material considered here, and partly due to $M_{\mathrm{G}}^{ij} \neq M_{\mathrm{F}}^{ij}$ in cases where $\mathscr{B} \neq 0$.

The incremental constitutive relations derived here could also directly be used in connection with other approximate yield conditions than (3.1), e.g., with (3.3). Some of the expressions (3.14)–(3.16) would have to be changed according to the different yield condition, but otherwise the constitutive relations would have the same form. It is noted that Shima and Oyane (1976) did discuss the incremental stress–strain relations associated with (3.3).

The uniaxial true stress natural strain curve for the matrix material, referred to in (3.7), is generally obtained from a uniaxial tensile test. In much of the following discussion, it will be assumed that the curve can be represented by a piecewise power law

$$\varepsilon = \begin{cases} \dfrac{\sigma}{E} & \text{for} \quad \sigma \leq \sigma_y, \\ \dfrac{\sigma_y}{E}\left(\dfrac{\sigma}{\sigma_y}\right)^n & \text{for} \quad \sigma > \sigma_y, \end{cases} \tag{3.21}$$

where σ_y is the uniaxial yield stress and n is the strain hardening exponent.

IV. Localization of Plastic Flow

During the deformation of ductile solids, it is frequently observed that at some stage a smoothly varying deformation pattern develops into a pattern involving highly localized deformations in the form of shear bands. Once localization has taken place, the strains inside the band become very large without contributing much to the overall deformation of the body. This can lead to shear fracture at an overall strain only slightly larger than that at the onset of localization.

For a homogeneously deformed rate-independent solid, the state at which bifurcation into a shear band mode is first possible coincides with loss of ellipticity of the equations governing incremental equilibrium (Hill, 1962; Rice, 1977). Thus, the analysis of this material instability follows the theoretical framework due to Hadamard (1903).

The value of the critical strain for loss of ellipticity is very sensitive to the constitutive law (Rudnicki and Rice, 1975; Rice, 1977). The classical elastic-plastic solid with a smooth yield surface and normality of the plastic flow rule is quite resistant to localization, but deviations from the classical model can have a strong effect. Thus, localization at a realistic strain level is predicted for a solid that develops a vertex on the yield surface, as arises in physical polycrystalline models based on the concept of single crystal slip. Also dilatational plastic flow and non-normality of the plastic flow rule, as induced by the porous ductile material model discussed in Section III, have a significant destabilizing effect (Rice, 1977; Needleman and Rice, 1978). Furthermore, the onset of localization is very sensitive to small material inhomogeneities.

Simple model studies for shear localization can be carried out by a procedure analogous to that developed by Marciniak and Kuczynski (1967) for the problem of plane stress sheet necking. In these analyses, an initial material inhomogeneity, such as a higher concentration of void nucleating particles, is assumed inside a plane slice of material, and the stress-states inside and outside this slice of material, respectively, are assumed to remain homogeneous throughout the deformation history.

In the model, illustrated in Figure 3, the principal directions of the stresses and strains outside the band are assumed to remain fixed, parallel to the x^i-axes in the Cartesian reference coordinate system. The major principal stress outside the band is taken to be in the x^1-direction, and the slice of material containing the initial inhomogeneity is assumed parallel with the x^3-axis, with the initial angle of inclination ψ_{I} and the unit normal vector n_i in the reference state. Then, the current angle of inclination ψ of the band, at any stage of the deformation, is given by

$$\tan \psi = \exp(\varepsilon_1^{\mathrm{o}} - \varepsilon_2^{\mathrm{o}}) \tan \psi_{\mathrm{I}}. \tag{4.1}$$

Here, ε_i are principal logarithmic strains, and $(\ )^{\mathrm{o}}$ denotes quantities outside the band while $(\ )^{\mathrm{b}}$ will denote quantities inside the band.

The field quantities outside the band are specified by the external loading, while the quantities inside the band have to satisfy compatibility and equilibrium over the band interface. Compatibility requires continuity of the tangential derivatives of the displacement components u_i

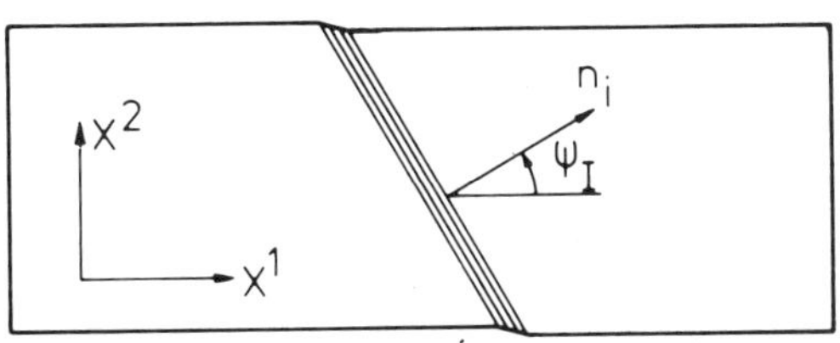

FIG. 3. Shear band in a homogeneously strained solid.

over the interface, so that the displacement gradients can be expressed by

$$u^{\mathrm{b}}_{i,j} = u^{\mathrm{o}}_{i,j} + c_i n_j, \tag{4.2}$$

where c_i are parameters to be determined. Equilibrium requires balance of the nominal tractions on each side of the interface

$$(T^i)^{\mathrm{b}} = (T^i)^{\mathrm{o}}, \tag{4.3}$$

where the nominal traction components T^i on a surface with reference normal n_j are given by (2.10). Now, a set of incremental equations for $\dot{c}_i$ are obtained by substituting the incremental constitutive relations (2.16) into the incremental form of (4.3), using (4.2) to express the strain increments $\dot{\eta}^{\mathrm{b}}_{ij}$.

In a case where there is no material inhomogeneity, the incremental equations for $\dot{c}_i$ are homogeneous, allowing only the trivial solution until bifurcation occurs. The first such bifurcation into a shear band (for any angle of inclination ψ of the band) marks the loss of ellipticity of the governing differential equations (Hill, 1962; Rice, 1977).

For porous ductile materials, described by the Gurson model, Yamamoto (1978) has analysed the bifurcation into shear bands, as a function of the initial void volume fraction f_{I}. In the limit of $f_{\mathrm{I}} \equiv 0$, where the material model reduces to J_2-flow theory, the critical strain for localization is essentially infinite, but the critical strain decays very rapidly for increasing f_{I}, so that just a few percent porosity is enough for the material instability to be predicted at a realistic strain level.

A detailed micromechanical study of shear band bifurcation that accounts for the interaction between neighbouring voids and the strongly nonhomogeneous stress distributions around each void has been carried out by Tvergaard (1981). A power-hardening elastic-plastic solid was considered, containing a doubly periodic array of circular cylindrical voids as shown in Figure 4. Due to symmetries, only the region hatched in Figure 4 needs to be analysed numerically, when this model material is subjected to plane strain tension in one of the coordinate directions. Bifurcation into another periodic pattern was also analysed by considering only the hatched region in Figure 4, making use of several symmetry properties of the repetitive pattern. This bifurcation mode was used to represent the deformations inside a shear band, as illustrated in Figure 5. The equilibrium conditions (4.3) on the band interface were satisfied exactly, while compatibility was only satisfied on the average, since the two periodic patterns are not locally compatible along the characteristic direction.

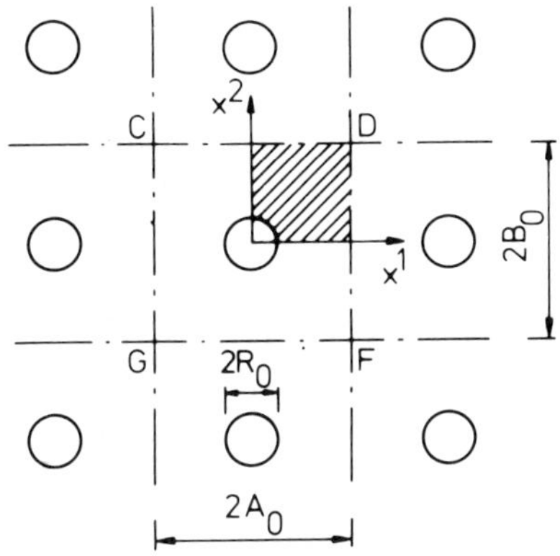

FIG. 4. Doubly periodic array of circular cylindrical voids.

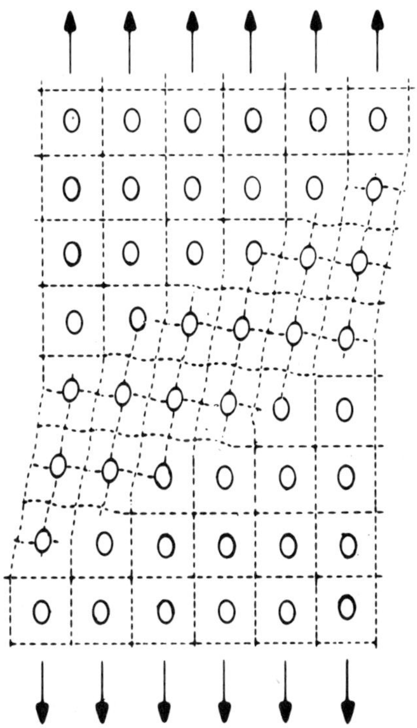

FIG. 5. Shear band bifurcation mode found numerically for model material containing a periodic array of cylindrical voids.

By comparison with the full numerical shear band bifurcation analyses described above, it was found that the Gurson model gives somewhat too large localization strains, but that reasonably good agreement is obtained by using $q_1 = 1.5$ in (3.1). It is noted that no loss of ellipticity of the governing differential equations happens in the full numerical bifurcation analysis, where the matrix material is described by isotropic hardening J_2-flow theory, but still there is good agreement with loss of ellipticity according to the approximate continuum model. This agreement with the more detailed micromechanical study gives good confidence in localization predictions obtained by the continuum model of a porous ductile solid, and such predictions do play an important role in studies of ductile fracture.

When there is a material inhomogeneity, the incremental equations for $\dot{c}_i$ obtained from (4.2) and (4.3) are inhomogeneous. Then the rate of deformation inside the band will gradually increase relative to that outside, and localization is defined by the onset of elastic unloading outside the band. For a given size of the inhomogeneity, the localization strain depends on the initial angle of inclination ψ_{I} of the slice containing the imperfection, and the variation of the localization strain with ψ_I must be determined to find the most critical angle of inclination. Such analyses have been carried out by Yamamoto (1978) for inhomogeneities in the form of a higher void volume fraction inside the band, and Saje, Pan, and Needleman (1982) have included the effect of an additional constant density of void nucleating particles.

Even if there is no initial porosity, $f_{\mathrm{I}}^{\mathrm{o}} = f_{\mathrm{I}}^{\mathrm{b}} = 0$, so that the initial response is that of a homogeneous bar, an initial inhomogeneity can be represented by a non-uniform distribution of void nucleating particles. Such cases have been studied by Tvergaard (1987a), with the inhomogeneity specified by Δf_{N}, so that the volume fractions of void nucleating particles inside and outside the band are related by

$$f_{\mathrm{N}}^{\mathrm{b}} = f_{\mathrm{N}}^{\mathrm{o}} + \Delta f_{\mathrm{N}}. \tag{4.4}$$

In all cases, the value of the initial angle of inclination ψ_{I} that leads to first localization tends to be much smaller for small imperfections than for larger imperfections. But independent of the value of the localization strain, the corresponding value of the current angle of inclination ψ at first localization is always about 43° in uniaxial plane strain tension, rather close to the critical inclination for shear band bifurcation in the homogeneous solid.

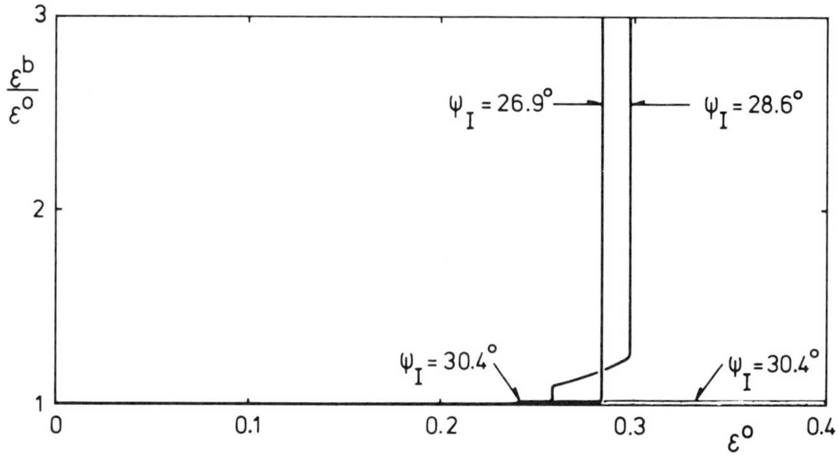

FIG. 6. Maximum principal strain ε^b inside shear band vs. strain ε^o outside. Initial voidage $f_I^b = 0.0001$, $f_I^o = 0$, and stress controlled nucleation $f_N = 0.04$, $\sigma_N = 2.1\sigma_y$, $s = 0.4\sigma_y$ (Tvergaard, 1982d).

The onset of localization is usually considered an indication of failure, since the subsequent severe straining inside the band contributes little to the overall strain. Several shear band computations have been continued beyond the onset of localization (Tvergaard, 1982c), and these results confirm that failure occurs by coalescence in a void-sheet inside the band, while the material outside the band remains elastically unloaded. However, the results shown in Figure 6 for a material with stress-controlled nucleation show that the behaviour can be somewhat more complex (Tvergaard, 1982d).

In Figure 6, the material is characterized by $f_N = 0.04$, $\sigma_N = 2.1\sigma_y$, and $s = 0.4\sigma_y$ in (3.9), with a very small inhomogeneity $f_I^b = 0.0001$, $f_I^o = 0$, so that the results are representative of the post-bifurcation behaviour corresponding to a completely homogeneous solid. For $\psi_I = 30.4°$, the band has rotated to the critical current angle $\psi = 42.8°$ for bifurcation at the corresponding critical strain $\varepsilon^o = 0.237$; however, elastic unloading occurs only briefly, and subsequently the deformations remain nearly uniform, $\varepsilon^b/\varepsilon^o \simeq 1$. For a smaller initial angle, $\psi_I = 28.6°$, localization occurs at a strain beyond the elliptic–hyperbolic interface, but then the material outside the band yields again, until final failure takes place at a secondary localization. A somewhat similar saturation of localized flow has been found by Hutchinson and Tvergaard (1981) for a material that forms a vertex on the yield surface. For an even smaller initial angle,

$\psi_I = 26.9°$, localization occurs later, but leads to direct failure. Thus, with stress controlled nucleation, where non-normality of the plastic flow rule (3.13) occurs, the localization leading to final failure may be a secondary localization.

For materials with stress controlled nucleation, Needleman and Rice (1978) and Saje, Pan, and Needleman (1982) have shown that a sudden burst of nucleation (e.g., a very small standard deviation s in (3.9)) causes very early localization. However, a computation continued beyond such an early onset of localization has shown that the localized strain inside the band increases only a little (during the burst of nucleation), and subsequently the material yields both inside and outside the band in a rather uniform manner (Tvergaard, 1982d). Thus, the early localizations due to a burst of nucleation appear not to have a strong effect on final material failure.

V. Cell Model Studies

Micromechanical investigations of the behaviour of ductile porous solids have mostly focussed on a characteristic volume element containing a single void. This includes an early numerical study by Needleman (1972) for a square array of cylindrical voids (Figure 4), the analyses of Gurson (1977a) for a spherical volume with a concentric spherical void, a numerical analysis by Nemat-Nasser and Taya (1977), and the localization studies by Tvergaard (1981, 1982a). Here, a few more recent cell model studies will be discussed that focus on void nucleation and on void growth to final failure.

A. Nucleation of Voids

The most obvious consequence of void nucleation in plastically deforming metals is the generation of damage, but another important consequence is a reduction of the macroscopic strain hardening capacity during the nucleation process. This softening due to nucleation has a dramatic effect on the prediction of localization and thus on the occurrence of ductile fracture.

As discussed in Section IV, a burst of stress controlled nucleation can result in very early localization according to the Gurson model (Needleman and Rice, 1978; Saje *et al.*, 1982), but strain-controlled nucleation or less sudden stress-controlled nucleation can also have a significant influence in reducing the critical strain for localization. The degree of nucleation-induced softening predicted by this material model is a result of assuming the nucleation model (3.6) and also assuming that (3.10) remains valid during nucleation.

A number of void nucleation studies have been carried out by Hutchinson and Tvergaard (1987) to get a more detailed understanding of the softening behaviour. Results for nucleation of an isolated spherical void in an infinite matrix under triaxial remote stressing were used to predict the overall stress–strain behaviour in cases where interaction between voids is negligible. Nucleation of the void, at stresses Σ_{ij} that have been applied proportionally, causes a redistribution of stress and additional straining of the matrix, above the straining that occurs in the absence of nucleation. This additional macroscopic straining during the nucleation of the void volume fraction $\dot{f}$ is denoted by $\dot{f}\Delta E_{ij}$, such that $\Sigma_{ij}\dot{f}\Delta E_{ij}$ is the extra work done by the remote stresses due to nucleation (here the macroscopic components refer to a small strain analysis on current Cartesian coordinates). Thus, if M_{ijkl} are the incremental compliances for $\dot{f}=0$, so that $\dot{E}_{ij}=M_{ijkl}\dot{\Sigma}_{kl}$ in the absence of nucleation, the macroscopic strain increments during first void nucleation are

$$\dot{E}_{ij}=M_{ijkl}\dot{\Sigma}_{kl}+\dot{f}\Delta E_{ij}. \tag{5.1}$$

For the nucleation of a spherical void in an isotropic hardening Mises material an excellent approximation of the computed results is

$$\Delta E_{ij}=\frac{1}{E_{\mathrm{t}}}F\left(\frac{\Sigma_{\mathrm{m}}}{\Sigma_{\mathrm{e}}}\right)\Sigma'_{ij}+\frac{1}{E}G\left(\frac{\Sigma_{\mathrm{m}}}{\Sigma_{\mathrm{e}}}\right)\Sigma_{\mathrm{m}}\delta_{ij}. \tag{5.2}$$

Here, Σ_{m}, Σ_{e}, and Σ'_{ij} are the macroscopic mean stress, Mises stress, and stress deviator, respectively. The functions F and G also depend implicitly on E_{t}/E and on the total strain increment during nucleation, but this dependence is weak. The variation of F and G with the stress triaxiality $\Sigma_{\mathrm{m}}/\Sigma_{\mathrm{e}}$, found in the computations, is illustrated in Figure 7.

For the Gurson model the strain contribution due to nucleation can be calculated from (3.4)–(3.12). For the first voids nucleated (when $f=0$) at Σ_{ij}, the result is

$$\Delta E_{ij}=\tfrac{3}{2}q_1\frac{1}{E_{\mathrm{t}}}\cosh\left(\frac{3q_2}{2}\frac{\Sigma_{\mathrm{m}}}{\Sigma_{\mathrm{e}}}\right)\Sigma'_{ij}, \tag{5.3}$$

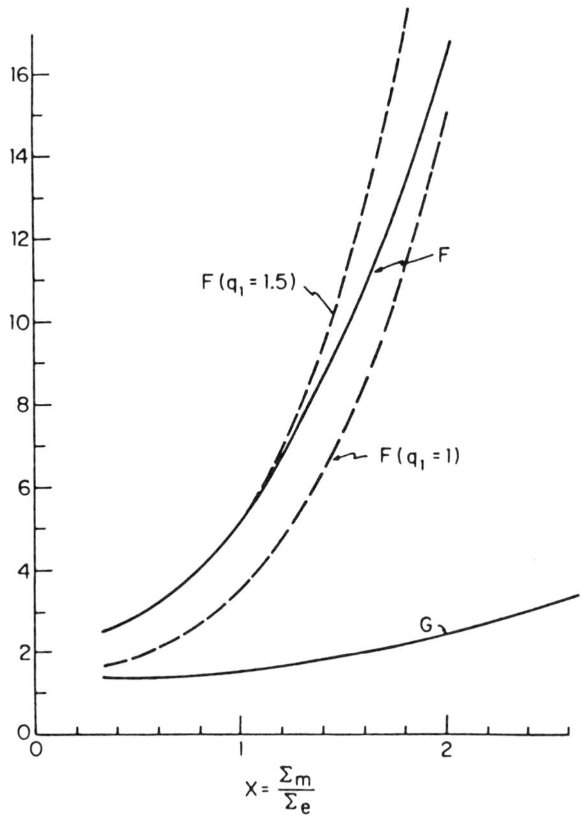

FIG. 7. Dependence of F and G in (5.2) on stress triaxiality (Hutchinson and Tvergaard, 1987).

assuming $E_t \ll E$. Then, in terms of (5.2), the Gurson model gives

$$F = \tfrac{3}{2} q_1 \cosh\left(\frac{3q_2}{2}\frac{\Sigma_m}{\Sigma_e}\right), \qquad G = 0, \tag{5.4}$$

which is also plotted in Figure 7. It is seen that in the low triaxiality range, the computed values of F are well represented by (5.4) for $q_1 = 1.5$ and $q_2 = 1$, while $q_1 \simeq 1.25$ gives a better approximation at higher triaxialities. The Gurson model neglects a dilatational contribution to ΔE_{ij}, but this contribution is inherently small anyway when $E_t \ll E$.

Nucleation is included in the Gurson model in a highly coupled manner, so that the quantitative effects of nucleation on macroscopic

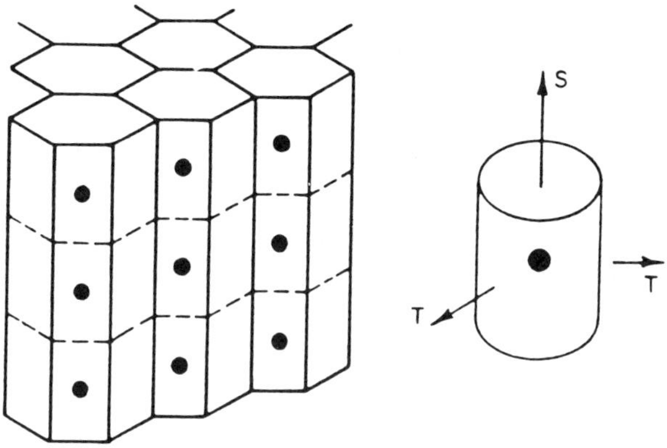

FIG. 8. Cylindrical cell model for periodic array of spherical void-nucleating particles (Hutchinson and Tvergaard, 1987).

behaviour are not transparent. However, the fact that the model is in good agreement with the micromechanical calculation of ΔE_{ij} increases the confidence in this model.

The effect of a uniform distribution of spherical particles that nucleate voids simultaneously has been investigated by numerical solution of the axisymmetric model problem illustrated in Figure 8 (Hutchinson and Tvergaard, 1987). Results for nucleation at constant marcoscopic stress are shown in Figure 9, and the same cases have been analysed for nucleation at constant macroscopic strain. The results for ΔE_{ij} obtained by these cell model calculations have been compared with (5.1) and (5.2), interpreting $\dot{E}_{ij}$ as logarithmic strain increments and $\dot{\Sigma}_{ij}$ as true stress increments. The values of F calculated from Figure 9 are not too much higher than the corresponding value shown in Figure 7, whereas the values found for G are a great deal larger. This agreement is quite reasonable since the nucleated void volumed fraction, $\dot{f} = 0.01$, in the cell model studies is not very small and the matrix material is assumed power-hardening while the tangent modulus E_t used in (5.2) is that at the onset of nucleation.

Both the calculations for an isolated spherical void in an infinite matrix and the cell model calculations have been repeated with a kinematic hardening description of the matrix material. These calculations show that the values of both F and G in (5.2) are significantly increased due to the yield surface translation. This is mainly an effect of the nonproportional stress history in the matrix material near the particle, due to

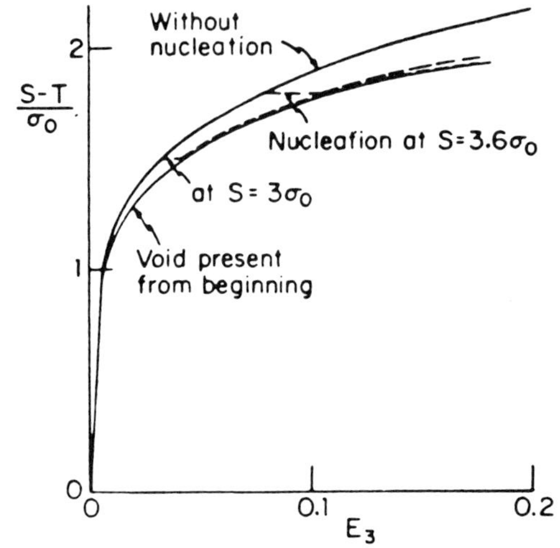

FIG. 9. Overall stress–strain behaviour for nucleation at constant stress, $\dot{f} = 0.01$, according to cylindrical cell model (Hutchinson and Tvergaard, 1987).

nucleation. In such circumstances, elastic-plastic material response is quite sensitive to the high curvature, or rounded vertex, that often develops at the loading point of the subsequent yield surface, and kinematic hardening was applied to represent this effect.

The reduction of the macroscopic hardening rate represented by the additional term $\dot{f}\Delta E_{ij}$ in (5.1) was estimated by Hutchinson and Tvergaard (1987). It was found that the hardening rate can even become negative, for a relatively small amount of void nucleation, if nucleation occurs over a relatively small range of straining.

Model studies of the nucleation process have also been carried out by Needleman (1987). This work has focussed on the process of interfacial decohesion, which has been described approximately by assuming that the interfacial tractions are derived from a potential function ϕ. In terms of the normal and tangential separations u_n and u_t, a characteristic length scale δ, and two other parameters σ_{max} and α, the potential function was defined as

$$\phi(u_n, u_t) = \frac{27}{4}\sigma_{max}\delta\left\{\frac{1}{2}\left(\frac{u_n}{\delta}\right)^2\left[1 - \frac{4}{3}\left(\frac{u_n}{\delta}\right) + \frac{1}{2}\left(\frac{u_n}{\delta}\right)^2\right]\right.$$

$$\left. + \frac{1}{2}\alpha\left(\frac{u_t}{\delta}\right)^2\left[1 - 2\left(\frac{u_n}{\delta}\right) + \left(\frac{u_n}{\delta}\right)^2\right]\right\} \tag{5.5}$$

with the normal and tangential tractions given by $T_n = \partial\phi/\partial u_n$ and

$T_t = \partial\phi/\partial u_t$, respectively. This procedure has made it possible to follow in detail the gradual separation of the interface, dependent on the stress state in the material. The procedure has also been used by Nutt and Needleman (1987) to study the separation between matrix and fibers in an aluminium alloy reinforced by SiC whiskers. The predicted patterns of separation at the fiber end are virtually identical to those found by direct observations using an electron microscope.

B. Growth and Coalescence of Voids

The axisymmetric model problem illustrated in Figure 8 has been used in a number of studies of the growth of an initially spherical void under various stress-states. Thus, Andersson (1977) has used the model to study void coalescence in an ideally plastic material in front of a crack tip, and Tvergaard (1982a) has used the model to study localization in a material containing spherical voids.

Quite recently Koplik and Needleman (1987) have made several analyses of this type to study void coalescence in an isotropically hardening material. The voids are assumed present from the beginning and void growth is studied for a fixed ratio of the radial and axial macroscopic true stresses. The computations are carried out for two different initial void volume fractions, $f_I = 0.0013$ and $f_I = 0.0104$, respectively. Initially, the rate of growth of the void volume fraction and the rate of macroscopic radial straining are rather constant, as functions of the macroscopic effective strain; but gradually the former increases while the latter decreases, until radial straining stops. Subsequently, the deformation takes place in a uniaxial straining mode, where all deformation is localized in the ligament between radially adjacent voids. During this final stage of localized flow, the void volume fraction grows rapidly, while the macroscopic true stress level decays rapidly.

Koplik and Needleman (1987) have compared the predictions of their cell model studies with the modification (3.2) in the Gurson model, introduced by Tvergaard and Needleman (1984), to account for final failure. Rather good agreement was achieved by using the parameter values $q_1 = 1.25$, $q_2 = 1$, and $f_F = 0.13$, while taking $f_C = 0.03$ for $f_I = 0.0013$ and $f_C = 0.055$ for $f_I = 0.0104$. The value of f_F was chosen to get a good approximation to the void volume fraction versus macroscopic effective strain curves at $f > f_C$. It was also found that q_1 should be increased for decreasing strain hardening.

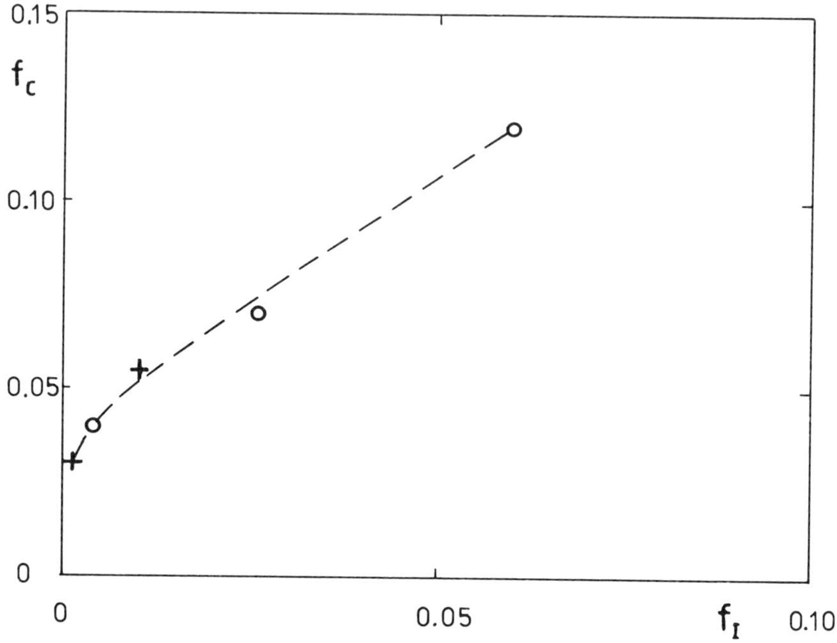

FIG. 10. Dependence of the critical value f_C in (3.2) on the initial void volume fraction f_I, according to various cell model studies.

Other cell model studies of the same type have been carried out by Becker *et al.* (1987a). Here the void volume fracture at final failure was chosen as $f_F = 0.25$ and the values of f_C were found to be 0.12, 0.07, and 0.04 for the initial void volume fractions 0.06, 0.026, and 0.004, respectively. Koplik and Needleman (1987) found that the value of f_C varies slowly with stress triaxiality and matrix strain hardening, but clearly there appears to be a rather strong dependence on the initial void volume fraction. The values of f_C, as a function of f_I, found in the two investigations are plotted in Figure 10. Thus, on the basis of these micromechanical investigations, it appears that the most realistic predictions are obtained by using an f_C value that depends on f_I (or on the volume fraction f_N of void nucleating particles). Clearly, the values in Figure 10 are smaller than the value $f_C = 0.15$ suggested by Tvergaard and Needleman (1984), based on the work of Brown and Embury (1973).

The results of these cell model studies have been used by Becker *et al.* (1987a) in a detailed comparison of numerical and experimental results for void growth and ductile failure in the non-uniform multiaxial stress fields of notched bars. The notched tensile specimens were machined from partially consolidated and sintered iron powder compacts, which had been made specifically to study the effects of porosity on mechanical behaviour. The pores were fairly random and nearly spherical in the higher density materials. Furthermore, the initial void volume fraction was large compared to the volume fraction of particles, so that these experiments allowed for a study of void growth and coalescence, independent of void nucleation issues. Quite good agreement between experiments and computations was obtained for several combinations of notch shapes and initial void volume fractions, which gives good confidence in the continuum model, and supports the use of the variable f_C indicated in Figure 10.

The values of f_C found by the micromechanical studies (Figure 10) are particularly interesting because the void (or particle) volume fractions in structural alloys are usually quite low, so that the relevant f_C value is significantly smaller than 0.15. This could be part of the explanation for the very low void volume fractions measured in the material near the fracture surface in the centre of the neck of round tensile test specimens (Cialone and Asaro, 1980; Moussy, 1985).

C. Two Size-Scales of Voids

Several structural alloys contain two size-scales of particles, i.e., a population of relatively large particles with low strength and a population of much stronger small particles. Thus, structural steels tested by Hancock and Mackenzie (1976) contain two populations of particles with average diameters around 100 μm and 1 μm, respectively, and aluminium alloys tested by Hahn and Rosenfield (1975) contain particles with average diameters of about 5 μm and 0.1 μm. In such materials, nucleation first occurs at the large weak particles, and during the subsequent growth of the large voids, a second population of small scale voids starts to nucleate, primarily in the regions of stress and strain concentrations near the larger voids.

A cell model study of the interaction between two size-scales of voids has been carried out by Tvergaard (1982b). The model material initially contains a periodic array of circular cylindrical voids, as shown in Figure 11, so that, due to symmetries, only the rectangular region DEFG in Figure 11 needs to be analysed numerically. The matrix material around

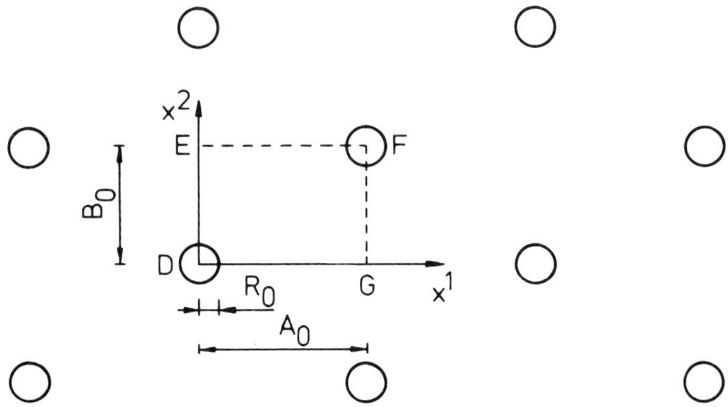

FIG. 11. Periodic array of circular cylindrical holes representing large-scale voids (Tvergaard, 1982b).

the cylindrical voids is described by the Gurson model, and thus the cylindrical holes represent the large-scale voids, while the small-scale voids are represented in terms of the continuum model of a porous ductile material. The larger voids are present from the beginning, while the small voids nucleate either by a strain criterion or a stress criterion. The material is subjected to generalized plane strain conditions, so that the strain in the x^3-direction is uniform.

The mesh used for this numerical analysis is chosen so that it allows for localization of plastic flow in a shear band between two larger voids. In the early stage of the deformation, nucleation starts to occur in the highly stressed and strained material near the cylindrical holes, and the void volume fraction varies smoothly along the surface of the holes. Subsequently, the nucleation and growth of voids starts to concentrate in a narrow band that grows into the material, and a shear band forms. Final fracture involves the formation of a void-sheet between two larger voids, as has been observed in a number of experiments (e.g., Cox and Low (1974) and Stone and Psioda (1975)).

This model analysis was used to predict final failure under a number of different macroscopic stress states including plane strain, $\varepsilon_{\mathrm{III}}=0$, or axisymmetric conditions approximated by taking either $\varepsilon_{\mathrm{III}}=\varepsilon_{\mathrm{I}}$ or $S_{\mathrm{III}}=S_{\mathrm{I}}$. Here, ε_{I} and S_{I} are the macroscopic logarithmic strain and true stress in the x^1-direction, etc., while S_{m}, S_{e}, and ε_{e} are the macroscopic mean stress, Mises stress, and effective strain, respectively, and S_{II} is the major principal stress. Values of the stress triaxiality parameter $S_{\mathrm{m}}/S_{\mathrm{e}}$ vs.

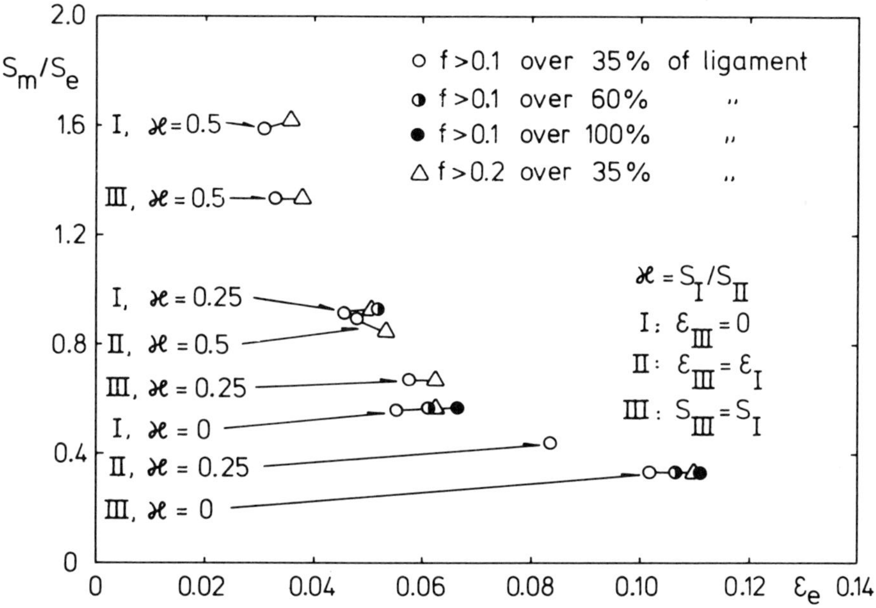

FIG. 12. Stress triaxiality S_m/S_e vs. effective strain ε_e near final failure in stress and strain fields ranging from plane strain to axisymmetry (Tvergaard, 1982b).

ε_e at stages near final failure are plotted in Figure 12 for the different stress states considered, corresponding to one particular set of material parameters.

In experimental investigations by Hancock and Mackenzie (1976) and Hancock and Brown (1983) for notched tensile specimens made of high strength steels, it has been found that the results can be represented approximately by a single failure locus in a plot of stress triaxiality vs. effective plastic strain at fracture initiation. This experimental observation does not agree with localization predictions for uniformly strained specimens, where is is known that materials are generally far more resistant to shear band formation when strained under axisymmetric conditions than under plane strain conditions (Needleman and Rice, 1978; Saje *et al.*, 1982). However, the interaction of two size-scales of voids does lead to agreement with the experimental observations, as shown in Figure 12, even though localization is also here an important part of the failure mechanism.

It should be noted that a reasonable agreement with the experimentally observed failure locus has also been found by Needleman and Tvergaard

(1984a) in a numerical analysis of notched bars, where the interaction of different void populations was not accounted for. Part of the reason for this agreement is that the notch geometries tend to suppress localization.

In a couple of more recent cell model studies, nucleation at both the large particles and the small particles is incorporated into the analysis, by representing both types of particles in terms of the Gurson model (Tvergaard, 1987c,d). The small scale particles are assumed to be uniformly distributed, with nucleation governed by a strain controlled criterion, while the larger inclusions are represented as "islands" of increased density of the amplitude of the void nucleation function. Nucleation in the "islands" is taken to be stress controlled, so that nucleation starts here, leading first to a soft-spot and subsequently to a large void. For a plane strain analysis (Tvergaard, 1987c), Figure 13 shows deformed meshes and curves of constant void volume fraction at

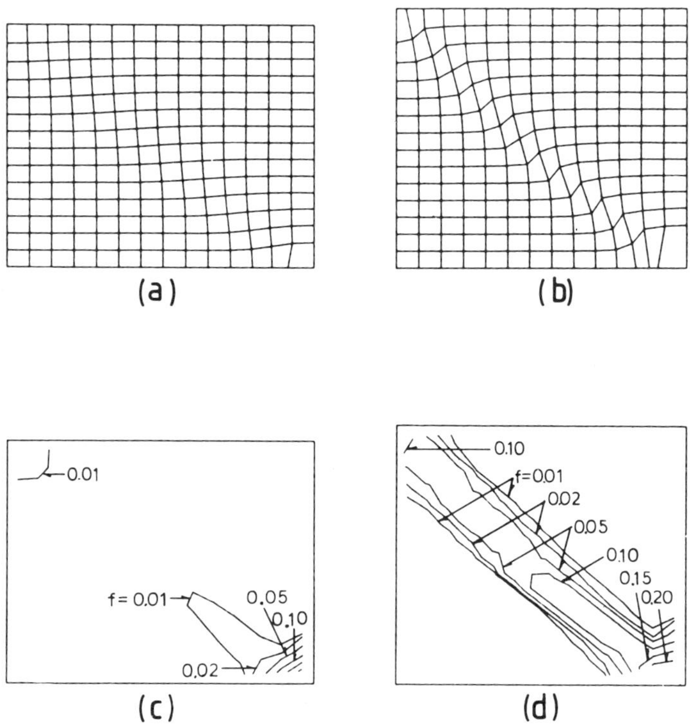

FIG. 13. Deformed meshes and curves of constant void volume fraction at two stages for material in uniaxial plane strain tension. Larger inclusions represented as "islands" of increased particle density (Tvergaard, 1987c).

two stages. At the first stage, the large cylindrical void has formed and localization is starting, although the mesh is still rather uniform, while at the second stage void-sheet failure is about to happen.

The same type of analysis has been carried out for a three-dimensional array of "islands" representing larger inclusions, where a full 3D numerical computation is necessary to study the final failure mode (Tvergaard, 1987d). This analysis includes a numerical representation of a 3D shear band, initiating in complex non-uniform stress and strain fields.

VI. Formation and Growth of Cracks

The analysis of material failure by shear localization between two larger voids, described in the previous section, was brought close to final failure, as indicated in Figure 12. However, predicting the development of an open crack requires some special modelling, which was not included in this void interaction analysis.

A. Crack Formation

A full description of final failure has been included in a numerical finite element analysis of ductile shear fracture at the free surface of a specimen subject to uniaxial plane strain tension (Tvergaard, 1982d). Here, loss of material stress carrying capacity as the voids start to coalesce was modelled by an additional term in (3.4), as suggested by Tvergaard (1982c), and an element vanish technique was introduced to describe the final formation of an open crack. When the failure condition is met in an element (the yield surface (3.1) has shrunk to a point), the element vanishes in that it no longer contributes to the virtual work integral (2.11). For numerical stability, the elements are actually taken to vanish slightly before the failure condition is met, and the nodal forces arising from the small remaining stresses in the nearly failed elements are gradually released in subsequent increments.

The surface region analysed corresponds to a semi-infinite solid occupying the half-space $x^2 \leq 0$, with a slight initial surface waviness of the form $w = -\frac{1}{2}\delta_0 \cos(\pi x^1/l_0)$. The x^2-axis is normal to the surface, and the x^1-axis is parallel to the surface. For this solid under tension in the

x^1-direction, solutions are available that satisfy symmetry conditions at the planes $x^1 = 0$ and $x^1 = l_0$, and such solutions are investigated by numerical analyses for a rectangular region of width l_0 and depth h_0, so that $h_0/l_0 \gg 1$. The material has no porosity initially, but voids nucleate, and at some stage localization of plastic flow occurs. The voids keep growing inside the shear bands, leading to void-sheet failure. The formation of open shear cracks predicted by the numerical analysis is illustrated in Figure 14. It should be noted here that the accurate prediction of plastic flow localization in a numerical analysis requires a careful mesh design, so that the mesh agrees with the critical orientation of shear bands at the strain where localization wants to occur (Tvergaard, Needleman, and Lo, 1981; Needleman and Tvergaard, 1984b).

The ductile fracture process in a round tensile test specimen has been analysed by Tvergaard and Needleman (1984). Here, a more realistic model of the void volume fraction evolution during coalescence was used, by introducing the function $f^*(f)$ defined in (3.2). Initially the bar has uniform thickness with no porosity; then necking occurs, and subsequently voids nucleate and grow rapidly inside the neck region, where increased triaxial tension develops. Figure 15 shows the predicted void volume fraction distribution at three stages, i.e., just before coalescence starts in the centre of the neck, at a later stage when an open

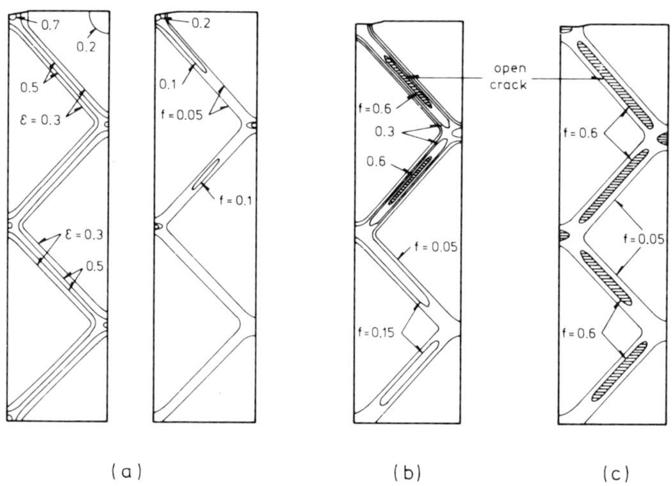

FIG. 14. Contours of constant maximum logarithmic strain ε and void volume fraction f in a surface region, where the top is the surface. Nucleation is strain controlled, with no initial voids. The average strain ε_a along the surface is (a) $\varepsilon_a = 0.268$, (b) $\varepsilon_a = 0.307$, (c) $\varepsilon_a = 0.332$ (Tvergaard, 1982d).

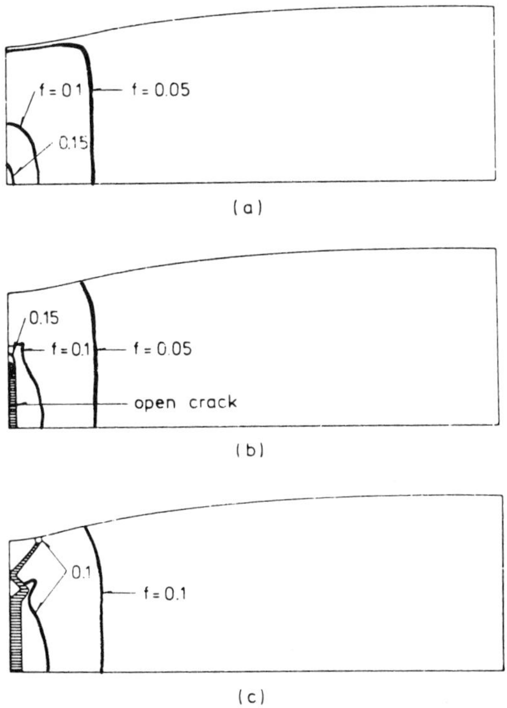

FIG. 15. Contours of constant void volume fraction in the neck of a round bar tensile test. The load P decays with increasing crack growth: (a) $P/P_{max} = 0.731$, (b) $P/P_{max} = 0.521$, (c) $P/P_{max} = 0.032$ (Tvergaard and Needleman, 1984).

penny-shaped crack has developed, and finally when zig-zag growth of the crack has given the conical void-sheet fracture known as the cup-cone.

Figures 14 and 15 illustrate the significant difference between the fracture mode in a plane strain tensile test and that in a round tensile test specimen, as is also observed experimentally (Speich and Spitzig, 1982). The analysis leading to Figure 14 did not account for necking. But in the neck region of a plane strain tensile test specimen, final fracture occurs in a shear mode induced by localization (Saje, Pan, and Needleman, 1982; Tvergaard, Needleman, and Lo, 1981), while the void volume fraction outside the band is still quite small. On the other hand, in the round bar tensile test specimen, fracture by void coalescence in the centre of the neck is predicted prior to localization (Figure 15).

B. Crack Tip Fields

The ductile fracture process at a crack tip differs significantly from that in more homogeneously strained circumstances, such as the plane strain or axisymmetric tensile test specimens described above (Rice, 1976). The large gradients of the stress and strain field near the crack tip typically result in critical conditions for failure being reached in part of the near-tip material at loads far less than the fracture load. Here, a characteristic length scale of the material plays a role, such as the particle spacing.

Rice and Johnson (1970) have studied the interaction of the crack-tip with a neighbouring void, using a slip line field analysis together with results of Rice and Tracey (1969) for the growth of an isolated spherical void to estimate the critical crack tip opening displacement for crack growth. An analogous study has been made by McMeeking (1977) on the basis of a full finite element solution for the blunting crack tip fields. The results of both analyses tend to overestimate critical crack tip opening displacements observed experimentally.

The interaction between a single cylindrical void and a plane strain tensile crack has been analysed numerically by Aoki *et al.* (1984) and Aravas and McMeeking (1985), accounting also for small scale voids in the surrounding material. The small scale voids are represented in terms of the porous ductile material model as suggested by Tvergaard (1982b) in the computation illustrated in Figures 11 and 12, and the crack growth criteria used refer to failure of the porous material.

In a similar analysis, also based on small scale yielding at a mode I plane strain crack, Needleman and Tvergaard (1987) consider the influence of a whole array of uniformly spaced larger inclusions in the material around the crack tip. In this investigation, no pre-existing voids are assumed present, and the large weak inclusions are represented by an array of "islands" of increased density of the amplitude of the void nucleation function, with stress-controlled nucleation inside the "islands." During crack tip blunting, self-similar stress and strain fields sweep over the near-tip material (McMeeking, 1977), so that an inclusion at initial distance R from the tip is reached by the stress peak when $J/\sigma_0 R \simeq 0.5$, while the strains are still very small (J and σ_0 denote the J-integral and the yield stress). Thus, large voids close to the crack tip nucleate rather early, when the stress peak sweeps by, whereas the small voids nucleate much later in the material close to the blunting crack tip, where large strains develop.

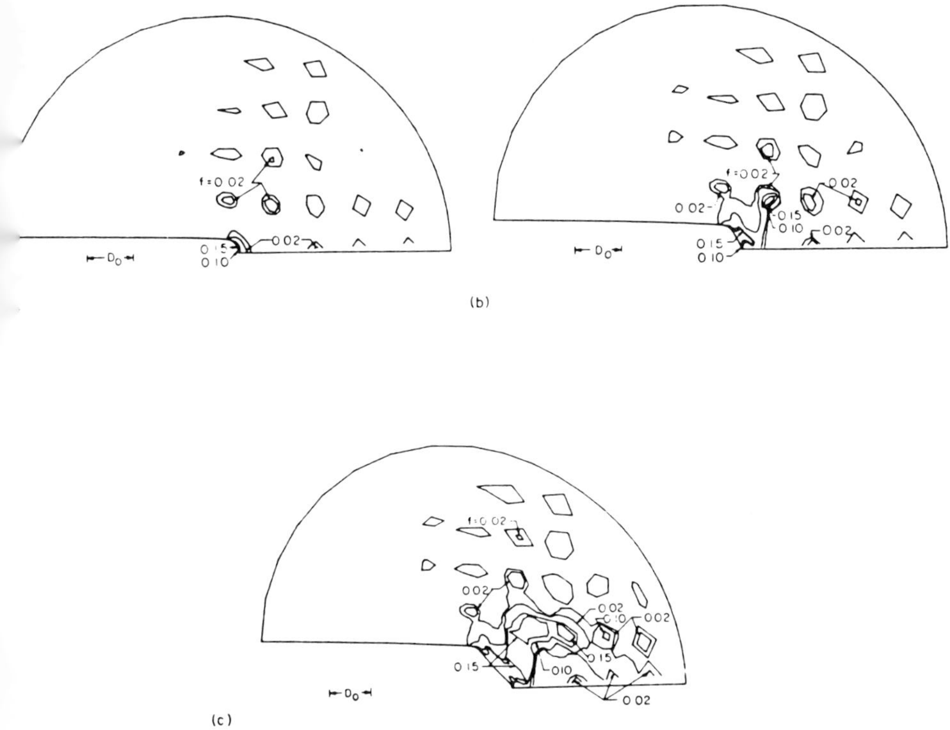

FIG. 16. Contours of constant void volume fraction near the tip of a blunting crack in a material containing two size-scales of voids. (a) $J/\sigma_0 D_0 = 0.856$, (b) $J/\sigma_0 D_0 = 1.89$, (c) $J/\sigma_0 D_0 = 3.89$ (Needleman and Tvergaard, 1987).

Four different distributions of the larger inclusions near the crack tip have been investigated by Needleman and Tvergaard (1987). Contours of constant void volume fraction in the near tip region are shown for one of the cases in Figure 16, where D_0 is the initial inclusion spacing. In this case, the inclusion nearest the crack tip is off the crack line. The large void at about 45° from the crack tip develops most rapidly, and crack growth involves localized shearing leading to void-sheet fracture in the material between this large void and the crack tip. Subsequently, the new crack tip interacts with the next larger void, and thus the crack keeps growing (Figure 16c).

The numerical results have been used to estimate the tearing modulus, which was introduced by Paris *et al.* (1979) to evaluate the limits of stable crack growth. The tearing modulus is defined in terms of the J-resistance

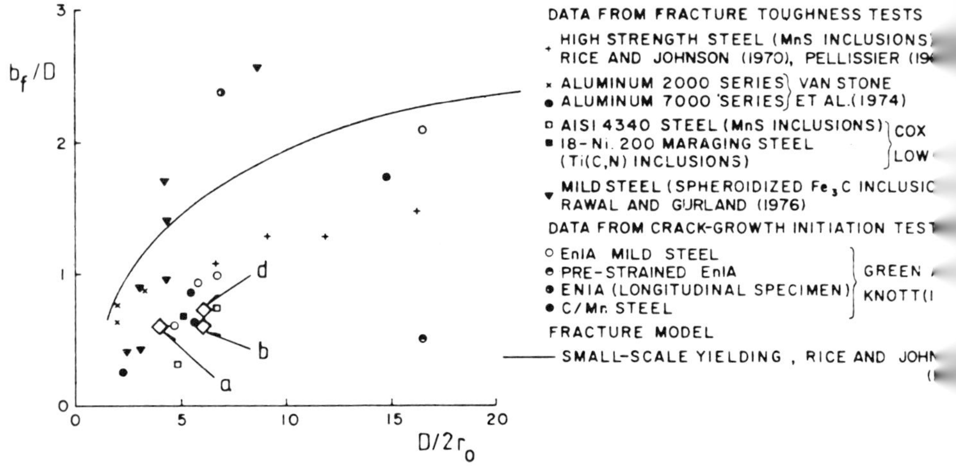

FIG. 17. Crack-tip opening b_f at fracture initiation, related to particle spacing D and particle size $2r_0$, plotted among the experimental results summarized by McMeeking (1977).

curve $J(\Delta a)$ as

$$T = \frac{E}{\sigma_0^2}\frac{dJ}{da}. \tag{6.1}$$

Subsequently, using the numerically calculated slopes dJ/da, the critical value J_{Ic} for the onset of crack growth was estimated in three of the computations. Thus, the numerical analyses discussed here have made it possible to relate the microstructure, represented by the particle distribution and nucleation criteria, to fracture mechanics parameters such as J_{Ic}, T, or the crack opening displacement. The values found for the crack opening displacement at fracture initiation b_f are compared with experimental results summarized by McMeeking (1977) in Figure 17. This figure shows the ratio of the crack-tip opening displacement and the particle spacing vs. the ratio of the particle spacing and diameter. It is seen that the computed values, marked *a*, *b*, and *d* in Figure 17, are in reasonable agreement with the experimental results.

A rather different mechanism of crack growth by void coalescence has been studied by Becker *et al.* (1987b), both experimentally and theoretically. The material considered is an Al–Li alloy with coarse grain boundary particles, which give rise to void nucleation at the grain boundaries, so that the final fracture is predominantly intergranular. This phenomenon of ductile grain boundary fracture is readily identified by the presence of microvoid dimples on the fracture surface and is distinct from the brittle grain boundary fracture observed in other cases.

In the numerical analysis, the grains near the crack tip are modelled as roughly circular regions free of void nucleating particles, while the nucleation and growth of voids in the bands between these grains is represented in terms of the ductile porous material model. Figure 18 shows grain distributions ahead of the crack tip for two of the cases analysed, corresponding to the grain sizes 200 μm and 60 μm, respectively, where the grains are outlined by curves of constant volume fraction of void nucleating particles. If the voids develop as a string of pearls along the grain boundary, then the most realistic initial width of the grain boundary porous zone, in the present type of model, is of the order of the void spacing (see comparisons made by Tvergaard (1982a)). This width is chosen somewhat too large in the distributions shown in Figure 18 in order to avoid the use of an extremely fine mesh. The initial part of the crack growth has also been analysed with a more realistic small width of the grain boundary porous zone.

The development of damage and crack growth found in the numerical analyses for the different grain diameters show that voids nucleate at the grain boundaries in the near vicinity of the crack tip, and that crack growth by void coalescence occurs on a zig-zag path following grain boundaries near the crack line.

Also these computations have been used to draw the J-resistance curves, J vs. Δa, and the slopes of these curves are used to estimate the tearing modulus T according to (6.1) for each case. Then, the critical value J_{Ic} for the onset of crack growth is estimated by back extrapolation using the slope dJ/da, and the critical value K_{Ic} of the stress intensity

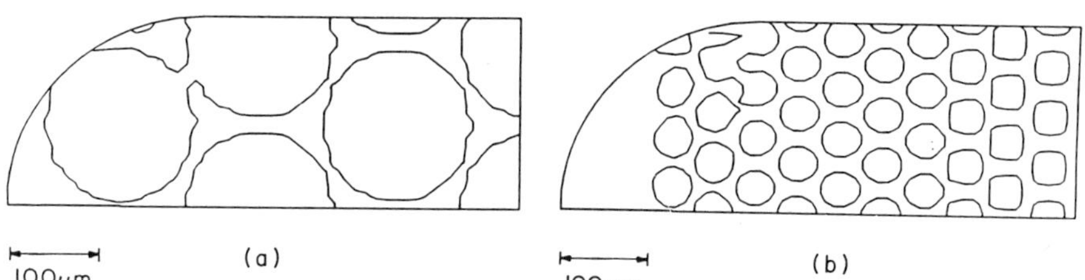

FIG. 18. Material with void nucleating particles in the grain boundary modelled in terms of porous material model, by considering the grains as roughly circular regions free of particles (Becker *et al.*, 1987b).

factor is obtained from the plane strain small scale yielding relation

$$K_{\mathrm{Ic}} = \left(\frac{J_{\mathrm{Ic}}E}{1-\nu^2}\right)^{1/2}. \tag{6.2}$$

The predicted fracture toughness decreases significantly as the width of the grain boundary porous zone is decreased. However, in spite of the rather crude description of the grain boundary porous zone used in these analyses, the quantitative agreement between the model predictions and the experimentally measured values is quite reasonable (within a factor of two), both for K_{Ic} and for the tearing modulus.

VII. Effect of Yield Surface Curvature

Predictions of plastic flow localization are very sensitive to the formation of a vertex on subsequent yield surfaces (Rice, 1977). A rounded vertex, or a relatively high local curvature of the yield surface near the loading point, is often found in experiments, and kinematic hardening can be used to approximately model such a rounded vertex. Both for necking in biaxially stretched sheets (Tvergaard, 1978) and for the onset of shear bands under plane strain conditions (Hutchinson and Tvergaard, 1981), it has been found that predictions of kinematic hardening are rather similar to those found for a solid that develops a sharp vertex on the yield surface.

In order to account for the effect of increased yield surface curvature during ductile fracture, Mear and Hutchinson (1985) have suggested a kinematic hardening model for a porous ductile material. The model has subsequently been extended by Tvergaard (1987a) to account for void nucleation.

A. Kinematic Hardening Porous Material

The model makes use of a family of isotropic/kinematic hardening yield surfaces of the form $\Phi(\sigma^{ij}, \alpha^{ij}, \sigma_{\mathrm{F}}, f) = 0$, where α^{ij} denotes the center of the yield surface, and σ_{F} is the radius of the yield surface for the matrix material. This radius is taken to be given by

$$\sigma_{\mathrm{F}} = (1-b)\sigma_{\mathrm{y}} + b\sigma_{\mathrm{M}}, \tag{7.1}$$

where σ_y and σ_M are the initial yield stress and the matrix flow stress, respectively, and the parameter b is a constant in the range $[0, 1]$. The constitutive relations are formulated such that for $b = 1$ they reduce to the isotropic hardening Gurson model, whereas a pure kinematic hardening model appears for $b = 0$.

The approximate yield condition to be used here is of the form

$$\Phi = \frac{\tilde{\sigma}_e^2}{\sigma_F^2} + 2q_1 f^* \cosh\left\{\frac{\tilde{\sigma}_k^k}{2\sigma_F}\right\} - 1 - (q_1 f^*)^2 = 0, \tag{7.2}$$

where $\tilde{\sigma}^{ij} = \sigma^{ij} - \alpha^{ij}$, $\tilde{\sigma}_e = (3\tilde{s}^{ij}\tilde{s}_{ij}/2)^{1/2}$, and $\tilde{s}^{ij} = \tilde{\sigma}^{ij} - G^{ij}\tilde{\sigma}_k^k/3$. For $f^* = f$ and $q_1 = 1$, the expression (7.2) is that proposed by Mear and Hutchinson (1985), which coincides with that of Gurson (1977a) if $b = 1$. The parameter q_1 is that also used in (3.1), and the function $f^*(f)$ is specified by (3.2).

The plastic part of the macroscopic strain increment $\dot{\eta}_{ij}^P$ and the effective plastic strain increment $\dot{\varepsilon}_M^P$ for the matrix material are taken to be related by

$$\tilde{\sigma}^{ij}\dot{\eta}_{ij}^P = (1 - f)\sigma_F\dot{\varepsilon}_M^P. \tag{7.3}$$

For $f = 0$, (7.3) is an exact relationship for the classical kinematic hardening solid, and for $b = 1$, the expression reduces to the equivalent plastic work expression (3.11). Substituting the uniaxial true stress natural strain curve (3.7) for the matrix material into (7.3) gives

$$\dot{\sigma}_M = \frac{EE_t}{E - E_t} \frac{\tilde{\sigma}^{ij}\dot{\eta}_{ij}^P}{(1 - f)\sigma_F}. \tag{7.4}$$

Furthermore, the rate of growth of the void volume fraction is still taken to be given by (3.4)–(3.6), with void nucleation specified by either (3.8) or (3.9).

A fictitious Gurson yield surface $\Phi_G = \Phi_G(\sigma_G^{ij}, \sigma_M, f)$ was used by Tvergaard (1987a) to formulate the constitutive relations, where σ_M and f are the current values and σ_G^{ij} are a set of fictitious stress components chosen such that

$$\frac{\sigma_G^{ij}}{\sigma_M} = \frac{\tilde{\sigma}^{ij}}{\sigma_F}. \tag{7.5}$$

With this assumption, $\Phi_G = 0$ is a direct consequence of $\Phi = 0$. In most cases the fictitious stresses σ_G^{ij} will differ from the actual stresses σ^{ij} at every point of the current yield surface (see Figure 19).

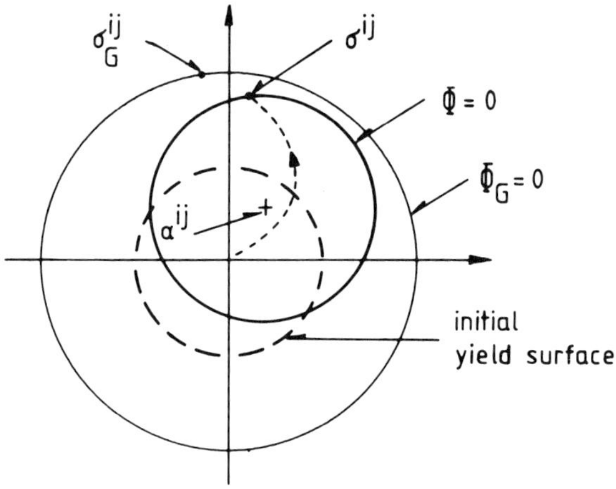

FIG. 19. Schematic representation of the current yield surface $\Phi = 0$ for the kinematic hardening model, and the fictitious yield surface $\Phi_G = 0$. The current stresses are σ^{ij}, while σ^{ij}_G are fictitious stresses.

The expression for $\dot{\eta}^P_{ij}$ in a point of the yield surface $\Phi = 0$ is chosen identical to that given by the Gurson model at the point σ^{ij}_G of the fictitious surface $\Phi_G = 0$. Thus, the plastic part of the macroscopic strain increment is taken to be of the form (3.13). Here, the expressions for the tensors m^G_{ij} and m^F_{ij} and the hardening H are

$$m^G_{ij} = \frac{3}{2}\frac{\tilde{s}_{ij}}{\sigma_F} + \alpha G_{ij}, \qquad m^F_{ij} = \frac{3}{2}\frac{\tilde{s}_{ij}}{\sigma_F} + \beta G_{ij}, \tag{7.6}$$

$$\alpha = \frac{f^*}{2} q_1 \sinh\left(\frac{\tilde{\sigma}^k_k}{2\sigma_F}\right), \qquad \beta = \alpha + \frac{\mathscr{B}\sigma_M}{6}\frac{\partial \Phi}{\partial f}, \tag{7.7}$$

$$H = \frac{\sigma_M}{2}\left[-3\alpha(1-f)\frac{\partial \Phi}{\partial f} - \left(\frac{\partial \Phi}{\partial f}\mathscr{A} + \frac{\sigma_F}{\sigma_M}\frac{\partial \Phi}{\partial \sigma_F}\right)\frac{EE_t}{E - E_t}\frac{1}{1-f}\left(\frac{\tilde{\sigma}_e^2}{\sigma_F^2} + \alpha\frac{\tilde{\sigma}^k_k}{\sigma_F}\right)\right] \tag{7.8}$$

Then, the instantaneous moduli to be used in the incremental stress–strain relationship of the form (2.16) are still obtained from (3.17)–(3.20).

The evolution equation for the yield surface centre during a plastic increment is taken to be

$$\overset{\nabla}{\alpha}{}^{ij} = \dot{\mu}\tilde{\sigma}^{ij}, \qquad \dot{\mu} \geq 0, \tag{7.9}$$

which is a finite strain generalization of Ziegler's (1959) hardening rule.

The value of the parameter $\dot{\mu}$ is determined so that the consistency condition, $\dot{\Phi} = 0$, is satisfied:

$$\dot{\mu} = (1-b)\frac{1}{2}\left[\frac{\tilde{\sigma}_e^2}{\sigma_F^2} + \alpha\frac{\tilde{\sigma}_k^k}{\sigma_F}\right]^{-1}\left[\frac{2}{\sigma_F} m_{kl}^G \overset{\nabla}{\sigma}{}^{kl} + \frac{\sigma_y}{\sigma_F}\frac{\partial\Phi}{\partial f}\frac{\mathscr{B}}{3} G_{kl}\overset{\nabla}{\sigma}{}^{kl}\right.$$

$$\left. + \frac{\sigma_y}{\sigma_F}\frac{1}{H}\frac{\partial\Phi}{\partial f}\left\{3\alpha(1-f) + \mathscr{A}\frac{EE_t}{E-E_t}\frac{1}{1-f}\left(\frac{\tilde{\sigma}_e^2}{\sigma_F^2} + \alpha\frac{\tilde{\sigma}_k^k}{\sigma_F}\right)\right\} m_{kl}^F \overset{\nabla}{\sigma}{}^{kl}\right]. \quad (7.10)$$

In cases where large rotations of the principal stress axes occur relative to the material, the formulation of (7.9) in terms of the Jaumann rate may give a poor representation of material behaviour, and other finite strain generalizations using other corotational rates may be preferred (Dafalias, 1983; Lee *et al.*, 1983). However, in shear localization studies the rotations of the principal stress axes prior to localization are quite small, and for such studies, Mear and Hutchinson (1985) have found that using the Dienes rate makes little difference.

B. Localization Predictions

A basic assumption in the development of the kinematic hardening porous material model has been that in proportional stressing the response should be identical to that for isotropic hardening. Clearly, the instantaneous moduli remain identical in such circumstances, and therefore bifurcation predictions are not changed by using the kinematic hardening model for a proportionally stressed solid. However, small initial inhomogeneities give rise to non-proportional stressing, and then the yield surface curvature has a strong effect on the imperfection-sensitivity.

Mear and Hutchinson (1985) have used the M-K-type model (Figure 3) with a larger initial porosity inside the band than outside the band to illustrate the influence of kinematic hardening on plastic flow localization. Similar studies have been carried out by Tvergaard (1987a) for cases where there is no initial porosity, but where voids nucleate from particles, which have a larger initial concentration inside the band, as described by (4.4). Figure 20 shows results for a solid subject to uniaxial plane strain tension with plastic strain controlled nucleation inside the band and no nucleation outside the band $f_N^o = 0$. It is seen that

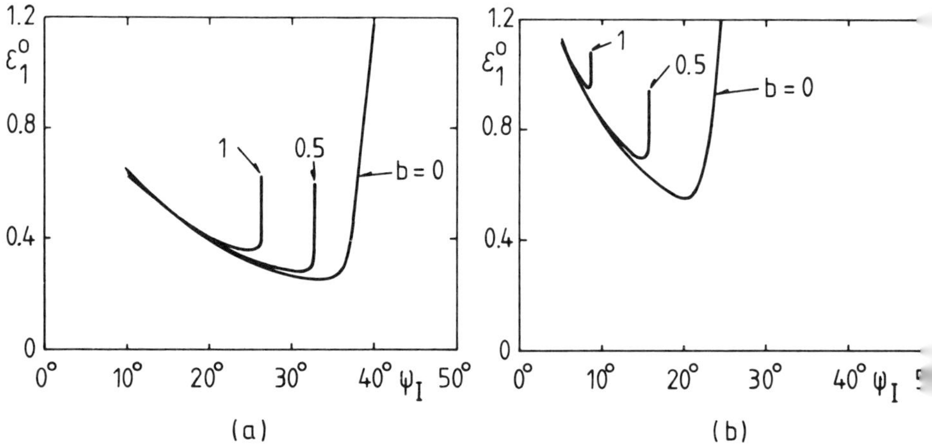

FIG. 20. Shear band localization strain vs. initial band orientation for uniaxial plane strain tension. No initial porosity, but strain controlled nucleation with $f_N^o = 0$ and (a) $\Delta f_N = 0.01$, (b) $\Delta f_N = 0.001$ (Tvergaard, 1987a).

localization occurs much earlier for kinematic hardening ($b = 0$) than for isotropic hardening ($b = 1$), particularly in the case of the very small initial inhomogeneity $\Delta f_N = 0.001$. The same type of results were found for solids under axisymmetric conditions.

The kinematic hardening material model makes it possible to study the question discussed by various authors (for example, see Rice (1977)) that even though failure occurs by shear localization with the final rupture surface made up of a void sheet, it is not certain that localization was initiated by porosity. In the case of no porosity, $f \equiv 0$, the Gurson model ($b = 1$) reduces to J_2-flow theory, which is very resistant to localization, whereas the localization predictions of the kinematic hardening model ($b = 0$) are rather similar to those found for a solid that develops a vertex on the yield surface (Hutchinson and Tvergaard, 1981).

Figure 21 shows localization predictions for a case in which the inhomogeneity is represented by a slightly lower initial yield stress inside the band than that outside, $\sigma_y^b = 0.99\sigma_y^o$, while the concentration of void nucleating particles is constant, $f_N^o = 0.01$ and $\Delta f_N = 0$, and nucleation occurs at very large strains, specified by $\varepsilon_N = 0.9$ and $s = 0.1$ in (3.8). For the kinematic hardening solid, the first critical localization ($\psi_I \simeq 27°$) occurs while there is essentially no porosity, so that here localization occurs due to the increased yield surface curvature. When the computation is continued beyond the onset of localization, nucleation occurs

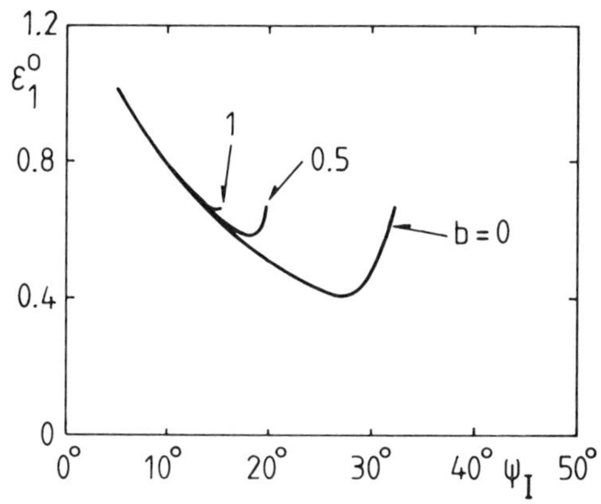

FIG. 21. Shear band localization vs. initial band orientation for uniaxial plane strain tension. Inhomogeneity in the initial yield stress, $\sigma_y^b = 0.99\sigma_y^o$, while void nucleating particles are uniformly distributed, $f_N^o = 0.01$, $\Delta f_N = 0$, $\varepsilon_N = 0.9$, $s = 0.1$, and there is no initial porosity (Tvergaard, 1987a).

inside the shear band as the localized strains grow large, and final failure occurs by the void sheet mechanism. Thus, for the kinematic hardening solid, there is a competition between two effects on localization. If voids nucleate late, vertex-type effects tend to dominate the onset of localization, while porosity has a strong influence if the voids nucleate earlier; but in both cases final failure occurs in a void sheet. On the other hand, for the isotropic hardening material a significant amount of porosity has to appear before the localization predicted in Figure 21.

The results in Figures 20 and 21 correspond to a homogeneously stressed solid, for which localization can be analysed by the simple model analysis (4.1)–(4.3). Shear band development in a solid subject to a non-uniform state of deformation is a more complex problem, which requires a full numerical analysis. Such numerical studies involving localization in strongly non-uniform strain fields have been discussed above, for example, see Figures 11 and 16. The strain field at the stretched surface of a bent specimen represents a less extreme non-uniformity, and this case has been used by Tvergaard (1987b) to study the basic influence of nonhomogeneous deformation on the development of localized shearing and shear failure.

An imperfection in the form of an initial surface waviness is assumed, and periodic solutions are considered analogous with those illustrated in Figure 14. The surface waviness has little effect initially but gives a

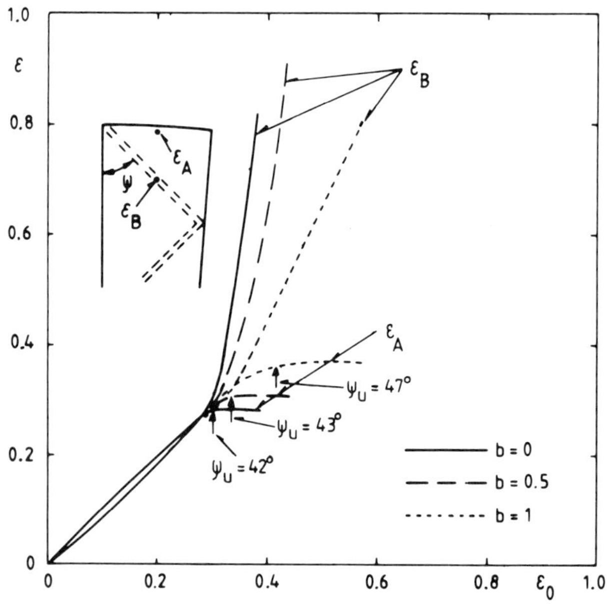

FIG. 22. Maximum principal logarithmic strains at two material points near the surface of a bent specimen. The material has strain-controlled nucleation and no initial porosity (Tvergaard, 1987b).

slightly non-uniform porosity distribution. Shear localization starts to develop from the wave bottoms, leading finally to a shear crack that grows into the material from the surface. For a kinematic hardening material with no initial voids and strain-controlled nucleation specified by $f_N = 0.04$, $\varepsilon_N = 0.3$, and $s = 0.1$ in (3.8), Figure 22 shows the corresponding development of the maximum principal logarithmic strain at two material points vs. an average surface strain measure ε_0, compared with isotropic hardening predictions. As expected, localization occurs earlier for the kinematic hardening model, but in all cases localization is significantly delayed relative to the first loss of ellipticity near the surface.

The stability of a periodic pattern of shear cracks, such as that implied by the results in Figure 22, has been investigated by a few computations. Figure 23 illustrates the results of one of these computations. At the first stage in Figure 23, two shear bands are developing nearly symmetrically from the central wave-bottom on the surface, whereas at the two following stages one of the bands has stopped growing while the other band develops into void sheet fracture (painted black in Figure 23). This

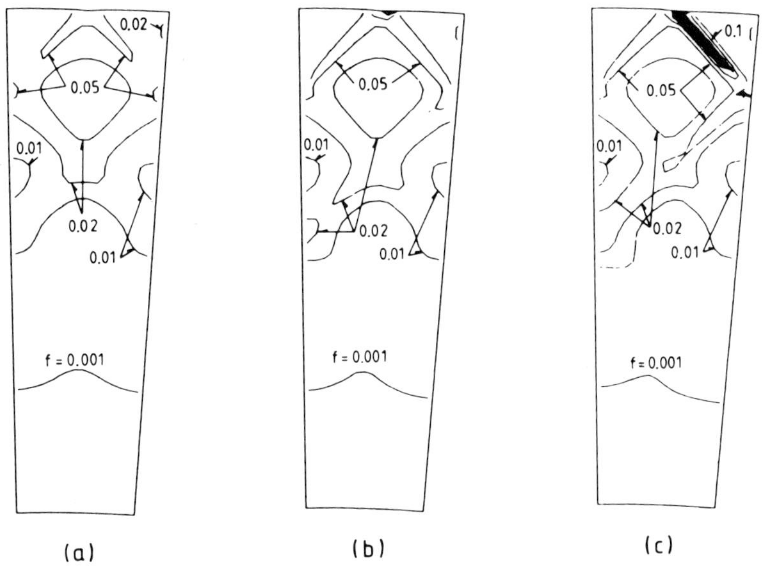

FIG. 23. Contours of constant void volume fraction calculated for a surface region of double length $2l_0$ in a bent specimen. The material has strain-controlled nucleation with no initial porosity. (a) $\varepsilon_0 = 0.323$, (b) $\varepsilon_0 = 0.340$, (c) $\varepsilon_0 = 0.354$ (Tvergaard, 1987b).

instability of the periodic growth pattern can explain a characteristic spacing between the shear cracks that is observed on the surface of a bent plate (for example, see Figure 10 in Hutchinson and Tvergaard, 1980). This behaviour is analogous to the instability of a system of straight edge cracks in a brittle solid found by Nemat-Nasser *et al.* (1980).

The kinematic/isotropic hardening model of a porous ductile solid has also been used to study the interaction of two size-scales of voids, where the fracture mechanism involves the formation of a void sheet between two larger voids. Both for a planar array of cylindrical large-scale inclusions (Tvergaard, 1987c) and for a full 3D array of spherical inclusions (Tvergaard, 1987d), it is found that an increased yield surface curvature gives rise to earlier final failure.

VIII. Strain Rate Sensitive Material

The strain rate sensitivity of plastic yielding is often introduced in material models by representing the inelastic part of the deformations in terms of a nonlinear viscous behaviour. Such a rate sensitive version of the Gurson model has been used by Pan, Saje, and Needleman (1983) and by Needleman and Tvergaard (1984c) to study the influence of the strain rate on plastic flow localization, and it has been found that localization is significantly delayed by the material rate sensitivity. Subsequently, Tvergaard and Needleman (1986, 1987) have incorporated a simple description of cleavage fracture in this material model and have used the model to study the brittle–ductile transition in the Charpy V-notch test.

A. Viscoplastic Material Model

The strain rate sensitivity is included in the porous ductile material model by taking the response of the matrix material to be elastic-viscoplastic. Thus, the microscopic effective plastic strain rate $\dot{\varepsilon}_M^P$ is here taken to be given by the power law relation

$$\dot{\varepsilon}_M^P = \dot{\varepsilon}_0 \left[\frac{\sigma_M}{g(\varepsilon_M^P)} \right]^{1/m}, \tag{8.1}$$

where m is the strain rate hardening exponent and $\dot{\varepsilon}_0$ is a reference strain rate. It is noted that in (8.1) and subsequently in this section $(\dot{\ })$ denotes the time derivative, since the material model is time-dependent.

The function $g(\varepsilon_M^P)$ in (8.1) represents the effective tensile flow stress in the matrix material in a tensile test carried out at a strain rate such that $\dot{\varepsilon}_M^P = \dot{\varepsilon}_0$. Thus, for a power hardening material with uniaxial stress strain behaviour as that given by (3.21), the function $g(\varepsilon_M^P)$ can be found from the relationship

$$\varepsilon_M^P = \frac{\sigma_0}{E} \left\{ \left[\frac{g(\varepsilon_M^P)}{\sigma_0} \right]^n - \frac{g(\varepsilon_M^P)}{\sigma_0} \right\}, \qquad g(0) = \sigma_0. \tag{8.2}$$

The concept of elastic unloading is not directly incorporated in the present elastic-viscoplastic material model. Consequently, there is no yield surface; but the function $\Phi(\sigma^{ij}, \sigma_M, f) = 0$ given by (3.1) is used as a plastic potential, so that the plastic part of the macroscopic strain rate is given by the expression

$$\dot{\eta}_{ij}^P = \Lambda \frac{\partial \Phi}{\partial \sigma^{ij}}. \tag{8.3}$$

The equivalent plastic work expression (3.11) is still assumed valid. Then, with $\dot{\varepsilon}_{\mathrm{M}}^{\mathrm{P}}$ specified by (8.1), the value of the parameter Λ is directly given by

$$\Lambda = (1-f)\sigma_{\mathrm{M}}\dot{\varepsilon}_{\mathrm{M}}^{\mathrm{P}}\left[\sigma^{ij}\frac{\partial \Phi}{\partial \sigma^{ij}}\right]^{-1}. \tag{8.4}$$

As in the time-independent material model, the total macroscopic strain rate is taken to be the sum of the elastic and plastic parts, $\dot{\eta}_{ij} = \dot{\eta}_{ij}^{\mathrm{E}} + \dot{\eta}_{ij}^{\mathrm{P}}$. Thus, using the elastic, incremental stress-strain relationship, the constitutive relations for the elastic-viscoplastic material can be written in the rate form (2.12).

For the viscoplastic material, the microscopic effective plastic strain rate $\dot{\varepsilon}_{\mathrm{M}}^{\mathrm{P}}$ is not proportional to $\dot{\sigma}_{\mathrm{M}}$, but is given directly by (8.1). Therefore, in the expression (3.4) for the rate of increase of the void volume fraction, the contribution due to void nucleation is here taken to be given by

$$(\dot{f})_{\text{nucleation}} = \mathscr{B}(\dot{\sigma}_{\mathrm{M}} + (\sigma_k^k)^{\cdot}/3) + \mathscr{D}\dot{\varepsilon}_{\mathrm{M}}^{\mathrm{P}}, \tag{8.5}$$

and plastic strain controlled nucleation is specified by

$$\mathscr{D} = \frac{f_{\mathrm{N}}}{s\sqrt{2\pi}}\exp\left[-\frac{1}{2}\left(\frac{\varepsilon_{\mathrm{M}}^{\mathrm{P}} - \varepsilon_{\mathrm{N}}}{s}\right)^2\right], \qquad \mathscr{B} = 0. \tag{8.6}$$

Since $\Phi = 0$ is used as the potential function, the consistency condition $\dot{\Phi} = 0$ must be satisfied. This equation together with (3.4), (3.5), and (8.5) determines the values of $\dot{f}$ and $\dot{\sigma}_{\mathrm{M}}$.

In an incremental numerical solution for this elastic-viscoplastic porous material, the stable step size can be significantly increased by using a forward gradient method proposed by Peirce *et al.* (1984). At time t, the microscopic effective strain rate $\dot{\varepsilon}_{\mathrm{M}}^{\mathrm{P}}$ to be used in (8.4) is expressed by a linear interpolation between the rates at time t and $t + \Delta t$, respectively,

$$\dot{\varepsilon}_{\mathrm{M}}^{\mathrm{P}} = (1+\theta)\dot{\varepsilon}_{\mathrm{M}}^{\mathrm{P}(t)} + \theta\dot{\varepsilon}_{\mathrm{M}}^{\mathrm{P}(t+\Delta t)}, \tag{8.7}$$

and a Taylor series expansion is employed to estimate the value $\dot{\varepsilon}_{\mathrm{M}}^{\mathrm{P}(t+\Delta t)}$ of the rate at time $t + \Delta t$. Then the constitutive relations based on (2.12), (8.3), and (8.4) can be rewritten in the form (2.17), where analogous to (3.17) the tangent moduli are given by

$$L_*^{ijkl} = \mathscr{L}^{ijkl} - \mu_* M_{\mathrm{G}}^{ij} M_{\mathrm{F}}^{kl} \tag{8.8}$$

with the value μ_* given by

$$\mu_* = \sqrt{\frac{G}{g}}\frac{\theta\,\Delta t}{1 - Q_1\theta\,\Delta t}Q_2 \tag{8.9}$$

and the initial stress rate type term given by

$$\dot{\tau}_*^{ij} = -\sqrt{\frac{G}{g}}\, Q_3 \frac{\dot{\varepsilon}_{\mathrm{M}}^{\mathrm{P}(t)}}{1 - Q_1 \theta\, \Delta t} M_{\mathrm{G}}^{ij}. \tag{8.10}$$

The functions Q_1, Q_2, and Q_3 are rather lengthy expressions that will not be repeated here. It is noted that $\theta = 0$ (no forward gradient method) gives $\mu_* = 0$, so that the tangent moduli (8.8) reduce to their elastic part.

A viscoplastic, kinematic hardening model of a porous ductile material has been introduced by Becker and Needleman (1985). This model, based on the time-independent model of Mear and Hutchinson (1985), has been used to study the effect of material strain rate sensitivity in plane strain and round tensile test specimens.

Failure by cleavage is included in the material model in a relatively simple manner, by assuming that failure occurs if the maximum principal tensile stress exceeds a critical value σ_{C}. Several investigations of cleavage fracture in body-centered-cubic (b.c.c.) metals have shown that such a constant critical stress is a realistic criterion for slip induced cleavage failure in the low temperature range (Cottrell, 1958; Petch, 1958; Smith, 1968; Hahn, 1984). At somewhat higher temperatures, near the transition to fibrous fracture, cleavage micro-cracks stretching over one grain each are not large enough to be immediately unstable by the Griffith criterion, and thus the final link-up of micro-cracks requires some further plastic deformations. However, it has been assumed by Tvergaard and Needleman (1986, 1987) that a constant critical value σ_{C} is a sufficiently good criterion in the whole temperature range considered.

The effect of the elastic-viscoplastic material model is illustrated in Figure 24 by a few results for a homogeneously strained material element under plane strain tension ($\varepsilon_3 = 0$), subject to a fixed prescribed logarithmic strain rate $\dot{\varepsilon}_1$. The material considered has the rate hardening exponent $m = 0.01$, the strain hardening exponent $n = 10$, and plastic strain controlled nucleation (8.6). It is seen that the peak value of the maximum principal tensile stress σ_1 is much higher for $\sigma_2/\sigma_1 = 0.6$ (where the stress triaxiality is high, $\sigma_k^k/(3\sigma_{\mathrm{e}}) \simeq 2.2$) than for $\sigma_2 = 0$. Furthermore, in both cases the peak value of σ_1 is increased by about 15%, when the strain rate is increased by a factor 10^6 (note $10^{0.06} = 1.148$). Thus, cleavage micro-cracks are most likely to nucleate in regions of high triaxiality, and more likely at high strain rates. For example, if the critical stress for cleavage is $\sigma_{\mathrm{C}} = 4.3\sigma_0$, failure at the low strain rate occurs by void coalescence, both for $\sigma_2/\sigma_1 = 0.6$ and for $\sigma_2 = 0$; but at the high

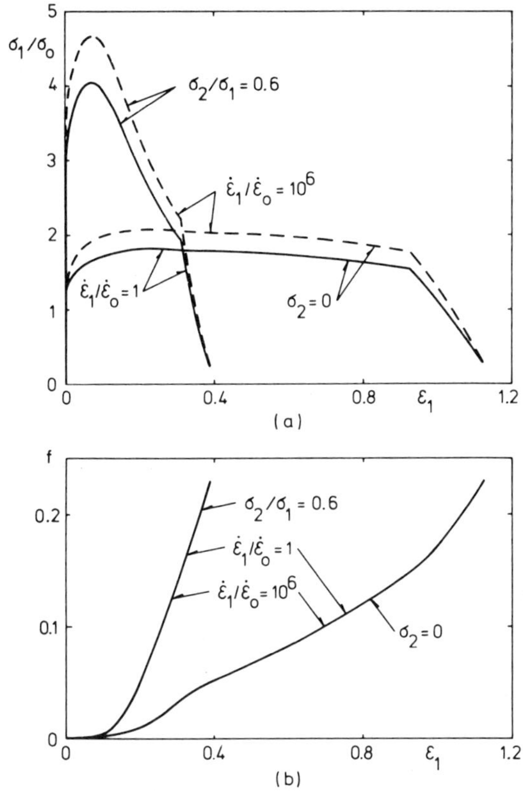

FIG. 24. Effect of stress triaxiality and strain rate in plane strain tension. Material with $n = 10$, $m = 0.01$, strain controlled nucleation and no initial porosity. (a) Maximum principal stress. (b) Void volume fraction (Tvergaard and Needleman, 1986).

strain rate fracture occurs by cleavage in the high triaxiality case, while the void volume fraction is still very low.

B. The Charpy V-Notch Test

The energy absorbed to fracture in the Charpy V-notch test is widely used to evaluate the fracture resistance of materials. For b.b.c. metals, fracture occurs by cleavage when the temperature is sufficiently low, whereas at higher temperatures fracture occurs by ductile hole growth and the absorbed energy is much larger. A similar transition can take place at a fixed temperature, from cleavage fracture at high strain rates to fibrous fracture at low strain rates.

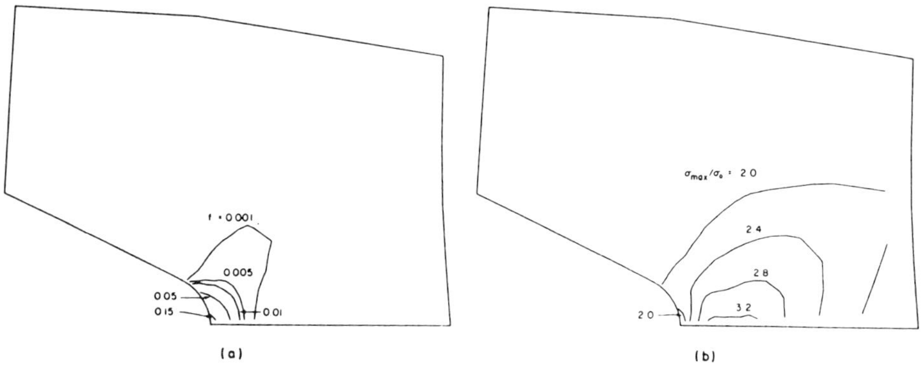

FIG. 25. Contours of constant void volume fraction and maximum principal stress in a Charpy V-notch test with $\dot{U}/B\dot{\varepsilon}_0 = 10^3$, at $U/B = 0.125$ (Tvergaard and Needleman, 1986).

The competition between the two failure mechanisms in the Charpy V-notch test at a fixed temperature has been analysed numerically by Tvergaard and Needleman (1986). For a material with $m = 0.01$, $n = 10$, and the critical stress for cleavage $\sigma_C = 3.35\sigma_0$, while nucleation is plastic strain-controlled, $f_N = 0.04$, $\varepsilon_N = 0.3$, $s = 0.1$, calculations have been carried out at three different rates of loading. Figure 25 shows distributions of the void volume fraction f and the maximum principal stress σ_{max} in a case with $\dot{U}/B\dot{\varepsilon}_0 = 10^3$, where B is the specimen thickness and $\dot{U}$ is the rate of displacement of the central point at which the hammer hits the specimen. At the stage illustrated in Figure 25, large plastic strains at the notch tip have given so much voidage that ductile failure is about to initiate, but simultaneously the critical stress σ_C has been reached in the material somewhat below the notch, and after this stage cleavage fracture takes over completely. For $\dot{U}/B\dot{\varepsilon}_0 = 10^5$, cleavage occurs earlier and the absorbed energy is significantly lower, while for $\dot{U}/B\dot{\varepsilon}_0 = 0.1$ the ductile failure mechanism is more dominant, in agreement with experimental observations (see Rolfe and Barsom (1977)).

In a subsequent investigation (Tvergaard and Needleman, 1987), the computational model has been used to directly analyse the temperature dependence of the absorbed energy for a particular material: the high-nitrogen mild steel investigated by Ritchie, Knott, and Rice (1973). This tempeature dependence is generally a result of the variation of material parameters with temperature, in particular the decreasing initial yield stress for increasing temperature. Uniaxial tensile test results for the high-nitrogen mild steel show that in addition to the decreasing yield stress, the degree of strain hardening increases with temperature, and it turns out that also this temperature dependence has a strong effect on the failure mode transition.

Curves of absorbed energies vs. temperature have been computed for two different imposed velocities, $\dot{U} = 5 \times 10^{-4}$ m/sec and $\dot{U} = 5$ m/sec, corresponding to slow bending and impact loading, respectively. These curves show a brittle/ductile transition in absorbed energy at $T = -85°C$ for the slow loading and at a somewhat higher temperature, $T = -50°C$, for impact loading. The transition in the fracture initiation mode from cleavage initiation to ductile void growth initiation is predicted at somewhat higher temperatures for both curves. In the range of temperatures below $T = 45°C$, the general variation of the absorbed energies with temperature and rate of deformation is found to be rather typical of many steels. However, a sharp reduction in absorbed energy is found at higher temperatures, which is untypical of most steels. This behaviour results from an unusually strong increase in strain hardening, which is characterized as "blue brittleness" for the particular material considered here.

The computations discussed so far are quasi-static, neglecting the effect of inertia, so that time dependence enters only through the viscoplastic material model. A transient analysis that does not account for material failure (Norris, 1979) indicates rather little effect of inertia at the loading rates typically encountered in the Charpy impact test. To check this in the presence of material failure Tvergaard and Needleman (1987) have repeated a few of the quasi-static computations as transient analyses, taking full account of material inertia. Figure 26 shows the dynamic force-displacement curve for $\dot{U} = 5$ m/sec at $T = -55°C$ compared with the response found by the quasistatic analysis. Plasticity damps the bending mode oscillations rather quickly, and the absorbed energy found by the transient analysis is only slightly higher than that found by neglecting the effect of inertia. It might be expected that wave effects have a significant influence on cleavage initiation, but in fact the magnitude of the maximum principal stress oscillations are very small once there is extensive plasticity.

Application of the elastic-viscoplastic material model is necessary in studies of the brittle/ductile transition, since the difference between the behaviour at slow loading and impact loading is entirely dependent on material strain-rate sensitivity. It is noted that also the investigations of

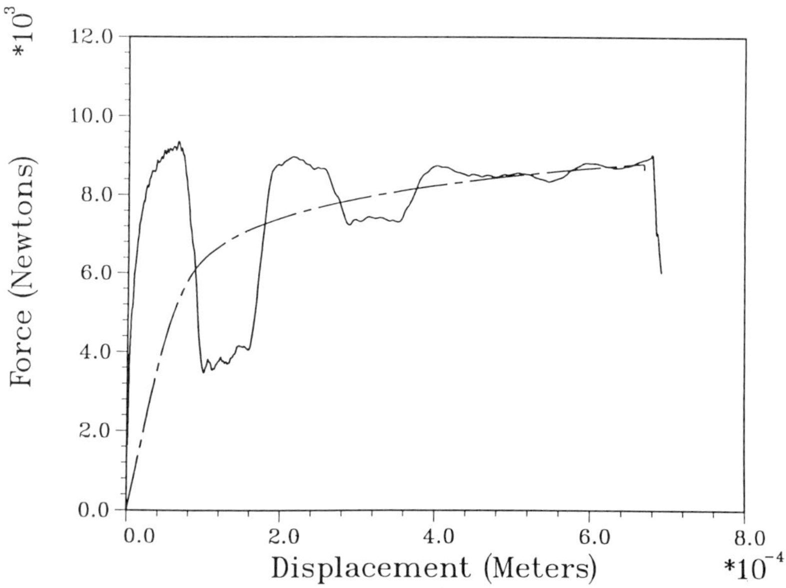

FIG. 26. Computed force-displacement curves for Charpy test of a high-nitrogen mild steel at $\dot{U} = 5\,\text{m/sec}$, $T = -55°\text{C}$. Full transient analysis compared with quasi-static result (Tvergaard and Needleman, 1987).

crack growth by Needleman and Tvergaard (1987) and Becker *et al.* (1987b) have been based on the viscoplastic version of the porous material model, although only one rate of loading was considered. Material strain-rate sensitivity does have an influence in these cases, due to the non-uniform distribution of strain rates in the fields near the crack tip; but here the predictions are not expected to differ much from those of the time-independent material model.

IX. Creep Failure by Grain Boundary Cavitation

In a polycrystalline metal subject to creep at elevated temperatures, the nucleation and growth of voids to coalescence also plays a major role in the failure process. In these circumstances, the voids occur primarily at the grain boundaries and the growth mechanism differs significantly from that at ductile fracture, discussed in the previous sections. Here, in addition to dislocation creep of the grains, grain boundary diffusion plays a significant role.

Experimental results show that high temperature cavitation occurs mainly on grain boundary facets normal to the maximum principal tensile stress direction (Hull and Rimmer, 1959; Cocks and Ashby, 1982; Argon, 1982). Coalescence of these cavities leads to micro-cracks, and the final intergranular creep fracture occurs as the micro-cracks link up. In cases where diffusion gives the dominant contribution to the growth of cavities, the rate of growth is often constrained by the rate of dislocation creep of the surrounding material, as has been noted by Dyson (1976). Several aspects of creep failure have been treated recently by Riedel (1986) and Cocks and Leckie (1987).

A set of constitutive relations for creep with grain boundary cavitation has been proposed by Tvergaard (1984b), as an extension of work by Rice (1981) and Hutchinson (1983). For the rate of growth of a single cavity both the contributions of diffusion and dislocation creep are accounted for, and furthermore, the model incorporates the creep constraint on the rate of cavity growth.

A. Creep with Grain Boundary Cavity Growth

The constitutive relations are based on the assumption that microscopic cavities nucleate and grow on a certain number of grain boundary facets, while the grains deform by dislocation creep, represented as power law creep (Tvergaard, 1984b). The model makes use of the idea of Rice (1981) that the cavitating grain boundary facets can be represented as penny-shaped cracks. Also an expression derived by Hutchinson (1983) for the rate of creep of a material containing traction free micro-cracks is applied.

The cavities are assumed to be uniformly distributed on each grain boundary facet, with average spacing $2b$, as indicated in Figure 27. Then, the average separation between two grains adjacent to a cavitating facet is $\delta = V/\pi b^2$, and the rate of growth of the average separation is

$$\dot{\delta} = \frac{\dot{V}}{\pi b^2} - \frac{2V}{\pi b^2}\frac{\dot{b}}{b}, \tag{9.1}$$

where V, $\dot{V}$, and $\dot{b}$ are the cavity volume, the rate of growth of this volume, and the rate of change of the cavity half-spacing, respectively (see Figure 27b).

When the cavitating facets are modelled as penny-shaped cracks, the average rate of separation $\dot{\delta}$ of the adjacent grains is also given as the

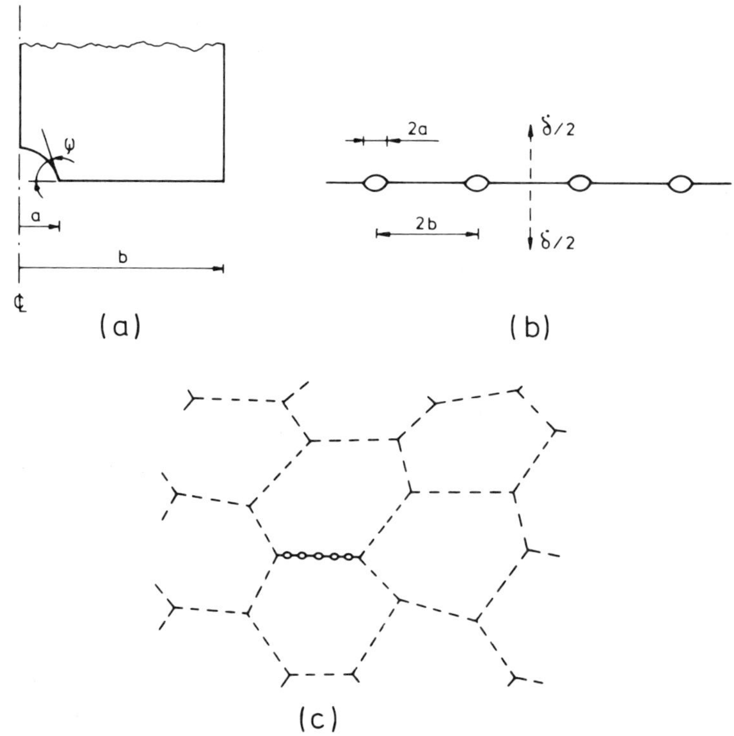

FIG. 27. (a) Spherical-caps shape of a single cavity. (b) Equally spaced cavities on a grain boundary. (c) An isolated, cavitated grain boundary facet in a polycrystalline material.

average rate of opening of a crack. This average rate of opening has been determined by He and Hutchinson (1981), and their expression, modified to account for a non-zero normal tensile stress σ_n on the crack surfaces (Tvergaard, 1984a), gives

$$\dot{\delta} = \beta \frac{S - \sigma_n}{\sigma_e} \dot{\varepsilon}_e^C 2R. \tag{9.2}$$

Here, R is the current radius of the crack, and $S = \sigma^{ij}\bar{n}_i\bar{n}_j$ represents the value of the normal stress on the facet in the absence of cavitation, where σ^{ij} is the macroscopic Cauchy stress tensor and $\bar{n}_i$ is the facet normal in the current configuration. The macroscopic effective Mises stress corresponding to σ^{ij} is σ_e (as in (3.1)), and the effective creep strain rate for an uncracked material is taken to be $\dot{\varepsilon}_e^C = \dot{\varepsilon}_T(\sigma_e/\sigma_0)^n$, where σ_0 is a reference stress quantity, n is the creep exponent, and $\dot{\varepsilon}_T$ is a temperature-dependent reference strain rate.

For this power law creeping material, the value of the parameter β is given by the asymptotic expression (He and Hutchinson, 1981)

$$\beta \simeq \frac{4}{\pi}\left(1+\frac{3}{n}\right)^{-1/2}, \tag{9.3}$$

which is highly accurate for all n as long as $|S/\sigma_e| \leq 2$, while for higher triaxialities the rate of opening of the micro-cracks is increasingly underestimated by using (9.3) in (9.2).

The two expressions (9.1) and (9.2) for the rate of separation $\dot{\delta}$ must be identical. This requirement determines the value of the normal stress σ_n on the facet and the cavity growth rate $\dot{V}$, which is generally a function of σ_n.

Hutchinson (1983) has derived an expression for the macroscopic creep strain rates in a material containing a certain density of penny-shaped micro-cracks. The expression is obtained for traction-free micro-cracks, partly based on the results of He and Hutchinson (1981) for a single crack, and can also be modified to account for a non-zero traction σ_n on the crack surfaces (Tvergaard, 1984b). This modified expression for the macroscopic creep strain-rates is

$$\dot{\eta}_{ij}^{C} = \dot{\varepsilon}_T\left(\frac{\sigma_e}{\sigma_0}\right)^n\left\{\frac{3}{2}\frac{s_{ij}}{\sigma_e} + \rho\left[\frac{3}{2}\frac{n-1}{n+1}\frac{s_{ij}}{\sigma_e}\left(\frac{S-\sigma_n}{\sigma_e}\right)^2 + \frac{2}{n+1}\frac{S-\sigma_n}{\sigma_e}m_{ij}\right]\right\}, \tag{9.4}$$

where $m_{ij} = \bar{n}_i\bar{n}_j$, so that $S = \sigma^{ij}m_{ij}$ and ρ reflects the density of cavitating facets. In terms of the crack radius R and the number of cracks per unit volume Λ, Hutchinson (1983) found, using (9.3),

$$\rho = 4R^3\Lambda(n+1)\left(1+\frac{3}{n}\right)^{-1/2}. \tag{9.5}$$

The total strain-rate is taken to be the sum of the elastic part, the creep part, and the thermal expansion part, $\dot{\eta}_{ij} = \dot{\eta}_{ij}^{E} + \dot{\eta}_{ij}^{C} + \dot{\eta}_{ij}^{T}$. Then, analogous to (2.12) the macroscopic stress strain relationship can be written as

$$\overset{\nabla}{\sigma}{}^{ij} = \mathcal{R}^{ijkl}(\dot{\eta}_{kl} - \dot{\eta}_{kl}^{C} - \dot{\eta}_{kl}^{T}). \tag{9.6}$$

A forward gradient method has been used by Tvergaard (1984b) to increase the stable step size in a numerical solution based on this visco-plastic material model. The derivations follow along the lines indicated in connection with (8.7) and result in constitutive relations of the general form (2.17), although the expressions differ from those given by (8.8)–(8.10).

It is noted that equations (9.1) to (9.6) are independent of the particular mechanisms for cavity growth or nucleation to be considered. These mechanisms will result in expressions for $\dot{V}$ and $\dot{b}$ to be substituted in (9.1), and thus the material model is valid in cases where the cavities grow in the spherical-caps shape as well as cases where crack-like cavity shapes are favoured. The model has also been extended to account for the effect of grain boundary sliding (Tvergaard, 1984b, 1985), and to represent different nucleation or growth behaviour at different groups of grain boundary facets; but these extensions will not be discussed here.

The particular void shape considered by Tvergaard (1984b) is the spherical-caps shape (Figure 27a), which is the relevant shape in cases where the diffusion along the void surface is sufficiently rapid, relative to the diffusion along the grain boundary, to maintain this quasi-equilibrium state. Detailed micromechanical studies of the rate of growth of such cavities, by combined grain boundary diffusion and dislocation creep, have been carried out numerically by Needleman and Rice (1980) and Sham and Needleman (1983). At sufficiently low tensile stresses diffusive cavity growth is completely dominant, while at somewhat higher stresses dislocation creep also contributes noticeably to the growth rate.

Using an interpolation between the rigid grains model for purely diffusional growth and approximate results (Budiansky *et al.*, 1982) for the growth of a spherical void in a power law creeping material, it has been found (Sham and Needleman, 1983; Tvergaard, 1984a) that over a wide range of stress states, the numerical results for the growth of a single void are well approximated by the expression

$$\dot{V} = \dot{V}_1 + \dot{V}_2 \qquad \text{for } \frac{a}{L} \le 10, \qquad f = \max\left\{\left(\frac{a}{b}\right)^2, \left(\frac{a}{a+1.5L}\right)^2\right\}, \tag{9.7}$$

where

$$\dot{V}_1 = 4\pi\mathscr{D}\frac{\sigma_n - (1-f)\sigma_s}{\ln(1/f) - (3-f)(1-f)/2}, \tag{9.8}$$

$$\dot{V}_2 = \begin{cases} \pm 2\pi\dot{\varepsilon}_e^C a^3 h(\psi)\left(\alpha_n \left|\frac{\sigma_m}{\sigma_e}\right| + \beta_n\right)^n & \text{for } \pm\frac{\sigma_m}{\sigma_e} > 1, \\ 2\pi\dot{\varepsilon}_e^C a^3 h(\psi)(\alpha_n + \beta_n)^n \frac{\sigma_m}{\sigma_e} & \text{for } \left|\frac{\sigma_m}{\sigma_e}\right| \le 1. \end{cases} \tag{9.9}$$

Here, a is the cavity radius (Figure 27a), $2b$ is the average cavity spacing, and $\mathscr{D} = D_B\delta_B\Omega/kT$ is the grain boundary diffusion parameter, where $D_B\delta_B$ is the boundary diffusivity, Ω is the atomic volume, k is Boltzmann's constant, and T is the absolute temperature. Furthermore,

σ_n, σ_m, and σ_e are the average normal stress, mean stress, and Mises stress, respectively, in the vicinity of the void, and the constants are given by $\alpha_n = 3/(2n)$, $\beta_n = (n-1)(n+0.4319)/n^2$, and $h(\psi) = [(1+\cos\psi)^{-1} - \frac{1}{2}\cos\psi]/\sin\psi$. A relevant value of the angle ψ (see Figure 27a) is $\psi \simeq 70°$, and the sintering stress σ_s is often approximated by $\sigma_s \simeq 0$. When $\dot{V}$ is known, the rate of growth of the cavity radius is given by $\dot{a} = \dot{V}/[4\pi a^2 h(\psi)]$.

Cavity coalescence on grain boundary facets requires $a/b \simeq 1$ in cases where creep constrained cavitation results in very small normal stresses σ_n. But, in other cases, where σ_n is higher, failure of the remaining ligaments between the cavities, by ductile tearing or by cleavage, may occur at somewhat smaller values of a/b (Cocks and Ashby, 1982). Such coalescence on grain boundary facets, leading to open micro-cracks, is often used as a failure criterion; but is is noted that final fracture actually occurs somewhat later, as these micro-cracks link up.

The parameter L in (9.7), defined by

$$L = (\mathscr{D}\sigma_e/\dot{\varepsilon}_e^C)^{1/3} \tag{9.10}$$

was used by Needleman and Rice (1980) as a stress and temperature dependent length scale. For $a/L < 0.1$, the cavity growth rate is completely dominated by diffusion, so that $\dot{V}_2$ is negligible relative to $\dot{V}_1$; but dislocation creep growth plays an increasing role for higher values of a/L, so that here $\dot{V}_2$ adds significantly to $\dot{V}_1$ in (9.7).

Continuous nucleation of new cavities during the creep process is observed in many metals (Argon, 1982; Dyson, 1983) and experiments often indicate that the number of cavities is proportional to the effective strain. An approximate way of incorporating such continuous nucleation in the present material model has been discussed in some detail by Tvergaard (1984b). With these approximations, the ratio $\dot{b}/b$, to be substituted into (9.1), is simply taken to be

$$\frac{\dot{b}}{b} = \frac{1}{2}\left(\frac{\dot{\lambda}_I}{\lambda_I} + \frac{\dot{\lambda}_{II}}{\lambda_{II}}\right) - \frac{1}{2}\frac{\dot{N}}{N}, \qquad \dot{N} = \frac{dN}{d\varepsilon_e}\dot{\varepsilon}_e, \tag{9.11}$$

where λ_I and λ_{II} are the principal stretches on the plane of the facet and N is the number of cavities per unit initial area of the facet. The value $dN/d\varepsilon_e$ is taken to be constant in some cases, while in other cases this value is taken to be stress dependent. It is assumed that no nucleation occurs if N has reached a saturation value N_{max}, or if $\sigma_n < 0$.

The temperature dependence of the material model is incorporated through the dependence of the effective creep strain rate $\dot{\varepsilon}_e^C$ and the grain boundary diffusion parameter $\mathscr{D}$ on the absolute temperature T (Needleman and Rice, 1980; Cocks and Ashby, 1982)

$$\dot{\varepsilon}_e^C = \dot{\varepsilon}_0\left(\frac{\sigma_e}{\sigma_0}\right)^n \frac{T_0}{T} \exp\left[\frac{Q_V}{\mathscr{R}}\left(\frac{1}{T_0} - \frac{1}{T}\right)\right], \tag{9.12}$$

$$\mathscr{D} = \mathscr{D}_0 \frac{T_0}{T} \exp\left[\frac{Q_B}{\mathscr{R}}\left(\frac{1}{T_0} - \frac{1}{T}\right)\right]. \tag{9.13}$$

Here, $\dot{\varepsilon}_0$, $\mathscr{D}_0$, σ_0, and T_0 are reference values, Q_V and Q_B are activation energies, and $\mathscr{R}$ is the gas constant.

B. Creep Rupture in Structural Components

The material model has been used to analyse the creep failure of thick-walled furnace tubes (Tvergaard, 1984b). The initial internal and external radii are denoted by A_0 and B_0, respectively, and the tube is subjected to an internal pressure p and heated from the outside, so that the absolute temperatures on the inside and outside are T_A and T_B with the radial temperature distribution corresponding to steady heat conduction.

Neglecting temperature variations along the tube, the solutions are uniform in the axial direction, with a constant axial strain-rate determined so that the resultant axial force in the tube balances the axial component of the internal pressure. Thus, the fields vary only in the radial direction (the x^2-direction), and a one-dimensional finite element approximation is applied in (2.11).

The material parameters employed do not refer to a particular catalyst tube steel, but are generally consistent with the orders of magnitude of the parameters tabulated by Frost and Ashby (1982). The values used are

$$E = 10^{11}\,\text{N/m}^2, \qquad \nu = 0.3, \qquad n = 5,$$

$$\sigma_0 = 30 \times 10^6\,\text{N/m}^2, \qquad \dot{\varepsilon}_0 = 10^{-6}\,\text{h}^{-1}, \qquad T_0 = 1173\,\text{K},$$

$$\frac{Q_V}{\mathscr{R}} = 6 \times 10^4\,\text{K}, \qquad \frac{Q_B}{\mathscr{R}} = 3.5 \times 10^4\,\text{K}, \qquad \mathscr{D}_0 = 10^{-29}\,\text{m}^5/\text{Nh},$$

and the thermal expansion coefficient is $\alpha = 16.8 \times 10^{-6}\,\text{K}^{-1}$. The initial dimensions of the tube are taken to be $A_0 = 0.045$ m and $B_0 = 0.055$ m, and the grain size is taken to be such that $R_I = 50\,\mu$m is representative of the initial radius of the grain boundary facets. Furthermore, in the case illustrated in Figure 28, continuous nucleation is assumed with $dN/d\varepsilon_e = 10^4/A_I$ and $N_{max} = 400/A_I$, where $A_I = \pi R_I^2$ is the initial area of a cavitating grain boundary facet. Very few cavities are assumed present from the beginning, as specified by $a_I/b_I = 0.005$ and $b_I/R_I = 1$, with the density of cavitating facets $\rho = 0.5$. A reference time is defined by $t_R = \sigma_e/(E\dot{\varepsilon}_e^C)$, where σ_e is the Mises stress obtained from the average axial and circumferential stresses, and $\dot{\varepsilon}_e^C$ is the corresponding effective creep strain rate at temperature T_0.

At time $t = 0$, the tube is subjected to an internal pressure $p = 3.5 \times 10^6\ \mathrm{N/m^2}$ and a temperature gradient specified by $T_A = T_0 - 20\ \mathrm{K}$, $T_B = T_0 + 20\ \mathrm{K}$. Figure 28 shows that the thermal expansion gives rise to significant stress gradients, with compressive hoop stresses S near the outside of the tube wall, where the temperatures are highest. These elastic stress peaks are relaxed by creep in the initial stages, until the

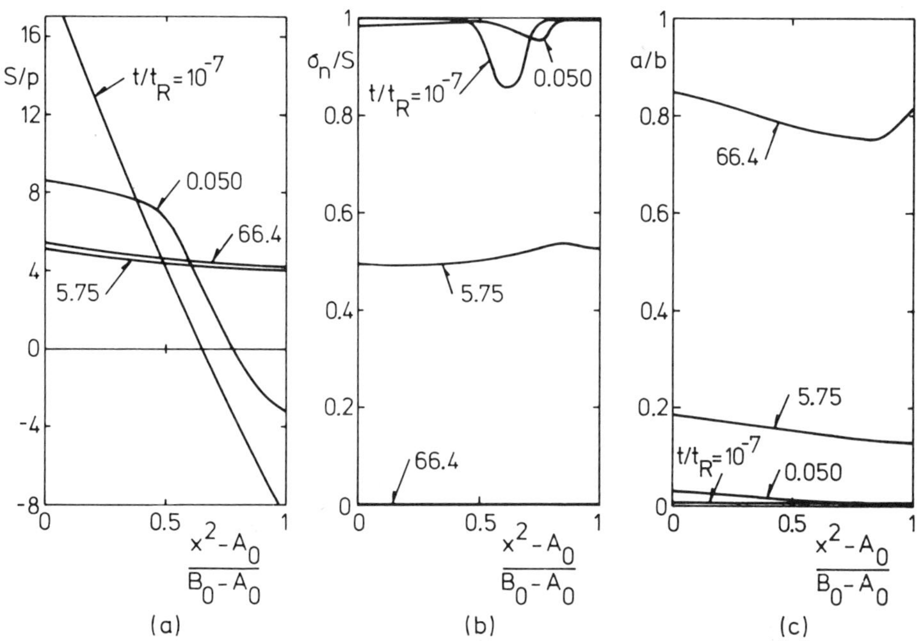

FIG. 28. Development of stresses and damage in a pressurized thick-walled tube with non-uniform temperature. The material has continuous nucleation. (a) Macroscopic hoop stress. (b) Normal stress on cavitating facets. (c) Cavity growth (Tvergaard, 1984b).

steady state stress distribution is reached at $t/t_R \simeq 5.75$. Initially, $\sigma_n/S \simeq 1$ (see Figure 28b), but the ratio σ_n/S decays rapidly towards zero, so that creep constrained cavity growth takes place during most of the lifetime. For tubes subject to uniform temperature, failure tends to occur first at the inside, where the strains are larger; but here the higher temperature at the outside results in nearly simultaneous failure at the inside and outside (Figure 28c).

Creep crack growth has also been analysed by application of this material model (Tvergaard, 1986). The particular configuration analysed is a center cracked panel subject to plane strain conditions, with the initial geometry specified by $A_0/B_0 = \frac{1}{3}$, where $2A_0$ is the initial crack length and $2B_0$ is the initial specimen width. In contrast to the ductile porous material description (Section II), the present creep failure model requires knowledge of the maximum principal stress directions, since the orientation of the facets, on which cavitation is considered, has to be chosen *a priori*. This is no problem in the thickwalled tube under internal pressure, where the principal stress directions remain fixed, but during crack growth the maximum principal stress directions rotate significantly. The field of angles ω_I used (Tvergaard, 1986) to define the reference normals of the cavitating facets has been computed as the maximum principal stress directions in a steady-state creep solution with no damage ($\rho = 0$). It has been checked in the computations that the angle between the current normal of a cavitating facet and the current maximum principal stress direction remains small in all material points where failure is going to occur subsequently.

Figures 29 and 30 are results obtained in a computation where all cavities are assumed to be present from the beginning, specified by $a_I/b_I = 0.1$, $b_I/R_I = 0.1$, and $\rho = 0.2$, while other material parameters are $n = 5$ and $\nu = 0.3$. The constant external load is specified by the initial nominal stress on the crack ligament, $\sigma_N = 0.0005E$, and the value of the grain boundary diffusion parameter is given by $a_I/L_N = 0.01$, where L_N appears by taking σ_e equal to σ_N in (9.10). In the figures, time t is normalized by the failure time t_f^o corresponding to uniaxial plane strain tension at stress level σ_N.

Initially, at $t = 0$ when the load is applied, the elastic singular stress field dominates near the crack tip; but already at $t/t_f^o \simeq 10^{-4}$, most of the initial elastic stress peak has been relaxed by creep, and soon after this stage the crack starts to grow. The creep damage in the material near the crack tip, measured by a/b, is shown in Figure 29 at two stages. To

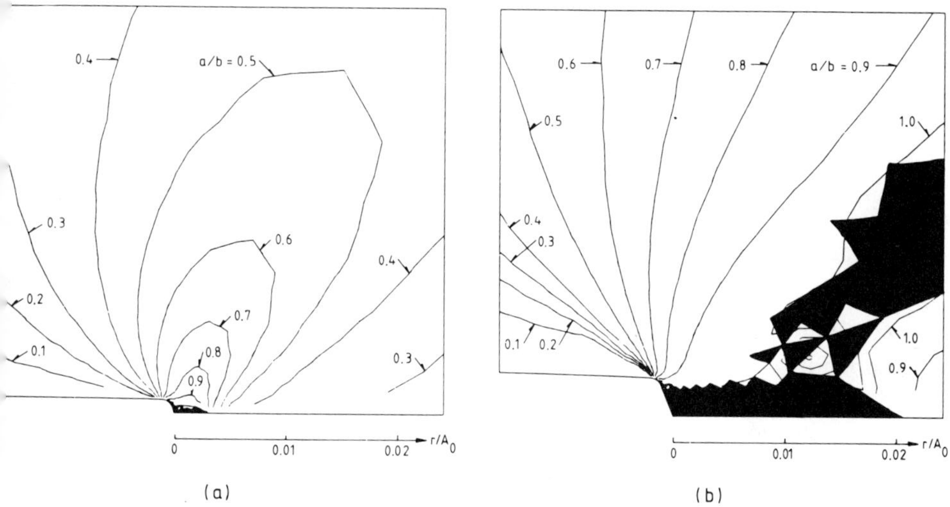

FIG. 29. Distribution of creep damage near the tip of a crack with initial length $2A_0$. (a) $t/t_f^o = 0.0042$, (b) $t/t_f^o = 0.0810$ (Tvergaard, 1986).

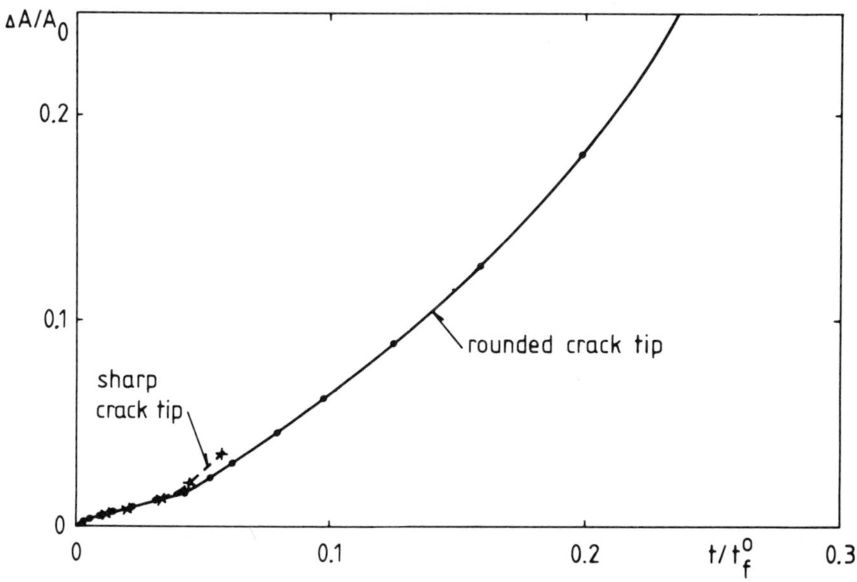

FIG. 30. Creep crack extension vs. time for crack with initial length $2A_0$ (Tvergaard, 1986).

illustrate the growth of the macroscopic crack, elements in which failure has occurred ($a/b = 1$) are painted black, and it is seen that initially the crack grows straight ahead ($\theta = 0°$). When the crack extension ΔA is about $0.014A_0$, the growth direction starts to deviate from the crack plane, and all subsequent crack growth occurs along a plane inclined about 40° to the initial crack plane.

The rate of crack growth is illustrated in Figure 30 by ΔA vs. t curves computed for two different near-tip meshes, where ΔA denotes the distance from the initial crack tip to the current crack tip in the reference configuration. Crack growth starts nearly immediately when the load is applied, with a relatively high growth rate that gradually decays as the elastic stress peak is relaxed by creep. The growth rate increases again when the growth direction starts to deviate from the crack plane, and subsequently the growth keeps accelerating as the remaining ligament is reduced by crack growth. Similar results for other sets of material parameters are also presented by Tvergaard (1986).

As an extension of this analysis, Tvergaard (1988) has investigated the effect of creep acceleration induced by degradation of the microstructure. These crack growth analyses are carried out for both center-cracked and double edge-cracked specimens, and it is found that creep acceleration reduces the crack growth rate due to a lower stress level at the tip.

The same material model has been used recently by Li, Needleman, and Shih (1988) to study creep crack growth under small scale yielding conditions. In this investigation, the effect of material parameters and load level on the dominant asymptotic stress and strain fields near the tip of the growing crack has been studied.

X. Discussion

A characteristic feature of the material models discussed here is that the damage parameter used is directly given by the material geometry on the micro-scale, either as a void volume fraction or as an area fraction of voids on grain boundary facets. This means that the current value of the damage parameter can be directly determined by the use of microscopy. Furthermore, this is the basis of the micromechanical studies that have been used extensively to improve the understanding of material behaviour. By contrast, the approach often characterized as continuum damage mechanics makes use of damage parameters relating to macroscopic material behaviour, with no direct relation to the cause of damage on the micro-level.

Although the void volume fraction can be measured, a given value of this scalar damage parameter can represent many different distributions and shapes of voids. Magnusen *et al.* (1987) have investigated the effect of the void distribution by carrying out a series of tensile tests for specimens with either random or regular arrays of equi-sized, cylindrical, through-thickness holes. The volume fraction of holes is the same in all cases, but different degrees of randomness are obtained by specifying different values of a minimum accepted spacing between adjacent holes. For each of steel, brass, and aluminium, it was found that the fracture strain is smaller the more randomness is allowed. The regular square array of holes that has often been used in micromechanical studies gives the maximum fracture strain for the given volume fraction of holes. Thus, the distribution of voids should certainly affect the material model, and this needs to be investigated further. It may be expected, based on the experiments, that the distribution will have a strong effect on the value of the critical void volume fraction f_C in (3.2), at which failure starts to occur rapidly, while little influence is expected at smaller void volume fractions.

Some materials contain long thin inclusions or rows of inclusions, which may be essentially parallel, dependent on previous metal working processes. For such materials, the influence of the voids on the stress strain relationship is anisotropic, both in the nucleation and growth phases, so that the isotropic material models discussed in the present paper are not adequate. Some work has been done to find approximate anisotropic yield conditions based on experimental results for various patterns of perforations in sheets, and there has also been work introducing a second order damage tensor f_{ij} for the description of ductile void growth. However, a full ductile porous material model that accounts for anisotropies due to void distributions or shapes has not yet been developed.

In spite of the problems raised above, it should be emphasized that in many cases a reasonable description is obtained by the isotropic material model, in which the void volume fraction is the only damage parameter. This has been strongly supported by the detailed comparison of theoretical predictions and experimental results for sintered iron powder compacts with known initial porosities, subjected to various non-uniform stress states (Becker *et al.*, 1987a). The comparison with these experiments supports the description of void growth and final failure incorporated in the material model. Nucleation models are not checked by these experiments, but will have to be fitted with the actual material behaviour for each particular application.

The size of the voids and the void spacing does not at all enter the ductile material model, where only the void volume fraction is known, and the same would be true if the void shapes and distribution patterns were also known. The only necessary assumption is that the voids are large enough relative to the dislocation cell size, so that the application of continuum plasticity is reasonable. This is different for the material subject to creep with grain boundary cavitation. Here, the grain size provides a natural length scale, and the material model requires knowledge of the number of cavities per grain boundary facet, so that in fact both the size and spacing of the voids need to be known.

In the porous ductile material models, the same damage parameter, the void volume fraction, applies to all possible stress states. This is more complicated in the model for creep with grain boundary cavitation, since here void growth occurs mainly on grain boundaries normal to the maximum principal tensile stress. Thus, with only one damage parameter (the area fraction of cavities on grain boundary facets), the maximum principal stress direction must be known in each material point, and it must be assumed that this field of directions remains approximately constant throughout the material lifetime. In the more general case, where the principal stress directions rotate during the lifetime, it is necessary to use a number of damage parameters, where each parameter has a direction associated with it (the facet normal). These damage parameters represent area fractions of cavities on different groups of grain boundary facets with different orientations of the normals.

Both the time-independent plasticity models in Sections III and VII and the viscous material models in Sections VIII and IX have been employed to analyse the gradually increasing damage and the mode of final failure in several test specimens or structural components. The predictions have been found to be at least qualitatively, and in some cases also quantitatively, in good agreement with experimental observations. Thus, although there is a need for further developments to make the descriptions more realistic in some respects, the material models in their current form appear to be quite powerful.

References

Andersson, H. (1977). Analysis of a model for void growth and coalescence ahead of a moving crack tip. *J. Mech. Phys. Solids* **25**, 217–233.

Aoki, S., Kishimoto, K., Takeya, A., and Sakata, M. (1984). Effects of microvoids on crack blunting and initiation in ductile materials. *Int. J. Fracture* **24**, 267–278.

Aravas, N., and McMeeking, R. M. (1985). Microvoid growth and failure in the ligament between a hole and a blunt crack. *Int. J. Fracture* **29**, 21–38.

Argon, A. S. (1982). Mechanisms and mechanics of fracture in creeping alloys. *In* "Recent Advances in Creep and Fracture of Engineering Materials and Structures" (B. Wilshire and D. R. J. Owen eds.), 1–52, Pinerage Press, Swansea, U.K.

Becker, R., and Needleman, A. (1986). Effect of yield surface curvature on necking and failure in porous plastic solids. *J. Appl. Mech.* **108**, 491–499.

Becker, R., Needleman, A., Richmond, O., and Tvergaard, V. (1987a). Void growth and failure in notched bars. Brown University, Division of Engineering.

Becker, R., Needleman, A., Suresh, S., Tvergaard, V., and Vasudevan, A. K. (1987b). An analysis of ductile failure by grain boundary void growth. Brown University, Division of Engineering.

Berg, C. A. (1970). Plastic dilatation and void interaction. In "Inelastic Behaviour of Solids" (M. F. Kanninen *et al.* eds.), pp. 171–210. McGraw–Hill.

Bishop, J. F. W., and Hill, R. (1951). A theory of the plastic distortion of a polycrystalline aggregate under combined stresses. *Phil. Mag.* **42**, 414–427.

Brown, L. M., and Embury, J. D. (1973). The initiation and growth of voids at second phase particles. *In* "Proc. 3rd Int. Conf. on Strength of Metals and Alloys," pp. 164–169. Inst. of Metals, London.

Budiansky, B. (1969). Remarks on theories of solid and structural mechanics. In "Problems of Hydrodynamics and Continuum Mechanics" (M. A. Lavrent'ev *et al.* eds.). SIAM, Philadelphia.

Budiansky, B., Hutchinson, J. W., and Slutsky, S. (1982). Void growth and collapse in viscous solids. *In* "Mechanics of Solids, The Rodney Hill 60th Anniversary Volume" (H. G. Hopkins and M. J. Sewell eds.), pp. 13–45. Pergamon Press.

Chu, C. C., and Needleman, A. (1980). Void nucleation effects in biaxially stretched sheets. *J. Eng. Materials Technol.* **102**, 249–256.

Cialone, H., and Asaro, R. J. (1980). Hydrogen assisted fracture of spheroidized plain carbon steels. *Met. Trans.* **12A**, 1373–1387.

Cocks, A. C. F., and Ashby, M. F. (1982). On creep fracture by void growth. *Prog. Materials Science* **27**, 189–244.

Cocks, A. C. F., and Leckie, F. A. (1987). Creep constitutive equations for damaged materials. *Adv. Appl. Mech.* **25**, 239–294.

Cottrell, A. H. (1958). Theory of brittle fracture in steel and similar metals. *Transactions AIME* **212**, 192–203.

Cox, T. B., and Low, J. R. (1974). An investigation of the plastic fracture of AISI 4340 and 18 nickel-200 grade maraging steels. *Met. Trans.* **5**, 1457–1470.

Dafalias, Y. F. (1983). Corotational rates for kinematic hardening at large plastic deformations. *J. Appl. Mech.* **50**, 561–565.

Doraivelu, S. M., Gegel, H. L., Gunaselsera, J. S., Malas, J. C., Morgan, J. T., and Thomas, J. F. (1984). A new yield function for compressible P/M materials. *Int. J. Mech. Sci.* **26**, 527–535.

Dyson, B. F. (1976). Constraints on diffusional cavity growth rates. *Metal Science* **10**, 349–353.

Dyson, B. F. (1983). Continuous cavity nucleation and creep fracture. *Scripta Metall.* **17**, 31–37.

Frost, H. J., and Ashby, M. F. (1982). "Deformation Mechanism Maps: The Plasticity and Creep of Metals and Ceramics." Pergamon Press, Oxford.

Goods, S. H., and Brown, L. M. (1979). The nucleation of cavities by plastic deformation. *Acta Metall.* **27**, 1–15.

Green, A. E., and Zerna, W. (1968). "Theoretical Elasticity." Oxford University Press, Oxford.

Guennouni, T., and Francois, D. (1987). Constitutive equations for rigid plastic or viscoplastic materials containing voids. *Fatigue Fract. Engng. Mater. Struct.* **10**, 399–418.

Gurson, A. L. (1977a). Continuum theory of ductile rupture by void nucleation and growth—I. Yield criteria and flow rules for porous ductile media. *J. Engng. Materials Technol.* **99**, 2–15.

Gurson, A. L. (1977b). Porous rigid-plastic materials containing rigid inclusions—Yield function, plastic potential, and void nucleation. *In* "Proc. Int. Conf. Fracture" (D. M. R. Taplin ed.), Vol. 2A, 357–364, Pergamon Press.

Hadamard, J. (1903). "Lecons sur la propagation des ondes et les Âquations de l'hydrodynamique." Libraire Scientifique A., Hermann, Paris.

Hahn, G. T. (1984). The influence of microstructure on brittle fracture toughness. *Met. Trans.* **15A**, 947–959.

Hahn, G. T., and Rosenfield, A. R. (1975). Metallurgical factors affecting fracture toughness of aluminum alloys. *Met. Trans.* **6A**, 653–668.

Hancock, J. W. (1982). Plasticity of porous metals. In "Yield, Flow and Fracture of Polycrystals" (T. N. Baker ed.), pp. 69–79. Appl. Sci. Publishers.

Hancock, J. W., and Brown, D. K. (1983). On the role of strain and stress state in ductile failure. *J. Mech. Phys. Solids* **31**, 1–24.

Hancock, J. W., and Mackenzie, A. C. (1976). On the mechanisms of ductile failure in high strength steels subjected to multiaxial stress-states. *J. Mech. Phys. Solids* **24**, 147–169.

He, M. Y., and Hutchinson, J. W. (1981). The penny-shaped crack and the plane strain crack in an infinite body of power-law material. *J. Appl. Mech.* **48**, 830–840.

Hill, R. (1962). Acceleration waves in solids. *J. Mech. Phys. Solids* **10**, 1–16.

Hull, D., and Rimmer, D. E. (1959). The growth of grain-boundary voids under stress. *Phil. Mag.* **4**, 673–687.

Hutchinson, J. W. (1973). Finite strain analysis of elastic-plastic solids and structures. In "Numerical Solution of Nonlinear Structural Problems" (R. F. Hartung ed.), p. 17, ASME, New York.

Hutchinson, J. W. (1983). Constitutive behaviour and crack tip fields for materials undergoing creep-constrained grain boundary cavitation. *Acta Metall.* **31**, 1079–1088.

Hutchinson, J. W., and Tvergaard, V. (1980). Surface instabilities on statically strained plastic solids. *Int. J. Mech. Sci.* **22**, 339–354.

Hutchinson, J. W., and Tvergaard, V. (1981). Shear band formation in plane strain. *Int. J. Solids Structures* **17**, 451–470.

Hutchinson, J. W., and Tvergaard, V. (1987). Softening due to void nucleation in metals. Harvard University, Division of Appl. Sci., Mech-99.

Koplik, J., and Needleman, A. (1987). Void growth and coalescence in porous plastic solids. Brown University, Division of Engineering.

Lee, E. H., Mallett, R. L., and Wertheimer, T. B. (1983). Stress analysis for kinematic hardening in finite-deformation plasticity. *J. Appl. Mech.* **50**, 554–560.

Li, F. Z., Needleman, A., and Shih, C. F. (1988). Creep crack growth by grain boundary cavitation: Crack tip fields and crack growth rates under transient conditions. Brown University Report.

Magnusen, P. E., Koss, D. A., and Dubensky, E. M. (1987). Ductile fracture and random vs. regular hole/void arrays. Office of Naval Research, N00014-86-K-0381, Techn. Rep. No. 5.

Marciniak, K., and Kuczynski, K. (1967). Limit strains in the process of stretch forming sheet metals. *Int. J. Mech. Sci.* **9**, 609–620.

McClintock, F. A. (1968). A criterion for ductile fracture by growth of holes. *J. Appl. Mech.* **35**, 363–371.

McMeeking, R. M. (1977). Finite deformation analysis of crack tip opening in elastic-plastic materials and implications for fracture initiation. *J. Mech. Phys. Solids* **25**, 357–381.

Mear, M. E., and Hutchinson, J. W. (1985). Influence of yield surface curvature on flow localization in dilatant plasticity. *Mech. Materials* **4**, 395–407.

Moussy, F. (1985). Les differentes echelles du developpement de l'endommagement dans del aciers. Influence sur la localization de la deformation a l'echelle microscopique. *In* "Proc. Int. Symp. on Plastic Instability; Considere Memorial" (J. Salencon ed.), pp. 263–272. Presses Ponts et ChaussÂes.

Needleman, A. (1972). Void growth in an elastic-plastic medium. *J. Appl. Mech.* **41**, 964–970.

Needleman, A. (1987). A numerical study of void nucleation at carbides. Brown University, Division of Engineering.

Needleman, A., and Rice, J. R. (1978). Limits to ductility set by plastic flow localization. *In* "Mechanics of Sheet Metal Forming" (D. P. Koistinen *et al.* eds.), pp. 237–267. Plenum Publishing.

Needleman, A., and Rice, J. R. (1980). Plastic creep flow effects in the diffusive cavitation of grain boundaries. *Acta Metall.* **28**, 1315–1332.

Needleman, A., and Tvergaard, V. (1984a). An analysis of ductile rupture in notched bars. *J. Mech. Phys. Solids* **32**, 461–490.

Needleman, A., and Tvergaard, V. (1984b). Finite element analysis of localization in plasticity. In "Finite Elements, Special Problems in Solid Mechanics" (J. T. Oden and G. F. Carey eds.), Vol. V, pp. 94–157. Prentice–Hall.

Needleman, A., and Tvergaard, V. (1984c). Limits to formability in rate-sensitive metal sheets. In "Mechanical Behavior of Materials—IV" (J. Carlsson and N. G. Ohlson eds.), pp. 51–65. Pergamon Press, Oxford.

Needleman, A., and Tvergaard, V. (1987). An analysis of ductile rupture modes at a crack tip. *J. Mech. Phys. Solids* **35**, 151–183.

Nemat-Nasser, S., and Taya, M. (1977). Model studies of ductile fracture. In "Continuum Models of Discrete Systems," pp. 387–405. University of Waterloo Press.

Nemat-Nasser, S., Sumi, Y., and Keer, L. M. (1980). Unstable growth of tension cracks in brittle solids: Stable and unstable bifurcations, snap-through, and imperfection sensitivity. *Int. J. Solids Structures* **16**, 1017–1035.

Norris, D. M. (1979). Computer simulation of the Charpy V-notch toughness test. *Engineering Fracture Mechanics* **11**, 261–274.

Nutt, S. R., and Needleman, A. (1987). Void nucleation at fiber ends in Al-SiC composites. *Scripta Metallurgica* **21**, 705–710.

Pan, J., Saje, M., and Needleman, A. (1983). Localization of deformation in rate sensitive porous plastic solids. *Int. J. Fracture* **21**, 261–278.

Paris, P. C., Tada, H., Zahoor, A., and Ernst, H. (1979). Initial experimental investigation of tearing instability theory. *In* "Elastic-Plastic Fracture," ASTM STP 668.

Peirce, D., Shih, C. F., and Needleman, A. (1984). A tangent modulus method for rate dependent solids. *Comput. Struct.* **18**, 875–887.

Petch, N. J. (1958). The ductile-brittle transition in the fracture of α-iron: I. *Phil. Mag.* **3**, 1089–1097.

Puttick, D. E. (1960). Ductile fracture in metals. *Phil. Mag.* **4**, 964–969.

Rice, J. R. (1976). Elastic-plastic fracture mechanics. In "The Mechanics of Fracture" (F. Erdogan ed.), pp. 23–53. ASME AMD 19.
Rice, J. R. (1977). The localization of plastic deformation. *In* "Theoretical and Appl. Mech." (W. T. Koiter ed.), pp. 207–220. North–Holland.
Rice, J. R. (1981). Constraints on the diffusive cavitation of isolated grain boundary facets in creeping polycrystals. *Acta Metall.* **29**, 675–681.
Rice, J. R., and Johnson, M. A. (1970). The role of large crack tip geometry changes in plane strain fracture. In "Inelastic Behaviour of Solids" (M. F. Kanninen *et al.* eds.), pp. 641–672. McGraw–Hill, New York.
Rice, J. R., and Tracey, D. M. (1969). On the ductile enlargement of voids in triaxial stress fields. *J. Mech. Phys. Solids* **17**, 201–217.
Riedel, H. (1986). "Fracture at High Temperatures." Springer-Verlag, Berlin.
Ritchie, R. O., Knott, J. F., and Rice, J. R. (1973). On the relationship between critical tensile stress and fracture toughness in mild steel. *J. Mech. Phys. Solids* **21**, 395–410.
Rolfe, S. T., and Barsom, J. M. (1977). "Fracture and Fatigue Control in Structures—Applications of Fracture Mechanics." Prentice–Hall, Englewood Cliffs, New Jersey.
Rudnicki, J. W., and Rice, J. R. (1975). Conditions for the localization of deformation in pressure-sensitive dilatant materials. *J. Mech. Phys. Solids* **23**, 371–394.
Saje, M., Pan, J., and Needleman, A. (1982). Void nucleation effects on shear localization in porous plastic solids. *Int. J. Fracture* **19**, 163–182.
Sham, T.-L., and Needleman, A. (1983). Effects of triaxial stressing on creep cavitation of grain boundaries. *Acta Metall.* **31**, 919–926.
Shima, S., and Oyane, M. (1976). Plasticity theory for porous metals. *Int. J. Mech. Sci.* **18**, 285–291.
Smith, E. (1968). Cleavage fracture in mild steel. *Int. J. Fracture Mechanics* **4**, 131–145.
Speich, G. R., and Spitzig, W. A. (1982). Effect of volume fraction and shape of sulfide inclusions on through-thickness ductility and impact energy of high-strength 4340 plate steels. *Met. Trans.* **13A**, 2239–2258.
Stone, R. H., and Psioda, J. A. (1975). Discussion of: Metallurgical factors affecting fracture toughness of aluminum alloys. *Met. Trans.* **6A**, 668–670.
Tvergaard, V. (1978). Effect of kinematic hardening on localized necking in biaxially stretched sheets. *Int. J. Mech. Sci.* **20**, 651–658.
Tvergaard, V. (1981). Influence of voids on shear band instabilities under plane strain conditions. *Int. J. Fracture* **17**, 389–407.
Tvergaard, V. (1982a). On localization in ductile materials containing spherical voids. *Int. J. Fracture* **18**, 237–252.
Tvergaard, V. (1982b). Ductile fracture by cavity nucleation between larger voids. *J. Mech. Phys. Solids* **30**, 265–286.
Tvergaard, V. (1982c). Material failure by void coalescence in localized shear bands. *Int. J. Solids Structures* **18**, 659–672.
Tvergaard, V. (1982d). Influence of void nucleation on ductile shear fracture at a free surface. *J. Mech. Phys. Solids* **30**, 399–425.
Tvergaard, V. (1984a). On the creep constrained diffusive cavitation of grain boundary facets. *J. Mech. Phys. Solids* **32**, 373–393.
Tvergaard, V. (1984b). Constitutive relations for creep in polycrystals with grain boundary cavitation. *Acta Metall.* **32**, 1977–1990.
Tvergaard, V. (1985). Effect of grain boundary sliding on creep constrained diffusive cavitation. *J. Mech. Phys. Solids* **33**, 447–469.
Tvergaard, V. (1986). Analysis of creep crack growth by grain boundary cavitation. *Int. J. Fracture* **31**, 183–209.

Tvergaard, V. (1987a). Effect of yield surface curvature and void nucleation on plastic flow localization. *J. Mech. Phys. Solids* **35**, 43–60.

Tvergaard, V. (1987b). Ductile shear fracture at the surface of a bent specimen. *Mech. Materials* **6**, 53–69.

Tvergaard, V. (1987c). Effect of anisotropic hardening on ductile failure of a material containing two size-scales of particles. Technical University of Denmark, DCAMM Report No. 354.

Tvergaard, V. (1987d). 3D-analysis of localization failure in a ductile material containing two size-scales of spherical particles. Technical University of Denmark, DCAMM Report No. 365.

Tvergaard, V. (1988). Effect of microstructure degradation on creep crack growth. Dept. of Solid Mech., Technical University of Denmark.

Tvergaard, V., and Needleman, A. (1984). Analysis of the cup-cone fracture in a round tensile bar. *Acta Metall.* **32**, 157–169.

Tvergaard, V., and Needleman, A. (1986). Effect of material rate sensitivity on failure modes in the Charpy V-notch test. *J. Mech. Phys. Solids* **34**, 213–241.

Tvergaard, V., and Needleman, A. (1987). Temperature and rate dependence of Charpy V-notch energies for a high nitrogen steel. Brown University, Division of Engineering.

Tvergaard, V., Needleman, A., and Lo, K. K. (1981). Flow localization in the plane strain tensile test. *J. Mech. Phys. Solids* **29**, 115–142.

Yamamoto, H. (1978). Conditions for shear localization in the ductile fracture of void-containing materials. *Int. J. Fracture* **14**, 347–365.

Ziegler, H. (1959). A modification of Prager's hardening rule. *Quart. Appl. Math.* **17**, 55–65.

ADVANCES IN APPLIED MECHANICS, VOLUME 27

Complete Solutions to Some Mixed Boundary Value Problems in Elasticity

V. I. FABRIKANT

Department of Mechanical Engineering
Concordia University
Montreal, Canada

I. Introduction

The great majority of solved punch and crack problems deal with the stresses and displacements in the plane $z = 0$ only. There are just a few

ISBN 0-12-002027-0

complete solutions published (Sneddon, 1951; Elliott, 1949; Westmann, 1965) where explicit expressions are given for the field of displacements and stresses for the most simple axisymmetric problems (a circular punch and a penny-shaped crack). Some explicit expressions for the field of displacements for an elliptic crack can be found in (Kassir and Sih, 1975). We are unaware of any nonaxisymmetric punch or crack problem for a transversely isotropic elastic half-space solved completely in elementary functions. Knowledge of complete solutions is indispensable for consideration of more complicated problems of crack interactions, influence of external loads on punches and cracks, etc.

The integral transform method combined with coupled integral equations is the most widely used for the solution of mixed boundary value problems. Some quite remarkable results were achieved by Uflyand (1977), nevertheless, one has to admit that our inability to solve problems directly is the main reason for a wide use of various integral transformations. Here we present some hints indicating that a direct method of solving some problems in elasticity does exist and is elementary. Consider the simplest case of a penny-shaped crack of radius a subjected to axisymmetric pressure $p(\rho)$. The corresponding potential function f can be found in Kassir and Sih (1975) and is as follows:

$$f(\rho, z) = \int_0^\infty \frac{A(s) J_0(\rho s) \exp(-sz)}{s}\, ds, \tag{1.1}$$

where

$$A(s) = -\frac{1}{\pi\mu} \int_0^a \sin st\, dt \int_0^t \frac{r p(r)\, dr}{(t^2 - r^2)^{1/2}}. \tag{1.2}$$

The user had to substitute the explicit expression for p in (1.2) and to compute two consecutive integrals. The result is substituted in (1.1), and an infinite integral with Bessel function is computed. It seems natural to try to spare one integration by substituting (1.2) in (1.1), changing the order of integration and computing the integral (Gradshteyn and Ryzhik, 1963, formula 6.752.1)

$$\int_0^\infty \frac{\sin st\, J_0(\rho s) \exp(-sz)}{s}\, ds = \sin^{-1}\left(\frac{t}{l_2(t)}\right), \tag{1.3}$$

where the notation

$$l_2(t) = \tfrac{1}{2}\{[(\rho + t)^2 + z^2]^{1/2} + [(\rho - t)^2 + z^2]^{1/2}\} \tag{1.4}$$

was introduced. Now formulae (1.1) and (1.2) can be unified giving

$$f(\rho, z) = -\frac{1}{\pi\mu}\int_0^a \sin^{-1}\left(\frac{t}{l_2(t)}\right) dt \int_0^t \frac{p(\rho_0)\rho_0\, d\rho_0}{(t^2-\rho_0^2)^{1/2}}. \tag{1.5}$$

Note that expression (1.5) contains no trace of the integral transform, and thus gives us a hint that a direct and elementary solution is indeed possible. For example, in the case of a uniform loading $p = \text{const.}$, the potential function (1.5) will take the form

$$f(\rho, z) = -\frac{p}{\pi\mu}\int_0^a t \sin^{-1}\left(\frac{t}{l_2(t)}\right) dt. \tag{1.6}$$

The integral in (1.6) though looking formidable due to (1.4) can be computed in elementary functions as shown in Appendix B (B.13), namely,

$$f(\rho, z) = -\frac{p}{4\pi\mu}\left[(2a^2+2z^2-\rho^2)\sin^{-1}\left(\frac{a}{l_2}\right) + l_1(\rho^2-l_1^2)^{1/2} - 2z(a^2-l_1^2)^{1/2}\right]. \tag{1.7}$$

Here l_2 is defined by (1.4), and

$$l_1(t) = \tfrac{1}{2}\{[(\rho+t)^2+z^2]^{1/2} - [(\rho-t)^2+z^2]^{1/2}\}. \tag{1.8}$$

One can show that arbitary polynomial loading in (1.5) will lead to an elementary expression for the potential function, and thus to an elementary complete solution.

The considerations above give just a hint that a direct solution is possible. This article describes how a general nonaxisymmetric solution can be derived for various punch and crack problems. The solution is elementary for the case of a circular line of division of boundary conditions; no integral transforms or special function expansions are involved. One may just wonder why this solution was not discovered at least a century ago.

II. Theory

A. General Solution

Consider a transversely isotropic elastic body that is characterized by five elastic constants A_{ik} defining the following stress–strain relationships

$$\begin{aligned}
\sigma_x &= A_{11}\frac{\partial u_x}{\partial x} + (A_{11} - 2A_{66})\frac{\partial u_y}{\partial y} + A_{13}\frac{\partial w}{\partial z},\\
\sigma_y &= (A_{11} - 2A_{66})\frac{\partial u_x}{\partial x} + A_{11}\frac{\partial u_y}{\partial y} + A_{13}\frac{\partial w}{\partial z},\\
\sigma_z &= A_{13}\frac{\partial u_x}{\partial x} + A_{13}\frac{\partial u_y}{\partial y} + A_{33}\frac{\partial w}{\partial z},\\
\tau_{xy} &= A_{66}\left(\frac{\partial u_x}{\partial y} + \frac{\partial u_y}{\partial x}\right),\\
\tau_{yz} &= A_{44}\left(\frac{\partial u_y}{\partial z} + \frac{\partial w}{\partial y}\right),\\
\tau_{zx} &= A_{44}\left(\frac{\partial w}{\partial x} + \frac{\partial u_x}{\partial z}\right).
\end{aligned} \tag{2.1}$$

The equilibrium equations are

$$\begin{aligned}
\frac{\partial \sigma_x}{\partial x} + \frac{\partial \tau_{xy}}{\partial y} + \frac{\partial \tau_{zx}}{\partial z} &= 0,\\
\frac{\partial \tau_{xy}}{\partial x} + \frac{\partial \sigma_y}{\partial y} + \frac{\partial \tau_{yz}}{\partial z} &= 0,\\
\frac{\partial \tau_{zx}}{\partial x} + \frac{\partial \tau_{yz}}{\partial y} + \frac{\partial \sigma_z}{\partial z} &= 0.
\end{aligned} \tag{2.2}$$

Substitution of (2.1) in (2.2) yields

$$\begin{aligned}
A_{11}\frac{\partial^2 u_x}{\partial x^2} + A_{66}\frac{\partial^2 u_x}{\partial y^2} + A_{44}\frac{\partial^2 u_x}{\partial z^2} + (A_{11} - A_{66})\frac{\partial^2 u_y}{\partial x\,\partial y} + (A_{13} + A_{44})\frac{\partial^2 w}{\partial x\,\partial z} &= 0,\\
A_{66}\frac{\partial^2 u_y}{\partial x^2} + A_{11}\frac{\partial^2 u_y}{\partial y^2} + A_{44}\frac{\partial^2 u_y}{\partial z^2} + (A_{11} - A_{66})\frac{\partial^2 u_x}{\partial x\,\partial y} + (A_{13} + A_{44})\frac{\partial^2 w}{\partial y\,\partial z} &= 0,\\
A_{44}\left[\frac{\partial^2 w}{\partial x^2} + \frac{\partial^2 w}{\partial y^2}\right] + A_{33}\frac{\partial^2 w}{\partial z^2} + (A_{44} + A_{13})\left[\frac{\partial^2 u_x}{\partial x\,\partial z} + \frac{\partial^2 u_y}{\partial y\,\partial z}\right] &= 0.
\end{aligned} \tag{2.3}$$

Now we introduce the complex tangential displacements $u = u_x + iu_y$ and $\bar{u} = u_x - iu_y$. This will allow us to reduce the number of equations in (2.3) by one and to rewrite these equations in a more compact manner, namely

$$\frac{1}{2}(A_{11}+A_{66})\,\Delta u + A_{44}\frac{\partial^2 u}{\partial z^2} + \frac{1}{2}(A_{11}-A_{66})\,\Lambda^2\bar{u} + (A_{13}+A_{44})\,\Lambda\frac{\partial w}{\partial z} = 0$$
$$A_{44}\,\Delta w + A_{33}\frac{\partial^2 w}{\partial z^2} + \frac{1}{2}(A_{13}+A_{44})\frac{\partial}{\partial z}(\bar{\Lambda}u + \Lambda\bar{u}) = 0. \tag{2.4}$$

Here the following differential operators were used

$$\Delta = \frac{\partial^2}{\partial x^2} + \frac{\partial^2}{\partial y^2}, \qquad \Lambda = \frac{\partial}{\partial x} + i\frac{\partial}{\partial y}, \tag{2.5}$$

and the overbar everywhere indicates the complex conjugate value. Note also that $\Delta = \Lambda\bar{\Lambda}$. One can verify that equations (2.4) can be satisfied by

$$u = \Lambda(F_1 + F_2 + iF_3), \qquad w = m_1\frac{\partial F_1}{\partial z} + m_2\frac{\partial F_2}{\partial z}, \tag{2.6}$$

where all three functions F_k satisfy the equation (Elliott, 1948)

$$\Delta F_k + \gamma_k^2\frac{\partial^2 F_k}{\partial z^2} = 0, \qquad \text{for } k = 1, 2, 3. \tag{2.7}$$

The values of m_k and γ_k are related by the following expression (Elliott 1948)

$$\frac{A_{44} + m_k(A_{13}+A_{44})}{A_{11}} = \frac{m_k A_{33}}{m_k A_{44} + A_{13} + A_{44}} = \gamma_k^2, \qquad \text{for } k = 1, 2;$$
$$\gamma_3 = (A_{44}/A_{66})^{1/2}. \tag{2.8}$$

Introducing the notation $z_k = z/\gamma_k$, for $k = 1, 2, 3$, we may call the function $F_k = F(x, y, z_k)$ harmonic. Note the property $m_1 m_2 = 1$, which seems to have escaped the attention of previous researchers, and which will help us to simplify various expressions to follow. The other elastic constants that will be used throughout the chapter are

$$G_1 = \beta + \gamma_1\gamma_2 H, \qquad G_2 = \beta - \gamma_1\gamma_2 H,$$
$$H = \frac{(\gamma_1+\gamma_2)A_{11}}{2\pi(A_{11}A_{33} - A_{13}^2)}, \qquad \alpha = \frac{(A_{11}A_{33})^{1/2} - A_{13}}{A_{11}(\gamma_1+\gamma_2)}, \qquad \beta = \frac{\gamma_3}{2\pi A_{44}}. \tag{2.9}$$

Introduce the following parameters

$$\sigma_1 = \sigma_x + \sigma_y, \qquad \sigma_2 = \sigma_x - \sigma_y + 2i\tau_{xy}, \qquad \tau_z = \tau_{zx} + i\tau_{yz}. \tag{2.10}$$

This will simplify expressions (2.1), giving

$$\sigma_1 = (A_{11} - A_{66})(\bar{\Lambda}u + \Lambda\bar{u}) + 2A_{13}\frac{\partial w}{\partial z}, \qquad \sigma_2 = 2A_{66}\,\Lambda u,$$
$$\sigma_z = \frac{1}{2}A_{13}(\bar{\Lambda}u + \Lambda\bar{u}) + A_{33}\frac{\partial w}{\partial z}, \qquad \tau_z = A_{44}\left(\frac{\partial u}{\partial z} + \Lambda w\right). \tag{2.11}$$

We have now only four components of stress, instead of six, as it was in (2.1). The substitution of (2.6) in (2.11) yields

$$\begin{aligned}
\sigma_1 &= 2A_{66}\frac{\partial^2}{\partial z^2}\{[\gamma_1^2 - (1+m_1)\gamma_3^2]F_1 + [\gamma_2^2 - (1+m_2)\gamma_3^2]F_2\},\\
\sigma_2 &= 2A_{66}\,\Lambda^2(F_1 + F_2 + iF_3),\\
\sigma_z &= A_{44}\frac{\partial^2}{\partial z^2}[(1+m_1)\gamma_1^2F_1 + (1+m_2)\gamma_2^2F_2]\\
&= -A_{44}\,\Delta[(1+m_1)F_1 + (1+m_2)F_2],\\
\tau_z &= A_{44}\,\Lambda\frac{\partial}{\partial z}[(1+m_1)F_1 + (1+m_2)F_2 + iF_3].
\end{aligned} \tag{2.12}$$

Here we used the fact that each F_k satisfies equation (2.7), and the relation (for $k = 1, 2$) $A_{11}\gamma_k^2 - A_{13}m_k = A_{44}(1 + m_k)$, which is an immediate consequence of (2.8). Expressions (2.6) and (2.12) give a general solution, expressed in terms of three harmonic functions F_k. It is very attractive to express each function F_k through just *one* harmonic function as follows:

$$F_k(x, y, z) = c_k F(x, y, z_k), \tag{2.13}$$

where $z_k = z/\gamma_k$, and c_k is an as yet unknown complex constant. As we shall see further, this is possible indeed.

B. Flat Crack under Arbitrary Normal Loading

Consider a transversely isotropic elastic space weakened by a flat crack S in the plane $z = 0$, with arbitrary pressure p applied to the crack faces. Due to symmetry, the problem can be formulated as follows: find the solution to the set of differential equations (2.3) for a half-space $z \geq 0$, subject to the mixed boundary conditions on the plane $z = 0$

$$\sigma_z = -p(x, y) \quad \text{for} \quad (x, y) \in S, \qquad w = 0 \quad \text{for} \quad (x, y) \notin S,$$
$$\tau_z = 0 \quad \text{for} \quad -\infty < (x, y) < \infty. \tag{2.14}$$

These conditions can be satisfied by one harmonic function. Let us put, according to (2.13),

$$F_1(z) = c_1 F(z_1), \qquad F_2(z) = c_2 F(z_2), \qquad F_3(z) = 0. \tag{2.15}$$

Expressions of the type $F_1(z)$ and $F(z_1)$, etc., everywhere in this article should be understood as $F_1(x, y, z)$ and $F(x, y, z_1)$ respectively. The substitution of (2.15) and (2.12) in the third condition (2.14) yields

$$c_1 = -\frac{c_2 \gamma_1}{m_1 \gamma_2}. \tag{2.16}$$

We can present the function F through the potential of a simple layer, namely,

$$F(\rho, \phi, z) \equiv F(z) = \iint_S \frac{\omega(N)}{R(M, N)}\, dS, \tag{2.17}$$

where ω stands for the crack face displacement $w(x, y, 0)$, $R(M, N)$ is the distance between the points $M(\rho, \phi, z)$ and $N(r, \psi, 0)$, and the integration is taken over the crack domain S. Expression (2.17) satisfies the second condition (2.14) identically, due to the well-known property of the potential of a simple layer. Inside the crack the same property gives

$$\left.\frac{\partial F}{\partial z}\right|_{z=0} = -2\pi\omega = -2\pi w(x, y, 0). \tag{2.18}$$

Now expressions (2.15), (2.17), (2.18), and (2.6) give the second equation for c_1 and c_2

$$-\frac{m_1 c_1}{\gamma_1} - \frac{m_2 c_2}{\gamma_2} = \frac{1}{2\pi}. \tag{2.19}$$

The constants c_1 and c_2 are defined from (2.16) and (2.19) as

$$c_1 = -\frac{\gamma_1}{2\pi(m_1 - 1)}, \qquad c_2 = -\frac{\gamma_2}{2\pi(m_2 - 1)}. \tag{2.20}$$

The potential functions will be defined by

$$F_1(z) = -\frac{\gamma_1}{2\pi(m_1 - 1)} F(z_1), \qquad F_2(z) = -\frac{\gamma_2}{2\pi(m_2 - 1)} F(z_2). \tag{2.21}$$

The substitution of (2.21) and (2.12) in the first condition (2.14) leads to the governing integral equation

$$p(N_0) = -\frac{1}{4\pi^2 H}\Delta \iint_S \frac{\omega(N)}{R(N_0, N)}\,dS, \tag{2.22}$$

where, as before, $R(N_0, N)$ stands for the distance between two points N_0 and N, and both $N_0, N \in S$. The following identities were used

$$m_1 m_2 = 1, \qquad \frac{m_1 - 1}{m_1 + 1} = 2\pi A_{44} H(\gamma_1 - \gamma_2). \tag{2.23}$$

The case of a general crack under a uniform pressure was considered by Fabrikant (1987a), where an approximate analytical solution was presented. A similar method can be used in the case of arbitary polynomial loading. An exact solution in elementary functions is possible when the crack is circular. Let a be the radius of the crack. The governing integral equation (2.21) can be rewritten in polar coordinates as follows (see Fabrikant (1986a)):

$$p(\rho, \phi) = -\frac{1}{\pi^2 H \rho} L\left(\frac{1}{\rho}\right) \frac{d}{d\rho} \int_0^\rho \frac{x\,dx}{(\rho^2 - x^2)^{1/2}} L(x^2) \frac{d}{dx}$$
$$\times \int_x^a \frac{\rho_0\,d\rho_0}{(\rho_0^2 - x^2)^{1/2}} L\left(\frac{1}{\rho_0}\right) \omega(\rho_0, \phi). \tag{2.24}$$

Here the L-operator is defined by

$$L(k) f(\phi) = \frac{1}{2\pi} \int_0^{2\pi} \lambda(k, \phi - \phi_0) f(\phi_0)\,d\phi_0 \tag{2.25}$$

and

$$\lambda(k, \psi) = \frac{1 - k^2}{1 + k^2 - 2k\cos\psi}. \tag{2.26}$$

The integral operator inverse to (2.24) was derived by Fabrikant (1986a), and is

$$\omega(\rho, \phi) = 4H \int_\rho^a \frac{dx}{(x^2 - \rho^2)^{1/2}} \int_0^x \frac{\rho_0\,d\rho_0}{(x^2 - \rho_0^2)^{1/2}} L\left(\frac{\rho\rho_0}{x^2}\right) p(\rho_0, \phi). \tag{2.27}$$

Another form of solution was derived by Fabrikant (1987b):

$$\omega = \frac{2}{\pi} H \int_0^{2\pi} \int_0^a \frac{p(\rho_0, \phi_0)}{R} \tan^{-1}\left(\frac{\eta}{R}\right) \rho_0\,d\rho_0\,d\phi_0, \tag{2.28}$$

where

$$R = [\rho^2 + \rho_0^2 - 2\rho\rho_0\cos(\phi - \phi_0)]^{1/2}, \qquad \eta = \frac{1}{a}(a^2 - \rho^2)^{1/2}(a^2 - \rho_0^2)^{1/2} \tag{2.29}$$

In this article, we do not stop at considering the situation in the plane $z = 0$: our purpose is obtaining a *complete* solution. We shall call $F(\rho, \phi, z)$, as defined by (2.17), the main potential function since both functions F_1 and F_2 become easily available, when F is found. The substitution of (2.28) in (2.17) allows us to express the main potential function as follows:

$$F(\rho, \phi, z) = \frac{2}{\pi}H\int_0^{2\pi}\int_0^a K(\rho, \phi, z; \rho_0, \phi_0)p(\rho_0, \phi_0)\rho_0\, d\rho_0\, d\phi_0, \tag{2.30}$$

where the Green's function K reads

$$K(M; N_0) = K(\rho, \phi, z; \rho_0, \phi_0)$$
$$= \int_0^{2\pi}\int_0^a \frac{1}{R(N, N_0)}\tan^{-1}\left[\frac{(a^2 - r^2)^{1/2}(a^2 - \rho_0^2)^{1/2}}{aR(N, N_0)}\right]\frac{r\, dr\, d\psi}{R(M, N)}. \tag{2.31}$$

Here $R(\cdot, \cdot)$ stands for the distance between respective points: $M(\rho, \phi, z)$, $N(r, \psi, 0)$, and $N_0(\rho_0, \phi_0, 0)$. Although we can not compute the integral in (2.31) in elementary functions, all its derivatives can be expressed in elementary functions, due to the fundamental integral established in Appendix C. Making use of (C.19), one can write

$$\frac{\partial K}{\partial z} = -\frac{2\pi}{R(M, N_0)}\tan^{-1}\left[\frac{h}{R(M, N_0)}\right], \tag{2.32}$$

where

$$h = \frac{1}{a}(a^2 - l_1^2)^{1/2}(a^2 - \rho_0^2)^{1/2}. \tag{2.33}$$

and the contraction l_1 everywhere in this article stands for $l_1(a)$, as it is defined by (1.8). Note that h tends to η, as defined by (2.29), for $z \to 0$ and $\rho < a$. Expressions (2.30) and (2.32) allow us to write

$$\frac{\partial F}{\partial z} = -4H\int_0^{2\pi}\int_0^a \frac{1}{R(M, N_0)}\tan^{-1}\left[\frac{h}{R(M, N_0)}\right]p(\rho_0, \phi_0)\rho_0\, d\rho_0\, d\phi_0. \tag{2.34}$$

The integral in (2.34), although looking difficult to compute even for $p = \text{const}$, can be expressed in elementary functions for any polynomial loading. This becomes evident, if we use the equivalent representations through the L-operator (see Fabrikant (1986b) for details):

$$\frac{\partial F}{\partial z} = -4H\int_{l_2(0)}^{l_2} \frac{dx}{(x^2-\rho^2)^{1/2}} \int_0^{g(x)} \frac{\rho_0\, d\rho_0}{[g^2(x)-\rho_0^2]^{1/2}} L\left(\frac{\rho\rho_0}{x^2}\right) p(\rho_0, \phi). \quad (2.35)$$

Here

$$g(x) = x\left[1 + \frac{z^2}{\rho^2 - x^2}\right]^{1/2}, \quad (2.36)$$

and the contraction l_2 everywhere in this article stands for $l_2(a)$, as it is defined by (1.4). Using the changes of variables $x = l_2(t)$, $t = g(x)$, expression (2.35) can be rewritten as follows:

$$\frac{\partial F}{\partial z} = -4H\int_0^a \frac{dl_2(t)}{[l_2^2(t)-\rho^2]^{1/2}} \int_0^t \frac{\rho_0\, d\rho_0}{(t^2-\rho_0^2)^{1/2}} L\left(\frac{\rho\rho_0}{l_2^2(t)}\right) p(\rho_0, \phi). \quad (2.37)$$

Since the function F vanishes at infinity, it can be defined from (2.37):

$$F(\rho, \phi, z) = -4H\int_\infty^z dz \int_0^a \frac{dl_2(t)}{[l_2^2(t)-\rho^2]^{1/2}} \int_0^t \frac{\rho_0\, d\rho_0}{(t^2-\rho_0^2)^{1/2}} L\left(\frac{\rho\rho_0}{l_2^2(t)}\right) p(\rho_0, \phi). \quad (2.38)$$

By using the property

$$\frac{\partial l_2(t)}{\partial t} = -\frac{[l_2^2(t)-\rho^2]^{1/2}}{[\rho^2 - l_1^2(t)]^{1/2}} \frac{\partial l_1(t)}{\partial z},$$

which is a consequence of formulae (A.27) and (A.28), expression (2.38) can be modified:

$$F(\rho, \phi, z) = 4H\int_0^a dt \int_0^t \frac{\rho_0\, d\rho_0}{(t^2-\rho_0^2)^{1/2}} \int_0^{l_1(t)} \frac{dy}{(\rho^2-y^2)^{1/2}} L\left(\frac{y^2\rho_0}{t^2\rho}\right) p(\rho_0, \phi). \quad (2.39)$$

Expression (2.39) is convenient for exact evaluation of the potential function F and proves that it can be expressed in elementary functions for arbitrary polynomial loading. A simple change of variables gives another formula, equivalent to (2.39),

$$F(\rho, \phi, z) = 4H\int_0^a t\,dt\int_0^t \frac{\rho_0\,d\rho_0}{(t^2-\rho_0^2)^{1/2}}\int_{l_2(t)}^\infty \frac{dx}{x(x^2-t^2)^{1/2}}L\left(\frac{\rho\rho_0}{x^2}\right)p(\rho_0, \phi). \tag{2.40}$$

We can proceed now with the remaining derivatives of the Green's function K, defined by (2.31). Differentiation of (2.31) yields

$$\Lambda K(\rho, \phi, z; \rho_0, \phi_0) = -\int_0^{2\pi}\int_0^a \frac{\rho e^{i\phi} - re^{i\psi}}{R^3(M, N)} \times \tan^{-1}\left[\frac{(a^2-r^2)^{1/2}(a^2-\rho_0^2)^{1/2}}{aR(N, N_0)}\right]\frac{r\,dr\,d\psi}{R(N, N_0)}. \tag{2.41}$$

This integral is computed in Appendix E. By using (E.11), one can write

$$\Lambda K(\rho, \phi, z; \rho_0, \phi_0) = \frac{2\pi}{\bar{q}}\left\{\frac{z}{R_0}\tan^{-1}\left(\frac{h}{R_0}\right) - \frac{(a^2-\rho_0^2)^{1/2}}{\bar{s}}\tan^{-1}\left[\frac{\bar{s}}{(l_2^2-a^2)^{1/2}}\right]\right\}, \tag{2.42}$$

where h is defined by (2.33) and

$$\bar{q} = \rho e^{-i\phi} - \rho_0 e^{-i\phi_0}, \qquad \bar{s} = (a^2 - \rho\rho_0 e^{-i(\phi-\phi_0)})^{1/2},$$
$$R_0 = R(M, N_0) = [\rho^2 + \rho_0^2 - 2\rho\rho_0\cos(\phi-\phi_0) + z^2]^{1/2}. \tag{2.43}$$

The other derivatives, which will be needed for the complete solution, are

$$\frac{\partial^2}{\partial z^2}K(\rho, \phi, z; \rho_0, \phi_0) = 2\pi\left[\frac{z}{R_0^3}\tan^{-1}\left(\frac{h}{R_0}\right) - \frac{h}{z(R_0^2+h^2)}\left(\frac{\rho^2-l_1^2}{l_2^2-l_1^2} - \frac{z^2}{R_0^2}\right)\right], \tag{2.44}$$

$$\frac{\partial}{\partial z}\Lambda K(\rho, \phi, z; \rho_0, \phi_0) = 2\pi\left[\frac{\rho e^{i\phi} - \rho_0 e^{i\phi_0}}{R_0^3}\tan^{-1}\left(\frac{h}{R_0}\right) + \frac{h}{R_0^2+h^2}\left(\frac{\rho e^{i\phi}}{l_2^2-l_1^2} + \frac{\rho e^{i\phi}-\rho_0 e^{i\phi_0}}{R_0^2}\right)\right], \tag{2.45}$$

$$\begin{aligned}\Lambda^2 K(\rho, \phi, z; \rho_0, \phi_0) = 2\pi\Bigg\{&\frac{(a^2-\rho_0^2)^{1/2}}{\bar{q}\bar{s}}\left(\frac{2}{\bar{q}} - \frac{\rho_0 e^{i\phi_0}}{\bar{s}^2}\right)\tan^{-1}\left[\frac{\bar{s}}{(l_2^2-a^2)^{1/2}}\right] \\ &- \frac{z(3R_0^2-z^2)}{\bar{q}^2R_0^3}\tan^{-1}\left(\frac{h}{R_0}\right) \\ &+ \frac{(a^2-\rho_0^2)^{1/2}(l_2^2-a^2)^{1/2}\rho_0 e^{i\phi_0}}{\bar{q}\bar{s}^2[l_2^2-\rho\rho_0 e^{-i(\phi-\phi_0)}]} \\ &- \frac{zh}{R_0^2+h^2}\left[\frac{q}{\bar{q}R_0^2} - \frac{\rho^2 e^{2i\phi}}{(l_2^2-l_1^2)(l_2^2-\rho^2)}\right]\Bigg\}.\end{aligned} \tag{2.46}$$

This concludes the general solution to the problem of a penny-shaped crack subjected to arbitrary pressure. Formulae (2.32) and (2.39)–(2.46) are the main new results of this section.

C. Application 1: Point Force Loading

Consider a penny-shaped crack opened by two equal concentrated forces P applied in opposite directions at the point $(\rho_0, \phi_0, 0^{\pm})$, $\rho_0 < a$. Formulae (2.6), (2.12), (2.21), (2.32), and (2.39)–(2.42) give a complete solution for the field of displacements and stresses in elementary functions, namely

$$u = \frac{2}{\pi} HP \left[\frac{\gamma_1}{m_1 - 1} f_1(z_1) + \frac{\gamma_2}{m_2 - 1} f_1(z_2) \right], \tag{2.47}$$

$$w = \frac{2}{\pi} HP \left[\frac{m_1}{m_1 - 1} f_2(z_1) + \frac{m_2}{m_2 - 1} f_2(z_2) \right], \tag{2.48}$$

$$\sigma_1 = \frac{2P}{\pi^2(\gamma_1 - \gamma_2)} \left\{ \left[\frac{\gamma_1}{(m_1 + 1)\gamma_3^2} - \frac{1}{\gamma_1} \right] f_3(z_1) - \left[\frac{\gamma_2}{(m_2 + 1)\gamma_3^2} - \frac{1}{\gamma_2} \right] f_3(z_2) \right\}, \tag{2.49}$$

$$\sigma_2 = \frac{4}{\pi} HA_{66} P \left[\frac{\gamma_1}{m_1 - 1} f_4(z_1) + \frac{\gamma_2}{m_2 - 1} f_4(z_2) \right], \tag{2.50}$$

$$\sigma_z = \frac{P}{\pi^2(\gamma_1 - \gamma_2)} [\gamma_1 f_3(z_1) - \gamma_2 f_3(z_2)], \tag{2.51}$$

$$\tau_z = \frac{P}{\pi^2(\gamma_1 - \gamma_2)} [f_5(z_1) - f_5(z_2)], \tag{2.52}$$

where

$$f_1(z) = \frac{1}{\bar{q}} \left\{ \frac{(a^2 - \rho_0^2)^{1/2}}{\bar{s}} \tan^{-1}\left[\frac{\bar{s}}{(l_2^2 - a^2)^{1/2}} \right] - \frac{z}{R_0} \tan^{-1}\left(\frac{h}{R_0} \right) \right\}, \tag{2.53}$$

$$f_2(z) = \frac{1}{R_0} \tan^{-1}\left(\frac{h}{R_0} \right), \tag{2.54}$$

$$f_3(z) = -\frac{z}{R_0^3} \tan^{-1}\left(\frac{h}{R_0} \right) + \frac{h}{z(R_0^2 + h^2)} \left(\frac{\rho^2 - l_1^2}{l_2^2 - l_1^2} - \frac{z^2}{R_0^2} \right), \tag{2.55}$$

$$f_4(z) = \frac{(a^2-\rho_0^2)^{1/2}}{\bar{q}\bar{s}}\left(\frac{\rho_0 e^{i\phi_0}}{\bar{s}^2} - \frac{2}{\bar{q}}\right)\tan^{-1}\left(\frac{\bar{s}}{(l_2^2-a^2)^{1/2}}\right)$$
$$+\frac{z(3R_0^2-z^2)}{\bar{q}^2R_0^3}\tan^{-1}\left(\frac{h}{R_0}\right) - \frac{(a^2-\rho_0^2)^{1/2}(l_2^2-a^2)^{1/2}\rho_0 e^{i\phi_0}}{\bar{q}\bar{s}^2(l_2^2-\rho\rho_0 e^{-i(\phi-\phi_0)})}$$
$$+\frac{zh}{R_0^2+h^2}\left[\frac{q}{\bar{q}R_0^2} - \frac{\rho^2 e^{2i\phi}}{(l_2^2-l_1^2)(l_2^2-\rho^2)}\right], \tag{2.56}$$

$$f_5(z) = -\frac{\rho e^{i\phi}-\rho_0 e^{i\phi_0}}{R_0^3}\tan^{-1}\left(\frac{h}{R_0}\right) - \frac{h}{R_0^2+h^2}\left(\frac{\rho e^{i\phi}}{l_2^2-l_1^2} + \frac{\rho e^{i\phi}-\rho_0 e^{i\phi_0}}{R_0^2}\right). \tag{2.57}$$

Here $R_0 = [\rho^2+\rho_0^2-2\rho\rho_0\cos(\phi-\phi_0)+z^2]^{1/2}$. It is believed that this problem has not been solved before. The expression for σ_z (2.51) simplifies when $z=0$ and $\rho > a$, namely,

$$\sigma_z = \frac{P}{\pi^2}\frac{(a^2-\rho_0^2)^{1/2}}{(\rho^2-a^2)^{1/2}[\rho^2+\rho_0^2-2\rho\rho_0\cos(\phi-\phi_0)]}. \tag{2.58}$$

Defining the stress intensity factor

$$k_1 = \lim_{\rho\to a}[(\rho-a)^{1/2}\sigma_z],$$

the following result may be obtained from (2.58):

$$k_1 = \frac{P}{\pi^2(2a)^{1/2}}\frac{(a^2-\rho_0^2)^{1/2}}{a^2+\rho_0^2-2a\rho_0\cos(\phi-\phi_0)}. \tag{2.59}$$

One can write for arbitrarily distributed pressure,

$$k_1 = \frac{1}{\pi^2(2a)^{1/2}}\int_0^{2\pi}\int_0^a \frac{(a^2-\rho_0^2)^{1/2}p(\rho_0,\phi_0)\rho_0\,d\rho_0\,d\phi_0}{a^2+\rho_0^2-2a\rho_0\cos(\phi-\phi_0)},$$

which corresponds to the well-known result (Cherepanov, 1974).

D. Application 2: Concentrated Load outside a Circular Crack

Consider transversely isotroic space weakened by a penny-shaped crack of radius a in the plane $z=0$. Let a concentrated force P be applied at arbitrary point (ρ, ϕ, z) in the Oz direction. The crack faces are stress-free. The problem has not been considered before, even for an isotropic body. A complete solution is beyond the scope of this article and will be published separately. Here we find the crack opening displacement and the opening mode stress intensity factor k_1 only.

Consider the second system in equilibrium: two unit concentrated forces Q applied normally to the crack faces in opposite directions at the point $(\rho_0, \phi_0, 0^{\pm})$. Denote the normal displacement in the space due to the forces Q by w_Q; w_P is the crack opening displacement $w(\rho_0, \phi_0, 0^+) - w(\rho_0, \phi_0, 0^-)$ due to force P. Application of the reciprocal theorem yields

$$Qw_P = Pw_Q,$$

which gives the crack opening displacement

$$w_P(\rho_0, \phi_0) = \frac{2}{\pi} HP\left[\frac{m_1}{m_1 - 1} f_2(z_1) + \frac{m_2}{m_2 - 1} f_2(z_2)\right], \tag{2.60}$$

with f_2 defined by (2.54). The stress intensity factor can be defined by

$$k_1(\phi_0) = \frac{1}{8\pi H} \lim_{\rho_0 \to a} \frac{w_P(\rho_0, \phi_0)}{(a - \rho_0)^{1/2}} = \frac{P}{2(2a)^{1/2}\pi^2}\left[\frac{m_1}{m_1 - 1} f_6(z_1) + \frac{m_2}{m_2 - 1} f_6(z_2)\right],$$

where

$$f_6(z) = \frac{(a^2 - l_1^2)^{1/2}}{r_a^2}, \qquad r_a^2 = \rho^2 + a^2 - 2\rho a \cos(\phi - \phi_0) + z^2. \tag{2.61}$$

The stress intensity factor vanishes as z tends to zero for $\rho \geq a$.

In the case of an isotropic body, expression (2.60) transforms into

$$w_P(\rho_0, \phi_0) = \frac{P}{\pi^2 \mu}\left\{\frac{1 - \nu}{R_0} \tan^{-1}\left(\frac{h}{R_0}\right) - \frac{1}{2}\left[\frac{h}{R_0^2 + h^2}\left(\frac{\rho^2 - l_1^2}{l_2^2 - l_1^2} - \frac{z^2}{R_0^2}\right) - \frac{z^2}{R_0^3} \tan^{-1}\left(\frac{h}{R_0}\right)\right]\right\}. \tag{2.62}$$

Here μ is the shear modulus, and ν is Poisson's ratio. The corresponding expression for the stress intensity factor will take the form

$$k_1(\phi_0) = \frac{P(a^2 - l_1^2)^{1/2}}{(8a)^{1/2}\pi^2 r_a^2}\left[1 + \frac{1}{1 - \nu}\left(\frac{z^2}{r_a^2} - \frac{\rho^2 - l_1^2}{2(l_2^2 - l_1^2)}\right)\right]. \tag{2.63}$$

In the case of axial symmetry $\rho = 0$, and formulae (2.62) and (2.63) simplify as follows:

$$w_P(\rho_0, \phi_0) = \frac{P}{\pi^2\mu}\left\{\left[\frac{1-\nu}{(\rho_0^2+z^2)^{1/2}} + \frac{z^2}{(\rho_0^2+z^2)^{3/2}}\right]\tan^{-1}\left(\frac{a^2-\rho_0^2}{\rho_0^2+z^2}\right)^{1/2} + \frac{z^2(a^2-\rho_0^2)^{1/2}}{2(\rho_0^2+z^2)(a^2+z^2)}\right\},$$

$$k_1 = \frac{Pa^{1/2}}{2\pi^2\sqrt{2}\,(a^2+z^2)}\left[1 + \frac{1}{1-\nu}\frac{z^2}{a^2+z^2}\right],$$

which is in agreement with the results reported by Collins (1962), who considered the axisymmetric case only.

E. Contact Problem for a Smooth Punch

A smooth punch is pressed against a transversely isotropic elastic half-space $z \geq 0$ by a normal force P. Let S denote the domain of contact. The mixed boundary conditions on the plane $z = 0$ are

$$\sigma_z = 0 \quad \text{for} \quad (\rho, \phi) \notin S, \qquad w = \omega(\rho, \phi) \quad \text{for} \quad (\rho, \phi) \in S,$$
$$\tau_z = 0 \quad \text{for} \quad -\infty < (x, y) < \infty. \tag{2.64}$$

We may assume again that

$$F_1(z) = c_1 F(z_1), \qquad F_2(z) = c_2 F(z_2), \qquad F_3(z) = 0. \tag{2.65}$$

Substitution of (2.65) and (2.12) in the third condition (2.64) yields

$$c_1 = -\frac{c_2\gamma_1}{m_1\gamma_2}. \tag{2.66}$$

Now we need to define the main potential function F so that its second z-derivative vanishes at $z = 0$ outside the domain of contact. Comparison with (2.17) gives

$$F(\rho, \phi, z) = F(M) = \iint_S \ln[R(M, N) + z]\sigma(N)\, dS_N. \tag{2.67}$$

The simple layer potential property yields, inside the domain of contact,

$$\left.\frac{\partial^2 F}{\partial z^2}\right|_{z=0} = -2\pi\sigma, \tag{2.68}$$

which, combined with (2.12), gives the second equation for the constants c_1 and c_2

$$2\pi A_{44}[(1+m_1)c_1+(1+m_2)c_2]=1. \tag{2.69}$$

Equations (2.66) and (2.69) define the constants

$$c_1=\frac{H\gamma_1}{m_1-1}, \qquad c_2=\frac{H\gamma_2}{m_2-1}. \tag{2.70}$$

The simplifications made in (2.70) are due to the properties (2.23).

Finally, substitution of (2.65), (2.67), and (2.70) in the second equation of (2.6) yields for $z=0$ the well-known governing integral equation of elastic contact problem for a smooth punch:

$$\omega(N_0)=H\iint_S \frac{\sigma(N)}{R(N,N_0)}\,dS_N. \tag{2.71}$$

An approximate analytical solution of (2.71) for a flat punch of arbitrary shape can be found in Fabrikant (1986c, 1986d). The case of a curved punch was considered in Fabrikant (1987d). An exact solution in closed form is available for a circular punch of arbitrary profile (Fabrikant, 1986b). The integral equation (2.71) can be rewritten for a circular domain of contact of radius a as follows:

$$4H\int_0^\rho \frac{dx}{(\rho^2-x^2)^{1/2}}\int_x^a \frac{\rho_0\,d\rho_0}{(\rho_0^2-x^2)^{1/2}}L\left(\frac{x^2}{\rho\rho_0}\right)\sigma(\rho_0,\phi)=\omega(\rho,\phi).$$

Its exact closed form solution is (Fabrikant, 1986b)

$$\sigma(\rho,\phi)=-\frac{1}{\pi^2H\rho}L(\rho)\frac{d}{d\rho}\int_\rho^a \frac{x\,dx}{(x^2-\rho^2)^{1/2}}L\left(\frac{1}{x^2}\right)\frac{d}{dx}$$
$$\times\int_0^x \frac{\rho_0\,d\rho_0}{(x^2-\rho_0^2)^{1/2}}L(\rho_0)\omega(\rho_0,\phi). \tag{2.72}$$

The potential function F can be found in two stages. First of all, one has from (2.67)

$$\frac{\partial F}{\partial z}=\int_0^{2\pi}\int_0^a \frac{\sigma(r,\psi)r\,dr\,d\psi}{[\rho^2+r^2-2r\rho\cos(\phi-\psi)+z^2]^{1/2}}. \tag{2.73}$$

Substitution of (2.72) in (2.73) yields, after integration (see Appendix F for details),

$$\frac{\partial F}{\partial z}=\frac{1}{\pi^2 H}\int_0^{2\pi}\int_0^a\left[\frac{R_0}{h}+\tan^{-1}\left(\frac{h}{R_0}\right)\right]\frac{z}{R_0^3}\,\omega(\rho_0,\phi_0)\rho_0\,d\rho_0\,d\phi_0. \quad (2.74)$$

Here $R_0=[\rho^2+\rho_0^2-2\rho\rho_0\cos(\phi-\phi_0)+z^2]^{1/2}$, and h is defined by (2.33). The next integration of (2.74) with respect to z gives the expression for the potential function directly through the prescribed displacement under the punch, namely,

$$F(\rho,\phi,z)=\frac{1}{\pi^2 H}\int_0^{2\pi}\int_0^a \mathcal{K}(\rho,\phi,z;\rho_0,\phi_0)\omega(\rho_0,\phi_0)\rho_0\,d\rho_0\,d\phi_0, \quad (2.75)$$

where

$$\mathcal{K}(\rho,\phi,z;\rho_0,\phi_0)=-\frac{1}{R_0}\tan^{-1}\left(\frac{h}{R_0}\right)+\frac{1}{(a^2-\rho_0^2)^{1/2}}\left\{\ln[l_2+(l_2^2-\rho^2)^{1/2}]\right.$$
$$\left.-2\Re\left(\frac{1}{(\zeta-1)^{1/2}}\left[\tan^{-1}\left(\frac{a(\zeta-1)^{1/2}}{(a^2-l_1^2)^{1/2}}\right)-\tan^{-1}(\zeta-1)^{1/2}\right]\right)\right\},$$
$$\zeta=\frac{\rho}{\rho_0}e^{i(\phi-\phi_0)}. \quad (2.76)$$

The details of integation are presented in Appendix F. Now all the Green's functions, related to the field of displacements and stresses can be obtained from (2.76) by differentiation in elementary functions:

$$\frac{\partial\mathcal{K}}{\partial z}=\frac{z}{R_0^3}\left[\frac{R_0}{h}+\tan^{-1}\left(\frac{h}{R_0}\right)\right], \quad (2.77)$$

$$\Lambda\mathcal{K}=\frac{q}{R_0^3}\left[\frac{R_0}{h}+\tan^{-1}\left(\frac{h}{R_0}\right)\right]+\frac{1}{h\bar{q}}\left\{1-\frac{(a^2-l_1^2)^{1/2}}{a}\right.$$
$$\left.-\frac{(a^2-l_1^2)^{1/2}}{a(\bar{\zeta}-1)^{1/2}}\left[\tan^{-1}\left(\frac{a(\bar{\zeta}-1)^{1/2}}{(a^2-l_1^2)^{1/2}}\right)-\tan^{-1}(\bar{\zeta}-1)^{1/2}\right]\right\}, \quad (2.78)$$

$$\frac{\partial^2\mathcal{K}}{\partial z^2}=\frac{1}{R_0^3}\left(1-\frac{3z^2}{R_0^2}\right)\left[\frac{R_0}{h}+\tan^{-1}\left(\frac{h}{R_0}\right)\right]+\frac{1}{h(R_0^2+h^2)}\left[\frac{z^2}{R_0^2}-\frac{\rho^2-l_1^2}{l_2^2-l_1^2}\right], \quad (2.79)$$

$$\Lambda\frac{\partial\mathcal{K}}{\partial z}=-\frac{3zq}{R_0^5}\left[\frac{R_0}{h}+\tan^{-1}\left(\frac{h}{R_0}\right)\right]+\frac{z}{h(R_0^2+h^2)}\left[\frac{\rho e^{i\phi}}{l_2^2-l_1^2}+\frac{q}{R_0^2}\right], \quad (2.80)$$

$$\Lambda^2\mathcal{K}=-\frac{3q^2}{R_0^5}\tan^{-1}\left(\frac{h}{R_0}\right)+\frac{1}{h(R_0^2+h^2)}\left[\frac{q^2}{R_0^2}-\frac{l_2^2-\rho^2}{l_2^2-l_1^2}e^{2i\phi}\right]$$
$$+\frac{3}{\bar{q}^2}\left\{\frac{1}{h}-\frac{1}{(a^2-\rho_0^2)^{1/2}}\left(1+\frac{1}{(\bar{\zeta}-1)^{1/2}}\left[\tan^{-1}\left(\frac{a(\bar{\zeta}-1)^{1/2}}{(a^2-l_1^2)^{1/2}}\right)-\tan^{-1}(\bar{\zeta}-1)^{1/2}\right]\right)\right\}$$
$$+\frac{e^{i\phi}}{\rho\bar{q}(a^2-\rho_0^2)^{1/2}}\left[1-\frac{a^3\bar{\zeta}}{(a^2-l_1^2)^{1/2}(a^2\bar{\zeta}-l_1^2)}\right]. \quad (2.81)$$

The following identities were used in the simplification of (2.77)–(2.81):

$$(l_2^2-\rho_0^2)(a^2-l_1^2)+a^2q\bar{q}=a^2(R_0^2+h^2),$$

$$(a^2\zeta-l_1^2)(a^2\bar{\zeta}-l_1^2)=\frac{(R_0^2+h^2)l_1^2a^2}{\rho_0^2},$$

$$(l_2^2-\rho_0^2)(q\bar{q}+z^2)-q\bar{q}(a^2-\rho_0^2)=(l_2^2-a^2)(R_0^2+h^2).$$

Formulae (2.76)–(2.81) are the main new results of this section. They give a complete solution for the considered contact problem when combined with formulae (2.6), (2.12), (2.65), and (2.70). When the prescribed displacement can be expressed as a sum of powers in x and y, the complete solution is elementary. Some particular examples will be considered in Section III.

F. Plane Crack under Arbitrary Shear Loading

Consider a transversely isotropic elastic space weakened by a flat crack S in the plane $z=0$, with arbitrary shear loading applied to the crack faces antisymmetrically. The problem can be formulated as follows: find the solution to the set of differential equations (2.3) for a half-space $z\geq 0$, subject to the mixed boundary conditions on the plane $z=0$:

$$\tau_z=-\tau(x,y)\quad\text{for}\quad(x,y)\in S,\qquad u=0\quad\text{for}\quad(x,y)\notin S,$$
$$\sigma_z=0\quad\text{for}\quad-\infty<(x,y)<\infty. \tag{2.82}$$

It is no longer possible to present the solution in the form (2.15). A more complicated representation is necessary, namely,

$$F_1=c_1(\Lambda\bar{\chi}_1+\bar{\Lambda}\chi_1),\qquad F_2=c_2(\Lambda\bar{\chi}_2+\bar{\Lambda}\chi_2),\qquad F_3=c_3(\Lambda\bar{\chi}_3-\bar{\Lambda}\chi_3). \tag{2.83}$$

Here c_1, c_2, and c_3 are the as yet unknown constants; χ_1, χ_2, and χ_3 are the as yet unknown complex harmonic functions. Introducing the notation $z_k=z/\gamma_k$ for $k=1, 2, 3$, we assume also that

$$\chi_1(z)=\chi(z_1),\qquad\chi_2(z)=\chi(z_2),\qquad\chi_3(z)=\chi(z_3). \tag{2.84}$$

This assumption will allow us to reduce the problem to finding just *one* harmonic function, which is much easier than searching for three. By substituting (2.83) into the third equation of (2.12), we obtain the first equation for the constants, namely,

$$c_1 + m_2 c_2 = 0. \tag{2.85}$$

The third condition in (2.82) is thus satisfied. Substitution of (2.83) in (2.6) yields

$$u = c_1(\Lambda^2\bar{\chi}_1 + \Delta\chi_1) + c_2(\Lambda^2\bar{\chi}_2 + \Delta\chi_2) + ic_3(\Lambda^2\bar{\chi}_3 - \Delta\chi_3). \tag{2.86}$$

The differential operators Λ and Δ are defined by (2.5). When $z = 0$, equation (2.86) transforms into

$$u = (c_1 + c_2 + ic_3)\,\Lambda^2\bar{\chi} + (c_1 + c_2 - ic_3)\,\Delta\chi. \tag{2.87}$$

It is convenient to assume

$$c_1 + c_2 + ic_3 = 0. \tag{2.88}$$

This assumption simplifies (2.87) giving

$$u = (c_1 + c_2 - ic_3)\,\Delta\chi, \tag{2.89}$$

and makes it possible to represent

$$\chi(M) = \iint_S \ln[R(M, N) + z]\, u(N)\, dS_N. \tag{2.90}$$

The representation (2.90) satisfies identically the second condition (2.82), and inside the crack the following equation becomes valid

$$c_1 + c_2 - ic_3 = \frac{1}{2\pi}. \tag{2.91}$$

The solution of the set (2.85), (2.88), and (2.91) gives

$$c_1 = -\frac{1}{4\pi(m_1 - 1)}, \qquad c_2 = -\frac{1}{4\pi(m_2 - 1)}, \qquad c_3 = \frac{i}{4\pi}. \tag{2.92}$$

Substitution of (2.83) and (2.92) in the last expression (2.12) gives the following expression for the tangential stress:

$$\tau_z = -\frac{A_{44}}{4\pi}\frac{\partial}{\partial z}\left[\frac{m_1 + 1}{m_1 - 1}(\Lambda^2\bar{\chi}_1 + \Delta\chi_1) \right.$$
$$\left. + \frac{m_2 + 1}{m_2 - 1}(\Lambda^2\bar{\chi}_2 + \Delta\chi_2) + (\Lambda^2\bar{\chi}_3 - \Delta\chi_3)\right]. \tag{2.93}$$

Expression (2.93) simplifies, for $z = 0$,

$$\tau_z = -\frac{A_{44}}{4\pi}\left\{\left[\frac{m_1+1}{(m_1-1)\gamma_1}+\frac{1}{\gamma_3}\right]\Lambda^2\frac{\partial\bar{\chi}}{\partial z}+\left[\frac{m_2+1}{(m_2-1)\gamma_2}-\frac{1}{\gamma_3}\right]\Delta\frac{\partial\chi}{\partial z}\right\}. \quad (2.94)$$

Finally, satisfaction of the first condition (2.82) yields the governing integro-differential equation

$$\tau(N_0) = -\frac{1}{2\pi^2(G_1^2-G_2^2)}\left[G_1\,\Delta\iint_S \frac{u(N)}{R(N,N_0)}\,dS_N + G_2\,\Lambda^2\iint_S \frac{\bar{u}(N)}{R(N,N_0)}\,dS_N\right], \quad (2.95)$$

where the elastic constants G_1 and G_2 are defined by (2.9). Let the crack boundary be described in polar coordinates by equation $\rho = a(\phi)$, where $a(\phi)$ is a single-valued continuous function. We can always choose the coordinate axis orientation so that the first harmonic will vanish from the Fourier expansion of $a(\phi)$. An approximate analytical solution of (2.95) for a general crack can be obtained by the method used in Fabrikant (1987a). The method implies the following representation

$$\iint_S \frac{u(N)}{R(N,N_0)}\,dS_N = \int_0^\rho \frac{dx}{(\rho^2-x^2)^{1/2}}\int_0^{2\pi} d\phi_0 \times \int_x^{a(\phi_0)} \frac{\lambda(x^2/(\rho\rho_0),\,\phi-\phi_0)}{(\rho_0^2-x^2)^{1/2}}\,u(\rho_0,\phi_0)\rho_0\,d\rho_0. \quad (2.96)$$

Consider the case of a uniform shear loading. Let the tangential displacements have the form

$$u(\rho,\phi) = \frac{u_0[a^2(\phi)-\rho^2]^{1/2}}{a(\phi)}, \quad (2.97)$$

where u_0 is an as yet unknown complex constant. Substitution of (2.97) in (2.96) yields, after integration and retaining the first two harmonics only,

$$\iint_S \frac{u(N)}{R(N,N_0)}\,dS_N \approx \frac{u_0}{16\pi(G_1^2-G_2^2)} \times \int_0^{2\pi}\left\{2a(\phi_0)-\frac{\rho^2}{a(\phi_0)}[1+3\cos 2(\phi-\phi_0)]\right\}d\phi_0. \quad (2.98)$$

By substituting (2.98) in (2.95) and performing the necessary differentiation, we obtain the relationship between the shear loading and the amplitude of the tangential displacements, namely,

$$\tau = \frac{1}{4\pi(G_1^2 - G_2^2)}(u_0 G_1 B_1 + 3\bar{u}_0 G_2 B_2), \tag{2.99}$$

where

$$B_1 = \int_0^{2\pi} \frac{d\phi}{a(\phi)}, \qquad B_2 = \int_0^{2\pi} \frac{e^{2i\phi}}{a(\phi)}\, d\phi. \tag{2.100}$$

Equation (2.99) can be solved:

$$u_0 = \frac{4\pi(G_1^2 - G_2^2)}{G_1^2 B_1^2 - 9G_2^2 B_2^2}(G_1 B_1 \tau - 3G_2 B_2 \bar{\tau}). \tag{2.101}$$

The integrals in (2.100) can be computed easily for various crack shapes. For example, a rectangular crack with sides $2a_1$ and $2a_2$ is characterized by the values

$$B_1 = \frac{4(a_1^2 + a_2^2)^{1/2}}{a_1 a_2}, \qquad B_2 = \frac{4(a_2^2 - a_1^2)}{3a_1 a_2 (a_1^2 + a_2^2)^{1/2}}. \tag{2.102}$$

It is noteworthy that despite the fact that the integral representation (2.96) is valid inside a circle $\rho \le \min\{a(\phi)\}$ only, and despite the approximate nature of (2.98), the solution given by (2.99) and (2.100) is *exact* for an ellipse (see Fabrikant (1987b) for details). We did not find in the literature any reliable data related to a non-elliptical crack under shear loading, therefore it is difficult to say how accurate the solution is for various crack shapes. A complete investigation of a general crack under shear loading is beyond the scope of this article and will be published separately.

The integro-differential equation (2.95) can be solved exactly for a penny-shaped crack. The closed form solution is (see Fabrikant (1987b) for details)

$$\begin{aligned} u(\rho, \phi) = {} & \frac{G_1}{\pi}\int_0^{2\pi}\int_0^a \left[\frac{1}{R}\tan^{-1}\left(\frac{\eta}{R}\right) - \frac{G_2^2}{G_1^2}\frac{(3-\bar{t})\eta}{a^2(1-\bar{t})^2}\right]\tau(\rho_0, \phi_0)\rho_0\, d\rho_0\, d\phi_0 \\ & + \frac{G_2}{\pi}\int_0^{2\pi}\int_0^a \left[\frac{q}{R\bar{q}}\tan^{-1}\left(\frac{\eta}{R}\right) \right. \\ & \left. + \frac{\eta[(q/\bar{q}) - te^{2i\phi_0}]}{a^2(1-t)(1-\bar{t})}\right]\bar{\tau}(\rho_0, \phi_0)\rho_0\, d\rho_0\, d\phi_0, \end{aligned} \tag{2.103}$$

where R and η are defined by (2.29), q is defined by (2.43), and

$$t = \frac{\rho\rho_0}{a^2} e^{i(\phi-\phi_0)}. \tag{2.104}$$

The potential functions can be found by substitution of (2.103) and (2.90) in (2.83) and computation of the integrals involved. This looks at first somewhat difficult, nevertheless, it will be shown here that all the Green's functions can be expressed in elementary functions. Note the following property

$$\Lambda\left[\frac{1}{R}\tan^{-1}\left(\frac{\eta}{R}\right) - \frac{(3-t)\eta}{a^2(1-t)^2}\right] = -\bar{\Lambda}\left[\frac{q}{R\bar{q}}\tan^{-1}\left(\frac{\eta}{R}\right) + \frac{\eta[q/\bar{q}) - te^{2i\phi_0}]}{a^2(1-t)(1-\bar{t})}\right]. \tag{2.105}$$

Introduce the following notation

$$E_1(N, N_0) = \frac{1}{R(N, N_0)}\tan^{-1}\left[\frac{(a^2-r^2)^{1/2}(a^2-\rho_0^2)^{1/2}}{aR(N, N_0)}\right],$$

$$E_2(N, N_0) = \frac{(3a^2 - r\rho_0 e^{i(\psi-\phi_0)})(a^2-r^2)^{1/2}(a^2-\rho_0^2)^{1/2}}{a(a^2 - r\rho_0 e^{i(\psi-\phi_0)})^2},$$

$$E_3(N, N_0) = \frac{re^{i\psi} - \rho_0 e^{i\phi_0}}{R(N, N_0)(re^{-i\psi} - \rho_0 e^{-i\phi_0})}\tan^{-1}\left[\frac{(a^2-r^2)^{1/2}(a^2-\rho_0^2)^{1/2}}{aR(N, N_0)}\right]$$
$$+ \frac{a(a^2-r^2)^{1/2}(a^2-\rho_0^2)^{1/2}}{(a^2 - r\rho_0 e^{i(\psi-\phi_0)})(a^2 - r\rho_0 e^{-i(\psi-\phi_0)})}$$
$$\times\left[\frac{re^{i\psi} - \rho_0 e^{i\phi_0}}{re^{-i\psi} - \rho_0 e^{-i\phi_0}} - \frac{r\rho_0}{a^2} e^{i(\psi+\phi_0)}\right]. \tag{2.106}$$

Here the points N and N_0 are characterized by the cylindrical coordinates $(r, \psi, 0)$ and $(\rho_0, \phi_0, 0)$ respectively. Note the following relationships of symmetry:

$$E_1(N, N_0) = E_1(N_0, N), \quad E_2(N, N_0) = \bar{E}_2(N_0, N), \quad E_3(N, {}_0) = E_3(N_0, N). \tag{2.107}$$

Let $R(M, N)$ denote the distance between the points $M(\rho, \phi, z)$ and $N(r, \psi, 0)$. By using (2.105) one may write

$$\iint_S \Lambda[E_1(N, N_0) - E_2(N, N_0)]\frac{dS_N}{R(M, N)} = -\iint_S \bar{\Lambda}E_3(N, N_0)\frac{dS_N}{R(M, N)}. \tag{2.108}$$

Integration by parts in (2.108) leads to the important property

$$\iint_S [E_1(N, N_0) - E_2(N, N_0)]\, \Lambda\left(\frac{1}{R(M, N)}\right) dS_N$$
$$= -\iint_S E_3(N, N_0)\, \bar{\Lambda}\left(\frac{1}{R(M, N)}\right) dS_N. \tag{2.109}$$

Two more properties can be obtained by applying Λ and $\bar{\Lambda}$ to both sides of (2.109), namely,

$$\iint_S [E_1(N, N_0) - E_2(N, N_0)]\, \Lambda^2\left(\frac{1}{R(M, N)}\right) dS_N$$
$$= -\iint_S E_3(N, N_0)\, \Delta\left(\frac{1}{R(M, N)}\right) dS_N, \tag{2.110}$$

$$\iint_S [E_1(N, N_0) - E_2(N, N_0)]\, \Delta\left(\frac{1}{R(M, N)}\right) dS_N$$
$$= e - \iint_S E_3(N, N_0)\, \bar{\Lambda}^2\left(\frac{1}{R(M, N)}\right) dS_N. \tag{2.111}$$

The properties (2.109)–(2.111) will allow us to substitute the computation of various integrals involving E_3, which look very formidable, by computation of integrals involving expressions E_1 and E_2, some of which have already been computed (2.41)–(2.46), and the remaining ones can be computed relatively easy (see Appendix G).

Next, we introduce the notation

$$X = \Lambda\bar{\chi} + \bar{\Lambda}\chi, \qquad Y = \Lambda\bar{\chi} - \bar{\Lambda}\chi. \tag{2.112}$$

In order to obtain the complete solution, we shall need the following expressions for various derivatives of X and Y: the tangential displacements are defined by ΛX and ΛY; the normal displacements by $\partial X/\partial z$; and the field of stresses may be computed through $\partial^2 X/\partial z^2$, $\Lambda^2 X$, $\Lambda^2 Y$, $\Lambda(\partial X/\partial z)$, and $\Lambda(\partial Y/\partial z)$. All the Green's functions involved can be expressed as various derivatives of two fundamental functions, namely,

$$\begin{aligned} K_1(M, N_0) &= \iint_S E_1(N, N_0) \ln[R(M, N) + z]\, dS_N, \\ K_2(M, N_0) &= \iint_S E_2(N, N_0) \ln[R(M, N) + z]\, dS_N. \end{aligned} \tag{2.113}$$

Rewrite formula (2.103) as

$$u(N) = \frac{G_1}{\pi} \iint_S \left[E_1(N, N_0) - \frac{G_2^2}{G_1^2} \bar{E}_2(N, N_0) \right] \tau(N_0)\, dS_{N_0}$$

$$+ \frac{G_2}{\pi} \iint_S E_3(N, N_0) \bar{\tau}(N_0)\, dS_{N_0}. \tag{2.114}$$

By substituting (2.90) and (2.114) in (2.112) and using the properties (2.109)–(2.111), we get the following results:

$$X = \frac{G_1 - G_2}{\pi} \left[\bar{\Lambda} \iint_S \left(K_1 + \frac{G_2}{G_1} \bar{K}_2 \right) \tau\, dS + \Lambda \iint_S \left(K_1 + \frac{G_2}{G_1} K_2 \right) \bar{\tau}\, dS \right],$$

$$Y = \frac{G_1 + G_2}{\pi} \left[-\bar{\Lambda} \iint_S \left(K_1 - \frac{G_2}{G_1} \bar{K}_2 \right) \tau\, dS + \Lambda \iint_S \left(K_1 - \frac{G_2}{G_1} K_2 \right) \bar{\tau}\, dS \right].$$

We shall only need the following derivatives of X and Y for the complete solution:

$$\Lambda X = \frac{G_1 - G_2}{\pi} \left[-\frac{\partial^2}{\partial z^2} \iint_S \left(K_1 + \frac{G_2}{G_1} \bar{K}_2 \right) \tau\, dS \right.$$

$$\left. + \Lambda^2 \iint_S \left(K_1 + \frac{G_2}{G_1} K_2 \right) \bar{\tau}\, dS \right], \qquad (2$$

$$\Lambda Y = \frac{G_1 + G_2}{\pi} \left[\frac{\partial^2}{\partial z^2} \iint_S \left(K_1 - \frac{G_2}{G_1} \bar{K}_2 \right) \tau\, dS \right.$$

$$\left. + \Lambda^2 \iint_S \left(K_1 - \frac{G_2}{G_1} K_2 \right) \bar{\tau}\, dS \right], \qquad (2$$

$$\frac{\partial X}{\partial z} = \frac{G_1 - G_2}{\pi} \frac{\partial}{\partial z} \left[\bar{\Lambda} \iint_S \left(K_1 + \frac{G_2}{G_1} \bar{K}_2 \right) \tau\, dS \right.$$

$$\left. + \Lambda \iint_S \left(K_1 + \frac{G_2}{G_1} K_2 \right) \bar{\tau}\, dS \right], \qquad (2$$

$$\frac{\partial Y}{\partial z} = \frac{G_1 + G_2}{\pi} \frac{\partial}{\partial z} \left[-\bar{\Lambda} \iint_S \left(K_1 - \frac{G_2}{G_1} \bar{K}_2 \right) \tau\, dS \right.$$

$$\left. + \Lambda \iint_S \left(K_1 - \frac{G_2}{G_1} K_2 \right) \bar{\tau}\, dS \right], \qquad (2.$$

$$\frac{\partial}{\partial z}\Lambda X=\frac{G_1-G_2}{\pi}\frac{\partial}{\partial z}\left[-\frac{\partial^2}{\partial z^2}\iint\limits_S\left(K_1+\frac{G_2}{G_1}\bar{K}_2\right)\tau\,dS\right.$$

$$\left.+\Lambda^2\iint\limits_S\left(K_1+\frac{G_2}{G_1}K_2\right)\bar{\tau}\,dS\right], \tag{2.119}$$

$$\frac{\partial}{\partial z}\Lambda Y=\frac{G_1+G_2}{\pi}\frac{\partial}{\partial z}\left[\frac{\partial^2}{\partial z^2}\iint\limits_S\left(K_1-\frac{G_2}{G_1}\bar{K}_2\right)\tau\,dS\right.$$

$$\left.+\Lambda^2\iint\limits_S\left(K_1-\frac{G_2}{G_1}K_2\right)\bar{\tau}\,dS\right], \tag{2.120}$$

$$\frac{\partial^2 X}{\partial z^2}=\frac{G_1-G_2}{\pi}\frac{\partial^2}{\partial z^2}\left[\bar{\Lambda}\iint\limits_S\left(K_1+\frac{G_2}{G_1}\bar{K}_2\right)\tau\,dS\right.$$

$$\left.+\Lambda\iint\limits_S\left(K_1+\frac{G_2}{G_1}K_2\right)\bar{\tau}\,dS\right], \tag{2.121}$$

$$\Lambda^2 X=\frac{G_1-G_2}{\pi}\Lambda\left[-\frac{\partial^2}{\partial z^2}\iint\limits_S\left(K_1+\frac{G_2}{G_1}\bar{K}_2\right)\tau\,dS\right.$$

$$\left.+\Lambda^2\iint\limits_S\left(K_1+\frac{G_2}{G_1}K_2\right)\bar{\tau}\,dS\right], \tag{2.122}$$

$$\Lambda^2 Y=\frac{G_1+G_2}{\pi}\Lambda\left[\frac{\partial^2}{\partial z^2}\iint\limits_S\left(K_1-\frac{G_2}{G_1}\bar{K}_2\right)\tau\,dS\right.$$

$$\left.+\Lambda^2\iint\limits_S\left(K_1-\frac{G_2}{G_1}K_2\right)\bar{\tau}\,dS\right]. \tag{2.123}$$

All the Green's functions in (2.115)–(2.123) are computed in elementary functions in Appendix G and in (2.42)–(2.46). The solution may be considered complete. In order to avoid duplication, we do not write here the explicit expressions for the fields of displacements and stresses since the equivalent expressions for the relevant Green's functions are given in the next section.

G. Application: Concentrated Tangential Loading of a Penny-Shaped Crack

Consider an infinite transversely isotropic solid weakened in the plane $z=0$ by a penny-shaped crack of radius a. Let two equal concentrated forces $T = T_x + iT_y$ be applied to the crack faces antisymmetrically at the point $N_0(\rho_0, \phi_0, 0^\pm)$. The previously obtained results give the complete solution in elementary functions:

$$u = \frac{\gamma_1\gamma_2 H}{\pi} \sum_{k=1}^{2} \frac{1}{m_k - 1} \left\{ -\left[f_2(z_k) + \frac{G_2}{G_1} \bar{f}_7(z_k) \right] T + \left[f_{16}(z_k) + \frac{G_2}{G_1} f_8(z_k) \right] \bar{T} \right\}$$
$$+ \frac{\beta}{\pi} \left\{ \left[f_2(z_3) - \frac{G_2}{G_1} \bar{f}_7(z_3) \right] T + \left[f_{16}(z_3) - \frac{G_2}{G_1} f_8(z_3) \right] \bar{T} \right\}, \tag{2.124}$$

$$w = \frac{2}{\pi} H\gamma_1\gamma_2 \Re \sum_{k=1}^{2} \frac{m_k}{(m_k - 1)\gamma_k} \left[\bar{f}_1(z_k) + \frac{G_2}{G_1} \bar{f}_9(z_k) \right] T, \tag{2.125}$$

$$\sigma_1 = \Re \left\{ \frac{2\gamma_1\gamma_2}{\pi^2(\gamma_1 - \gamma_2)} \sum_{k=1}^{2} (-1)^{k+1} \left[\frac{1}{\gamma_3^2(m_k + 1)} - \frac{1}{\gamma_k^2} \right] \left[\bar{f}_5(z_k) + \frac{G_2}{G_1} \bar{f}_{10}(z_k) \right] T \right\}, \tag{2.126}$$

$$\sigma_2 = -\frac{2}{\pi} A_{66} H\gamma_1\gamma_2 \sum_{k=1}^{2} \frac{1}{m_k - 1}$$
$$\times \left\{ \left[f_5(z_k) + \frac{G_2}{G_1} \bar{f}_{13}(z_k) \right] T + \left[f_{11}(z_k) + \frac{G_2}{G_1} f_{12}(z_k) \right] \bar{T} \right\}$$
$$- \frac{1}{\pi^2\gamma_3} \left\{ \left[-f_5(z_3) + \frac{G_2}{G_1} \bar{f}_{13}(z_3) \right] T + \left[f_{11}(z_3) - \frac{G_2}{G_1} f_{12}(z_3) \right] \bar{T} \right\}, \tag{2.127}$$

$$\sigma_z = \Re \left\{ \frac{\gamma_1\gamma_2}{\pi^2(\gamma_1 - \gamma_2)} \sum_{k=1}^{2} (-1)^{k+1} \left[\bar{f}_5(z_k) + \frac{G_2}{G_1} \bar{f}_{10}(z_k) \right] T \right\}, \tag{2.128}$$

$$\tau_z = \frac{\gamma_1\gamma_2}{2\pi^2(\gamma_1 - \gamma_2)} \sum_{k=1}^{2} \frac{(-1)^k}{\gamma_k}$$
$$\times \left\{ \left[f_3(z_k) + \frac{G_2}{G_1} \bar{f}_{14}(z_k) \right] T + \left[-f_4(z_k) + \frac{G_2}{G_1} f_{15}(z_k) \right] \bar{T} \right\}$$
$$+ \frac{1}{2\pi^2} \left\{ \left[f_3(z_3) - \frac{G_2}{G_1} \bar{f}_{14}(z_3) \right] T + \left[f_4(z_3) + \frac{G_2}{G_1} f_{15}(z_3) \right] \bar{T} \right\}. \tag{2.129}$$

Here $\Re$ indicates the real part, the elastic coefficients are defined by (2.9), and the functions f with subindex less than 6 are defined by formulae (2.53)–(2.57), and the remaining functions are computed in

Appendix G, namely,

$$f_7(z) = \frac{ha^2}{s^2}\left\{\frac{3}{s^2} - \frac{t}{l_2^2 - a^2 t} - \frac{3(l_2^2 - a^2)^{1/2}}{s^3}\tan^{-1}\left[\frac{s}{(l_2^2 - a^2)^{1/2}}\right]\right\}, \tag{2.130}$$

$$f_8(z) = \frac{1}{\bar{q}}(a^2 - \rho_0^2)^{1/2}\left\{\frac{(\bar{\xi} - 1)^{1/2}}{\bar{q}}\left[\tan^{-1}\left(\frac{1}{(\bar{\xi} - 1)^{1/2}}\right) - \tan^{-1}\left(\frac{(a^2 - l_1^2)^{1/2}}{a(\bar{\xi} - 1)^{1/2}}\right)\right]\right.$$
$$\left. - \frac{e^{i\phi}}{\rho}\left[\frac{(a^2 - l_1^2)^{1/2}}{a}\left(1 + \frac{\rho^2}{l_2^2 - \rho\rho_0 e^{i(\phi - \phi_0)}}\right) - 1\right]\right\}, \tag{2.131}$$

$$f_9(z) = -\rho e^{i\phi}\frac{(a^2 - \rho_0^2)^{1/2}}{a^3}\left\{\frac{1}{t}\sin^{-1}\left(\frac{a}{l_2}\right) + \frac{a(l_2^2 - a^2)^{1/2}}{(1 - t)(l_2^2 - \rho\rho_0 e^{i(\phi - \phi_0)})}\right.$$
$$\left. - \frac{1}{t(1 - t)^{3/2}}\tan^{-1}\left(\frac{a(1 - t)^{1/2}}{(l_2^2 - a^2)^{1/2}}\right)\right\}, \tag{2.132}$$

$$f_{10}(z) = -\frac{h\rho e^{i\phi}(3l_2^2 - a^2 t)}{(l_2^2 - l_2^2)(l_2^2 - a^2 t)^2}, \tag{2.133}$$

$$f_{11}(z) = \frac{1}{\bar{q}}\left\{\frac{3R_0^4 + 6R_0^2 z^2 - z^4}{R_0^3\bar{q}^2}\tan^{-1}\left(\frac{h}{R_0}\right)\right.$$
$$- (a^2 - \rho_0^2)^{1/2}\left[\frac{z}{\bar{s}}\left(\frac{8}{\bar{q}^2} - \frac{4\rho_0 e^{i\phi_0}}{\bar{s}^2\bar{q}} + \frac{3\rho_0^2 e^{2i\phi_0}}{\bar{s}^4}\right)\tan^{-1}\left(\frac{\bar{s}}{(l_2^2 - a^2)^{1/2}}\right)\right.$$
$$- \frac{e^{i\phi}}{\rho}\left(\frac{2e^{i\phi}}{\rho} + \frac{3}{\bar{q}}\right) - \frac{3(\bar{\xi} - 1)^{1/2}}{\bar{q}^2}$$
$$\left.\times\left(\tan^{-1}\left[\frac{1}{(\bar{\xi} - 1)^{1/2}}\right] - \tan^{-1}\left[\frac{(a^2 - l_1^2)^{1/2}}{a(\bar{\xi} - 1)^{1/2}}\right]\right)\right]$$
$$+ \frac{ha^2 e^{i\phi}}{\rho\bar{s}^2}\left[\frac{2\rho_0 e^{i\phi_0}}{\bar{s}^2} - \frac{2e^{i\phi}}{\rho} - \frac{2}{\bar{q}} + \left(\frac{\rho_0 e^{i\phi_0}}{\bar{s}^2} - \frac{2}{\bar{q}}\right)\frac{(l_2^2 - a^2)\bar{t}}{l_2^2 - a^2\bar{t}}\right]$$
$$\left. - \frac{h}{R_0^2 + h^2}\left[\frac{\bar{q}\rho e^{3i\phi}}{l_2^2 - l_1^2} + \frac{e^{i\phi}(l_2^2 - \rho^2)}{\rho\bar{q}} - \frac{z^2 q}{R_0^2\bar{q}} + 2e^{2i\phi}\right]\right\}, \tag{2.134}$$

$$f_{12}(z) = \frac{1}{\bar{q}}(a^2 - \rho_0^2)^{1/2}\left\{\frac{3(\bar{\xi} - 1)^{1/2}}{\bar{q}^2}\left[\tan^{-1}\left(\frac{1}{(\bar{\xi} - 1)^{1/2}}\right) - \tan^{-1}\left(\frac{(a^2 - l_1^2)^{1/2}}{a(\bar{\xi} - 1)^{1/2}}\right)\right]\right.$$
$$- \frac{e^{2i\phi}(a^2 - l_1^2)^{1/2}}{a(l_2^2 - l_1^2)}\left[\frac{l_2^2 + \rho^2}{l_2^2 - \rho\rho_0 e^{i(\phi - \phi_0)}} + \frac{2\rho^2(l_2^2 - a^2)}{(l_2^2 - \rho\rho_0 e^{i(\phi - \phi_0)})^2} + 1\right]$$
$$\left. + \frac{e^{i\phi}}{\rho}\left[\frac{3}{\bar{q}} + \frac{2e^{i\phi}}{\rho} - \frac{(a^2 - l_1^2)^{1/2}}{a}\left(\frac{l_2^2 + 2\rho^2}{\bar{q}(l_2^2 - \rho\rho_0 e^{i(\phi - \phi_0)})} + 2\left[\frac{1}{\bar{q}} + \frac{e^{i\phi}}{\rho}\right]\right)\right]\right\}, \tag{2.135}$$

$$\bar{f}_{13}(z) = -h\left\{\frac{a^2}{\bar{s}^2}\rho_0 e^{i\phi_0}\left[\frac{15(l_2^2-a^2)^{1/2}}{\bar{s}^5}\tan^{-1}\left(\frac{\bar{s}}{(l_2^2-a^2)^{1/2}}\right)-\frac{15}{\bar{s}^4}\right.\right.$$
$$\left.\left.+\frac{5}{\bar{s}^2(l_2^2-a^2\bar{t})}+\frac{2\bar{t}}{(l_2^2-a^2\bar{t})^2}\right]+\frac{\rho e^{i\phi}(3l_2^2-a^2\bar{t})}{(l_2^2-l_1^2)(l_2^2-a^2\bar{t})^2}\right\}, \tag{2.136}$$

$$f_{14}(z) = \frac{(a^2-\rho_0^2)^{1/2}}{a^3(1-t)}\left\{\frac{a(l_2^2-a^2)^{1/2}}{(l_2^2-l_1^2)(l_2^2-\rho\rho_0 e^{i(\phi-\phi_0)})}\right.$$
$$\times\left[\frac{3(l_2^2-l_1^2t)}{1-t}+\frac{\rho\rho_0 e^{i(\phi-\phi_0)}(2l_2^2+l_1^2t-3\rho^2)}{l_2^2-\rho\rho_0 e^{i(\phi-\phi_0)}}\right]$$
$$\left.-\frac{3}{(1-t)^{3/2}}\tan^{-1}\left[\frac{a(1-t)^{1/2}}{(l_2^2-a^2)^{1/2}}\right]\right\}, \tag{2.137}$$

$$f_{15}(z) = \frac{\rho^2 e^{2i\phi}(a^2-\rho_0^2)^{1/2}(l_2^2-a^2)^{1/2}(3l_2^2-\rho\rho_0 e^{i(\phi-\phi_0)})}{l_2^2(l_2^2-l_1^2)(l_2^2-\rho\rho_0 e^{i(\phi-\phi_0)})^2} \tag{2.138}$$

$$f_{16}(z) = \frac{1}{\bar{q}}\left\{\frac{R_0^2+z^2}{R_0\bar{q}}\tan^{-1}\left(\frac{h}{R_0}\right)+(a^2-\rho_0^2)^{1/2}\right.$$
$$\times\left[\frac{z}{\bar{s}}\left(\frac{\rho_0 e^{i\phi_0}}{\bar{s}^2}-\frac{2}{\bar{q}}\right)\tan^{-1}\left(\frac{\bar{s}}{(l_2^2-a^2)^{1/2}}\right)+\frac{(\bar{\xi}-1)^{1/2}}{\bar{q}}\right.$$
$$\left.\left.\times\left(\tan^{-1}\left[\frac{1}{(\bar{\xi}-1)^{1/2}}\right]-\tan^{-1}\left[\frac{(a^2-l_1^2)^{1/2}}{a(\bar{\xi}-1)^{1/2}}\right]\right)+\frac{e^{i\phi}}{\rho}\right]-\frac{e^{i\phi}ha^2}{\rho\bar{s}^2}\right\}. \tag{2.139}$$

This problem was not considered before. The solution (2.124)–(2.139) presents, in fact, the explicit expressions for the Green's functions, and allows us to write a complete solution for the case of arbitrary tangential loading in quadratures. The general results simplify significantly for $z=0$, namely,

$$u = \frac{G_1}{\pi}\left[\frac{1}{R}\tan^{-1}\left(\frac{\eta}{R}\right)-\frac{G_2^2}{G_1^2}\frac{(3-\bar{t})\eta}{a^2(1-\bar{t})^2}\right]T$$
$$+\frac{G_2}{\pi}\left[\frac{q}{R\bar{q}}\tan^{-1}\left(\frac{\eta}{R}\right)+\frac{\eta[(q/\bar{q})-te^{2i\phi_0}]}{a^2(1-t)(1-\bar{t})}\right]\bar{T} \qquad \text{for } \rho<a, \tag{2.140}$$

$$w = H\alpha(a^2-\rho_0^2)^{1/2}\Re\left\{\left[\frac{1}{qs}+\frac{G_2}{G_1}\frac{e^{-i\phi_0}}{\rho_0}\left(\frac{a^2}{\bar{s}^3}-\frac{1}{\alpha}\right)\right]T\right\} \qquad \text{for } \rho\le a,$$

$$w = \frac{2}{\pi}H\alpha(a^2-\rho_0^2)^{1/2}\Re\left\{\left[\frac{1}{qs}\tan^{-1}\left(\frac{s}{(\rho^2-a^2)^{1/2}}\right)\right.\right.$$
$$+\frac{G_2}{G_1}\left(\frac{a^2e^{-i\phi_0}}{\rho_0\bar{s}^3}\tan^{-1}\left(\frac{\bar{s}}{(\rho^2-a^2)^{1/2}}\right)\right.$$
$$\left.\left.\left.-\frac{(\rho^2-a^2)^{1/2}}{\bar{s}^2q}-\frac{e^{-i\phi_0}}{a\rho_0}\sin^{-1}\left(\frac{a}{\rho}\right)\right)\right]T\right\} \qquad \text{for } \rho>a, \tag{2.141}$$

$$\sigma_1 = \frac{2}{\pi^2}\Re\bigg\{\left(2\pi HA_{66}\gamma_1\gamma_2 - \frac{\gamma_1+\gamma_2}{\gamma_1\gamma_2}\right)\left[\frac{1}{qR}\tan^{-1}\left(\frac{\eta}{R}\right)\right.$$

$$+\frac{\eta}{a^2(1-t)(1-\bar{t})}\left(\frac{\rho e^{-i\phi}}{a^2-\rho^2}+\frac{1}{q}\right)$$

$$\left.+\frac{G_2}{G_1}\frac{\eta\rho e^{-i\phi}(3-\bar{t})}{(a^2-\rho^2)a^2(1-\bar{t})^2}\right]T\bigg\} \qquad \text{for } \rho < a, \tag{2.142}$$

$$\sigma_1 = 0 \qquad \text{for } \rho > a,$$

$$\sigma_2 = \frac{1}{\pi^2}\bigg\{\left(2\pi A_{66}H\gamma_1\gamma_2+\frac{1}{\gamma_3}\right)\left[f_5(0)T+\frac{G_2}{G_1}f_{12}(0)\bar{T}\right]$$

$$+\left(2\pi A_{66}H\gamma_1\gamma_2-\frac{1}{\gamma_3}\right)\left[f_{11}(0)\bar{T}+\frac{G_2}{G_1}\bar{f}_{13}(0)T\right]\bigg\}, \tag{2.143}$$

$$\sigma_z = 0,$$

$$\tau_z = \frac{(a^2-\rho_0^2)^{1/2}}{\pi^2(\rho^2-a^2)^{1/2}}\left[\frac{T}{R^2}+\frac{G_2}{G_1}\frac{e^{2i\phi}(3\rho-\rho_0 e^{i(\phi-\phi_0)})\bar{T}}{\rho(\rho-\rho_0 e^{i(\phi-\phi_0)})^2}\right] \qquad \text{for } \rho > a. \tag{2.144}$$

Here R and η are defined by (2.29). The stress intensity factor of the second and the third mode can be expressed through the decomposition $\tau^{(n)} = \tau_{zn} + i\tau_{tz}$, which is related to τ_z by the relationship $\tau_z = \tau^{(n)}e^{i\phi}$. Introducing the complex stress intensity factor

$$k_2 + ik_3 = \lim_{\rho\to a}[(\rho-a)^{1/2}\tau_z e^{-i\phi}], \tag{2.145}$$

one gets from (2.144)

$$k_2+ik_3 = \frac{(a^2-\rho_0^2)^{1/2}}{\pi^2(2a)^{1/2}}\left[\frac{Te^{-i\phi}}{\rho_0^2+a^2-2a\rho\cos(\phi-\phi_0)}\right.$$

$$\left.+\frac{G_2}{G_1}\frac{e^{i\phi}(3a-\rho_0 e^{i(\phi-\phi_0)})\bar{T}}{a(a-\rho_0 e^{i(\phi-\phi_0)})^2}\right]. \tag{2.146}$$

In the general case of arbitrarily distributed loading, the stress intensity factor takes the form

$$k_2+ik_3 = \int_0^{2\pi}\int_0^a \frac{(a^2-\rho_0^2)^{1/2}}{\pi^2(2a)^{1/2}}\left[\frac{e^{-i\phi}\tau(\rho_0,\phi_0)}{\rho_0^2+a^2-2a\rho\cos(\phi-\phi_0)}\right.$$

$$\left.+\frac{G_2}{G_1}\frac{(3a-\rho_0 e^{i(\phi-\phi_0)})e^{i\phi}\bar{\tau}(\rho_0,\phi_0)}{a(a-\rho_0 e^{i(\phi-\phi_0)})^2}\right]\rho_0\,d\rho_0\,d\phi_0, \tag{2.147}$$

which is in agreement with the results of Fabrikant (1987b).

III. Illustrative Examples

Here we present the complete solutions to some simple punch and crack problems. The axisymmetric solutions were reported in the literature, so that we shall be able to compare the results. The non-axisymmetric ones seem to be new.

A. Example 1: Penny-Shaped Crack under Uniform Pressure

Let a penny-shaped crack of radius a be opened by the pressure $p = \text{const}$. In this case one gets from (2.27)

$$\omega(\rho, \phi) = 4Hp(a^2 - \rho^2)^{1/2}. \tag{3.1}$$

The potential function F can be obtained by substitution of (3.1) in (2.17). The integral can be computed in elementary functions (A.4):

$$F = 2\pi Hp\left[(2a^2 + 2z^2 - \rho^2)\sin^{-1}\left(\frac{a}{l_2}\right) - \frac{2a^2 - 3l_1^2}{a}(l_2^2 - a^2)^{1/2}\right]. \tag{3.2}$$

The complete solution can be expressed through various derivatives of the potential function, as prescribed by formulae (2.6) and (2.12). All the derivatives are given in Appendix A. The solution is:

$$u = 2Hp\rho e^{i\phi}\sum_{k=1}^{2}\frac{\gamma_k}{m_k - 1}\left[\sin^{-1}\left(\frac{a}{l_{2k}}\right) - \frac{a(l_{2k}^2 - a^2)^{1/2}}{l_{2k}^2}\right], \tag{3.3}$$

$$w = 4Hp\sum_{k=1}^{2}\frac{m_k}{m_k - 1}\left[(a^2 - l_{1k}^2)^{1/2} - z\sin^{-1}\left(\frac{a}{l_{2k}}\right)\right], \tag{3.4}$$

$$\sigma_1 = 8HpA_{66}\sum_{k=1}^{2}\frac{\gamma_k^2 - (m_k + 1)\gamma_3^2}{\gamma_k(m_k - 1)}\left[\frac{a(l_{2k}^2 - a^2)^{1/2}}{(l_{2k}^2 - l_{1k}^2)} - \sin^{-1}\left(\frac{a}{l_{2k}}\right)\right], \tag{3.5}$$

$$\sigma_2 = -8HpA_{66}ae^{2i\phi}\sum_{k=1}^{2}\frac{\gamma_k}{m_k - 1}\left[\frac{l_{1k}^2(l_{2k}^2 - a^2)^{1/2}}{l_{2k}^2(l_{2k}^2 - l_{1k}^2)}\right], \tag{3.6}$$

$$\sigma_z = \frac{2p}{\pi(\gamma_1 - \gamma_2)}\sum_{k=1}^{2}(-1)^{k+1}\gamma_k\left[\frac{a(l_{2k}^2 - a^2)^{1/2}}{l_{2k}^2 - l_{1k}^2} - \sin^{-1}\left(\frac{a}{l_{2k}}\right)\right], \tag{3.7}$$

$$\tau_z = \frac{2pa^2\rho e^{i\phi}}{\pi(\gamma_1 - \gamma_2)}\sum_{k=1}^{2}(-1)^k\frac{(a^2 - l_{1k}^2)^{1/2}}{l_{2k}^2(l_{2k}^2 - l_{1k}^2)}, \tag{3.8}$$

Here the following notation was introduced:

$$l_{1k}=\tfrac{1}{2}\{[(a+\rho)^2+z_k^2]^{1/2}-[(a-\rho)^2+z_k^2]^{1/2}\},$$
$$l_{2k}=\tfrac{1}{2}\{[(a+\rho)^2+z_k^2]^{1/2}+[(a-\rho)^2+z_k^2]^{1/2}\}, \tag{3.9}$$
$$z_k=\frac{z}{\gamma_k}, \qquad k=1,2.$$

The problem was first solved by Elliott (1949) by the integral transform method. Our results are essentially in agreement with those of Elliott, who expressed them in the form of integrals involving Bessel functions, namely,

$$C_n^m=\int_0^\infty x^{n-2}\cos x\,J_m\left(x\frac{\rho}{a}\right)\exp\left(-x\frac{z}{a}\right)dx,$$
$$S_n^m=\int_0^\infty x^{n-2}\sin x\,J_m\left(x\frac{\rho}{a}\right)\exp\left(-x\frac{z}{a}\right)dx.$$

These integrals can be computed in elementary functions

$$S_1^0=\sin^{-1}\left(\frac{a}{l_2}\right),\qquad S_2^0=\frac{a(a^2-l_1^2)^{1/2}}{l_2^2-l_1^2},\qquad C_2^0=\frac{a(l_2^2-a^2)^{1/2}}{l_2^2-l_1^2},$$
$$C_3^0=\frac{a(a^2-l_1^2)^{1/2}[l_2^4+a^2(\rho^2-2a^2-2z^2)]}{(l_2^2-l_1^2)^3},$$
$$S_3^0=\frac{a(l_2^2-a^2)^{1/2}[a^2(2a^2+2z^2-\rho^2)-l_1^4]}{(l_2^2-l_1^2)^3},$$
$$C_1^1=\frac{z[a-(a^2-l_1^2)^{1/2}]}{\rho(a^2-l_1^2)^{1/2}},\qquad S_1^1=\frac{a-(a^2-l_1^2)^{1/2}}{\rho},$$
$$S_1^2=\frac{z[a-(a^2-l_1^2)^{1/2}]^2}{\rho^2(a^2-l_1^2)^{1/2}},\qquad C_2^1=\frac{a-(a^2-l_1^2)^{1/2}}{\rho}-\frac{al_1(a^2-l_1^2)^{1/2}}{l_2(l_2^2-l_1^2)},$$
$$S_2^1=\frac{al_1(l_2^2-a^2)^{1/2}}{l_2(l_2^2-l_1^2)},$$
$$C_3^1=\frac{a^2\rho(l_2^2-a^2)^{1/2}(l_2^2+3l_1^2-4a^2)}{(l_2^2-l_1^2)^3},$$
$$S_3^1=\frac{a^2\rho(a^2-l_1^2)^{1/2}(3l_2^2+l_1^2-4a^2)}{(l_2^2-l_1^2)^3},$$
$$C_2^2=\frac{2a[(l_2^2-a^2)^{1/2}-z]}{\rho^2}-\frac{a(l_2^2-a^2)^{1/2}}{l_2^2-l_1^2},$$

$$S_2^2 = \frac{2a[a-(a^2-l_1^2)^{1/2})]}{\rho^2} - \frac{a(a^2-l_1^2)^{1/2}}{l_2^2-l_1^2},$$

$$C_3^2 = \frac{2a}{\rho}\left[\frac{a-(a^2-l_1^2)^{1/2}}{\rho} - \frac{al_1(a^2-l_1^2)^{1/2}}{l_2(l_2^2-l_1^2)}\right] - \frac{a(a^2-l_1^2)^{1/2}[l_2^4+a^2(\rho^2-2a^2-2z^2)]}{(l_2^2-l_1^2)^3}.$$

There are some misprints (or errors) in the Elliott's paper. For example, according to his formula (4.2.5), the tangential displacement u vanishes on the plane $z=0$, which can not be correct; there are missing terms and obvious misprints in formula (4.2.6).

Though formulae (3.3)–(3.8) are valid for a penny-shaped crack subjected to a uniform pressure only, they can be used to obtain some general results that are valid for an arbitrary crack with a smooth boundary subjected to arbitrary loading; namely, we can explore the asymptotic behavior of displacements and stresses near the rim of a general crack. Such expressions were derived first for the case of transversely isotropic body by Kassir and Sih (1975) who considered a more complicated case of an elliptical crack. Their derivation looks so complicated that nobody so far has been able to verify its accuracy. The case of a circular crack is much simpler, so that we can obtain the same results by elementary means. Introduce the local system of spherical coordinates (r, θ, ϕ), with the coordinate origin at the crack rim. The following asymptotics are valid for the main parameters used in (3.3)–(3.8):

$$\begin{gathered} \rho = a + r\cos\theta, \qquad z = r\sin\theta, \\ l_{1k} \approx a - \tfrac{1}{2}rS_k^2, \qquad l_{2k} \approx a + \tfrac{1}{2}rT_k^2, \\ l_{2k}^2 - l_{1k}^2 \approx 2arQ_k, \qquad (a^2-l_{1k}^2)^{1/2} \approx (ar)^{1/2}S_k, \\ (l_{2k}^2-a^2)^{1/2} \approx (ar)^{1/2}T_k, \qquad \sin^{-1}\left(\frac{a}{l_{2k}}\right) \approx -\left(\frac{r}{a}\right)^{1/2} T_k. \end{gathered} \tag{3.10}$$

Here the following notation was introduced:

$$Q_k = \left(\cos^2\theta + \frac{1}{\gamma_k^2}\sin^2\theta\right)^{1/2}, \quad S_k = (Q_k - \cos\theta)^{1/2},$$
$$T_k = (Q_k + \cos\theta)^{1/2}, \quad \text{for } k = 1, 2, 3. \tag{3.11}$$

Introducing the opening mode stress intensity factor

$$k_1 = \frac{p\sqrt{2a}}{\pi}, \tag{3.12}$$

the following asymptotic expressions can be derived by substitution of (3.10) and (3.12) in (3.3)–(3.8):

$$u = u_n = -2\pi H k_1 \sqrt{2r}\left[\frac{\gamma_1 T_1}{m_1 - 1} + \frac{\gamma_2 T_2}{m_2 - 1}\right] + O(1), \tag{3.13}$$

$$w = 2\pi H k_1 \sqrt{2r}\left[\frac{m_1 S_1}{m_1 - 1} + \frac{m_2 S_2}{m_2 - 1}\right] + O(r), \tag{3.14}$$

$$\sigma_1 = 2\pi A_{66} H k_1 \left(\frac{2}{r}\right)^{1/2} \sum_{k=1}^{2} \frac{\gamma_k^2 - (m_k + 1)\gamma_3^2}{\gamma_k(m_k - 1)} \frac{T_k}{Q_k} + O(1), \tag{3.15}$$

$$\sigma_2 = -2\pi A_{66} H k_1 \left(\frac{2}{r}\right)^{1/2} \sum_{k=1}^{2} \frac{\gamma_k T_k}{(m_k - 1) Q_k} + O(\sqrt{r}), \tag{3.16}$$

$$\sigma_z = \frac{k_1}{\sqrt{2r}\,(\gamma_1 - \gamma_2)}\left[\frac{\gamma_1 T_1}{Q_1} - \frac{\gamma_2 T_2}{Q_2}\right] + O(1), \tag{3.17}$$

$$\tau_z = \tau_{zn} = -\frac{k_1}{\sqrt{2r}\,(\gamma_1 - \gamma_2)}\left[\frac{S_1}{Q_1} - \frac{S_2}{Q_2}\right] + O(\sqrt{r}), \tag{3.18}$$

These results were computed for $\phi = 0$. This assumption allows us to avoid a cumbersome axis transformation, without loss of generality. The parameter σ_1 in this case is interpreted as the sum $\sigma_n + \sigma_t$, and $\sigma_2 = \sigma_n - \sigma_t + 2i\tau_{nt}$. By taking the sum and the difference of (3.15) and (3.16), one can get

$$\sigma_n = \frac{k_1}{\sqrt{2r}\,(\gamma_1 - \gamma_2)}\left[-\frac{T_1}{\gamma_1 Q_1} + \frac{T_2}{\gamma_2 Q_2}\right],$$

$$\sigma_t = \pi A_{66} H k_1 \left(\frac{2}{r}\right)^{1/2}\left[\frac{[2\gamma_1^2 - (m_1 + 1)\gamma_3^2] T_1}{\gamma_1(m_1 - 1) Q_1} + \frac{[\gamma_2^2 - (m_2 + 1)\gamma_3^2] T_2}{\gamma_2(m_2 - 1) Q_2}\right].$$

Formulae (3.13)–(3.18) are essentially in agreement with the results of Kassir and Sih (1975), except for some misprints; for example, one should read $\sqrt{n_1}$ and $\sqrt{n_2}$ instead of n_1 and n_2 in the denominator of the terms in curly brackets of formula (8.94c). In order to compare our results with those of Kassir and Sih, one should keep in mind that their definition of the stress intensity factor is $\sqrt{2}$ times greater than ours, and their notation n_k corresponds to our γ_k^2. Kassir and Sih seem to have not noticed the properties (2.23) and the relationship $S_k = (\sin\theta)/(\gamma_k T_k)$, which in some cases can simplify their results significantly. For example, they have an expression $[A_{13} m_k - A_{11}\gamma_k^2]/[A_{44}(m_k + 1)]$ in formula (8.95a), without realizing that it is equal to -1 for $k = 1, 2$.

Formulae (3.3)–(3.8) in the limiting case of $\gamma_1 \to \gamma_2 \to \gamma_3 \to 1$, $m_1 \to m_2 \to 1$, $H = (1-\nu)/2\pi\mu$, and $A_{44} = A_{66} = \mu$, give the complete solution for an isotropic body. By using the L'Hôpital rule, one obtains

$$u = \frac{p\rho e^{i\phi}}{2\pi\mu}\left\{(1-2\nu)\left[\frac{a(l_2^2-a^2)^{1/2}}{l_2^2} - \sin^{-1}\left(\frac{a}{l_2}\right)\right] + \frac{2a^2\,|z|\,(a^2-l_1^2)^{1/2}}{l_2^2(l_2^2-l_1^2)}\right\}, \tag{3.19}$$

$$w = \frac{2p}{\pi\mu}\left\{2(1-\nu)\left[\frac{z}{|z|}(a^2-l_1^2)^{1/2} - z\sin^{-1}\left(\frac{a}{l_2}\right)\right] + z\left[\sin^{-1}\left(\frac{a}{l_2}\right) - \frac{a(l_2^2-a^2)^{1/2}}{l_2^2-l_1^2}\right]\right\}, \tag{3.20}$$

$$\sigma_1 = \frac{2p}{\pi}\left\{(1+2\nu)\left[\frac{a(l_2^2-a^2)^{1/2}}{l_2^2-l_1^2} - \sin^{-1}\left(\frac{a}{l_2}\right)\right] + \frac{az^2[l_1^4 + a^2(2a^2+2z^2-3\rho^2)]}{(l_2^2-l_1^2)^3(l_2^2-a^2)^{1/2}}\right\}, \tag{3.21}$$

$$\sigma_2 = \frac{2p}{\pi}\frac{al_1^2 e^{2i\phi}(l_2^2-a^2)^{1/2}}{l_2^2(l_2^2-l_1^2)}\left\{1-2\nu + \frac{z^2[a^2(6l_2^2-2l_1^2+\rho^2)-5l_2^4]}{(l_2^2-l_1^2)^2(l_2^2-a^2)}\right\}, \tag{3.22}$$

$$\sigma_z = \frac{2p}{\pi}\left\{\frac{a(l_2^2-a^2)^{1/2}}{l_2^2-l_1^2} - \sin^{-1}\left(\frac{a}{l_2}\right) - \frac{az^2[l_1^4+a^2(2a^2+2z^2-3\rho^2)]}{(l_2^2-l_1^2)^3(l_2^2-a^2)^{1/2}}\right\}, \tag{3.23}$$

$$\tau_z = -\frac{2p}{\pi}\frac{zl_1 e^{i\phi}(l_2^2-a^2)^{1/2}[a^2(4l_2^2-5\rho^2)+l_1^4]}{l_2(l_2^2-l_1^2)^3}. \tag{3.24}$$

This problem was first solved by Sneddon (1951) by the integral transform method. He was seemingly unable to compute the potential function (3.2), so that he resorted to differentiation under the integral sign, with a subsequent computation of various integrals involving Bessel functions. His final results are given as elementary functions of four parameters, namely,

$$r = \left(1 + \left(\frac{z}{a}\right)^2\right)^{1/2}, \qquad R^2 = \left[\left(\frac{\rho}{a}\right)^2 + \left(\frac{z}{a}\right)^2 - 1\right]^2 + 4\left(\frac{z}{a}\right)^2,$$

$$\theta = \tan^{-1}\left(\frac{a}{z}\right), \qquad \phi = \tan^{-1}\left(\frac{2az}{\rho^2+z^2-a^2}\right).$$

This choice of parameters is not the best possible. Here is one illustration. The expression for S_1^0 in Sneddon's notation takes the form

(Sneddon, 1951, p. 497)

$$S_1^0 = \tan^{-1}\left(\frac{r \sin\theta + \sqrt{R}\sin(\phi/2)}{r\cos\theta + \sqrt{R}\cos(\phi/2)}\right),$$

with the limitation $\rho \neq 0$, and there is no indication of what the result would be if $\rho = 0$. Our result is $\sin^{-1}(a/l_2)$, with no limitations attached. Introduction of Sneddon's parameters r and θ seems to be unnecessary. There exist relationships between his parameters R and ϕ, and our l_1 and l_2, namely,

$$R^2 = \frac{l_2^2 - l_1^2}{a^2}, \qquad \sin\frac{\phi}{2} = \frac{(a^2 - l_1^2)^{1/2}}{(l_2^2 - l_1^2)^{1/2}}.$$

These relationships may be used to compare the solutions, which are in good agreement, except for some misprints: factor ζ is missing in formula (139, p. 496), and the last term in formula (145, p. 499) should read $-S_1^0$, rather than $+S_0^0$.

The asymptotic behavior of displacements and stresses near the crack edge in an isotropic body can be found from either (3.13)–(3.18) or (3.19)–(3.24). The result is

$$u = u_n = \frac{k_1\sqrt{r}}{2\mu}\cos\frac{\theta}{2}\left[2(1-\nu) - \cos^2\frac{\theta}{2}\right] + O(1), \tag{3.25}$$

$$w = \frac{k_1\sqrt{r}}{\mu}\sin\frac{\theta}{2}\left[2(1-\nu) - \cos^2\frac{\theta}{2}\right] + O(r), \tag{3.26}$$

$$\sigma_1 = \frac{k_1}{\sqrt{r}}\cos\frac{\theta}{2}\left[1 + 2\nu - \sin\frac{\theta}{2}\sin\frac{3\theta}{2}\right] + O(1), \tag{3.27}$$

$$\sigma_2 = \frac{k_1}{\sqrt{r}}\cos\frac{\theta}{2}\left[1 - 2\nu - \sin\frac{\theta}{2}\sin\frac{3\theta}{2}\right] + O(\sqrt{r}), \tag{3.28}$$

$$\sigma_z = \frac{k_1}{\sqrt{r}}\cos\frac{\theta}{2}\left[1 + \sin\frac{\theta}{2}\sin\frac{3\theta}{2}\right] + O(1), \tag{3.29}$$

$$\tau_z = \tau_{zn} = \frac{k_1}{2\sqrt{r}}\sin\theta\cos\frac{3\theta}{2} + O(\sqrt{r}), \tag{3.30}$$

which is in agreement with the results given in (Sih and Liebowitz, 1968).

B. Example 2: Flat Centrally Loaded Circular Punch

Consider a transversely isotropic elastic half-space $z \geq 0$, penetrated by a rigid circular punch of radius a. The punch loading is statically equivalent to a centrally applied normal force P. Denote the punch settlement by $\omega = \text{const}$. Solution of the integral equation (2.71) is given by (2.72), and in this particular case takes the form

$$\sigma(\rho, \phi) = \frac{\omega}{\pi^2 H(a^2 - \rho^2)^{1/2}}. \tag{3.31}$$

Integration of the last expression over the circle $\rho \leq a$ relates the punch settlement ω with the total force P as

$$P = \frac{2\omega a}{\pi H}.$$

The substitution of (3.31) in (2.67) leads to an integral which can be computed by differentiation of the expression for the main potential function (A.1) with respect to a. The result is

$$F = \frac{2\omega}{\pi H}\left\{z \sin^{-1}\left(\frac{a}{l_2}\right) - (a^2 - l_1^2)^{1/2} + a \ln[l_2 + (l_2^2 - \rho^2)^{1/2}]\right\}. \tag{3.32}$$

The appropriate differentiation of the potential function (3.32) gives the complete solution:

$$u = \frac{2\omega a e^{i\phi}}{\pi\rho} \sum_{k=1}^{2} \frac{\gamma_k}{m_k - 1}\left[1 - \frac{(a^2 - l_{1k}^2)^{1/2}}{a}\right], \tag{3.33}$$

$$w = \frac{2\omega}{\pi} \sum_{k=1}^{2} \frac{m_k}{m_k - 1} \sin^{-1}\left(\frac{a}{l_{2k}}\right), \tag{3.34}$$

$$\sigma_1 = -\frac{4\omega A_{66}}{\pi} \sum_{k=1}^{2} \frac{\gamma_k^2 - (m_k + 1)\gamma_3^2}{\gamma_k(m_k - 1)} \frac{(a^2 - l_{1k}^2)^{1/2}}{l_{2k}^2 - l_{1k}^2}, \tag{3.35}$$

$$\sigma_2 = \frac{4\omega A_{66} e^{2i\phi}}{\pi} \sum_{k=1}^{2} \frac{\gamma_k}{(m_k - 1)} \left\{\frac{(a^2 - l_{1k}^2)^{1/2}}{l_{2k}^2 - l_{1k}^2} - \frac{2a}{\rho^2}\left[1 - \frac{(a^2 - l_{1k}^2)^{1/2}}{a}\right]\right\}, \tag{3.36}$$

$$\sigma_z = \frac{\omega}{\pi^2 H(\gamma_1 - \gamma_2)} \sum_{k=1}^{2} (-1)^k \gamma_k \frac{(a^2 - l_{1k}^2)^{1/2}}{l_{2k}^2 - l_{1k}^2}, \tag{3.37}$$

$$\tau_z = \frac{\omega e^{i\phi}}{\pi^2 H(\gamma_1 - \gamma_2)} \sum_{k=1}^{2} (-1)^k \frac{l_{1k}(l_{2k}^2 - a^2)^{1/2}}{l_{2k}(l_{2k}^2 - l_{1k}^2)}. \tag{3.38}$$

This problem was first solved by Elliott (1949). His solution is essentially in agreement with ours. The fact that he uses the notation ρ as an elastic parameter and as a polar radius simultaneously, somewhat complicates the comparison.

Introduce the stress intensity factor as

$$k_1 = \lim_{\rho \to a} [(a - \rho)^{1/2} \sigma_z] \qquad \text{for } z = 0. \tag{3.39}$$

Substitution of (3.37) in (3.39) yields

$$k_1 = -\frac{\omega}{\pi^2 H \sqrt{2a}} = -\frac{P}{2\pi} \left(\frac{a}{2}\right)^{1/2}. \tag{3.40}$$

The asymptotical behavior of the field of stresses and displacements near the punch edge can be derived by substitution of (3.10) and (3.40) in (3.33)–(3.38). The result is

$$u = u_n = 2\pi H k_1 \sqrt{2r} \left(\frac{\gamma_1 S_1}{m_1 - 1} + \frac{\gamma_2 S_2}{m_2 - 1}\right) + O(1), \tag{3.41}$$

$$w = -2\pi H k_1 \sqrt{2r} \left(\frac{m_1 T_1}{m_1 - 1} + \frac{m_2 T_2}{m_2 - 1}\right) + O(1), \tag{3.42}$$

$$\sigma_1 = 2\pi A_{66} H k_1 \left(\frac{2}{r}\right)^{1/2} \sum_{k=1}^{2} \frac{\gamma_k^2 - (m_k + 1)\gamma_3^2}{\gamma_k(m_k - 1)} \frac{S_k}{Q_k} + O(1), \tag{3.43}$$

$$\sigma_2 = -2\pi A_{66} H k_1 \left(\frac{2}{r}\right)^{1/2} \sum_{k=1}^{2} \frac{\gamma_k S_k}{(m_k - 1) Q_k} + O(\sqrt{r}), \tag{3.44}$$

$$\sigma_z = \frac{k_1}{\sqrt{2r}\,(\gamma_1 - \gamma_2)} \left(\frac{\gamma_1 S_1}{Q_1} - \frac{\gamma_2 S_2}{Q_2}\right) + O(1), \tag{3.45}$$

$$\tau_z = \tau_{zn} = -\frac{k}{\sqrt{2r}\,(\gamma_1 - \gamma_2)} \left(\frac{T_1}{Q_1} - \frac{T_2}{Q_2}\right) + O(\sqrt{r}). \tag{3.46}$$

Comparison of (3.41)–(3.46) with (3.13)–(3.18) indicates that they become identical if one substitutes formally θ by $\pi - \theta$. This is easy to explain. The punch problem is mathematically equivalent to an external crack problem. There exists a notion that the asymptotic behavior of stresses and displacements near the edge of an arbitary flat crack with a smooth boundary is completely defined by three stress intensity factors. This means that the asymptotics of an internal crack and an external one should be the same. The system of local axes was introduced in previous paragraph so that the angle θ was measured from the direction outside the penny-shaped crack. In the case of the punch problem the axis *On* should be inverted inside the circle, this will make the expressions (3.41)–(3.46) identical to (3.13)–(3.18).

In the limiting case of an isotropic body formulae (3.33)–(3.38) transform into

$$u = \frac{\omega e^{i\phi}}{\pi(1-\nu)}\left[-(1-2\nu)\frac{a-(a^2-l_1^2)^{1/2}}{\rho} + \frac{zl_1(l_2^2-a^2)^{1/2}}{l_2(l_2^2-l_1^2)}\right], \tag{3.47}$$

$$w = \frac{2\omega}{\pi}\left[\sin^{-1}\left(\frac{a}{l_2}\right) + \frac{z(a^2-l_1^2)^{1/2}}{2(1-\nu)(l_2^2-l_1^2)}\right], \tag{3.48}$$

$$\sigma_1 = -\frac{2\omega\mu}{\pi(1-\nu)}\left[(1+2\nu)\frac{(a^2-l_1^2)^{1/2}}{l_2^2-l_1^2} + \frac{z^2[l_1^4-a^2(2a^2-\rho^2+2z^2)]}{(a^2-l_1^2)^{1/2}(l_2^2-l_1^2)^3}\right], \tag{3.49}$$

$$\sigma_2 = \frac{2\omega\mu e^{2i\phi}}{\pi(1-\nu)}\left\{-(1-2\nu)\left[\frac{(a^2-l_1^2)^{1/2}}{l_2^2-l_1^2} - \frac{2[a-(a^2-l_1^2)^{1/2}]}{\rho^2}\right]\right.$$
$$\left. - \frac{az(l_2^2-a^2)^{1/2}[2l_1^4+\rho^2(l_1^2+3l_2^2-6a^2)]}{l_2^2(l_2^2-l_1^2)^3}\right\}, \tag{3.50}$$

$$\sigma_z = \frac{2\omega\mu}{\pi(1-\nu)}\left\{-\frac{(a^2-l_1^2)^{1/2}}{l_2^2-l_1^2} + \frac{z^2[l_1^4+a^2(\rho^2-2a^2-2z^2)]}{(a^2-l_1^2)^{1/2}(l_2^2-l_1^2)^3}\right\}, \tag{3.51}$$

$$\tau_z = -\frac{2\omega\mu}{\pi(1-\nu)}\frac{z\rho e^{i\phi}(a^2-l_1^2)^{1/2}(3l_2^2+l_1^2-4a^2)}{(l_2^2-l_1^2)^3}. \tag{3.52}$$

The equivalent expressions for the components in polar coordinates will take the form:

$$u_\rho = \frac{\omega}{\pi(1-\nu)}\left[-(1-2\nu)\frac{a-(a^2-l_1^2)^{1/2}}{\rho} + \frac{zl_1(l_2^2-a^2)^{1/2}}{l_2(l_2^2-l_1^2)}\right], \tag{3.53}$$

$$\sigma_\rho = \frac{2\omega\mu}{\pi(1-\nu)}\left\{(1-2\nu)\frac{a-(a^2-l_1^2)^{1/2}}{\rho^2} - \frac{(a^2-l_1^2)^{1/2}}{l_2^2-l_1^2}\right.$$
$$\left. - \frac{az(l_2^2-a^2)^{1/2}[l_1^4-l_2^4+\rho^2(l_1^2+3l_2^2-4a^2)]}{l_2^2(l_2^2-l_1^2)^3}\right\}, \tag{3.54}$$

$$\sigma_\phi = -\frac{2\omega\mu}{\pi(1-\nu)}\left\{(1-2\nu)\frac{a-(a^2-l_1^2)^{1/2}}{\rho^2}\right.$$
$$\left. + \frac{(a^2-l_1^2)^{1/2}[a^2-(1-2\nu)l_2^2]}{l_2^2(l_2^2-l_1^2)}\right\}, \tag{3.55}$$

$$\tau_{\rho z} = -\frac{2\omega\mu}{\pi(1-\nu)}\frac{z\rho(a^2-l_1^2)^{1/2}(3l_2^2+l_1^2-4a^2)}{(l_2^2-l_1^2)^3}. \tag{3.56}$$

The integral transform solution can be found in (Sneddon, 1951). It is in agreement with ours, except for the missing factor ρ in the last expression of (56, p. 462). In the examples to follow we shall consider several nonaxisymmetric problems that have not been solved before.

C. Example 3: Inclined Punch on an Elastic Half-Space

The case of a flat circular punch pressed against a tansversely isotropic elastic half-space by a noncentrally applied force P can be considered as a superposition of two: that of a centrally loaded punch, which was considered earlier, and a punch subjected to a tilting moment $\mathcal{M}$, which will be considered here. Let the displacements under the punch be

$$\omega = b_x y - b_y x. \tag{3.57}$$

Here b_x and b_y are the tilting angles about the axes Ox and Oy respectively. Introduce the complex tilting angle

$$b = b_x + ib_y. \tag{3.58}$$

Expression (3.57) can be rewritten now as

$$\omega(\rho, \phi) = \Im\{\bar{b}\rho e^{i\phi}\}. \tag{3.59}$$

Here $\Im$ indicates the imaginary part. The substitution of (3.59) in (2.72) yields

$$\sigma(\rho, \phi) = \frac{2\Im\{\bar{b}\rho e^{i\phi}\}}{\pi^2 H(a^2 - \rho^2)^{1/2}}. \tag{3.60}$$

Evaluation of the potential function (2.67) leads to the integral

$$\int_0^{2\pi} \int_0^a \frac{\rho_0 e^{i\phi_0}}{(a^2 - \rho_0^2)^{1/2}} \ln(R_0 + z)\, \rho_0\, d\rho_0\, d\phi_0. \tag{3.61}$$

Taking into consideration that

$$\frac{\rho e^{i\phi}}{(a^2 - \rho^2)^{1/2}} = \Lambda(a^2 - \rho^2)^{1/2},$$

the integral in (3.61) can be computed by parts, with the result given by (A.5). The potential function takes the form

$$F = \frac{2\Im\{\bar{b}\rho e^{i\phi}\}}{\pi H}\left[z \sin^{-1}\left(\frac{a}{l_2}\right) - (a^2 - l_1^2)^{1/2}\left(1 - \frac{l_1^2 + 2a^2}{3\rho^2}\right) - \frac{2a^3}{3\rho^2}\right]. \tag{3.62}$$

All the necessary derivatives are readily available from Appendix A, and we can write the complete solution:

$$u = \frac{2i}{\pi} \sum_{k=1}^{2} \frac{\gamma_k}{m_k - 1} \left\{ b\left[z_k \sin^{-1}\left(\frac{a}{l_{2k}}\right) - (a^2 - l_{1k}^2)^{1/2} \right] \right.$$

$$\left. - \bar{b} e^{2i\phi} \frac{2a^3 - (l_{1k}^2 + 2a^2)(a^2 - l_{1k}^2)^{1/2}}{3\rho^2} \right\}, \tag{3.63}$$

$$w = \frac{2}{\pi}(b_x y - b_y x) \sum_{k=1}^{2} \frac{m_k}{m_k - 1} \left[\sin^{-1}\left(\frac{a}{l_{2k}}\right) - \frac{a(l_{2k}^2 - a^2)^{1/2}}{l_{2k}^2} \right], \tag{3.64}$$

$$\sigma_1 = -\frac{8A_{66}}{\pi}(b_x y - b_y x) a^2 \sum_{k=1}^{2} \frac{\gamma_k^2 - (m_k + 1)\gamma_3^2}{\gamma_k(m_k - 1)} \frac{(a^2 - l_{1k}^2)^{1/2}}{l_{2k}^2(l_{2k}^2 - l_{1k}^2)}, \tag{3.65}$$

$$\sigma_2 = \frac{2A_{66} i e^{i\phi}}{\pi} \sum_{k=1}^{2} \frac{\gamma_k}{m_k - 1} \left\{ b \frac{a l_{1k}(a^2 - l_{1k}^2)^{1/2}}{l_{2k}(l_{2k}^2 - l_{1k}^2)} \right.$$

$$\left. - \bar{b} e^{2i\phi} \left[\frac{4[(l_1^2 + 2a^2)(a^2 - l_1^2)^{1/2} - 2a^3]}{3\rho^3} + \frac{a l_1 (a^2 - l_1^2)^{1/2}}{l_2(l_2^2 - l_1^2)} \right] \right\}, \tag{3.66}$$

$$\sigma_z = \frac{2(b_x y - b_y x)}{\pi^2 H(\gamma_1 - \gamma_2)} \sum_{k=1}^{2} (-1)^k \gamma_k \frac{a^2(a^2 - l_{1k}^2)^{1/2}}{l_{2k}^2(l_{2k}^2 - l_{1k}^2)}, \tag{3.67}$$

$$\tau_z = \frac{i}{\pi^2 H(\gamma_1 - \gamma_2)} \sum_{k=1}^{2} \left\{ b\left[\sin^{-1}\left(\frac{a}{l_{2k}}\right) - \frac{a(l_{2k}^2 - a^2)^{1/2}}{l_{2k}^2 - l_{1k}^2} \right] \right.$$

$$\left. + \bar{b} e^{2i\phi} \frac{a l_{1k}^2 (l_{2k}^2 - a^2)^{1/2}}{l_{2k}^2(l_{2k}^2 - l_{1k}^2)} \right\}. \tag{3.68}$$

To the best of our knowledge, this problem has not been solved before.

D. Example 4: Penny-Shaped Crack under Uniform Shear Loading

Consider a penny-shaped crack of radius a in a transversely isotropic elastic space, subjected to a uniform shear loading τ, where τ is a complex constant. The solution of the integro-differential equation (2.95) in this case is

$$u = \frac{2(G_1^2 - G_2^2)}{G_1} \tau (a^2 - \rho^2)^{1/2} \qquad \text{for } z = 0 \text{ and } \rho \le a. \tag{3.69}$$

Substitution of (3.69) in (2.90) leads to the integral

$$\frac{2(G_1^2 - G_2^2)}{G_1} \tau \int_0^{2\pi} \int_0^a (a^2 - \rho_0^2)^{1/2} \ln(R_0 + z) \rho_0 \, d\rho_0 \, d\phi_0,$$

which has been computed in Appendix A, with all the necessary derivatives. The complete solution will take the form:

$$u = \frac{2(G_1^2 - G_2^2)}{G_1}\left\{\frac{1}{m_1 - 1}[f_{17}(z_1)\bar{\tau} + f_{18}(z_1)\tau]\right.$$

$$\left. + \frac{1}{m_2 - 1}[f_{17}(z_2)\bar{\tau} + f_{18}(z_2)\tau] + f_{17}(z_3)\bar{\tau} - f_{18}(z_3)\tau\right\}, \tag{3.70}$$

$$w = \frac{G_1^2 - G_2^2}{2G_1}(\bar{\tau}\rho e^{i\phi} + \tau\rho e^{-i\phi}) \sum_{k=1}^{2} \frac{m_k}{m_k - 1}\left[\sin^{-1}\left(\frac{a}{l_{2k}}\right) - \frac{a(l_{2k}^2 - a^2)^{1/2}}{l_{2k}^2}\right], \tag{3.71}$$

$$\sigma_1 = \frac{2(G_1^2 - G_2^2)A_{66}}{G_1} a(\bar{\tau}e^{i\phi} + \tau e^{-i\phi})$$

$$\times \sum_{k=1}^{2} \frac{\gamma_3^2(m_k + 1) - \gamma_k^2}{\gamma_k^2(m_k - 1)} \frac{l_{1k}(a^2 - l_{1k}^2)^{1/2}}{l_{2k}(l_{2k}^2 - l_{1k}^2)}, \tag{3.72}$$

$$\sigma_2 = \frac{2(G_1^2 - G_2^2)A_{66}}{G_1}\left\{\frac{1}{m_1 - 1}[f_{19}(z_1)\bar{\tau} + f_{20}(z_1)\tau]\right.$$

$$\left. + \frac{1}{m_2 - 1}[f_{19}(z_2)\bar{\tau} + f_{20}(z_2)\tau] + f_{19}(z_3)\bar{\tau} - f_{20}(z_3)\tau\right\}, \tag{3.73}$$

$$\sigma_z = \frac{2\gamma_1\gamma_2\beta(\tau e^{-i\phi} + \bar{\tau}e^{i\phi})}{\pi G_1(\gamma_1 - \gamma_2)} \sum_{k=1}^{2} (-1)^k \frac{al_{1k}(a^2 - l_{1k}^2)^{1/2}}{l_{2k}(l_{2k}^2 - l_{1k}^2)}, \tag{3.74}$$

$$\tau_z = \frac{2\gamma_1\gamma_2}{\pi G_1}\left\{\frac{\beta}{\gamma_1 - \gamma_2} \sum_{k=1}^{2} \frac{(-1)^k}{\gamma_k}[f_{21}(z_k)\tau + f_{22}(z_k)\bar{\tau}]\right.$$

$$\left. + H[f_{21}(z_3)\tau - f_{22}(z_3)\bar{\tau}]\right\}. \tag{3.75}$$

Here

$$f_{17}(z) = e^{2i\phi}\frac{2a^3 - (l_1^2 + 2a^2)(a^2 - l_1^2)^{1/2}}{3\rho^2}, \tag{3.76}$$

$$f_{18}(z) = z\sin^{-1}\left(\frac{a}{l_2}\right) - (a^2 - l_1^2)^{1/2}, \tag{3.77}$$

$$f_{19}(z) = e^{3i\phi}\frac{al_1(a^2 - l_1^2)^{1/2}}{l_2(l_2^2 - l_1^2)} - 4e^{i\phi}f_{17}(z), \tag{3.78}$$

$$f_{20}(z) = e^{i\phi}\frac{al_1(a^2 - l_1^2)^{1/2}}{l_2(l_2^2 - l_1^2)}, \tag{3.79}$$

$$f_{21}(z) = -\sin^{-1}\left(\frac{a}{l_2}\right) + \frac{a(l_2^2 - a^2)^{1/2}}{l_2^2 - l_1^2}, \tag{3.80}$$

$$f_{22}(z) = \frac{e^{2i\phi} a l_1^2 (l_2^2 - a^2)^{1/2}}{l_2^2(l_2^2 - l_1^2)}. \tag{3.81}$$

A complete solution to this problem for the case of isotropy can be found in Westmann (1965); the case of transverse isotropy does not seem to have been considered before.

We can derive again some results of general nature, namely, the asymptotic behavior of the field of stresses and displacements in the neighborhood of the edge of a flat crack with a smooth boundary. We recall that at $\phi = 0$ the decompositions $u = u_x + iu_y$ and $\tau_z = \tau_{zx} + i\tau_{yz}$ are equal to $u^{(n)} = u_n + iu_t$ and $\tau^{(n)} = \tau_{zn} + i\tau_{tz}$ respectively; σ_1 is understood as $\sigma_n + \sigma_t$, and $\sigma_2 = \sigma_n - \sigma_t + 2i\tau_{nt}$. This will allow us to avoid a cumbersome axis transformation. The complex stress intensity factor, introduced in (2.145), can be expressed through the prescribed shear loading τ as

$$k = k_2 + ik_3 = \frac{\sqrt{2a}}{\pi}\left[\tau + \frac{G_2}{G_1}\bar{\tau}\right], \tag{3.82}$$

and its inversion gives

$$\tau = \frac{\pi G_1(kG_1 - \bar{k}G_2)}{\sqrt{2a}\,(G_1^2 - G_2^2)}. \tag{3.83}$$

Substitution of (3.10) and (3.83) in (3.70)–(3.81) yields

$$u_n + iu_t = \frac{k_2\gamma_1\gamma_2\sqrt{2r}}{A_{44}(\gamma_1 - \gamma_2)}\left(-\frac{S_1}{m_1 + 1} + \frac{S_2}{m_2 + 1}\right) + \frac{ik_3\gamma_3\sqrt{2r}}{A_{44}} S_3, \tag{3.84}$$

$$w = \frac{k_2\gamma_1\gamma_2\sqrt{2r}}{A_{44}(\gamma_1 - \gamma_2)}\left[-\frac{m_1 T_1}{\gamma_1(m_1 + 1)} + \frac{m_2 T_2}{\gamma_2(m_2 + 1)}\right], \tag{3.85}$$

$$\sigma_1 = 2\pi k_2\gamma_1\gamma_2 H A_{66}\left(\frac{2}{r}\right)^{1/2} \sum_{k=1}^{2} \frac{\gamma_3^2(m_k + 1) - \gamma_k^2}{\gamma_k^2(m_k - 1)} \frac{S_k}{Q_k}, \tag{3.86}$$

$$\sigma_2 = 2\pi k_2\gamma_1\gamma_2 H A_{66}\left(\frac{2}{r}\right)^{1/2}\left[\frac{S_1}{(m_1 - 1)Q_1} + \frac{S_2}{(m_2 - 1)Q_2}\right] - \frac{ik_3\sqrt{2}\,S_3}{\gamma_3\sqrt{r}\,Q_3}, \tag{3.87}$$

$$\sigma_z = -\frac{k_2\gamma_1\gamma_2}{\sqrt{2r}\,(\gamma_1-\gamma_2)}\left(\frac{S_1}{Q_1}-\frac{S_2}{Q_2}\right), \tag{3.88}$$

$$\tau_{zn}+i\tau_{tz} = -\frac{k_2\gamma_1\gamma_2}{\sqrt{2r}\,(\gamma_1-\gamma_2)}\left(\frac{T_1}{\gamma_1 Q_1}-\frac{T_2}{\gamma_2 Q_2}\right)+\frac{ik_3T_3}{\sqrt{2r}\,Q_3}. \tag{3.89}$$

By taking the sum and the difference of (3.86) and (3.87), one gets

$$\sigma_n = \frac{k_2\gamma_1\gamma_2}{\sqrt{2r}\,(\gamma_1-\gamma_2)}\left(\frac{S_1}{\gamma_1^2 Q_1}-\frac{S_2}{\gamma_2^2 Q_2}\right), \tag{3.90}$$

$$\sigma_t = \frac{k_2\gamma_1\gamma_2}{\sqrt{2r}}\left\{\frac{1}{(\gamma_1-\gamma_2)}\left[\left(\frac{1}{\gamma_1^2}-\frac{1}{\gamma_3^2}\right)\frac{S_1}{Q_1}-\left(\frac{1}{\gamma_2^2}-\frac{1}{\gamma_3^2}\right)\frac{S_2}{Q_2}\right]\right.$$
$$\left.+\,2\pi HA_{66}\left(\frac{S_1}{Q_1}+\frac{S_2}{Q_2}\right)\right\}, \tag{3.91}$$

$$\tau_{nt} = -\frac{k_3S_3}{\sqrt{2r}\,\gamma_3Q_3}. \tag{3.92}$$

Our formulae (3.84)–(3.92) are in relatively good agreement with similar results of Kassir and Sih (1975), except for formula (8.96b, p. 371) for u_t, which should correspond to the imaginary part of our (3.84). Formula by Kassir and Sih (8.96b) seems to be in error because it implies that u_t depends on k_2, γ_1, and γ_2, which is wrong: our results relate u_t to k_3 and γ_3 only, as it should be. There are several misprints in their formulae (8.96a) and (8.96c). The remaining formulae are in agreement, though the formulae by Kassir and Sih (1975) look more complicated than ours, mainly because they did not notice the properties (2.23), which could make some expressions much simpler.

In the case of isotropy, formulae (3.70)–(3.75) transform into

$$u = \frac{1}{\pi\mu(2-\nu)}\left\{\left[(-5+4\nu)z\sin^{-1}\left(\frac{a}{l_2}\right)+4(1-\nu)(a^2-l_1^2)^{1/2}\right]\tau\right.$$
$$\left.+\frac{za(l_2^2-a^2)^{1/2}}{l_2^2-l_1^2}\left(\tau+\frac{l_1^2}{l_2^2}\bar{\tau}e^{2i\phi}\right)\right\}, \tag{3.93}$$

$$w = \frac{(\bar{\tau}e^{i\phi}+\tau e^{-i\phi})\rho}{\pi\mu(2-\nu)}\left\{\frac{1-2\nu}{2}\left[\sin^{-1}\left(\frac{a}{l_2}\right)-\frac{a(l_2^2-a^2)^{1/2}}{l_2^2}\right]+\frac{za^2(a^2-l_1^2)^{1/2}}{l_2^2(l_2^2-l_1^2)}\right\}, \tag{3.94}$$

$$\sigma_1 = \frac{2(\bar{\tau}e^{i\phi} + \tau e^{-i\phi})}{\pi(2-\nu)}\left\{-2(1+\nu)\frac{al_1(a^2-l_1^2)^{1/2}}{l_2(l_2^2-l_1^2)}\right.$$
$$\left. + \frac{zl_1(l_2^2-a^2)^{1/2}[a^2(4l_2^2-5\rho^2)+l_1^4]}{l_2(l_2^2-l_1^2)^3}\right\}, \tag{3.95}$$

$$\sigma_2 = -\frac{2e^{i\phi}}{\pi(2-\nu)}\left\{4(1-\nu)\frac{al_1(a^2-l_1^2)^{1/2}}{l_2(l_2^2-l_1^2)}\tau\right.$$
$$\left. + \frac{zl_1(l_2^2-a^2)^{1/2}}{l_2(l_2^2-l_1^2)}\left[\frac{4a^2}{l_2^2}\bar{\tau}e^{2i\phi} - \frac{a^2(4l_2^2-5\rho^2)+l_1^4}{(l_2^2-l_1^2)^2}(\tau+\bar{\tau}e^{2i\phi})\right]\right\}, \tag{3.96}$$

$$\sigma_z = -\frac{2(\bar{\tau}e^{i\phi}+\tau e^{-i\phi})}{\pi(2-\nu)}\frac{zl_1(l_2^2-a^2)^{1/2}[a^2(4l_2^2-5\rho^2)+l_1^4]}{l_2(l_2^2-l_1^2)^3}, \tag{3.97}$$

$$\tau_z = \frac{2}{\pi(2-\nu)}\left\{\left[(2-\nu)\left(\frac{a(l_2^2-a^2)^{1/2}}{l_2^2-l_1^2} - \sin^{-1}\left(\frac{a}{l_2}\right)\right)\right.\right.$$
$$\left. + \frac{z(a^2-l_1^2)^{1/2}[l_1^4+a^2(2a^2+2z^2-3\rho^2)]}{(l_2^2-l_1^2)^3}\right]\tau$$
$$\left. + \left[\nu a(l_2^2-a^2)^{1/2} + \frac{z(a^2-l_1^2)^{1/2}[a^2(6l_2^2-2l_1^2+\rho^2)-5l_2^4]}{(l_2^2-l_1^2)^2}\right]\frac{l_1^2e^{2i\phi}}{l_2^2(l_2^2-l_1^2)}\bar{\tau}\right\}. \tag{3.98}$$

The asymptotic behavior near the crack boundary for the case of isotropy will take the form

$$u_n + iu_t = \frac{\sqrt{r}}{\mu}\sin\frac{\theta}{2}\left\{\left[2(1-\nu)+\cos^2\frac{\theta}{2}\right]k_2 + 2ik_3\right\},$$
$$w = k_2\frac{\sqrt{r}}{\mu}\cos\frac{\theta}{2}\left[-(1-2\nu)+\sin^2\frac{\theta}{2}\right],$$
$$\sigma_1 = \frac{k_2}{\sqrt{r}}\sin\frac{\theta}{2}\left[-2(1+\nu)-\cos\frac{\theta}{2}\cos\frac{3\theta}{2}\right],$$
$$\sigma_2 = \frac{\sin(\theta/2)}{\sqrt{r}}\left\{\left[-2(1-\nu)-\cos\frac{\theta}{2}\cos\frac{3\theta}{2}\right]k_2 - 2ik_3\right\},$$
$$\sigma_z = \frac{k_2}{2\sqrt{r}}\sin\theta\cos\frac{3\theta}{2},$$
$$\tau_{zn} + i\tau_{tz} = \frac{\cos(\theta/2)}{\sqrt{r}}\left\{\left[1-\sin\frac{\theta}{2}\sin\frac{3\theta}{2}\right]k_2 + ik_3\right\}.$$

The last set of formulae is in agreement with the results presented in Sih and Liebowitz (1968).

The case of nonuniform shear loading can be treated in a similar manner. For example, in the case when the loading is proportional to the term $\rho e^{i\phi}$, the main potential function will be proportional to the integral

$$I = \int_0^{2\pi} \int_0^a \ln(R_0 + z)\,(a^2 - \rho_0^2)^{1/2} e^{i\phi_0} \rho_0^2 \,d\rho_0\, d\phi_0$$

$$= \frac{\pi}{8} \rho e^{i\phi} \left\{ (a^2 - l_1^2)^{1/2} \left[\frac{4}{3} a^2 + 7\rho^2 - \frac{19}{3} l_1^2 - 4 l_2^2 + \frac{2(8a^4 + 4a^2 l_1^2 + 3 l_1^4)}{15\rho^2} \right] \right.$$

$$\left. + z(4a^2 - 3\rho^2 + 4z^2) \sin^{-1}\left(\frac{a}{l_2}\right) - \frac{16a^5}{15\rho^2} \right\}.$$

It might be a good exercise for the reader to do the differentiation and write a complete solution. As an example, here are the first two derivatives

$$\frac{\partial I}{\partial z} = \frac{\pi}{8} \rho e^{i\phi} \left\{ a(l_2^2 - a^2)^{1/2} \left[\frac{15 l_1^2}{a^2} - 12 - \frac{2 l_1^2}{l_2^2} \right] + \sin^{-1}\left(\frac{a}{l_2}\right) [4a^2 - 3\rho^2 + 12z^2] \right\},$$

$$\frac{\partial^2 I}{\partial z^2} = \pi \rho e^{i\phi} \left[(a^2 - l_1^2)^{1/2} \left(\frac{a^2}{l_2^2} - 3 \right) + 3z \sin^{-1}\left(\frac{a}{l_2}\right) \right].$$

It is noteworthy that $\partial I / \partial z$ is proportional to the main potential function for the case of linear *normal* loading of a penny-shaped crack.

IV. Discussion

Here we discuss some immediate and future applications of the method presented in this paper. Comparison is made also with some results reported in the literature.

The expressions for the energy release rate can be obtained from (3.46), (3.47), (3.88), and (3.89) by the procedure similar to the one employed by Kassir and Sih (1968). The result is

$$\mathscr{G}_1 = 2\pi^2 H k_1^2, \qquad \mathscr{G}_2 = 2\pi^2 H \gamma_1 \gamma_2 k_2^2, \qquad \mathscr{G}_3 = 2\pi^2 \beta k_3^2,$$

which is in agreement with similar results of Kassir and Sih (1968), and gives an additional reason for the introduction of the parameters H and β.

Investigation of various interactions (loads with cracks and punches, between cracks, etc.) becomes readily available due to the new theory. For example, consider the interaction of a concentrated load P_0, applied at an arbitrary point (ρ, ϕ, z) in the z-direction, with a flat circular punch of radius a. One can deduce from (3.34) that the normal displacement w at the point (ρ, ϕ, z), due to a unit force applied to the punch, is

$$w = \frac{H}{a} \sum_{k=1}^{2} \frac{m_k}{m_k - 1} \sin^{-1}\left(\frac{a}{l_{2k}}\right).$$

Application of the reciprocal theorem immediately gives the punch settlement ω due to the load P_0:

$$\omega = \frac{H}{a} P_0 \sum_{k=1}^{2} \frac{m_k}{m_k - 1} \sin^{-1}\left(\frac{a}{l_{2k}}\right).$$

The tilting angle δ of the punch can be obtained in a similar manner from (3.64). The result is

$$\delta = \frac{3H}{2a^3} P_0 \sum_{k=1}^{2} \frac{m_k}{m_k - 1} \left[\rho \sin^{-1}\left(\frac{a}{l_{2k}}\right) - \frac{l_{1k}(l_{2k}^2 - a^2)^{1/2}}{l_{2k}}\right].$$

Some well-known results can be simplified significantly by using the method of computation of various integrals involving distances between several points. For example, here is the set of governing integral equations derived by Panasiuk *et al.* (1986, p. 83) for the problem of interaction of N coplanar thin spheroidal inclusions:

$$\begin{aligned} w_n(x_n, y_n) + H\Gamma w_n(x_n, y_n) - \frac{1}{4\pi^2} \sum_{\substack{k=1 \\ k\neq n}}^{N} \iint_{S_k} w_k(x_k, y_k)\Gamma[(x_n - x_k)^2 \\ + (y_n - y_k)^2]^{-3/2}\, dx_k\, dy_k \\ = f_n(x_n, y_n) \qquad \text{for } n = 1, 2, \ldots, N. \end{aligned} \tag{4.1}$$

Here f_n is a known function, w_n is the normal displacement at the boundary of the n-th inclusion, S_n is its median crossection, $(x_n, y_n) \in S_n$, and the integral operator Γ is defined by

$$\begin{aligned} \Gamma\Phi(x_k, y_k) = \iint_{S_k} \frac{\Phi(\xi, \eta)}{[(x_k - \xi)^2 + (y_k - \eta)^2]^{1/2}}\, d\xi\, d\eta \\ - \frac{1}{\pi^2} \iint_{\bar{S}_k} \frac{d\xi\, d\eta}{\{[\xi^2 + \eta^2 - a_k^2][(x_k - \xi)^2 + (y_k - \eta)^2]\}^{1/2}} \\ \times \iint_{S_k} \frac{\Phi(\xi_1, \eta_1)(a_k^2 - \xi_1^2 - \eta_1^2)^{1/2}}{(\xi - \xi_1)^2 + (\eta - \eta_1)^2}\, d\xi_1\, d\eta_1. \end{aligned} \tag{4.2}$$

Here a_k is the radius of median crosssection of the k-th inclusion, $\bar{S}_k$

indicates the area outside S_k. The double and quadruple integrals in (4.2), which is a kernel of the integral equation (4.1), make any numerical solution next to impossible. Panasiuk *et al.* (1986) have managed to give an approximate solution for the case when the inclusions are far apart, which is of little practical value, since there is almost no interaction at such distances between the inclusions. Let us show that (4.1) can be simplified so that its kernel be presented in elementary functions. Making use of the integral (compare with (C.31))

$$\int_0^{2\pi}\int_a^{\infty}\frac{(a^2-\rho_0^2)^{1/2}}{(r^2-a^2)^{1/2}}\frac{1}{\rho_0^2+r^2-2\rho_0 r\cos(\phi_0-\psi)}\frac{r\,dr\,d\psi}{[\rho^2+r^2-2\rho r\cos(\phi-\psi)]^{1/2}}$$
$$=\frac{2\pi}{R}\left\{\frac{\pi}{2}-\tan^{-1}\left[\frac{(a^2-\rho^2)^{1/2}(a^2-\rho_0^2)^{1/2}}{aR}\right]\right\},$$

one can change the order of integration in the second integral of (4.2), perform the integration in $\bar{S}_k$, and the integral operator Γ simplifies in polar coordinates significantly, namely,

$$\Gamma\Phi(\rho,\phi)=\frac{2}{\pi}\iint_{S_k}\frac{1}{R}\tan^{-1}\left(\frac{\eta}{R}\right)\Phi(\rho_0,\phi_0)\rho_0\,d\rho_0\,d\phi_0,$$

which is much simpler than (4.2). It is reminded that R and η are defined by (2.29). We can also compute $\Gamma[(x_n-x_k)^2+(y_n-y_k)^2]^{-3/2}$ in elementary functions. Indeed, expression (C.19) in the particular case of $z\to 0$ gives

$$\int_0^{2\pi}\int_0^{a}\frac{1}{R}\tan^{-1}\left(\frac{\eta}{R}\right)\frac{\rho_0\,d\rho_0\,d\phi_0}{[\rho_0^2+r^2-2r\rho_0\cos(\phi_0-\psi)]^{3/2}}$$
$$=\frac{2\pi(a^2-\rho^2)^{1/2}}{(r^2-a^2)^{1/2}[r^2+\rho^2-2r\rho\cos(\phi-\psi)]}\qquad\text{for } r>a.$$

The results above simplify (4.1) so significantly that now it can be easily solved by iteration numerically or analytically. The reader is referred for more detail to Fabrikant (1987c). A full investigation of this and other interaction problems is beyond the scope of this article and will be published separately.

A complete solution to the problem of a semi-infinite plane crack in an isotropic body was obtained by Uflyand (1965), who used a very complicated Kontorovich–Lebedev integral transform. To the best of our knowledge, the same problem for a transversely isotropic solid was not considered as yet. All these results can now be obtained as a limiting case of the penny-shaped crack solutions, when the radius $a \to \infty$, and the coordinate origin moves from the circle centre to its boundary. No integral transform is necessary. All the Green's functions are elementary.

Some of the integrals, involving special functions, can now be computed just by comparison of the solutions obtained by the integral transform method with the similar solution, given by the present method. For example, comparison of (2.39) with formulae (1.42) and (1.43) from Kassir and Sih (1975) leads to

$$\int_0^\infty \frac{J_{n+1/2}(as)J_n(\rho s)e^{-sz}}{\sqrt{s}}\,ds = \left(\frac{2}{\pi a}\right)^{1/2}(a\rho)^{-n}\int_0^{l_1}\frac{x^{2n}}{(\rho^2-x^2)^{1/2}}\,dx.$$

We have verified numerically that the last formula is correct for noninteger n as well. An enormous amount of material on Bessel functions exists in the literature, so it is difficult to claim that the last result is new, but there is some chance that it is new, since our notation l_1 and l_2 does not seem to have been used before. Here is another example of just how useful this notation is. Suppose that we need to compute the integral

$$\int_0^\infty \frac{\sin ax\, J_1(\rho x)e^{-zx}}{x^2}\,dx. \tag{4.3}$$

We could not find this integral in the tables, but we have found another one (Gradshtein and Ryzhik, 1963, formula 6.752.2)

$$\int_0^\infty \frac{\sin ax\, J_1(\rho x)e^{-zx}}{x}\,dx = \frac{a}{\rho}(1-r), \tag{4.4}$$

where the parameter r is a positive root of the equation

$$a^2 = \frac{\rho^2}{1-r^2} - \frac{z^2}{r^2}. \tag{4.5}$$

The original integral (4.3) can be computed by integration of both sides of (4.4) with respect to z. We need to get an explicit expression of r from the fourth order algebraic equation (4.5), substitute the result in (4.4), and integrate the result with respect to z, which does not seem possible at first. The introduction of the parameters l_1 and l_2 allows us to find the positive root of (4.5) in a very simple form, namely, $r = (a^2 - l_1^2)^{1/2}/a$. The z-integration can be performed by using (B.3), and the final result is

$$\int_0^\infty \frac{\sin ax\, J_1(\rho x)e^{-zx}}{x^2}\,dx = \frac{(2a^2-l_1^2)(l_2^2-a^2)^{1/2} - 2a^2 z}{2a\rho} - \frac{\rho}{2}\sin^{-1}\left(\frac{a}{l_2}\right). \tag{4.6}$$

It is of interest to indicate several solutions to the governing integral equation (2.71) in the case of a circular punch that are equivalent to (2.72). One such solution was obtained in Fabrikant (1986b):

$$\sigma(\rho, \phi) = \frac{1}{\pi^2 H}\left\{\frac{1}{(a^2-\rho^2)^{1/2}}\frac{\partial}{\partial a}\int_0^a \frac{\rho_0\, d\rho_0}{(a^2-\rho_0^2)^{1/2}} L\left(\frac{\rho\rho_0}{a^2}\right)\omega(\rho_0\phi)\right.$$
$$\left. -\frac{1}{2\pi}\int_0^{2\pi}\int_0^a \frac{1}{R}\tan^{-1}\left(\frac{\eta}{R}\right)\Delta\omega(\rho_0, \phi_0)\rho_0\, d\rho_0\, d\phi_0\right\}.$$

Here R and η are defined by (2.29), and Δ is a two-dimensional Laplace operator. Taking into consideration that σ can be expressed through the limiting case of (2.79) for $z \to 0$, the following expression can be obtained

$$\sigma(\rho, \phi) = -\frac{1}{4\pi^2 H}\left\{\Delta\int_0^{2\pi}\int_0^a \frac{\omega(\rho_0, \phi_0)}{R}\rho_0\, d\rho_0\, d\phi_0\right.$$
$$\left. +\frac{2}{\pi}\int_0^{2\pi}\int_0^a \left[\frac{R}{\eta} - \tan^{-1}\left(\frac{R}{\eta}\right)\right]\frac{\omega(\rho_0, \phi_0)}{R^3}\rho_0\, d\rho_0\, d\phi_0\right\}.$$

This result was first derived by Leonov (1953). Another type of solution can be expressed through the second z-derivative of (2.75) for $z = 0$. Taking into consideration that for a harmonic function the second z-derivative is equivalent to $-\Delta$, the following result is valid:

$$\sigma(\rho, \phi) = -\frac{1}{2\pi^3 H}\left\{-\Delta\int_0^{2\pi}\int_0^a \frac{1}{R}\tan^{-1}\left(\frac{\eta}{R}\right)\omega(\rho_0, \phi_0)\rho_0\, d\rho_0\, d\phi_0\right.$$
$$\left. +\frac{a}{(a^2-\rho^2)^{3/2}}\int_0^{2\pi}\int_0^a \frac{1-t\bar{t}}{(1-t)(1-\bar{t})}\frac{\omega(\rho_0, \phi_0)}{(a^2-\rho_0^2)^{1/2}}\rho_0\, d\rho_0\, d\phi_0\right\}$$
$$= -\frac{1}{\pi^2 H}\left\{-\Delta\int_\rho^a \frac{dx}{(x^2-\rho^2)^{1/2}}\int_0^x \frac{\rho_0\, d\rho_0}{(x^2-\rho_0^2)^{1/2}} L\left(\frac{\rho\rho_0}{x^2}\right)\omega(\rho_0, \phi)\right.$$
$$\left. +\frac{a}{(a^2-\rho^2)^{3/2}}\int_0^a \frac{\rho_0\, d\rho_0}{(a^2-\rho_0^2)^{1/2}} L\left(\frac{\rho\rho_0}{a^2}\right)\omega(\rho_0, \phi_0)\right\}.$$

The last formula corresponds to the solution by Mossakovskii *et al.* (1985).

V. Conclusion

The new method is proven capable of solving various non-axisymmetric problems as easy as axisymmetric ones. Some of the problems solved are still beyond the reach of the existing methods, and many more new solutions are to come. The fact that the new method does not use any integral transforms or special function expansions makes it available for use by the average engineer. It is believed that some of the results presented are of fundamental value not only to elasticity but to any branch of engineering science where potential theory is being used.

Appendix A

Here the main potential function is given, together with selected partial derivatives. Define the potential function by

$$\Psi = \int_0^{2\pi} \int_0^a (a^2 - \rho_0^2)^{1/2} \ln(R_0 + z)\rho_0 \, d\rho_0 \, d\phi_0 ,$$

where $R_0 = [\rho^2 + \rho_0^2 - 2\rho\rho_0 \cos(\phi - \phi_0) + z^2]^{1/2}$. The integral can be computed in elementary functions:

$$\Psi = \frac{\pi}{2}\left\{z\left(2a^2 - \rho^2 + \frac{2}{3}z^2\right)\sin^{-1}\left(\frac{a}{l_2}\right) + \frac{1}{3}(a^2 - l_1^2)^{1/2} \right.$$
$$\left. \times\left(5\rho^2 - \frac{10}{3}a^2 - 2l_2^2 - \frac{11}{3}l_1^2\right) + \frac{4}{3}a^3 \ln[l_2 + (l_2^2 - \rho^2)^{1/2}]\right\}. \quad (A.1)$$

The following derivatives may be computed:

$$\frac{\partial\Psi}{\partial x} = \pi x\left[-z\sin^{-1}\left(\frac{a}{l_2}\right) + (a^2 - l_1^2)^{1/2}\left(1 - \frac{l_1^2 + 2a^2}{3\rho^2}\right) + \frac{2a^3}{3\rho^2}\right], \quad (A.2)$$

$$\frac{\partial\Psi}{\partial y} = \pi y\left[-z\sin^{-1}\left(\frac{a}{l_2}\right) + (a^2 - l_1^2)^{1/2}\left(1 - \frac{l_1^2 + 2a^2}{3\rho^2}\right) + \frac{2a^3}{3\rho^2}\right], \quad (A.3)$$

$$\frac{\partial\Psi}{\partial z} = \frac{\pi}{2}\left[(2a^2 + 2z^2 - \rho^2)\sin^{-1}\left(\frac{a}{l_2}\right) - \frac{2a^2 - 3l_1^2}{a}(l_2^2 - a^2)^{1/2}\right], \quad (A.4)$$

$$\Lambda\Psi = \pi\rho e^{i\phi}\left[-z\sin^{-1}\left(\frac{a}{l_2}\right) + (a^2 - l_1^21)^{1/2}\left(1 - \frac{l_1^2 + 2a^2}{3\rho^2}\right) + \frac{2a^3}{3\rho^2}\right], \quad (A.5)$$

$$\frac{\partial^2\Psi}{\partial x^2} = \pi\left[-z\sin^{-1}\left(\frac{a}{l_2}\right) + (a^2 - l_1^2)^{1/2} + \left(1 - \frac{2x^2}{\rho^2}\right)\frac{2a^3 - (l_1^2 + 2a^2)(a^2 - l_1^2)^{1/2}}{3\rho^2}\right], \tag{A.6}$$

$$\frac{\partial^2\Psi}{\partial y^2} = \pi\left[-z\sin^{-1}\left(\frac{a}{l_2}\right) + (a^2 - l_1^2)^{1/2} + \left(1 - \frac{2y^2}{\rho^2}\right)\frac{2a^3 - (l_1^2 + 2a^2)(a^2 - l_1^2)^{1/2}}{3\rho^2}\right], \tag{A.7}$$

$$\frac{\partial^2\Psi}{\partial x\,\partial y} = -2\pi xy\,\frac{2a^3 - (l_1^2 + 2a^2)(a^2 - l_1^2)^{1/2}}{3\rho^4}, \tag{A.8}$$

$$\Lambda^2\Psi = -2\pi e^{2i\phi}\,\frac{2a^3 - (l_1^2 + 2a^2)(a^2 - l_1^2)^{1/2}}{3\rho^2}, \tag{A.9}$$

$$\Delta\Psi = 2\pi\left[-z\sin^{-1}\left(\frac{a}{l_2}\right) + (a^2 - l_1^2)^{1/2}\right], \tag{A.10}$$

$$\frac{\partial^2\Psi}{\partial z^2} = 2\pi\left[z\sin^{-1}\left\{\frac{a}{l_2}\right\} - (a^2 - l_1^2)^{1/2}\right], \tag{A.11}$$

$$\frac{\partial}{\partial z}\Lambda\Psi = \pi\rho e^{i\phi}\left[-\sin^{-1}\left(\frac{a}{l_2}\right) + \frac{a(l_2^2 - a^2)^{1/2}}{l_2^2}\right], \tag{A.12}$$

$$\frac{\partial^3\Psi}{\partial x^2\,\partial z} = \pi\left\{-\sin^{-1}\left(\frac{a}{l_2}\right) + \frac{a(l_2^2 - a^2)^{1/2}}{l_2^2}\left[1 + \frac{2a^2x^2}{l_2^2(l_2^2 - l_1^2)}\right]\right\}, \tag{A.13}$$

$$\frac{\partial^3\Psi}{\partial y^2\,\partial z} = \pi\left\{-\sin^{-1}\left(\frac{a}{l_2}\right) + \frac{a(l_2^2 - a^2)^{1/2}}{l_2^2}\left[1 + \frac{2a^2y^2}{l_2^2(l_2^2 - l_1^2)}\right]\right\}, \tag{A.14}$$

$$\frac{\partial^3\Psi}{\partial z^3} = 2\pi\left\{\sin^{-1}\left(\frac{a}{l_2}\right) - \frac{a(l_2^2 - a^2)^{1/2}}{l_2^2 - l_1^2}\right\}, \tag{A.15}$$

$$\frac{\partial^3\Psi}{\partial x\,\partial y\,\partial z} = 2\pi\,\frac{a^3xy(l_2^2 - a^2)^{1/2}}{l_2^4(l_2^2 - l_1^2)}, \tag{A.16}$$

$$\frac{\partial^3\Psi}{\partial x\,\partial z^2} = 2\pi\,\frac{a^2x(a^2 - l_1^2)^{1/2}}{l_2^2(l_2^2 - l_1^2)}, \tag{A.17}$$

$$\frac{\partial^3\Psi}{\partial y\,\partial z^2} = 2\pi\,\frac{a^2y(a^2 - l_1^2)^{1/2}}{l_2^2(l_2^2 - l_1^2)}, \tag{A.18}$$

$$\frac{\partial}{\partial z}\Lambda^2\Psi = 2\pi\rho^2e^{2i\phi}\,\frac{a^3(l_2^2 - a^2)^{1/2}}{l_2^4(l_2^2 - l_1^2)}, \tag{A.19}$$

$$\frac{\partial}{\partial z}\Delta\Psi = 2\pi\left[-\sin^{-1}\left(\frac{a}{l_2}\right) + \frac{a(l_2^2-a^2)^{1/2}}{l_2^2-l_1^2}\right], \tag{A.20}$$

$$\frac{\partial^2}{\partial z^2}\Lambda\Psi = 2\pi a^2\rho e^{i\phi}\frac{(a^2-l_1^2)^{1/2}}{l_2^2(l_2^2-l_1^2)}, \tag{A.21}$$

$$\Lambda\Delta\Psi = -2\pi a^2\rho e^{i\phi}\frac{(a^2-l_1^2)^{1/2}}{l_2^2(l_2^2-l_1^2)}, \tag{A.22}$$

$$\Lambda^3\Psi = -2\pi e^{3i\phi}\left\{\frac{4[(l_1^2+2a^2)(a^2-l_1^2)^{1/2}-2a^3]}{3\rho^3} + \frac{al_1(a^2-l_1^2)^{1/2}}{l_2(l_2^2-l_1^2)}\right\}, \tag{A.23}$$

$$\frac{\partial^4\Psi}{\partial z^4} = -2\pi\frac{za[l_1^4+a^2(2a^2+2z^2-3\rho^2)]}{(l_2^2-l_1^2)^3(l_2^2-a^2)^{1/2}}, \tag{A.24}$$

$$\frac{\partial^4\Psi}{\partial\rho\,\partial z^3} = -2\pi\frac{l_1(l_2^2-a^2)^{1/2}}{l_2(l_2^2-l_1^2)^3}[a^2(4l_2^4-5\rho^2)+l_1^4], \tag{A.25}$$

$$\frac{\partial^2}{\partial z^2}\Lambda^2\Psi = 2\pi\rho^2e^{2i\phi}a^3z\frac{a^2(6l_2^2-2l_1^2+\rho^2)-5l_2^4}{l_2^4(l_2^2-l_1^2)^3(l_2^2-a^2)^{1/2}}. \tag{A.26}$$

The following identities were used during the derivation of (A.1–A.26):

$$\begin{gathered} l_1l_2 = a\rho, \qquad l_1^2+l_2^2 = a^2+\rho^2+z^2, \\ (l_2^2-\rho^2)^{1/2}(l_2^2-a^2)^{1/2} = zl_2, \qquad (a^2-l_1^2)^{1/2}(\rho^2-l_1^2)^{1/2} = zl_1, \\ (a^2-l_1^2)^{1/2}(l_2^2-a^2)^{1/2} = za, \qquad (l_2^2-\rho^2)^{1/2}(\rho^2-l_1^2)^{1/2} = z\rho. \end{gathered} \tag{A.27}$$

$$\begin{gathered} \frac{\partial l_1}{\partial z} = -\frac{zl_1}{l_2^2-l_1^2}, \qquad \frac{\partial l_2}{\partial z} = \frac{zl_2}{l_2^2-l_1^2}, \\ \frac{\partial l_1}{\partial\rho} = \frac{al_2-\rho l_1}{l_2^2-l_1^2} = \frac{\rho(a^2-l_1^2)}{l_1(l_2^2-l_1^2)}, \qquad \frac{\partial l_2}{\partial\rho} = \frac{\rho l_2-al_1}{l_2^2-l_1^2} = \frac{\rho(l_2^2-a^2)}{l_2(l_2^2-l_1^2)}. \end{gathered} \tag{A.28}$$

Appendix B

Here we present some indefinite integrals of expressions containing l_1 and l_2.

$$\int (l_2^2-a^2)^{1/2}\,dz = (a^2-l_1^2)^{1/2}\frac{l_2^2-2a^2}{2a} + \frac{\rho^2}{2}\ln[l_2+(l_2^2-\rho^2)^{1/2}], \tag{B.1}$$

$$\int (l_2^2-a^2)^{1/2}l_1^2\,dz = -a(a^2-l_1^2)^{1/2}\frac{l_1^2+2a^2}{3} + a^2\rho^2\ln[l_2+(l_2^2-\rho^2)^{1/2}], \tag{B.2}$$

$$\int (a^2 - l_1^2)^{1/2}\, dz = \frac{2a^2 - l_1^2}{2a}(l_2^2 - a^2)^{1/2} + \frac{\rho^2}{2}\sin^{-1}\left(\frac{a}{l_2}\right), \tag{B.3}$$

$$\int (a^2 - l_1^2)^{1/2} l_1^2\, dz = -\frac{l_1^2(2l_1^2 + 3\rho^2)}{8a}(l_2^2 - a^2)^{1/2} + \rho^2\left(\frac{3}{8}\rho^2 - a^2\right)\sin^{-1}\left(\frac{a}{l_2}\right), \tag{B.4}$$

$$\int (l_2^2 - a^2)^{1/2}\frac{l_1^2}{l_2^2}\, dz = a(a^2 - l_1^2)^{1/2}\left[1 - \frac{8a^2}{15\rho^2} - \frac{4a^2 + 3l_1^2}{15l_2^2}\right], \tag{B.5}$$

$$\int \frac{(a^2 - l_1^2)^{1/2}}{l_2^2 - l_1^2}\, dz = -\sin^{-1}\left(\frac{a}{l_2}\right), \tag{B.6}$$

$$\int \frac{(a^2 - l_1^2)^{1/2}}{l_2^2(l_2^2 - l_1^2)}\, dz = \frac{1}{2a^2}\left[\frac{a(l_2^2 - a^2)^{1/2}}{l_2^2} - \sin^{-1}\left(\frac{a}{l_2}\right)\right], \tag{B.7}$$

$$\int \sin^{-1}\left(\frac{a}{l_2}\right) dz = z\sin^{-1}\left(\frac{a}{l_2}\right) - (a^2 - l_1^2)^{1/2} + a\ln[l_2 + (l_2^2 - \rho^2)^{1/2}], \tag{B.8}$$

$$\int z\sin^{-1}\left(\frac{a}{l_2}\right) dz = \frac{1}{4}(2a^2 + 2z^2 + \rho^2)\sin^{-1}\left(\frac{a}{l_2}\right) + (l_2^2 - a^2)^{1/2}\frac{2a^2 + l_1^2}{4a}, \tag{B.9}$$

$$\int z^2\sin^{-1}\left(\frac{a}{l_2}\right) dz = \frac{1}{3}z^3\sin^{-1}\left(\frac{a}{l_2}\right) + \frac{1}{18}(a^2 - l_1^2)^{1/2}(3l_2^2 + 6\rho^2 + 8a^2 - 2l_1^2) - \frac{1}{6}a(3\rho^2 + 2a^2)\ln[l_2 + (l_2^2 - \rho^2)^{1/2}]. \tag{B.10}$$

The integration in (B.1)–(B.10) was performed by parts, with a consequent change of variables: $z = (a^2 - l_1^2)^{1/2}(\rho^2 - l_1^2)^{1/2}/l_1$ or $z = (l_2^2 - a^2)^{1/2}(l_2^2 - \rho^2)^{1/2}/l_2$.

$$\int \rho\sin^{-1}\left(\frac{a}{l_2}\right) d\rho = \frac{\rho^2}{2}\sin^{-1}\left(\frac{a}{l_2}\right) + \frac{z(2a^2 - l_1^2)}{2(a^2 - l_1^2)^{1/2}}, \tag{B.11}$$

$$\int \rho^2\sin^{-1}\left(\frac{a}{l_2}\right) d\rho = \frac{\rho^3}{3}\sin^{-1}\left(\frac{a}{l_2}\right) + \frac{z\rho(2a^2 - l_1^2)}{6(a^2 - l_1^2)^{1/2}} + \frac{1}{6}a(a^2 - 3z^2)\cosh^{-1}\left[\frac{l_2}{(a^2 + z^2)^{1/2}}\right] - \frac{1}{6}z(3a^2 - z^2)\sin^{-1}\left[\frac{l_1}{(a^2 + z^2)^{1/2}}\right], \tag{B.12}$$

$$\int a \sin^{-1}\left(\frac{a}{l_2}\right) da = \frac{1}{4}\left[(2a^2 + 2z^2 - \rho^2) \sin^{-1}\left(\frac{a}{l_2}\right)\right.$$

$$\left. + l_1(\rho^2 - l_1^2)^{1/2} - 2z(a^2 - l_1^2)^{1/2}\right]. \tag{B.13}$$

The integration in (B.11) and (B.12) was performed by parts, with a consequent change of variables: $\rho = y[1 + z^2/(a^2 - y^2)]^{1/2}$, which corresponds to a substitution $l_2 = y$. A similar remark is valid for formula (B.13).

Appendix C

Several integrals, used in this article, are computed here. Consider three points in the system of cylindrical coordinates, namely, $M(\rho, \phi, z)$, $M_0(\rho_0, \phi_0, z_0)$, and $N(r, \psi, 0)$. The following notation is introduced:

$$l_1(t) = \tfrac{1}{2}\{[(\rho + t)^2 + z^2]^{1/2} - [(\rho - t)^2 + z^2]^{1/2}\}, \tag{C.1}$$

$$l_2(t) = \tfrac{1}{2}\{[(\rho + t)^2 + z^2]^{1/2} + [(\rho - t)^2 + z^2]^{1/2}\}, \tag{C.2}$$

$$l_{10}(t) = \tfrac{1}{2}\{[(\rho_0 + t)^2 + z_0^2]^{1/2} - [(\rho_0 - t)^2 + z_0^2]^{1/2}\}, \tag{C.3}$$

$$l_{20}(t) = \tfrac{1}{2}\{[(\rho_0 + t)^2 + z_0^2]^{1/2} + [(\rho_0 - t)^2 + z_0^2]^{1/2}\}. \tag{C.4}$$

According to the earlier convention, l_{10} stands as an abbreviation for $l_{10}(a)$, etc.; $R(\cdot, \cdot)$ denotes the distance between two points.

Consider the following integral:

$$\int_0^{2\pi}\int_0^a \frac{z}{R^3(M, N)}\frac{1}{R(N, M_0)} \tan^{-1}\left[\frac{h_0}{R(N, M_0)}\right] r\, dr\, d\psi, \tag{C.5}$$

where

$$h_0 = (a^2 - l_{10})^{1/2}(a^2 - r^2)^{1/2}/a. \tag{C.6}$$

Make use of the integral representation (Fabrikant, 1986b):

$$\frac{1}{R(N, M_0)} \tan^{-1}\left[\frac{h_0}{R(N, M_0)}\right]$$

$$= \int_r^a \frac{dl_{20}(x)}{(l_{20}^2(x) - \rho_0^2)^{1/2}(x^2 - r^2)^{1/2}} \lambda\left(\frac{\rho_0 r}{l_{20}^2(x)}, \psi - \phi_0\right), \tag{C.7}$$

where $\lambda(\cdot, \cdot)$ is defined by (2.26). The substitution of (C.7) in (C.5) yields, after changing the order of integration,

$$\int_0^{2\pi} d\psi \int_0^a \frac{dl_{20}(x)}{[l_{20}^2(x) - \rho_0^2]^{1/2}} \int_0^x \frac{r\,dr}{(x^2 - r^2)^{1/2}} \lambda\left(\frac{\rho_0 r}{l_{20}^2(x)}, \psi - \phi_0\right) \frac{z}{R^3(M, N)}. \quad \text{(C.8)}$$

By substituting the integral representation (D.6)

$$\frac{z}{R^3(M, N)} \equiv \frac{z}{[\rho^2 + r^2 - 2r\rho\cos(\phi - \psi) + z^2]^{3/2}}$$
$$= \frac{2}{\pi r} L\left(\frac{1}{r}\right) \frac{d}{dr} \int_0^r \frac{t\,dt}{(r^2 - t^2)^{1/2}} \frac{[t^2 - l_1^2(t)]^{1/2}}{l_2^2(t) - l_1^2(t)} \lambda\left(\frac{l_1(t)t}{l_2(t)}, \phi - \psi\right) \quad \text{(C.9)}$$

in (C.8), the following result can be obtained, after integration with respect to ψ:

$$4\int_0^a \frac{dl_{20}(x)}{[l_{20}(x) - \rho_0^2]^{1/2}} \int_0^x \frac{dr}{(x^2 - r^2)^{1/2}}$$
$$\times \frac{d}{dr} \int_0^r \frac{t\,dt}{(r^2 - t^2)^{1/2}} \frac{[t^2 - l_1^2(t)]^{1/2}}{l_2^2(t) - l_1^2(t)} \lambda\left(\frac{l_1(t)t\rho_0}{l_{20}^2(x) l_2(t)}, \phi - \phi_0\right). \quad \text{(C.10)}$$

Here the following property of the L-operators was used:

$$L(k)\lambda(k_1, \cdot) = \lambda(kk_1, \cdot) \qquad \text{for } k, k_1 < 1. \quad \text{(C.11)}$$

The well-known property of the Abel-type operators, namely,

$$\int_0^x \frac{dr}{(x^2 - r^2)^{1/2}} \frac{d}{dr} \int_0^{r} \frac{f(t)t\,dt}{(r^2 - t^2)^{1/2}} = \frac{\pi}{2} f(x), \quad \text{(C.12)}$$

allows us to simplify (C.10) significantly:

$$2\pi \int_0^a \frac{[x^2 - l_1^2(x)]^{1/2}}{l_2^2(x) - l_1^2(x)} \frac{[x^2 - l_{10}^2(x)]^{1/2}}{l_{20}^2(x) - l_{10}^2(x)} \lambda\left(\frac{l_1(x) l_{10}(x)}{l_2(x) l_{20}(x)}, \phi - \phi_0\right) dx. \quad \text{(C.13)}$$

It is noteworthy that the integrand in (C.13) is symmetric with respect to the points M and M_0 while it did not look so in the original expression (C.5). The integrand in (C.13) is a perfect differential, so that the integral can be computed as indefinite:

$$\int \frac{[x^2 - l_1^2(x)]^{1/2}}{l_2^2(x) - l_1^2(x)} \frac{[x^2 - l_{10}^2(x)]^{1/2}}{l_{20}^2(x) - l_{10}^2(x)} \lambda\left(\frac{l_1(x) l_{10}(x)}{l_2(x) l_{20}(x)}, \phi - \phi_0\right) dx$$
$$= \frac{1}{2R_1} \tan^{-1}\left(\frac{\Theta_1(x)}{R_1}\right) + \frac{1}{2R_2} \tan^{-1}\left(\frac{\Theta_2(x)}{R_2}\right), \quad \text{(C.14)}$$

where

$$R_1 = [\rho^2 + \rho_0^2 - 2\rho\rho_0\cos(\phi - \phi_0) + (z - z_0)^2]^{1/2},$$
$$R_2 = [\rho^2 + \rho_0^2 - 2\rho\rho_0\cos(\phi - \phi_0) + (z + z_0)^2]^{1/2},$$
$$\Theta_1(x) = \theta(x) + \frac{zz_0}{\theta(x)}, \qquad \Theta_2(x) = \theta(x) - \frac{zz_0}{\theta(x)}, \tag{C.15}$$
$$\theta(x) = \frac{1}{x}[x^2 - l_1^2(x)]^{1/2}[x^2 - l_{10}^2(x)]^{1/2}.$$

Correctness of the integral in (C.14) can be verified by differentiation. The algebra involved is not trivial. Here we present some intermediary transformations:

$$\frac{\partial}{\partial x}\theta(x) = \frac{\theta(x)[l_2^2(x)l_{20}^2(x) - l_1^2(x)l_{10}^2(x)]}{x[l_2^2(x) - l_1^2(x)][l_{20}^2(x) - l_{10}^2(x)]}, \tag{C.16}$$

$$\lambda\left(\frac{l_1(x)l_{10}(x)}{l_2(x)l_{20}(x)}, \phi - \phi_0\right) = \frac{l_2^2(x)l_{20}^2(x) - l_1^2(x)l_{10}^2(x)}{2x^2} \times \left[\frac{1}{R_1^2 + \Theta_1^2(x)} + \frac{1}{R_2^2 + \Theta_2^2(x)}\right]. \tag{C.17}$$

Formula (C.14) allows to to compute the integral (C.5):

$$\int_0^{2\pi}\int_0^a \frac{z}{R^3(M, N)}\frac{1}{R(N, M_0)}\tan^{-1}\left[\frac{h_0}{R(N, M_0)}\right] r\,dr\,d\psi$$
$$= \pi\frac{|z|}{z}\left\{\frac{1}{R_1}\left[\tan^{-1}\left(\frac{\Theta_1}{R_1}\right) - \frac{\pi}{2}\frac{|zz_0|}{zz_0}\right] + \frac{1}{R_2}\left[\tan^{-1}\left(\frac{\Theta_2}{R_2}\right) + \frac{\pi}{2}\frac{|zz_0|}{zz_0}\right]\right\}, \tag{C.18}$$

where the contractions Θ_1 and Θ_2 stand for $\Theta_1(a)$ and $\Theta_2(a)$ respectively. Note an important particular case when $z_0 = 0$. Formula (C.18) in this case transforms into

$$\int_0^{2\pi}\int_0^a \frac{z}{R^3(M, N)}\frac{1}{R(N, N_0)}\tan^{-1}\left[\frac{(a^2 - r^2)^{1/2}(a^2 - \rho_0^2)^{1/2}}{aR(N, N_0)}\right] r\,dr\,d\psi$$
$$= \frac{2\pi}{R(M, N_0)}\tan^{-1}\left[\frac{h}{R(M, N_0)}\right] \qquad \text{for } \rho_0 < a, \tag{C.19}$$

and the integral vanishes when $\rho_0 \geq a$. Here the point N_0 has the cylindrical coordinates $(\rho_0, \phi_0, 0)$, and h is defined by (2.33).

The second fundamental integral to be considered is:

$$\int_0^{2\pi}\int_0^a \frac{z_0}{R^3(N, M_0)}\left[\frac{R(N, M_0)}{h_0} + \tan^{-1}\left(\frac{h_0}{R(N, M_0)}\right)\right]\frac{r\,dr\,d\psi}{R(M, N)}, \quad \text{(C.20)}$$

where h_0 is defined by (C.6). Make use of the integral representation for the reciprocal distance

$$\frac{1}{R(M, N)} = \frac{2}{\pi}\int_0^r \frac{dx}{(r^2 - x^2)^{1/2}}\frac{[l_2^2(x) - x^2]^{1/2}}{l_2^2(x) - l_1^2(x)}\lambda\left(\frac{l_1(x)x}{l_2(x)r}, \phi - \psi\right), \quad \text{(C.21)}$$

which is a variation of the result established in Fabrikant (1986b). By substituting (C.21) in (C.20), the following expression results, after changing the order of integration:

$$\frac{2}{\pi}\int_0^{2\pi} d\psi \int_0^a \frac{[l_2^2(x) - x^2]^{1/2}}{l_2^2(x) - l_2^2(x)}\,dx \int_x^a \frac{r\,dr}{(r^2 - x^2)^{1/2}}$$
$$\times \lambda\left(\frac{l_1(x)x}{l_2(x)r}, \phi - \psi\right)\frac{z_0}{R^3(N, M_0)}\left[\frac{R(N, M_0)}{h_0} + \tan^{-1}\left(\frac{h_0}{R(N, M_0)}\right)\right], \quad \text{(C.22)}$$

We use the integral representation (Fabrikant, 1986b):

$$\frac{z_0}{R^3(N, M_0)}\left[\frac{R(N, M_0)}{h_0} + \tan^{-1}\left(\frac{h_0}{R(N, M_0)}\right)\right]$$
$$= -\frac{L(r)}{r}\frac{d}{dr}\int_r^a \frac{t\,dt}{(t^2 - r^2)^{1/2}}\frac{[l_{20}^2(t) - t^2]^{1/2}}{l_{20}^2(t) - l_{10}^2(t)}$$
$$\times \lambda\left(\frac{l_{10}(t)}{tl_{20}(t)}, \phi_0 - \psi\right). \quad \text{(C.23)}$$

The substitution of (C.23) in (C.22) yields, after integration with respect to ψ:

$$-4\int_0^a \frac{[l_2^2(x) - x^2]^{1/2}}{l_2^2(x) - l_1^2(x)}\,dx \int_x^a \frac{dr}{(r^2 - x^2)^{1/2}}\frac{d}{dr}\int_r^a \frac{t\,dt}{(t^2 - r^2)^{1/2}}$$
$$\times \frac{[l_{20}^2(t) - t^2]^{1/2}}{l_{20}^2(t) - l_{10}^2(t)}\lambda\left(\frac{l_1(x)l_{10}(t)x}{l_2(x)l_{20}(t)t}, \phi - \phi_0\right). \quad \text{(C.24)}$$

We recall another well-known property of the Abel operators:

$$\int_x^a \frac{dr}{(r^2 - x^2)^{1/2}}\frac{d}{dr}\int_r^a \frac{f(t)t\,dt}{(t^2 - r^2)^{1/2}} = -\frac{\pi}{2}f(x). \quad \text{(C.25)}$$

Application of (C.25) to (C.24) yields

$$2\pi\int_0^a \frac{[l_2^2(x)-x^2]^{1/2}}{l_2^2(x)-l_1^2(x)}\frac{[l_{20}^2(x)-x^2]^{1/2}}{l_{20}^2(x)-l_{10}^2(x)}\lambda\left(\frac{l_1(x)l_{10}(x)}{l_2(x)l_{20}(x)},\ \phi-\phi_0\right)dx. \quad (C.26)$$

Note certain similarity between (C.26) and (C.13). The integrand in (C.26) is a perfect differential and can be computed in elementary functions:

$$\int \frac{[l_2^2(x)-x^2]^{1/2}}{l_2^2(x)-l_1^2(x)}\frac{[l_{20}^2(x)-x^2]^{1/2}}{l_{20}^2(x)-l_{10}^2(x)}\lambda\left(\frac{l_1(x)l_{10}(x)}{l_2(x)l_{20}(x)},\ \phi-\phi_0\right)dx$$
$$= -\frac{1}{2R_1}\tan^{-1}\left(\frac{\Xi_1(x)}{R_1}\right)-\frac{1}{2R_2}\tan^{-1}\left(\frac{\Xi_2(x)}{R_2}\right), \quad (C.27)$$

where R_1 and R_2 are defined by (C.15) and

$$\Xi_1(x)=\xi(x)+\frac{zz_0}{\xi(x)}, \qquad \Xi_2(x)=\xi(x)-\frac{zz_0}{\xi(x)},$$
$$\xi(x)=\frac{1}{x}[l_2^2(x)-x^2]^{1/2}[l_{20}^2(x)-x^2]^{1/2}. \quad (C.28)$$

As before, correctness of the integration can be verified by differentiation, using (C.17) and the property

$$\frac{\partial}{\partial x}\xi(x) = -\frac{\xi(x)[l_2^2(x)l_{20}^2(x)-l_1^2(x)l_{10}^2(x)]}{x[l_2^2(x)-l_1^2(x)][l_{20}^2(x)-l_{10}^2(x)]}, \quad (C.29)$$

Finally, the integral (C.20) can be computed as follows:

$$\int_0^{2\pi}\int_0^a \frac{z_0}{R^3(N,M_0)}\left[\frac{R(N,M_0)}{h_0}+\tan^{-1}\left(\frac{h_0}{R(N,M_0)}\right)\right]\frac{r\,dr\,d\psi}{R(M,N)}$$
$$=\pi\frac{|z_0|}{z_0}\left\{\frac{1}{R_1}\left[\frac{\pi}{2}-\tan^{-1}\left(\frac{\Xi_1}{R_1}\right)\right]+\frac{1}{R_2}\left[\frac{\pi}{2}-\tan^{-1}\left(\frac{\Xi_2}{R_2}\right)\right]\right\}. \quad (C.30)$$

According to our convention, Ξ_1 and Ξ_2 denote $\Xi_1(a)$ and $\Xi_2(a)$ respectively.

Consider a particular case, when $z_0=0$ and $\rho_0>a$. Due to the relationship (see (A.27))

$$\frac{az_0}{(a^2-l_{10}^2)^{1/2}}\rightarrow(\rho_0^2-a^2)^{1/2} \qquad \text{for } z_0\rightarrow 0,$$

the integral (C.30) will take the form

$$\int_0^{2\pi}\int_0^a \frac{(\rho_0^2-a^2)^{1/2}}{(a^2-r^2)^{1/2}}\frac{1}{\rho_0^2+r^2-2\rho_0 r\cos(\phi_0-\psi)}$$

$$\times\frac{r\,dr\,d\psi}{[\rho^2+r^2-2\rho r\cos(\phi-\psi)+z^2]^{1/2}}$$

$$=\frac{2\pi}{R_0}\left\{\frac{\pi}{2}-\tan^{-1}\left[\frac{(l_2^2-a^2)^{1/2}(\rho_0^2-a^2)^{1/2}}{aR_0}\right]\right\}. \tag{C.31}$$

Here $R_0=[\rho^2+\rho_0^2-2\rho\rho_0\cos(\phi-\phi_0)+z^2]^{1/2}$. The integration in (C.30) for $z_0=0$ and $\rho_0<a$ yields π^2/R_0. The case of $z=0$ can be considered in a similar manner.

Appendix D

An integral representation for z/R^3 is presented here, for $R=[\rho^2+r^2-2\rho r\cos\phi+z^2]^{1/2}$. Consider the integral

$$\frac{1}{r}L\left(\frac{1}{r}\right)\frac{d}{dr}\int_0^r\frac{x\,dx}{(r^2-x^2)^{1/2}}\frac{[x^2-l_1^2(x)]^{1/2}}{l_2^2(x)-l_1^2(x)}\lambda\left(\frac{l_1(x)x}{l_2(x)},\phi\right). \tag{D.1}$$

By using the rule of differentiation under the integral sign

$$\frac{d}{dr}\int_0^r\frac{f(x)\,dx}{(r^2-x^2)^{1/2}}=\lim_{x\to 0}\left[\frac{f(x)}{x}\right]+r\int_0^r\frac{dx}{(r^2-x^2)^{1/2}}\frac{d}{dx}\left[\frac{f(x)}{x}\right],$$

the integral (D.1) can be rewritten as

$$\int_0^r\frac{dx}{(r^2-x^2)^{1/2}}\frac{d}{dx}\left[\frac{[x^2-l_1^2(x)]^{1/2}}{l_2^2(x)-l_1^2(x)}\lambda\left(\frac{l_1(x)x}{l_2(x)r},\phi\right)\right]. \tag{D.2}$$

Introduce the new variable

$$t=\frac{(r^2-x^2)^{1/2}[l_2^2(x)-x^2]^{1/2}}{x},$$

$$t'=\frac{dt}{dx}=-\frac{[l_2^2(x)-x^2]^{1/2}[r^2l_2^2(x)-x^2l_1^2(x)]}{x^2[l_2^2(x)-l_1^2(x)](r^2-x^2)^{1/2}} \tag{D.3}$$

Let us transform the expression for λ:

$$\lambda\left(\frac{l_1(x)x}{l_2(x)r}, \phi\right) = \lambda\left(\frac{l_1^2(x)}{\rho r}, \phi\right) = -\frac{l_1^2(x) - \rho^2 r^2/l_1^2(x)}{l_1^2(x) + \rho^2 r^2/l_1^2(x) - 2\rho r \cos\psi}$$

$$= -\frac{l_1^2(x) - \rho^2 r^2/l_1^2(x)}{l_1^2(x) + \rho^2 r^2/l_1^2(x) - 2\rho r \cos\psi + \rho^2 + r^2 + z^2 - l_1^2(x) - l_2^2(x) + x^2 - r^2}$$

$$= -\frac{x^2 l_1^2(x) - r^2 l_2^2(x)}{x^2(R^2 + t^2)} = -\frac{[l_2^2(x) - l_1^2(x)](r^2 - x^2)^{1/2}}{[l_2^2(x) - x^2]^{1/2}} \frac{t'}{R^2 + t^2}. \tag{D.4}$$

The substitution of (D.3) and (D.4) in (D.2) allows us to continue the transformations:

$$-\int_0^r \frac{dx}{(r^2 - x^2)^{1/2}} \frac{d}{dx}\left[\frac{[x^2 - l_1^2(x)]^{1/2}(r^2 - x^2)^{1/2}}{[l_2^2(x) - x^2]^{1/2}} \frac{t'}{R^2 + t^2}\right]$$

$$= -\int_0^r \frac{dx}{(r^2 - x^2)^{1/2}} \frac{d}{dx}\left[\frac{xz(r^2 - x^2)^{1/2}}{l_2^2(x) - x^2} \frac{t'}{R^2 + t^2}\right]$$

$$= -z\int_0^r \frac{dx}{(r^2 - x^2)^{1/2}} \frac{d}{dx}\left[\frac{(r^2 - x^2)^{3/2} t'}{xt^2(R^2 + t^2)}\right]$$

$$= -z\left\{\frac{(r^2 - x^2)t'}{xt^2(R^2 + t^2)}\bigg|_0^r - \int_0^r \frac{dt(x)}{t^2(R^2 + t^2)}\right\}$$

$$= -z\left\{\frac{(r^2 - x^2)t'}{xt^2(R^2 + t^2)} + \frac{1}{R^2 t} + \frac{1}{R^3}\tan^{-1}\left(\frac{t}{R}\right)\right\}\bigg|_0^r = \frac{\pi}{2}\frac{z}{R^3}. \tag{D.5}$$

Finally, (D.5) allows us to write the required representation

$$\frac{z}{R^3} = \frac{2}{\pi r} L\left(\frac{1}{r}\right) \frac{d}{dr} \int_0^r \frac{x\,dx}{(r^2 - x^2)^{1/2}} \frac{[x^2 - l_1^2(x)]^{1/2}}{l_2^2(x) - l_1^2(x)} \lambda\left(\frac{l_1(x)x}{l_2(x)}, \phi\right). \tag{D.6}$$

Appendix E

The following integral is computed here:

$$\frac{1}{2\pi}\int_0^{2\pi}\int_0^a \frac{\rho e^{i\phi} - re^{i\psi}}{R^3(M, N)} \tan^{-1}\left[\frac{(a^2 - r^2)^{1/2}(a^2 - \rho_0^2)^{1/2}}{aR(N, N_0)}\right] \frac{r\,dr\,d\psi}{R(N, N_0)}. \tag{E.1}$$

The integral (E.1) can be computed indirectly by using (2.32), which leads to an equivalent expression:

$$\int_\infty^z \Lambda\left[\frac{1}{R(M, N_0)} \tan^{-1}\left(\frac{h}{R(M, N_0)}\right)\right] dz. \tag{E.2}$$

Let us make use of the following identities:

$$\Lambda h = -\frac{h\rho e^{i\phi}}{l_2^2 - l_1^2}, \tag{E.3}$$

$$\Lambda\left[\frac{1}{R(M, N_0)}\tan^{-1}\left(\frac{h}{R(M, N_0)}\right)\right]$$
$$= -\frac{\rho e^{i\phi} - \rho_0 e^{i\phi_0}}{R_0^3}\tan^{-1}\left(\frac{h}{R_0}\right) - \frac{h}{R_0^2 + h^2}\left[\frac{\rho e^{i\phi}}{l_2^2 - l_1^2} + \frac{\rho e^{i\phi} - \rho_0 e^{i\phi_0}}{R_0^2}\right]. \tag{E.4}$$

The notation R_0 in this appendix is used as a contraction for $R(M, N_0)$. The substitution of (E.4) in (E.2) yields, after integration by parts,

$$\int_\infty^z \Lambda\left[\frac{1}{R(M, N_0)}\tan^{-1}\left(\frac{h}{R(M, N_0)}\right)\right] dz$$
$$= -\frac{1}{\rho e^{-i\phi} - \rho_0 e^{-i\phi_0}}\left\{\frac{z}{R_0}\tan^{-1}\left(\frac{h}{R_0}\right) - \int_\infty^z \frac{z}{R_0}\frac{\partial}{\partial z}\left[\tan^{-1}\left(\frac{h}{R_0}\right)\right] dz\right\}$$
$$-\int_\infty^z \left[\frac{\rho e^{i\phi}}{l_2^2 - l_1^2} + \frac{\rho e^{i\phi} - \rho_0 e^{i\phi_0}}{R_0^2}\right]\frac{h}{R_0^2 + h^2}\, dz. \tag{E.5}$$

The following identities are now used:

$$\frac{\partial h}{\partial z} = \frac{h(\rho^2 - l_1^2)}{z(l_2^2 - l_1^2)}, \tag{E.6}$$

$$\frac{\partial}{\partial z}\tan^{-1}\left(\frac{h}{R_0}\right) = \frac{hR_0}{z(R_0^2 + h^2)}\left[\frac{\rho^2 - l_1^2}{l_2^2 - l_1^2} - \frac{z^2}{R_0^2}\right]. \tag{E.7}$$

The substitution of (E.7) in (E.5) allows us to proceed:

$$-\frac{1}{\rho e^{-i\phi} - \rho_0 e^{-i\phi_0}}\left\{\frac{z}{R_0}\tan^{-1}\left(\frac{h}{R_0}\right) - \int_\infty^z \frac{h}{R_0^2 + h^2}\left[\frac{\rho^2 - l_1^2}{l_2^2 - l_1^2} - \frac{z^2}{R_0^2}\right] dz\right\}$$
$$-\int_\infty^z \left[\frac{\rho e^{i\phi}}{l_2^2 - l_1^2} + \frac{\rho e^{i\phi} - \rho_0 e^{i\phi_0}}{R_0^2}\right]\frac{h\, dz}{R_0^2 + h^2}$$
$$= -\frac{z}{(\rho e^{-i\phi} - \rho_0 e^{-i\phi_0})R_0}\tan^{-1}\left(\frac{h}{R_0}\right)$$
$$+\int_0^z \frac{h\, dz}{(R_0^2 + h^2)(l_2^2 - l_1^2)}\left[\frac{\rho^2 - l_1^2}{\rho e^{-i\phi} - \rho_0 e^{-i\phi_0}} - \rho e^{i\phi}\right]$$
$$-\int_\infty^z \frac{h\, dz}{R_0^2(R_0^2 + h^2)}\left[\frac{z^2}{\rho e^{-i\phi} - \rho_0 e^{-i\phi_0}} + \rho e^{i\phi} - \rho_0 e^{i\phi_0}\right]$$

$$= \frac{1}{\rho e^{-i\phi} - \rho_0 e^{-i\phi_0}} \left\{ -\frac{z}{R_0} \tan^{-1}\left(\frac{h}{R_0}\right) + \int_\infty^z \left[\frac{\rho\rho_0 e^{i(\phi-\phi_0)} - l_1^2}{l_2^2 - l_1^2} - 1\right] \frac{h\,dz}{R_0^2 + h^2} \right\}$$

$$= \frac{1}{\rho e^{-i\phi} - \rho_0 e^{-i\phi_0}} \left\{ -\frac{z}{R_0} \tan^{-1}\left(\frac{h}{R_0}\right) + \int_\infty^z \left[\frac{\rho\rho_0 e^{i(\phi-\phi_0)} - l_2^2}{l_2^2 - l_1^2}\right] \frac{h\,dz}{R_0^2 + h^2} \right\}. \tag{E.8}$$

Taking into consideration the identity

$$R_0^2 + h^2 = \frac{1}{l_2^2}(l_2^2 - \rho\rho_0 e^{i(\phi-\phi_0)})(l_2^2 - \rho\rho_0 e^{-i(\phi-\phi_0)}), \tag{E.9}$$

the integral in (E.8) can be transformed as follows:

$$\int \frac{h l_2^2\,dz}{(l_2^2 - l_1^2)(l_2^2 - \rho\rho_0 e^{-i(\phi-\phi_0)})} = \frac{(a^2 - \rho_0^2)^{1/2}}{a} \int \frac{(a^2 - l_1^2)^{1/2} l_2^2\,dl_2}{z l_2 (l_2^2 - \rho\rho_0 e^{-i(\phi-\phi_0)})}$$

$$= (a^2 - \rho_0^2)^{1/2} \int \frac{l_2\,dl_2}{(l_2^2 - a^2)^{1/2}(l_2^2 - \rho\rho_0 e^{-i(\phi-\phi_0)})}$$

$$= \frac{(a^2 - \rho_0^2)^{1/2}}{\bar{s}} \tan^{-1}\left[\frac{(l_2^2 - a^2)^{1/2}}{\bar{s}}\right], \tag{E.10}$$

where $\bar{s} = (a^2 - \rho\rho_0 e^{-i(\phi-\phi_0)})^{1/2}$. Finally, formulae (E.8) and (E.10) allow us to compute the original integral (E.1):

$$\frac{1}{2\pi} \int_0^{2\pi} \int_0^a \frac{\rho e^{i\phi} - r e^{i\psi}}{R^3(M, N)} \tan^{-1}\left[\frac{(a^2 - r^2)^{1/2}(a^2 - \rho_0^2)^{1/2}}{a R(N, N_0)}\right] \frac{r\,dr\,d\psi}{R(N, N_0)}$$

$$= \frac{1}{\rho e^{-i\phi} - \rho_0 e^{-i\phi_0}} \left\{ \frac{(a^2 - \rho_0^2)^{1/2}}{\bar{s}} \tan^{-1}\left[\frac{\bar{s}}{(l_2^2 - a^2)^{1/2}}\right] - \frac{z}{R(M, N_0)} \tan^{-1}\left[\frac{h}{R(M, N_0)}\right] \right\}. \tag{E.11}$$

It is reminded that h is defined by (2.33), and $R(M, N_0) = [\rho^2 + \rho_0^2 - 2\rho\rho_0 \cos(\phi - \phi_0) + z^2]^{1/2}$. The right hand side in (E.11) simplifies in the limiting case of $\rho_0 \to \rho$ and $\phi_0 \to \phi$, namely,

$$\frac{\rho e^{i\phi}}{2(a^2 - \rho^2)^{1/2}} \left\{ \frac{1}{(a^2 - \rho^2)^{1/2}} \tan^{-1}\left[\frac{(a^2 - \rho^2)^{1/2}}{(l_2^2 - a^2)^{1/2}}\right] - \frac{(l_2^2 - a^2)^{1/2}}{l_2^2 - \rho^2} \right\}. \tag{E.12}$$

Appendix F

Some integrals, related to elastic contact problems, are computed here. The first integral to compute is (2.73)

$$I_1 = \int_0^{2\pi} \int_0^a \frac{\sigma(r, \psi) r \, dr \, d\psi}{[\rho^2 + r^2 - 2r\rho \cos(\phi - \psi) + z^2]^{1/2}}, \tag{F.1}$$

where, according to (2.72),

$$\sigma(r, \psi) = -\frac{1}{\pi^2 H r} L(r) \frac{d}{dr} \int_r^a \frac{t \, dt}{(t^2 - r^2)^{1/2}} L\left(\frac{1}{t^2}\right) \frac{d}{dt} \\ \times \int_0^t \frac{\rho_0 \, d\rho_0}{(t^2 - \rho_0^2)^{1/2}} L(\rho_0) \omega(\rho_0, \psi). \tag{F.2}$$

By substituting the integral representation for the reciprocal distance (C.21) in (F.1), we get, after changing the order of integration

$$\frac{2}{\pi} \int_0^{2\pi} d\psi \int_0^a \frac{[l_2^2(x) - x^2]^{1/2}}{l_2^2(x) - l_1^2(x)} \lambda\left(\frac{l_1(x) x}{l_2(x) r}, \phi - \psi\right) dx \int_x^a \frac{\sigma(r, \psi) r \, dr}{(r^2 - x^2)^{1/2}}. \tag{F.3}$$

We substitute (F.2) in (F.3) and use the property (C.11) to obtain

$$-\frac{2}{\pi^3 H} \int_0^{2\pi} d\psi \int_0^a \frac{[l_2^2(x) - x^2]^{1/2}}{l_2^2(x) - l_1^2(x)} \lambda\left(\frac{l_1(x) x}{l_2(x)}, \phi - \psi\right) dx \\ \times \int_x^a \frac{dr}{(r^2 - x^2)^{1/2}} \frac{d}{dr} \int_r^a \frac{t \, dt}{(t^2 - r^2)^{1/2}} L\left(\frac{1}{t^2}\right) \frac{d}{dt} \int_0^t \frac{\rho_0 \, d\rho_0}{(t^2 - \rho_0^2)^{1/2}} L(\rho_0) \omega(\rho_0, \psi). \tag{F.4}$$

The property (C.25) allows us to simplify (F.4) as follows

$$\frac{1}{\pi^2 H} \int_0^{2\pi} d\psi \int_0^a \frac{[l_2^2(x) - x^2]^{1/2}}{l_2^2(x) - l_1^2(x)} \lambda\left(\frac{\rho}{l_2^2(x)}, \phi - \psi\right) dx \frac{d}{dx} \\ \times \int_0^x \frac{\rho_0 \, d\rho_0}{(x^2 - \rho_0^2)^{1/2}} L(\rho_0) \omega(\rho_0, \psi). \tag{F.5}$$

One can establish the following rule of changing the order of integration:

$$\int_0^a F(x) \, dx \frac{d}{dx} \int_0^x \frac{f(\rho_0) \rho_0 \, d\rho_0}{(x^2 - \rho_0^2)^{1/2}} = -\int_0^a f(\rho_0) \, d\rho_0 \frac{d}{d\rho_0} \int_{\rho_0}^a \frac{F(x) x \, dx}{(x^2 - \rho_0^2)^{1/2}}. \tag{F.6}$$

The rule (F.6), applied to (F.5), yields

$$-\frac{1}{\pi^2 H} \int_0^{2\pi} \int_0^a \left\{ \frac{d}{d\rho_0} \int_{\rho_0}^a \frac{[l_2^2(x) - x^2]^{1/2}}{l_2^2(x) - l_1^2(x)} \lambda\left(\frac{\rho \rho_0}{l_2^2(x)}, \phi - \psi\right) \right. \\ \left. \times \frac{x \, dx}{(x^2 - \rho_0^2)^{1/2}} \right\} \omega(\rho_0, \phi_0) \, \rho_0 \, d\rho_0 \, d\phi_0. \tag{F.7}$$

Now we need to compute the integral in curly brackets (F.7). Introduce a new variable

$$h(x)=\frac{[x^2-l_1^2(x)]^{1/2}(x^2-\rho_0^2)^{1/2}}{x},$$
$$h'(x)=\frac{dh(x)}{dx}=\frac{xh(x)[l_2^4(x)-\rho^2\rho_0^2]}{l_2^2(x)[l_2^2(x)-l_1^2(x)](x^2-\rho_0^2)}. \tag{F.8}$$

Formulae (F.8) may be considered as a particular case of (C.15) and (C.16) for $z_0=0$ and $x\geq\rho_0$. By using the analogy with (D.4), one can write

$$\lambda\left(\frac{\rho\rho_0}{l_2^2(x)},\ \phi-\psi\right)=\frac{[l_2^4(x)-\rho^2\rho_0^2]}{l_2^2(x)[R_0^2+h^2(x)]}. \tag{F.9}$$

Here $R_0=[\rho^2+\rho_0^2-2\rho\rho_0\cos(\phi-\phi_0)+z^2]^{1/2}$. Equations (F.8) and (F.9) justify the following identity:

$$\frac{[l_2^2(x)-x^2]^{1/2}}{l_2^2(x)-l_1^2(x)}\lambda\left(\frac{\rho\rho_0}{l_2^2(x)},\ \phi-\psi\right)=\frac{z(x^2-\rho_0^2)^{3/2}h'(x)}{xh^2(x)[R_0^2+h^2(x)]}. \tag{F.10}$$

We can now compute the integral in (F.7).

$$\begin{aligned}&\frac{d}{d\rho_0}\int_{\rho_0}^{a}\frac{[l_2^2(x)-x^2]^{1/2}}{l_2^2(x)-l_1^2(x)}\lambda\left(\frac{\rho\rho_0}{l_2^2(x)},\ \phi-\psi\right)\frac{x\,dx}{(x^2-\rho_0^2)^{1/2}}\\
&\quad=-\frac{(l_2^2-a^2)^{1/2}\rho_0}{a(a^2-\rho_0^2)^{1/2}(l_2^2-l_1^2)}+\rho_0\int_{\rho_0}^{a}\frac{dx}{(x^2-\rho_0^2)^{1/2}}\frac{d}{dx}\\
&\qquad\times\left[\frac{[l_2^2(x)-x^2]^{1/2}}{l_2^2(x)-l_1^2(x)}\lambda\left(\frac{\rho\rho_0}{l_2^2(x)},\ \phi-\psi\right)\right]\\
&\quad=-\frac{(l_2^2-a^2)^{1/2}\rho_0}{a(a^2-\rho_0^2)^{1/2}(l_2^2-l_1^2)}+\rho_0\int_{\rho_0}^{a}\frac{dx}{(x^2-\rho_0^2)^{1/2}}\frac{d}{dx}\left[\frac{z(x^2-\rho_0^2)^{3/2}h'(x)}{xh^2(x)[R_0^2+h^2(x)]}\right].\end{aligned}$$

The last integral can be computed by parts:

$$\begin{aligned}&-\frac{(l_2^2-a^2)^{1/2}\rho_0}{a(a^2-\rho_0^2)^{1/2}(l_2^2-l_1^2)}+\frac{z(x^2-\rho_0^2)h'(x)}{xh^2(x)[R_0^2+h^2(x)]}\bigg|_{\rho_0}^{a}+z\int_{\rho_0}^{a}\frac{h'(x)\,dx}{h^2(x)[R_0^2+h^2(x)]}\\
&\quad=-\frac{(l_2^2-a^2)^{1/2}\rho_0}{a(a^2-\rho_0^2)^{1/2}(l_2^2-l_1^2)}+\left\{\frac{z(x^2-\rho_0^2)h'(x)}{xh^2(x)[R_0^2+h^2(x)]}\right.\\
&\qquad\left.-z\left[\frac{1}{R_0^2h(x)}+\frac{1}{R_0^3}\tan^{-1}\left(\frac{h(x)}{R_0}\right)\right]\right\}\bigg|_{\rho_0}^{a}.\end{aligned} \tag{F.11}$$

Computation of the upper and lower limits in (F.11) gives the final result

$$\int_0^{2\pi}\int_0^a \frac{\sigma(r,\,\psi) r\,dr\,d\psi}{[\rho^2+r^2-2r\rho\cos(\phi-\psi)+z^2]^{1/2}}$$
$$=\frac{1}{\pi^2 H}\int_0^{2\pi}\int_0^a \frac{z}{R_0^3}\left[\frac{R_0}{h}+\tan^{-1}\left(\frac{h}{R_0}\right)\right]\omega(\rho_0,\,\phi_0)\rho_0\,d\rho_0\,d\phi_0. \quad \text{(F.12)}$$

According to our convention, h stands as an abbreviation for $h(a)$. Expression (F.12) validates (2.74).

The second integral to be computed is

$$I_2=\int \frac{z}{R_0^3}\left[\frac{R_0}{h}+\tan^{-1}\left(\frac{h}{R_0}\right)\right]dz. \quad \text{(F.13)}$$

We proceed with integration by parts. The result is

$$\int \frac{z\,dz}{R_0^2 h}-\frac{1}{R_0}\tan^{-1}\left(\frac{h}{R_0}\right)+\int \frac{dz}{R_0}\frac{d}{dz}\tan^{-1}\left(\frac{h}{R_0}\right). \quad \text{(F.14)}$$

We modify (E.7):

$$\frac{d}{dz}\tan^{-1}\left(\frac{h}{R_0}\right)=\frac{R_0}{R_0^2+h^2}\left[\frac{h(\rho^2-l_1^2)}{z(l_2^2-l_1^2)}+\frac{z}{h}\right]-\frac{z}{hR_0}$$
$$=\frac{R_0(l_2^2-a^2)^{1/2}(l_2^4-\rho^2\rho_0^2)}{(a^2-\rho_0^2)^{1/2}(l_2^2-l_1^2)l_2^2(R_0^2+h^2)}-\frac{z}{hR_0}. \quad \text{(F.15)}$$

By substituting (F.15) in (F.14), and taking into consideration (E.9) and (A.28), we obtain

$$-\frac{1}{R_0}\tan^{-1}\left(\frac{h}{R_0}\right)+\frac{1}{(a^2-\rho_0^2)^{1/2}}$$
$$\times\int \frac{(l_2^4-\rho^2\rho_0^2)\,dl_2}{(l_2^2-\rho^2)^{1/2}(l_2^2-\rho\rho_0 e^{i(\phi-\phi_0)})(l_2^2-\rho\rho_0 e^{-i(\phi-\phi_0)})}. \quad \text{(F.16)}$$

The integral in (F.16) is elementary:

$$-\frac{1}{R_0}\tan^{-1}\left(\frac{h}{R_0}\right)+\frac{1}{(a^2-\rho_0^2)^{1/2}}\left\{\ln[l_2+(l_2^2-\rho^2)^{1/2}]\right.$$
$$+\frac{1}{(\zeta-1)^{1/2}}\tan^{-1}\left[\frac{(a^2-l_1^2)^{1/2}}{a(\zeta-1)^{1/2}}\right]$$
$$\left.+\frac{1}{(\bar{\zeta}-1)^{1/2}}\tan^{-1}\left[\frac{(a^2-l_1^2)^{1/2}}{a(\bar{\zeta}-1)^{1/2}}\right]\right\}, \qquad \zeta=\frac{\rho}{\rho_0}e^{i(\phi-\phi_0)}. \quad \text{(F.17)}$$

Since the integration was indefinite, we might have lost a function of the variables, other than z. This function can be found from the condition that the result of the integration does not have a logarithmic singularity at $\rho = 0$. The function eliminating such a singularity is $\tan^{-1}(\zeta - 1)^{1/2}$. Now the final result can be represented by

$$\int \frac{z}{R_0^3}\left[\frac{R_0}{h} + \tan^{-1}\left(\frac{h}{R_0}\right)\right] dz$$

$$= -\frac{1}{R_0}\tan^{-1}\left(\frac{h}{R_0}\right) + \frac{1}{(a^2-\rho_0^2)^{1/2}}\left\{\ln[l_2 + (l_2^2-\rho^2)^{1/2}]\right.$$

$$- 2\Re\left(\frac{1}{(\zeta-1)^{1/2}}\left[\tan^{-1}\left(\frac{a(\zeta-1)^{1/2}}{(a^2-l_1^2)^{1/2}}\right)\right.\right.$$

$$\left.\left.\left. - \tan^{-1}(\zeta-1)^{1/2}\right]\right)\right\}, \qquad \zeta = \frac{\rho}{\rho_0}e^{i(\phi-\phi_0)}$$

The last expression proves correctness of formula (2.76).

Appendix G

Some integrals related to the problem of a penny-shaped crack under shear loading are presented here, without derivation. The first integral, which can be computed directly, is

$$\int_0^{2\pi}\int_0^a \frac{(3a^2 - r\rho_0 e^{i(\psi-\phi_0)})}{(a^2 - r\rho_0 e^{i(\psi-\phi_0)})^2}\frac{(a^2-r^2)^{1/2}(a^2-\rho_0^2)^{1/2}}{aR(M,N)} r\,dr\,d\psi$$

$$= \pi\frac{(a^2-\rho_0^2)^{1/2}}{a^3}\left\{\frac{\rho^2}{t}\sin^{-1}\left(\frac{a}{l_2}\right) + \frac{(l_2^2-a^2)^{1/2}[l_1^2(4-t)-3a^2]}{a(1-t)^2}\right.$$

$$\left. + \frac{1}{(1-t)^{3/2}}\left[\frac{3z^2}{1-t} + a^2(3-2t) - \frac{\rho^2}{t}\right]\tan^{-1}\left(\frac{a(1-t)^{1/2}}{(l_2^2-a^2)^{1/2}}\right)\right\}. \quad \text{(G.1)}$$

Here t is defined by (2.104). Application of the operator Λ to both sides of (G.1) yields

$$\int_0^{2\pi}\int_0^a \frac{(3a^2 - r\rho_0 e^{i(\psi-\phi_0)})}{(a^2 - r\rho_0 e^{i(\psi-\phi_0)})^2}\frac{(a^2-r^2)^{1/2}(a^2-\rho_0^2)^{1/2}}{aR^3(M,N)}(\rho e^{i\phi} - re^{i\psi}) r\,dr\,d\psi$$

$$= -2\pi\rho e^{i\phi}\frac{(a^2-\rho_0^2)^{1/2}}{a^3}\left\{\frac{1}{t}\sin^{-1}\left(\frac{a}{l_2}\right) + \frac{a(l_2^2-a^2)^{1/2}}{(1-t)(l_2^2-\rho\rho_0 e^{i(\phi-\phi_0)})}\right.$$

$$\left. - \frac{1}{t(1-t)^{3/2}}\tan^{-1}\left(\frac{a(1-t)^{1/2}}{(l_2^2-a^2)^{1/2}}\right)\right\}. \quad \text{(G.2)}$$

Another application of Λ to both sides of (G.2) gives

$$\int_0^{2\pi}\int_0^a \frac{(3a^2 - r\rho_0 e^{i(\psi-\phi_0)})}{(a^2 - r\rho_0 e^{i(\psi-\phi_0)})^2}\frac{(a^2-r^2)^{1/2}(a^2-\rho_0^2)^{1/2}}{aR^5(M, N)} 3(\rho e^{i\phi} - re^{i\psi})^2 r\,dr\,d\psi$$
$$= 2\pi \frac{\rho^2 e^{2i\phi}(a^2-\rho_0^2)^{1/2}(l_2^2-a^2)^{1/2}(3l_2^2 - \rho\rho_0 e^{i(\phi-\phi_0)})}{l_2^2(l_2^2-l_1^2)(l_2^2-\rho\rho_0 e^{i(\phi-\phi_0)})^2}. \tag{G.3}$$

Differentiation with respect to z of both sides of (G.1) results in

$$\int_0^{2\pi}\int_0^a \frac{(3a^2 - r\rho_0 e^{i(\psi-\phi_0)})}{(a^2 - r\rho_0 e^{i(\psi-\phi_0)})^2}\frac{z(a^2-r^2)^{1/2}(a^2-\rho_0^2)^{1/2}}{aR^3(M, N)} r\,dr\,d\psi$$
$$= 2\pi \frac{ha^2}{s^2}\left[\frac{3}{s^2} - \frac{t}{l_2^2-a^2 t} - \frac{3(l_2^2-a^2)^{1/2}}{s^3}\tan^{-1}\left(\frac{s}{(l_2^2-a^2)^{1/2}}\right)\right]. \tag{G.4}$$

Here h is defined by (2.33). Another differentiation of both sides of (G.4) with respect to z yields

$$\int_0^{2\pi}\int_0^a \frac{(3a^2 - r\rho_0 e^{i(\psi-\phi_0)})}{(a^2 - r\rho_0 e^{i(\psi-\phi_0)})^2}\frac{(a^2-r^2)^{1/2}(a^2-\rho_0^2)^{1/2}}{aR^3(M, N)}\left(1 - \frac{3z^2}{R^2(M, N)}\right) r\,dr\,d\psi$$
$$= 2\pi \frac{(a^2-\rho_0^2)^{1/2}}{a^3(1-t)}\left\{\frac{a(l_2^2-a^2)^{1/2}}{(l_2^2-l_1^2)(l_2^2-\rho\rho_0 e^{i(\phi-\phi_0)})}\left[\frac{3(l_2^2-l_1^2 t)}{1-t}\right.\right.$$
$$\left.\left. + \frac{\rho\rho_0 e^{i(\phi-\phi_0)}(2l_2^2 + l_1^2 t - 3\rho^2)}{l_2^2 - \rho\rho_0 e^{i(\phi-\phi_0)}}\right] - \frac{3}{(1-t)^{3/2}}\tan^{-1}\left(\frac{a(1-t)^{1/2}}{(l_2^2-a^2)^{1/2}}\right)\right\}. \tag{G.5}$$

Application of the operator Λ to both sides of (G.4) yields

$$\int_0^{2\pi}\int_0^a \frac{(3a^2 - r\rho_0 e^{i(\psi-\phi_0)})}{(a^2 - r\rho_0 e^{i(\psi-\phi_0)})^2}\frac{z(a^2-r^2)^{1/2}(a^2-\rho_0^2)^{1/2}}{aR^5(M, N)} 3(\rho e^{i\phi} - re^{i\psi}) r\,dr\,d\psi$$
$$= 2\pi \frac{h\rho e^{i\phi}(3l_2^2 - a^2 t)}{(l_2^2-l_1^2)(l_2^2-a^2 t)^2}. \tag{G.6}$$

A different result appears if Λ is applied to a complex conjugate of expression (G.4), namely,

$$\int_0^{2\pi}\int_0^a \frac{(3a^2 - r\rho_0 e^{-i(\psi-\phi_0)})}{(a^2 - r\rho_0 e^{-i(\psi-\phi_0)})^2}\frac{z(a^2-r^2)^{1/2}(a^2-\rho_0^2)^{1/2}}{aR^5(M, N)} 3(\rho e^{i\phi} - re^{i\psi}) r\,dr\,d\psi$$
$$= 2\pi h\left\{\frac{a^2}{\bar{s}^2}\rho_0 e^{i\phi_0}\left[\frac{15(l_2^2-a^2)^{1/2}}{\bar{s}^5}\tan^{-1}\left(\frac{\bar{s}}{(l_2^2-a^2)^{1/2}}\right) - \frac{15}{\bar{s}^4}\right.\right.$$
$$\left.\left. + \frac{5}{\bar{s}^2(l_2^2-a^2\bar{t})} + \frac{2\bar{t}}{(l_2^2-a^2\bar{t})^2}\right] + \frac{\rho e^{i\phi}(3l_2^2-a^2\bar{t})}{(l_2^2-l_1^2)(l_2^2-a^2\bar{t})^2}\right\}. \tag{G.7}$$

Integration of both sides of (G.1) with respect to z gives

$$\int_0^{2\pi}\int_0^a \frac{(3a^2 - r\rho_0 e^{i(\psi-\phi_0)})}{(a^2 - r\rho_0 e^{i(\psi-\phi_0)})^2}\frac{(a^2-r^2)^{1/2}(a^2-\rho_0^2)^{1/2}}{a}\ln[R(M, N)+z]r\,dr\,d\psi$$

$$= \pi(a^2-\rho_0^2)^{1/2}\Big\{2\ln[l_2 + (l_2^2-\rho^2)^{1/2}] - 2$$

$$+ \frac{(a^2-l_1^2)^{1/2}}{a^3(1-t)}\left[-\frac{z^2}{1-t}+\rho^2-\frac{1}{3}(l_1^2+2a^2)\right] + \frac{z\bar{\xi}}{a}\sin^{-1}\left(\frac{a}{l_2}\right)$$

$$+ \frac{z}{a^3(1-t)^{3/2}}\left[\frac{z^2}{1-t}+a^2(3-2t-\bar{\xi})\right]\tan^{-1}\left[\frac{a(1-t)^{1/2}}{(l_2^2-a^2)^{1/2}}\right]$$

$$+ 2(\bar{\xi}-1)^{1/2}\left[\tan^{-1}\left(\frac{1}{(\bar{\xi}-1)^{1/2}}\right) - \tan^{-1}\left(\frac{(a^2-l_1^2)^{1/2}}{a(\bar{\xi}-1)^{1/2}}\right)\right]\Big\}. \qquad \text{(G.8)}$$

Here ξ is defined by (2.76), and the bar, as usually, indicates the complex conjugate value. A similar integration with respect to z of (G.3) yields

$$\int_0^{2\pi}\int_0^a \frac{(3a^2 - r\rho_0 e^{i(\psi-\phi_0)})}{(a^2 - r\rho_0 e^{i(\psi-\phi_0)})^2}\frac{(a^2-r^2)^{1/2}(a^2-\rho_0^2)^{1/2}}{a}\Lambda^2$$

$$\times \{\ln[R(M, N)+z]\}r\,dr\,d\psi$$

$$= \frac{2\pi}{\bar{q}}(a^2-\rho_0^2)^{1/2}\Big\{-\frac{(\bar{\xi}-1)^{1/2}}{\bar{q}}$$

$$\times\left[\tan^{-1}\left(\frac{1}{(\bar{\xi}-1)^{1/2}}\right) - \tan^{-1}\left(\frac{(a^2-l_1^2)^{1/2}}{a(\bar{\xi}-1)^{1/2}}\right)\right]$$

$$+ \frac{e^{i\phi}}{\rho}\left[\frac{(a^2-l_1^2)^{1/2}}{a}\left(1+\frac{\rho^2}{l_2^2-\rho\rho_0 e^{i(\phi-\phi_0)}}\right) - 1\right]\Big\}. \qquad \text{(G.9)}$$

It is reminded that q is defined by (2.43) and that

$$\Lambda^2\ln[R(M, N)+z] = -(\rho e^{i\phi} - re^{i\psi})^2\frac{2R(M, N)+z}{R^3(M, N)[R(M, N)+z]^2}. \qquad \text{(G.10)}$$

And yet another application of the operator Λ to (G.9) gives

$$\int_0^{2\pi}\int_0^a \frac{(3a^2-r\rho_0 e^{i(\psi-\phi_0)}}{(a^2-r\rho_0 e^{i(\psi-\phi_0)})^2}\frac{(a^2-r^2)^{1/2}(a^2-\rho_0^2)^{1/2}}{a}\Lambda^3\{\ln[R(M,N)+z]\}r\,dr\,d\psi$$

$$=\frac{2\pi}{\bar q}(a^2-\rho_0^2)^{1/2}\left\{\frac{3(\bar\xi-1)^{1/2}}{\bar q^2}\left[\tan^{-1}\left(\frac{1}{(\bar\xi-1)^{1/2}}\right)-\tan^{-1}\left(\frac{(a^2-l_1^2)^{1/2}}{a(\bar\xi-1)^{1/2}}\right)\right]\right.$$

$$-\frac{e^{2i\phi}(a^2-l_1^2)^{1/2}}{a(l_2^2-l_1^2)}\left[\frac{l_2^2+\rho^2}{l_2^2-\rho\rho_0 e^{i(\phi-\phi_0)}}+\frac{2\rho^2(l_2^2-a^2)}{(l_2^2-\rho\rho_0 e^{i(\phi-\phi_0)})^2}+1\right]$$

$$\left.+\frac{e^{i\phi}}{\rho}\left[\frac{3}{\bar q}+\frac{2e^{i\phi}}{\rho}-\frac{(a^2-l_1^2)^{1/2}}{a}\left(\frac{l_2^2+2\rho^2}{\bar q(l_2^2-\rho\rho_0 e^{i(\phi-\phi_0)})}+2\left(\frac{1}{\bar q}+\frac{e^{i\phi}}{\rho}\right)\right)\right]\right\}. \tag{G.11}$$

Here

$$\Lambda^3\ln[R(M,N)+z]=(\rho e^{i\phi}-re^{i\psi})^3\frac{8R^2(M,N)+9R(M,N)z+3z^2}{R^3(M,N)[r(m,n)+z]^3}. \tag{G.12}$$

Formula (2.42) can be used to obtain some additional results. Integration of (2.42) with respect to z gives

$$\int_0^{2\pi}\int_0^a \frac{\rho e^{i\phi}-re^{i\psi}}{R(M,N)[R(M,N)+z]}\tan^{-1}\left(\frac{(a^2-r^2)^{1/2}(a^2-\rho_0^2)^{1/2}}{aR(N,N_0)}\right)\frac{r\,dr\,d\psi}{R(N,N_0)}$$

$$=\frac{2\pi}{\bar q}\left\{R_0\tan^{-1}\left(\frac{h}{R_0}\right)-(a^2-\rho_0^2)^{1/2}\left[\frac{z}{\bar s}\tan^{-1}\left(\frac{\bar s}{(l_2^2-a^2)^{1/2}}\right)\right.\right.$$

$$\left.\left.-(\bar\xi-1)^{1/2}\left(\tan^{-1}\left(\frac{1}{(\bar\xi-1)^{1/2}}\right)-\tan^{-1}\left(\frac{(a^2-l_1^2)^{1/2}}{a(\bar\xi-1)^{1/2}}\right)\right)\right]\right\}. \tag{G.13}$$

The following indefinite integrals were used here

$$\int\tan^{-1}\left(\frac{\bar s}{(l_2^2-a^2)^{1/2}}\right)dz=z\tan^{-1}\left(\frac{\bar s}{(l_2^2-a^2)^{1/2}}\right)+\bar s\left[\ln[l_2+(l_2^2-\rho^2)^{1/2}]\right.$$

$$\left.+(\zeta-1)^{1/2}\tan^{-1}\left(\frac{a(\zeta-1)^{1/2}}{(a^2-l_1^2)^{1/2}}\right)\right],$$

$$\int\frac{z}{R_0}\tan^{-1}\left(\frac{h}{R_0}\right)dz=R_0\tan^{-1}\left(\frac{h}{R_0}\right)-(a^2-\rho_0^2)^{1/2}\left[-\ln[l_2+(l_2^2-\rho^2)^{1/2}]\right.$$

$$+(\bar\xi-1)^{1/2}\tan^{-1}\left(\frac{(a^2-l_1^2)^{1/2}}{a(\bar\xi-1)^{1/2}}\right)$$

$$\left.+(\zeta-1)^{1/2}\tan^{-1}\left(\frac{(a^2-l_1^2)^{1/2}}{a(\zeta-1)^{1/2}}\right)\right]. \tag{G.14}$$

By applying the operator Λ to both sides of (G.13), one gets

$$\int_0^{2\pi}\int_0^a \frac{(\rho e^{i\phi}-re^{i\psi})^2[2R(M,N)+z]}{R^3(M,N)[R(M,N)+z]^2}\tan^{-1}\left(\frac{(a^2-r^2)^{1/2}(a^2-\rho_0^2)^{1/2}}{aR(N,N_0)}\right)\frac{r\,dr\,d\psi}{R(N,N_0)}$$

$$=\frac{2\pi}{\bar{q}}\left\{\frac{R_0^2+z^2}{R_0\bar{q}}\tan^{-1}\left(\frac{h}{R_0}\right)+(a^2-\rho_0^2)^{1/2}\left[\frac{z}{\bar{s}}\left(\frac{\rho_0 e^{i\phi_0}}{\bar{s}^2}-\frac{2}{\bar{q}}\right)\right.\right.$$
$$\times\tan^{-1}\left(\frac{\bar{s}}{(l_2^2-a^2)^{1/2}}\right)+\frac{(\bar{\xi}-1)^{1/2}}{\bar{q}}\left(\tan^{-1}\left(\frac{1}{(\bar{\xi}-1)^{1/2}}\right)\right.$$
$$\left.\left.\left.-\tan^{-1}\left(\frac{(a^2-l_1^2)^{1/2}}{a(\bar{\xi}-1)^{1/2}}\right)\right)+\frac{e^{i\phi}}{\rho}\right]-\frac{e^{i\phi}ha^2}{\rho\bar{s}^2}\right\}. \quad \text{(G.15)}$$

And yet another application of Λ to (G.15) yields

$$\int_0^{2\pi}\int_0^a \frac{(\rho e^{i\phi}-re^{i\psi})^3[8R^2(M,N)+9R(M,N)z+3z^2]}{R^5(M,N)[R(M,N)+z]^3}$$
$$\times\tan^{-1}\left(\frac{(a^2-r^2)^{1/2}(a^2-\rho_0^2)^{1/2}}{aR(N,N_0)}\right)\frac{r\,dr\,d\psi}{R(N,N_0)}$$
$$=\frac{2\pi}{\bar{q}}\left\{\frac{3R_0^4+6R_0^2z^2-z^4}{R_0^3\bar{q}^2}\tan^{-1}\left(\frac{h}{R_0}\right)-(a^2-\rho_0^2)^{1/2}\right.$$
$$\times\left[\frac{z}{\bar{s}}\left(\frac{8}{\bar{q}^2}-\frac{4\rho_0 e^{i\phi_0}}{\bar{s}^2\bar{q}}+\frac{3\rho_0^2e^{2i\phi_0}}{\bar{s}^4}\right)\tan^{-1}\left(\frac{\bar{s}}{(l_2^2-a^2)^{1/2}}\right)\right.$$
$$\left.-\frac{3(\bar{\xi}-1)^{1/2}}{\bar{q}^2}\left(\tan^{-1}\left(\frac{1}{(\bar{\xi}-1)^{1/2}}\right)-\tan^{-1}\left(\frac{(a^2-l_1^2)^{1/2}}{a(\bar{\xi}-1)^{1/2}}\right)\right)\right]$$
$$+\frac{ha^2e^{i\phi}}{\rho\bar{s}^2}\left[\frac{2\rho_0 e^{i\phi_0}}{\bar{s}^2}-\frac{2e^{i\phi}}{\rho}-\frac{2}{\bar{q}}+\left(\frac{\rho_0 e^{i\phi_0}}{\bar{s}^2}-\frac{2}{\bar{q}}\right)\frac{(l_2^2-a^2)\bar{t}}{l_2^2-a^2\bar{t}}\right]$$
$$+(a^2-\rho_0^2)^{1/2}\frac{e^{i\phi}}{\rho}\left(\frac{2e^{i\phi}}{\rho}+\frac{3}{\bar{q}}\right)-\frac{h}{R_0^2+h^2}$$
$$\left.\times\left[\frac{\bar{q}\rho e^{3i\phi}}{l_2^2-l_1^2}+\frac{e^{i\phi}(l_2^2-\rho^2)}{\rho\bar{q}}-\frac{z^2q}{R_0^2\bar{q}}+2e^{2i\phi}\right]\right\}. \quad \text{(G.16)}$$

Acknowledgment

The reported research was supported by a grant from the Natural Sciences and Engineering Research Council of Canada.

References

Collins, W. D. (1962). Some axially symmetric stress distributions in elastic solids containing penny-shaped cracks. I. Cracks in an infinite solid and a thick plate. *Proc. Roy. Soc. London Ser. A* **266**, 359–386.

Cherepanov, G. P. (1974). "Mechanics of Brittle Fracture." Nauka, Moscow (in Russian). English translation: McGraw–Hill (1979).

Elliott, H. A. (1948). Three-dimensional stress distributions in hexagonal aeolotropic crystals. *Proc. Cambridge Philos. Soc.* **44**, 522–533.

Elliott, H. A. (1949). Axial symmetric stress distributions in aeolotropic hexagonal crystals. The problem of the plane and related problems. *Proc. Cambridge Philos. Soc.* **45**, 621–630.

Fabrikant, V. I. (1986a). Inverse crack problem in elasticity. *Acta Mech.* **61**, 29–36.

Fabrikant, V. I. (1986b). A new approach to some problems in potential theory. *Z. Angew. Math. Mech.* **66**, 363–368.

Fabrikant, V. I. (1986c). Flat punch of arbitrary shape on an elastic half-space. *Internat. J. Engineering Sci.* **24**, 1731–1740.

Fabrikant, V. I. (1986d). Inclined flat punch of arbitrary shape on an elastic half-space. *ASME J. of App. Mech.* **53**, 798–806.

Fabrikant, V. I. (1987a). Flat crack of arbitrary shape in elastic body: Analytical approach. *Philos. Magazine A* **56**, 175–189.

Fabrikant, V. I. (1987b). Penny-shaped crack revisited: Closed form solutions. *Philos. Magazine A* **56**, 191–207.

Fabrikant, V. I. (1987c). Close interaction of coplanar circular cracks in an elastic medium. *Acta Mech.* **67**, 39–59.

Fabrikant, V. I. (1987d). Frictionless elastic contact problem for a curved rigid punch of arbitrary shape. *Acta Mech.* **67**, 1–25.

Gradshteyn, I. S., and Ryzhik, I. M. (1963). "Tables of Integrals, Series and Products." Moscow. English translation: Academic Press, New York (1965).

Kassir, M. K., and Sih, G. (1968). Three-dimensional stresses around elliptical cracks in transversely isotropic solids. *Engineering Fracture Mech.* **1**, 327–345.

Kassir, M. K., and Sih, G. (1975). "Three-Dimensional Crack Problems." Noordhoff International, Leyden.

Leonov, M. Ya. (1953). General problem of a circular punch pressed against an elastic half-space. *Prikl. Mat. Mekh.* **17**, 87–98 (in Russian).

Mossakovskii, V. I., Kachalovskaya, N. E., and Golikova, S. S. (1985). "Contact Problems of the Mathematical Theory of Elasticity." Naukova Dumka, Kiev (in Russian).

Panasiuk, V. V., Stadnik, M. M., and Silovaniuk, V. P. (1986). "Stress Concentration in Three-Dimensional Bodies with Thin Inclusions." Naukova Dumka, Kiev (in Russian).

Sih, G. C., and Liebowitz, H. (1968). Mathematical theories of brittle fracture. *In* "Fracture", Vol. 2 Academic Press, New York.

Sneddon, I. N. (1951). Fourier Transforms." McGraw–Hill, New York.

Uflyand, Ya. S. (1965). Survey of articles on the application of integral transforms in the theory of elasticity. North Carolina State University, Applied Mathematics Research Group, File No. PSR-24/6.

Uflyand, Ya. S. (1977). "Method of Coupled Equations in Mathematical Physics Problems." Academy of Sciences of the U.S.S.R., Nauka, Leningrad (in Russian).

Westmann, R. A. (1965). Asymmetric mixed boundary-value problem of the elastic half-space. *J. Appl. Mech.* **32**, 411–417.

ADVANCES IN APPLIED MECHANICS, VOLUME 27

Convective and Radiative Heat Transfer in Porous Media

CHANG-LIN TIEN†

Department of Mechanical Engineering
University of California,
Berkeley, California

and

KAMBIZ VAFAI

Department of Mechanical Engineering
Ohio State University
Columbus, Ohio

† Present affiliation: University of California, Irvine.

ISBN 0-12-002027-0

I. Introduction

Convective and radiative heat transfer and multiphase transport processes in porous media, both with or without phase change, have gained extensive attention in recent years. This is due to the wide range of applicability of these research areas in contemporary technology. These applications include, but are not restricted to, areas such as geothermal engineering, building thermal insulation, chemical catalytic reactors, packed cryogenic microsphere insulations, petroleum reservoirs, direct-contact heat exchangers, coal combustors, nuclear waste repositories, and heat pipe technology.

The reviews by Cheng (1978a) and Combarnous and Bories (1975) provide some well-thought views on a number of different issues in heat transfer in fluid-saturated porous media. The present work aims at providing a comprehensive review that will cover most of the aspects of interest in convective and radiative heat transfer and multiphase transport in porous media.

Several applications related to porous media require a detailed analysis of convective heat transfer in different geometrical shapes, orientations, and configurations. Based on the specific application, the flow in the porous medium may be internal or external. Most of the studies in porous media carried out until now are based on the Darcy flow model, which in turn is based on the assumption of creeping flow through an infinitely extended uniform medium. However, it is now generally recognized that the non-Darcian effects are quite important for certain applications. Different models have been introduced for studying and accounting for such non-Darcian effects as the inertial, boundary, and variable porosity effects. The ultimate goal of studies in convective heat transfer in porous media is to determine the dimensionless heat transfer coefficient, the Nusselt number. A considerable amount of research has been carried out to accomplish this, and empirical correlations for the Nusselt number for a variety of configurations and boundary conditions have been established, with certain limitations.

In keeping with the evolution of the non-Darcian model, it has become increasingly important to consider and study the effects of multiphase transport and phase change processes in porous media. Such an analysis has become, in recent years, a more pronounced area of research. A survey of the work performed in this area reveals several simplifying assumptions that are commonly made in the modeling of multiphase transport phenomena. These include assumptions of constant transport coefficients, absence of noncondensible gas effects, Darcian flow in the porous matrix, and local thermodynamic equilibrium between the phases of the porous system. It should be noted that such simplifying assumptions may be used for a limited number of phenomena, such as the drying of different porous materials and moisture migration due to a temperature gradient in a porous material, without significant loss of accuracy. However, for certain applications the assumption of Darcian flow and local thermodynamic equilibrium will not be valid. At present there is a lack of multiphase transport models for porous media that account for the non-Darcian effects in momentum transfer, and also a lack of any comprehensive analysis for multiphase transport in the absence of local thermodynamic equilibrium.

Some modern technologies and industrial processes utilizing porous and fluidized beds of solid particles operate at temperatures high enough for thermal radiation to be a significant mechanism compared to convective heat transfer (Flamant and Arnaud, 1984; Saxena *et al.*, 1978). Some examples are coal combustors, chemical reactors, and nuclear fuel rods. Other types of packed beds in which thermal radiation is important, even though the temperatures may not be high, are those where other modes of heat transfer have been suppressed, such as packed cryogenic microsphere insulations (Tien and Cunnington, 1973). For these reasons, radiative heat transfer is considered to be a significant mode of heat transfer for these types of porous systems.

The present work delivers a comprehensive and updated review of the research performed in the aforementioned areas and serves as a conduit for better understanding of the current research directions for convective and radiative heat transfer as well as multiphase transport processes in porous media. While providing a general and useful guide for future research, this work primarily deals with the problems that have engineering applications. To make this work a convenient reference for researchers, an attempt has been made to, as much as possible, subdivide and categorize the pertinent research areas and the available literature. Furthermore, by concentrating on the infrastructure of each research area and minimizing the peripheral information to the bare essentials, it is hoped that this exposition will serve as an objective, accurate, and readily accessible source of information for researchers in these areas. To this end, a conscious effort has been made to avoid duplicating equation after equation but instead concentrate on more physical and precise presentation with a practical point of view. However, every effort was made to provide the reader with a precise path to the most pertinent peripheral information related to a particular infrastructure. In what follows, the aforementioned subjects are divided into four sections for convenience. These are

II. Forced convection in porous media,
III. Natural convection in porous media,
IV. Multiphase transport in porous media,
V. Radiative heat transfer in porous and fluidized beds.

In Sections II and III the subject areas were well suited to be classified according to their geometric configurations as shown later in Figures 1–3.

II. Forced Convection in Porous Media

A. Flow over External Boundaries

Forced convection over external boundaries in the presence of a porous medium (Figure 1) constitutes a very important subject area. This is due to the very fundamental and generic nature of this type of problem which makes it pertinent to a wide variety of applications.

Consider the flow over a semi-infinite flat plate first. Analysis of convective heat transfer from this type of external boundary embedded in a porous medium has many important applications, e.g., drying of agricultural products, geothermal engineering, etc. Cheng (1977) and Bejan (1984) documented the local Nusselt number for forced convection over a constant temperature solid wall as

$$\mathrm{Nu}_x = 0.504\,\mathrm{Pe}_x^{1/2}, \tag{2.1}$$

where $\mathrm{Nu}_x = hx/k$, $\mathrm{Pe}_x = U_\infty x/\alpha$, h the heat transfer coefficient, k the thermal conductivity, U_∞ the free-stream velocity, and α is the thermal diffusivity. They also documented the local Nusselt number for a solid wall with constant heat flux as

$$\mathrm{Nu}_x = 0.886\,\mathrm{Pr}_x^{1/2}, \tag{2.2}$$

where $\mathrm{Nu}_x = qx/k(T_w - T_\infty)$, q is the supplied heat flux, and T_w is the temperature of the solid boundary. These two equations are valid if $\mathrm{Pe}_x \gg 1$, that is, when the longitudinal heat transfer is dominated by

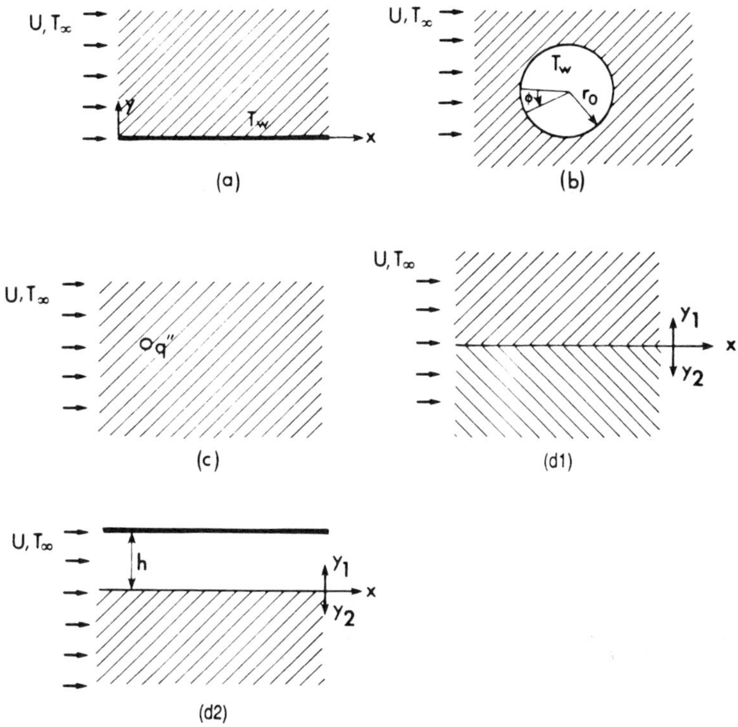

FIG. 1. Geometric configuration for forced convection heat transfer.
(a) Flat plate.
(b) Horizontal cylinder or sphere.
(c) Horizontal line source or point source.
(d1) Interface region between two different porous media.
(d2) Interface region between a fluid region and a porous medium.

convection. These results are based on Darcy's law, which neglects the effects of a solid boundary, inertia forces, and variable porosity on fluid flow and heat transfer through porous media.

However, in many cases of practical interest, it becomes quite important to consider the non-Darcian effects. In such cases, the fluid velocity is high, the porous medium is bounded, and the porosity is variable, thus rendering the Darcy flow model ineffective. Vafai and Tien (1981) analyzed the effects of a solid boundary and inertial forces on forced convection through constant porosity media, after establishing the governing equations by local volume-averaging technique. The governing equations established by Vafai and Tien (1981) and used in their analysis were

$$\nabla \cdot \langle \mathbf{v} \rangle = 0, \tag{2.3}$$

$$(\rho_{\mathrm{f}}/\delta)\langle(\mathbf{v}\cdot\nabla)\mathbf{v}\rangle = -\nabla\langle P\rangle^{\mathrm{f}} + (\mu_{\mathrm{f}}/\delta)\nabla^2\langle\mathbf{v}\rangle - (\mu_{\mathrm{f}}/K)\langle\mathbf{v}\rangle - \rho_{\mathrm{f}}F\delta^{1/2}\gamma[\langle\mathbf{v}\rangle\cdot\langle\mathbf{v}\rangle]\mathbf{J}, \tag{2.4}$$

$$(\langle\mathbf{v}\rangle\cdot\nabla)\langle T\rangle = \alpha_{\mathrm{e}}\nabla^2\langle T\rangle, \tag{2.5}$$

where $\mathbf{v}$ represents the velocity vector, $\langle\ \rangle$ denotes the local volume average of a quantity, ρ_{f} the fluid density, δ the porosity, K the permeability, T the temperature, $\mathbf{J}$ a unit vector oriented along the velocity vector, $\langle P\rangle^{\mathrm{f}}$ the intrinsic phase average of the pressure, μ_{f} the fluid viscosity, $\gamma = (\delta/K)^{1/2}$ the porous medium shape parameter, and F an empirical function which depends primarily on the microstructure of the porous medium.

It was shown that for the flow field the boundary effect is confined within a thin momentum boundary layer which often plays an insignificant role in the overall flow consideration. The effect of the boundary on heat transfer, however, was shown to be quite important and was more pronounced for the thermal boundary layer with a thickness less than or of the same order as that of the momentum boundary layer. Recently, Kaviany (1987) obtained Karman–Pohlhausen solutions for forced convection from a semi-infinite flat plate embedded in a porous medium on the basis of the Brinkman–Forschheimer-extended equation. These boundary and inertia effects were numerically and experimentally investigated by Vafai and Tien (1982) for transient mass transfer in the vicinity of an external impermeable boundary. It should be mentioned that the concepts and the general results obtained from these studies are equally valid for forced as well as natural convection over an external impermeable boundary in the vicinity of a porous medium.

In many applications such as fixed-bed catalytic reactors, packed bed heat exchangers, drying, chemical reaction engineering, and metal processing, the constant-porosity assumption does not hold because of the influence of an impermeable boundary. Therefore, there is a need to focus on the variable-porosity effects on forced convection in the vicinity of an impermeable boundary. Vafai (1984, 1986) presented in investigation of the channeling effect and its influence on heat transfer and fluid flow through variable-porosity media using the method of matched asymptotic expansions. The variable porosity effects were shown to be important for most heat transfer problems and the theoretical results were found to be in good agreement with the available experimental data. Furthermore, Vafai, Alkire, and Tien (1985) performed an experimental investigation on the effects of a solid impermeable boundary and variable porosity on forced convection in porous media. Their results confirm the qualitative and to a lesser degree the quantitative agreement between the experimental and the theoretical results that account for the variable porosity effects.

For flow over horizontal cylinders and spheres Cheng (1982) obtained similarity solutions using the thermal boundary layer approximation for forced convection around these types of geometries embedded in a porous medium. The local Nusselt numbers for an isothermal cylinder and an isothermal sphere are given respectively by

$$\mathrm{Nu}_\phi = 0.564\,\mathrm{Pe}_\phi^{1/2}\sqrt{2\phi}\,\sin(1-\cos\phi)^{-1/2}, \tag{2.6}$$

$$\mathrm{Nu}_\phi = 0.564\,\mathrm{Pe}_\phi^{1/2}\sqrt{\frac{3\phi}{2}}\,\sin^2\!\left(\cos^3\frac{\phi}{3}-\cos\phi+\frac{2}{3}\right)^{-1/2}, \tag{2.7}$$

where $\mathrm{Nu}_\phi = hr_0\phi/k$ and $\mathrm{Pe}_\phi = U_\infty r_0\phi/\alpha$ with ϕ denoting the angle measured from the stagnation point and r_0 denoting the radius of the cylinder or the sphere. These results are based on Darcy's law as the governing momentum equation and are valid when $\mathrm{Pe}_\phi \gg 1$.

For the cases of a horizontal line source and a point source placed in a uniform flow Bejan (1984) presented the resultant temperature distributions. Also Cheng and Zheng (1985) have reported the inertia and thermal dispersion effects on flow and temperature fields for a horizontal line heat source in a porous medium.

Another configuration of engineering importance is the interface region between a porous medium and another medium. In general, the other medium could be a fluid, a solid or another porous medium. A specific example of the interface region can be cited from petroleum reservoirs wherein the oil flow encounters different layers of sand, rock, shale, limestone, etc. Similar situations are encountered in many other cases of practical interest such as geothermal operations, nuclear waste repositories, water reservoirs, underground coal gasification, ground water hydrology, iron blast furnaces, and solid matrix heat exchangers. Vafai and Thiyagaraja (1987) analyzed a general class of problems involving interface interactions on flow and heat transfer for three different types of interface zones, the interface region between two different porous media, the interface region between a fluid region and a porous medium where the fluid layer is sandwiched between an impermeable wall and a porous medium, and the interface region between an impermeable medium and a porous medium. They obtained analytical solutions for both the velocity and temperature distributions as well as analytical expressions for the Nusselt number for all of the cases that were investigated.

First for the interface region between an impermeable medium and a porous medium, the Nusselt number was derived for the low and high Prandtl number Pr fluids. For the low Prandtl number fluids, the Nusselt number is

$$\mathrm{Nu}_\xi = \left(\frac{\mathrm{Pr}\,\delta}{\mathrm{Da}}\right)^{1/2} \sqrt{\Omega}\,\frac{\Gamma(p+1)}{\Gamma(p+\frac{1}{2})}\,\xi^{1/2}, \tag{2.8}$$

where Γ is the gamma function, ξ the nondimensional longitudinal distance, and p is related to the temperature variation. For the high Prandtl number fluids, the Nusselt number is

$$\mathrm{Nu}_\xi = \frac{3^{1/3}\Gamma(\frac{2}{3})}{\Gamma(\frac{1}{3})}\,\frac{\Gamma(p+1)}{\Gamma(p+\frac{2}{3})}\,\lambda^{1/3}\left(\frac{\mathrm{Pr}\,\delta}{\mathrm{Da}}\right)^{1/2} \xi^{2/3}, \tag{2.9}$$

where λ is a parameter that is explicitly related to the friction factor, Reynolds number, porosity, Darcy number, and the Prandtl number, whereas Ω is not dependent on the Prandtl number. The analysis shows that for high Prandtl number fluids, the Nusselt number is proportional to $\mathrm{Pr}^{1/3}$, and for low Prandtl number fluids, the Nusselt number is proportional to $\mathrm{Pr}^{1/2}$. These Nusselt number expressions are extremely accurate even for Prandtl numbers where the analytical temperature distribution shows some deviations. The Nusselt number for the interface region between two different porous media with different free stream temperatures was also derived for the low and high Prandtl number fluids based on the temperature distributions. For the low Prandtl number fluids, the Nusselt number is

$$\mathrm{Nu}_\xi = \left(\frac{\mathrm{Pr}_1\delta_1}{\mathrm{Da}_1}\right)^{1/2} \sqrt{\frac{\Omega_1}{\pi}}\,\xi^{1/2}. \tag{2.10}$$

For the high Prandtl number fluids the Nusselt number for the interface region between two different porous media with different free stream temperatures is

$$\mathrm{Nu}_\xi = \frac{\Delta_0\psi_1^{1/2}\,\mathrm{Pr}_1^{1/2} + [\psi_1^{1/2}\omega/2 - \Lambda_2\Delta_0/(4\psi_1)]\pi\xi^{1/2}}{[\omega\,\mathrm{Pr}_1^{-1/2} + \Delta_0(\pi/\xi)^{1/2}](\delta_1\,\mathrm{Da}_1)^{1/2}}. \tag{2.11}$$

For the interface region between a high Prandtl number fluid and a porous medium the Nusselt number is found to be

$$\mathrm{Nu}_\xi = \frac{\Delta_{c0}\psi_{c1}^{*1/2}\,\mathrm{Pr}_1^{1/2} + [\psi_{c1}^{*1/2}\Omega/2 - \Lambda_{c2}\Delta_{c0}/(4\psi_{c1}^{*})]\pi\xi^{1/2}}{[\Omega\,\mathrm{Pr}_1^{-1/2} + \Delta_{c0}(\pi/\xi)^{1/2}](\delta\,\mathrm{Da})^{1/2}}. \quad (2.12)$$

In the above expressions ψ_1, Δ_0, ω, Λ_2, Λ_{c1}^*, Λ_{c2}, and Λ_{c0} are given explicitly by Vafai and Thiyagaraja (1987) in terms of the derived interfacial velocity and temperature.

It should also be noted that their results for the interface between a fluid region and a porous medium are found to be in good agreement with the experimental hypothesis of Beavers and Joseph (1967).

B. Confined Flows

For forced convection in confined structures two important and interrelated cases are considered. First the convective heat transfer process for porous channels is discussed. The importance of this configuration is due to the fact that convective cooling can be augmented by insertion of a high-conductivity porous material in a coolant passage. Koh and Colony (1974) performed a numerical analysis of the cooling effectiveness of a heat exchanger containing a conductive porous medium while Koh and Stevens (1975) conducted an experimental investigation for the same problem. It was shown for the case with fixed allowable wall temperature that the heat flux at the channel wall can be increased by over three times by using a porous material in the channel. For the case with prescribed heat flux at the channel wall, a significant reduction in the wall temperature was observed when a porous material was used in the channel. This scheme has important practical applications, a typical example being the cooling of rocket nozzles. Here, an augmentation of regenerative cooling in the nozzle can be effectively achieved when the convective cooling of the nozzle has reached its limit.

The Nusselt number for the fully developed regime in a porous medium bounded by two parallel plates, based on the Darcy flow model (slug flow) can be represented by (Rohsenow and Hartnett, 1973)

$$\mathrm{Nu} = \begin{cases} 4.93 & \text{for constant wall temperature,} \\ 6.0 & \text{for constant wall heat flux,} \end{cases} \quad (2.13)$$

where $\mathrm{Nu} = qH/k(T_W - T_m)$, T_m is the bulk mean temperature of the fluid in the porous channel, and H the width of the channel.

To account for the effect of a solid boundary, Kaviany (1985a) performed a numerical study of laminar flow through a porous channel bounded by isothermal plates based on the Brinkman-extended Darcy model for constant porosity media. Recently Poulikakos and Renken (1987) have investigated the effects of flow inertia, variable porosity, and a solid boundary on fluid flow and heat transfer through porous media bounded by constant-temperature parallel plates or a circular pipe. They found that the general flow model predicts an overall enhancement in heat transfer between the fluid-saturated porous medium and the walls, compared to the predictions of the widely used Darcy flow model.

The convective heat transfer process through packed beds constitutes the second case. Many industrial operations in the areas of chemical and metallurgical engineering involve the passage of a fluid stream through a packed bed of particulate solids to obtain extended solid/fluid interfacial areas or good fluid mixing. Typical applications involving such systems include catalytic and chromatographic reactions, packed absorption and distillation towers, ion exchange columns, packed filters, pebble-type heat exchangers, etc. The design of these systems is governed by the pressure drop, fluid flow, and heat and mass transfer processes in the packed bed arrangement. Considerable attention has been paid to these aspects because of their direct influence on the optimiation and stability of these systems. Since most packed and fluidized beds are characterized by either high volume fractions of particles or large particles, or both, many features are unique for analyzing such systems.

In the past, a one-dimensional model has been used to describe the heat exchange process between the packed bed and its walls. This continuum model assumes a uniform fluid temperature for a section perpendicular to the flow direction. However, this condition of radial uniformity is usually not satisfied, especially in chemical reactors with important radial heat transfer effects. It has also been shown that the mean temperature predicted by the one-dimensional model deviates significantly from the actual radial mean temperature (De Wasch and Froment, 1972). These considerations led to the use of a two-dimensional model that accounts for the transverse heat transfer as well as the longitudinal heat transfer by convection. In such a model, the heat transfer in transverse direction can be characterized by the effective thermal conductivity and the wall heat transfer coefficient. The latter parameter was introduced to account for the higher thermal resistance near the wall. These two parameters were estimated from the temperature profile measurements performed in fluids flowing through packed beds, heated or cooled from the wall (Lerou and Froment, 1977). However, caution must be exercised while using many of the empirical and semi-empirical correlations, especially while extrapolating into regions where measurements are not available. The results of different investigations show many discrepancies, the causes for which have been attributed to the experimental techniques and the various definitions for the parameters employed by different researchers (Balakrishnan and Pei, 1974). This type of analysis was also shown to be valid only after a certain entrance length, thereby necessitating the introduction of a length effect in these studies. The inclusion of the length effect in these studies does not completely remedy this situation. This makes much of the data from the above-mentioned investigations unsuitable for use in chemical reactor design (Li and Finlayson, 1977).

Recently published research work emphasizes the importance of porosity changes and the corresponding velocity changes near the container wall and has shown, by comparison with experiments, that the theoretical prediction of chemical reactor performance can be improved by assuming a nonuniform velocity distribution within the packed bed (Kalthoff and Vortmeyer, 1980). Vortmeyer and Schuster (1983) solved the Brinkman-extended Darcy equation by a variational method and explained the large deviations between calculated and measured profiles. Vafai (1986) performed a boundary layer analysis for a variable-porosity medium using the method of matched asymptotic expansions on the Brinkman–Forschheimer-extended Darcy equation and the energy equation. The concept of the triple momentum boundary layer, to incorporate the channeling effect in variable porosity media, was introduced in this work. This type of analysis relies directly on the fundamental principles, thus enabling a more thorough analysis of the problem. Cheng and Hsu (1986) performed a numerical analysis for a fully developed forced convective flow through a packed bed, confined by concentric walls maintained at different temperatures. Their analysis was based on the Brinkman model with variable permeability and the mixing-length theory proposed recently by Cheng and Vortmeyer (1988) for transverse thermal dispersion.

Hunt and Tien (1987) also investigated the non-Darcian convection in cylindrical packed beds on the basis of Brinkman–Forschheimer-extended Darcy equation. To account for the transverse thermal dispersion, they used the effective thermal conductivity varying with radial direction in the volume-averaged steady energy equation. Their results agree well with experimental data for chemical catalytic reactors and are independent of the extensive experimental relations needed in conventional reactor models. However, most of the different approaches for incorporating the thermal dispersion effects are essentially reduced to the determination of the effective thermal conductivity by the use of more adjustable parameters. A more fundamental approach for determination of the thermal dispersion effects is needed for future research work.

III. Natural Convection in Porous Media

A. Natural Convection over External Boundaries

Vertical flat plate boundaries are the most analyzed configuration for natural convection over external boundaries (Figure 2). Cheng and Minkowycz (1977) performed an analysis on natural convection from a heated impermeable surface embedded in fluid-saturated porous media. Their investigation was primarily motivated by their aim to simulate the groundwater heating of an aquifer by a dike. They used a boundary layer formulation based on Darcy's law and the energy equation to obtain similarity solutions. The local Nusselt number for a vertical isothermal wall was found to be

$$\mathrm{Nu}_x = 0.444\,\mathrm{Ra}_x^{1/2}, \tag{3.1}$$

where $\mathrm{Nu}_x = hx/k$ and $\mathrm{Ra}_x = Kg\beta x(T_w - T_\infty)/\alpha\nu$, g the gravitational constant, ν the kinematic viscosity, and β is the volumetric thermal expansion coefficient. This equation is valid in the boundary layer regime, where $\mathrm{Ra}_x^{1/2} \gg 1$. The local Nusselt number for a vertical wall with constant heat flux can be written as

$$\mathrm{Nu}_x = 0.772\,\mathrm{Ra}_x^{*1/3}, \tag{3.2}$$

where $\mathrm{Ra}_x^* = Kg\beta x^2 q/\alpha\nu k$. This relation is valid in the boundary-layer regime, $\mathrm{Ra}_x^{*1/3} \gg 1$. Cheng and Pop (1984) also considered transient free convection about a heated vertical wall embedded in a porous medium.

To extend the range of applicability of boundary layer analysis to relatively lower values of Rayleigh number, higher-order effects such as entrainments from the edge of the boundary layer, axial heat conduction and normal pressure gradients have to be included in the boundary layer

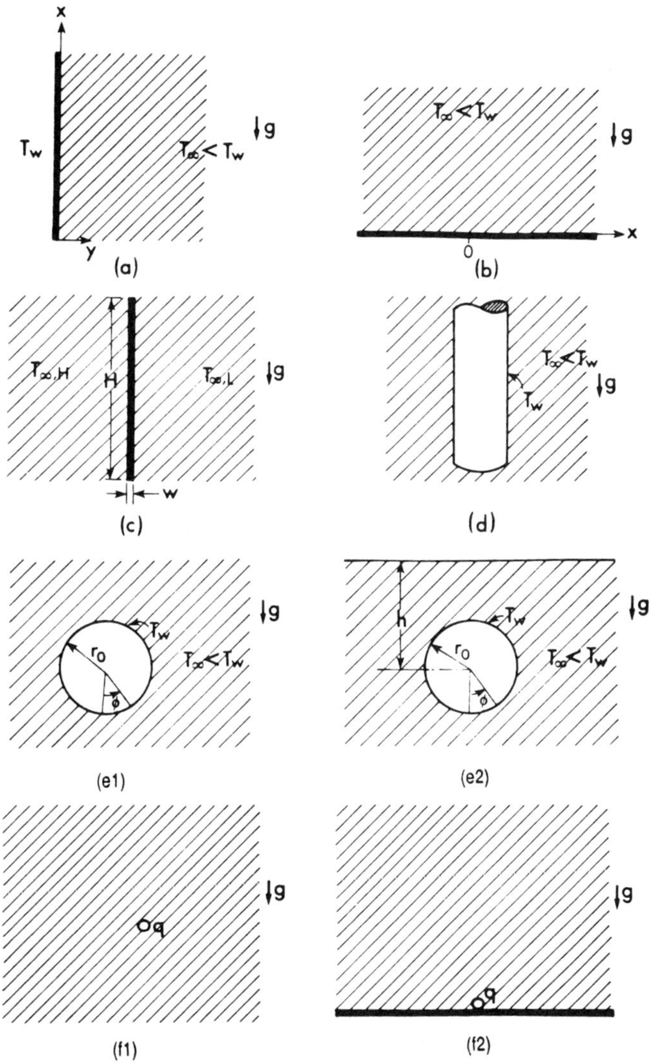

FIG. 2. Geometric configuration for natural convection heat transfer (boundary layer).
(a) Vertical flat plate.
(b) Horizontal flat plate.
(c) Conjugate boundary layer.
(d) Vertical cylinder.
(e1) Horizontal cylinder or sphere.
(e2) Horizontal cylinder buried beneath an impermeable horizontal surface.
(f1) Point source or horizontal line source in an infinite porous medium.
(f2) Point source or horizontal line source located on the lower boundary of a semi-infinite porous medium.

analysis. Cheng and Chang (1979) and Cheng and Hsu (1984) examined the magnitude of these effects using the method of matched asymptotic expansions for the problem of natural convection in a porous medium near a vertical plate with a power law variation of wall temperature. It was found that higher-order theory has a significant effect on the velocity profiles but a relatively small effect on the temperature profiles. However, their results have been found to have questionable validity. Joshi and Gebhart (1984) have pointed out that those results are valid only for a surface temperature variation specified *a priori,* such as an isothermal surface condition, and that the problem with uniform heat flux condition at the wall has to be treated separately. This is required because the surface temperature variation is improved successively at each order and is not imposed externally in the perturbation analysis. In order to remedy the above problem, they developed consistent higher-order approximations including second-order solutions using the method of matched asymptotic expansions.

A number of recent studies have considered the various non-Darcian effects for the problem of natural convection about a vertical flat plate embedded in a porous medium; Plumb and Huenefeld (1981) and Bejan and Poulikakos (1984) used Forschheimer's equation to account for inertial effects and obtained similarity solutions with the boundary layer approximations. Based on Brinkman's equation Evans and Plumb (1978) investigated numerically the boundary effect on heat and fluid flow, and Hsu and Cheng (1985) used the method of matched asymptotic expansions for the same problem. Recently, Kim and Vafai (1989) obtained analytical and numerical solutions for buoyancy-driven fluid flow and heat transfer about a vertical flat plate embedded in a porous medium. The governing equations were solved analytically using the method of matched asymptotic expansions along with the modified Oseen method. Their analysis, which included the boundary effects, was performed for two general cases, the constant wall temperature and the constant wall heat flux cases. For each case, two important categories were considered. For the constant wall temperature case these categories were shown to be physically related to either $\mathrm{Da}_L^{1/2} \ll \mathrm{Ra}_L^{-1/2}$ ($\mathrm{Da}_L^{1/2} \ll \mathrm{Ra}_L^{*-1/3}$ for the constant wall heat flux) or $\mathrm{Da}_L^{1/2} \gg \mathrm{Ra}_L^{-1/2}$ ($\mathrm{Da}_L^{1/2} \gg \mathrm{Ra}_L^{*-1/3}$ for the constant wall heat flux). Hong, Tien, and Kaviany (1985) included both boundary and inertia effects as well as the convective term in the momentum equation. Also Kaviany (1985b) reported a Karman–Pohlhausen solution to the same problem and Chen, Hung, and Cleaver

(1987) studied numerically the non-Darcian effect of a vertical flat plate embedded in a high-porosity medium on transient natural convection. These investigations have shown that both inertia and boundary effects decrease the heat transfer rate as the Rayleigh number is increased. However, the influence of boundary effects on the Rayleigh number is less pronounced than the influence of inertial effects. Another important non-Darcian effect is the thermal dispersion effect. It has been found that this effect needs to be considered when the inertia effects are prevalent. Cheng (1981) and Plumb (1983) examined the thermal dispersion effects in non-Darcian convective flows. Their analyses neglected the boundary effect, which was subsequently included in the recent study by Hong and Tien (1987). It is shown that the heat transfer rate is enhanced as a result of the thermal dispersion effects.

Evans and Plumb (1978), Cheng, Ali, and Verma (1981), Cheng and Ali (1981) and Kaviany and Mittal (1987) have performed experimental investigations of natural convection around a vertical and an inclined isothermal plate. Also Huenefeld and Plumb (1981) experimentally investigated natural convection about a vertical plate with constant heat flux.

The results reviewed in this section are also applicable to the case of an inclined flat plate if the gravitational acceleration appearing in the Rayleigh number is replaced by the gravitational acceleration component that acts along the inclined surface. However, using the appropriate component of the gravitational acceleration produces valid results only up to a certain critical inclination angle. Little work has been done on determining this critical inclination angle, where apparently the convection mode changes character from one type to another.

For natural convection over a horizontal flat plate, which has important applications in heat transfer around a geothermal reservoir, Cheng and Chang (1976) obtained a similarity solution leading to a local Nusselt number given by

$$\mathrm{Nu}_x = 0.420\,\mathrm{Ra}_x^{1/3}, \tag{3.3}$$

where $\mathrm{Nu}_x = hx/k$ and $\mathrm{Ra}_x = Kg\beta x(T_W - T_\infty)/\alpha\nu$. This equation is valid only in the thermal boundary layer regime, i.e., $\mathrm{Ra}_x^{1/3} \gg 1$. Also the local Nusselt number for a horizontal wall with uniform heat flux was found to be

$$\mathrm{Nu}_x = 0.859\,\mathrm{Ra}_x^{*1/4}, \tag{3.4}$$

where $\mathrm{Ra}_x^* = Kg\beta x^2 q/k\alpha\nu$. This relation holds also only in the thermal boundary layer regime, $\mathrm{Ra}_x^{*1/4} \gg 1$. In a subsequent paper, Cheng (1978b) obtained Karman–Pohlhausen type integral solutions for the same problem. Also Pop and Cheng (1983) reported a Karman–Pohlhausen solution for the growth of a thermal boundary layer in a porous medium adjacent to a suddenly heated semi-infinite horizontal surface.

To extend the applicability of the boundary layer theory to lower Rayleigh number ranges, Chang and Cheng (1983) refined the boundary layer analysis by including the higher-order effects such as the upward-drift induced friction and fluid entrainment, using the method of matched asymptotic expansions.

The upward-facing cold plate with finite length, another horizontal wall configuration, was examined by Kimura, Bejan, and Pop (1985). They have shown that the overall heat transfer rate between the porous medium and the flat plate varies as

$$\mathrm{Nu} \sim \mathrm{Ra}^{1/3}, \tag{3.5}$$

where Ra is the Darcy-modified Rayleigh number based on plate half-length.

The analysis of the buoyancy induced flow occurring on both sides of a vertical impermeable wall facing two fluid-saturated porous reservoirs maintained at different temperatures could be useful for calculations in the design of thermal insulations. This type of configuration falls under a category which is usually referred to as "conjugate boundary layers." Bejan and Anderson (1981) analyzed such a configuration. They found that a boundary layer exists on each side of the impermeable wall. The overall Nusselt number correlation, correct to within 1%, is

$$\mathrm{Nu} = 0.382(1 + 0.615\,\omega)^{-0.875}\,\mathrm{Ra}_H^{1/2}, \tag{3.6}$$

where $\mathrm{Nu} = q_{0\text{-}H}H/(T_{\infty,H} - T_{\infty,L})k$, $\omega = (Wk/Hk_w)\,\mathrm{Ra}_H^{1/2}$ is the parameter describing the size of the wall thermal resistance relative to the convective resistance, k_w is the thermal conductivity of the wall, and descriptions for $T_{\infty,H}$, $T_{\infty,L}$, H, and W are given in Figure 2c.

Bejan and Anderson (1983) examined the natural convection flow between a fluid-saturated porous medium and a fluid reservoir separated by a vertical impermeable wall when the two fluids are maintained at different temperatures. For this case, it was also shown that a boundary layer exists on each side of the partition. The overall Nusselt number may be represented as

$$\mathrm{Nu} = [(0.638)^{-1} + (0.888\,\mathrm{B})^{-1}]^{-1}\,\mathrm{Ra}_{H,f}^{1/4}, \tag{3.7}$$

where $\mathrm{Ra}_{H,f} = g(\beta/\alpha\nu)_f H^3(T_{\infty,H} - T_{\infty,L})$, and $B = k\,\mathrm{Ra}_H^{1/2}/k_f\,\mathrm{Ra}_{H,f}^{1/4}$.

For the case of a vertical cylinder, which has practical application in predicting the convective flow generated in the ground water adjacent to hot intrusives like magma, Minkowycz and Cheng (1976) obtained an approximate solution using the local nonsimilarity method. Their solution included different power law variations for the surface temperature of the cylinder. In case of an isothermal cylinder, the local Nusselt number is given by

$$\mathrm{Nu}_x = 0.444\,\mathrm{Ra}_x^{1/2}\left[1 + 0.6\left(\frac{x}{r_0}\right)\mathrm{Ra}_x^{-1/2}\right], \tag{3.8}$$

where r_0 is the radius of the cylinder and x refers to the curvilinear coordinate along the surface of the cylinder. By comparing the above result with the Nusselt number for the vertical flat plate, it becomes apparent that for the same surface area, a vertical cylinder transfers more heat than a vertical plate.

For horizontal cylinders and spheres the flow and temperature fields are part of the information that can be used in the design of underground electrical cables, thermal insulation of power plants, and steam and water distribution lines. Merkin (1979) obtained similarity solutions for natural convection boundary layers on axisymmetric and two-dimensional bodies of arbitrary shape embedded in a saturated porous medium. Cheng (1982), using a similarity transformation similar to Merkin's, showed that the local Nusselt numbers for boundary layer convection about a horizontal isothermal cylinder or sphere embedded in an infinite porous medium are respectively given by

$$\mathrm{Nu}_\phi = 0.444\,\mathrm{Ra}_\phi^{1/2}\sqrt{2\phi}\,\sin(1 - \cos\phi)^{-1/2}, \tag{3.9}$$

$$\mathrm{Nu}_\phi = 0.444\,\mathrm{Ra}_\phi^{1/2}\sqrt{\frac{3\phi}{2}}\,\sin^2\!\left(\cos^3\frac{\phi}{3} - \cos\phi + \frac{2}{3}\right)^{-1/2}, \tag{3.10}$$

where $\mathrm{Nu}_\phi = q_w r_0/k(T_w - T_\infty)$ and $\mathrm{Ra}_\phi = Kg\beta r_0\phi(T_w - T_\infty)/\alpha\nu$ with ϕ denoting the angle measured from the downward vertical (as shown in Figure 2e) and r_0 denoting the radius of the cylinder or sphere. It should be noted that Cheng neglected the buoyancy force normal to the heated surface in order to obtain the above equations. Recently Fand, Steinberger, and Cheng (1986) performed an experimental investigation of heat transfer by natural convection from a horizontal cylinder embedded in a porous medium. They have shown that the overall range of Rayleigh number can be divided into two subregions, called "low" and "high," in each of which the Nusselt number behaves differently. It is shown that the low-Rayleigh number region corresponds to Darcy flow and the high-Rayleigh number region to Forschheimer flow.

Natural convection around a cylinder buried beneath an impermeable horizontal surface is another important configuration. Schrock, Fernandez, and Kesavan (1970) and Fernandez and Schrock (1982) performed experimental investigations of this geometry. Recently Farouk and Shayer (1985) conducted a numerical study of natural convection from a heated cylinder buried in a semi-infinite liquid-saturated porous medium. It should be mentioned that this geometry is the same configuration studied by Schrock *et al.*

Natural convection about a line source or a point source, which is the limiting case for a horizontal cylinder or sphere, respectively, has an important application to heat transfer problems in a nuclear-waste repository. Bejan (1978) considered the problem of transient and steady-state natural convection around a concentrated point heat source in an infinite porous medium at low Rayleigh numbers and obtained a perturbation solution for the transient case up to the first-order Rayleigh number and for the steady-state case up to the third-order Rayleigh number. For the same point source geometry, Hickox and Watts (1980) obtained a numerical solution using a similarity transformation for the steady-state natural convection. They also obtained a numerical solution for a point source located on the lower, insulated boundary of a semi-infinite region for a range of Rayleigh numbers from 0.1 to 100. In a subsequent paper, Hickox (1981) has shown that in the limit of small Rayleigh numbers, superposition of transient and steady convection solutions for a concentrated heat source, a finite vertical line source and a constant temperature sphere can be used for obtaining solutions for some other geometries.

Prior to these works, Wooding (1963) had proposed a boundary layer model for large-Rayleigh number natural convection about a point and line source located on the lower boundary of a semi-infinite fluid-saturated porous medium. Furthermore, Wooding was able to present an analytical solution for the temperature and flow field around a horizontal line source in a porous medium at high Rayleigh numbers. Bejan (1984) presented an analytical solution using a boundary layer analysis for the flow and temperature field in the high-Rayleigh number regime for the plume above a point source.

B. Natural Convection in Confined Structures

Convective heat transfer in a rectangular porous cavity in which vertical walls are maintained at two different temperatures is a fundamental problem (Figure 3). It has important applications to building insulation and geothermal operations. The boundary layer regime that develops at large Rayleigh numbers was studied by Weber (1975a) using boundary layer approximations with the modified Oseen method. Darcy's equation was used as the governing momentum equation, and the Nusselt number, defined as the ratio of the total heat transport to the heat transferred by pure conduction, was found to be

$$\mathrm{Nu} = 0.577\left(\frac{L}{H}\right)\mathrm{Ra}_H^{1/2}. \tag{3.11}$$

By replacing Weber's impermeable horizontal boundary condition with average zero energy flux conditions along the horizontal walls, Bejan (1979) was able to get better agreement with the experimental results of Klarsfeld (1963) as well as with the numerical calculations of Bankvall (1974) and Burns, Chow, and Tien (1977). Bejan presented the Nusselt number as

$$\mathrm{Nu} = 0.508\left(\frac{L}{H}\right)\mathrm{Ra}_H^{1/2}. \tag{3.12}$$

Simpkins and Blythe (1980) arrived at similar results by using an integral method.

Bories and Combarnous (1973) and Seki, Fukusako, and Inaba (1978) performed experimental work to investigate the effects of the cavity aspect ratio and the physical properties of the porous medium on the overall heat transfer rates. Shiralkar, Haajizadeh, and Tien (1983) performed a numerical study of high-Rayleigh number convection for the same problem using a stable exponential differencing scheme. To account for the non-Darcian effect on fluid flow and heat transfer, Tong and Subramanian (1985) used the Brinkman-extended model based on boundary layer approximations and the modified Oseen method. They

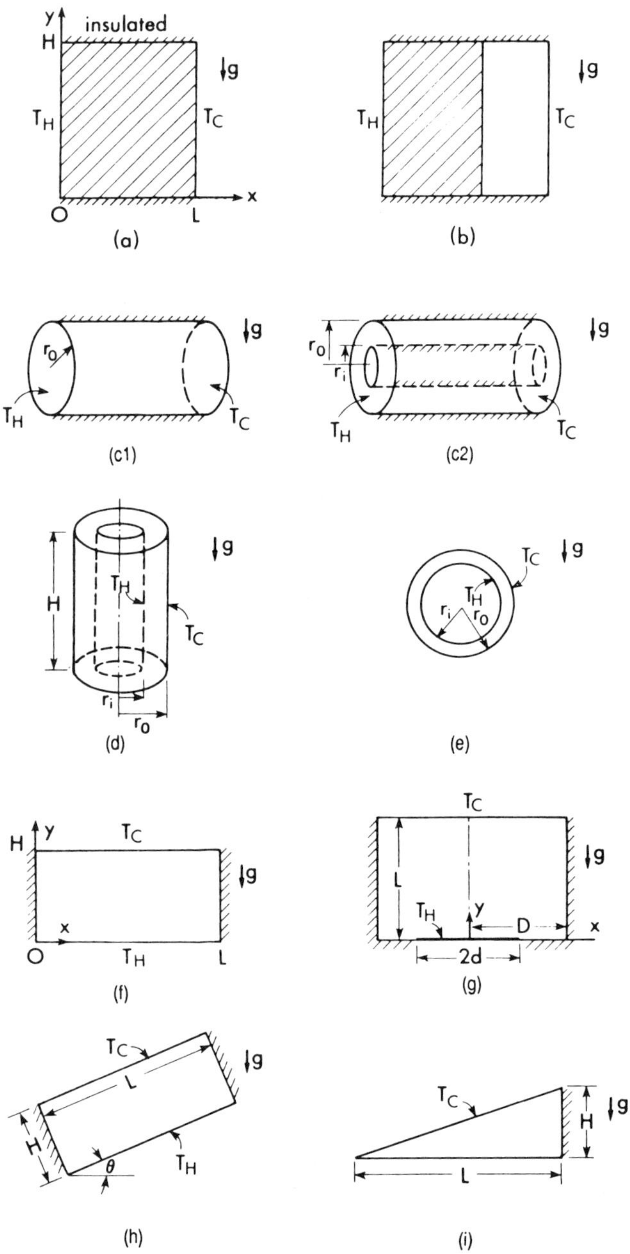

insulated
y
H
T_H
T_C
g
x
O
L
(a)
(b)
r_o
r_i
(c1)
(c2)
(d)
(e)
(f)
D
2d
(g)
θ
(h)
(i)

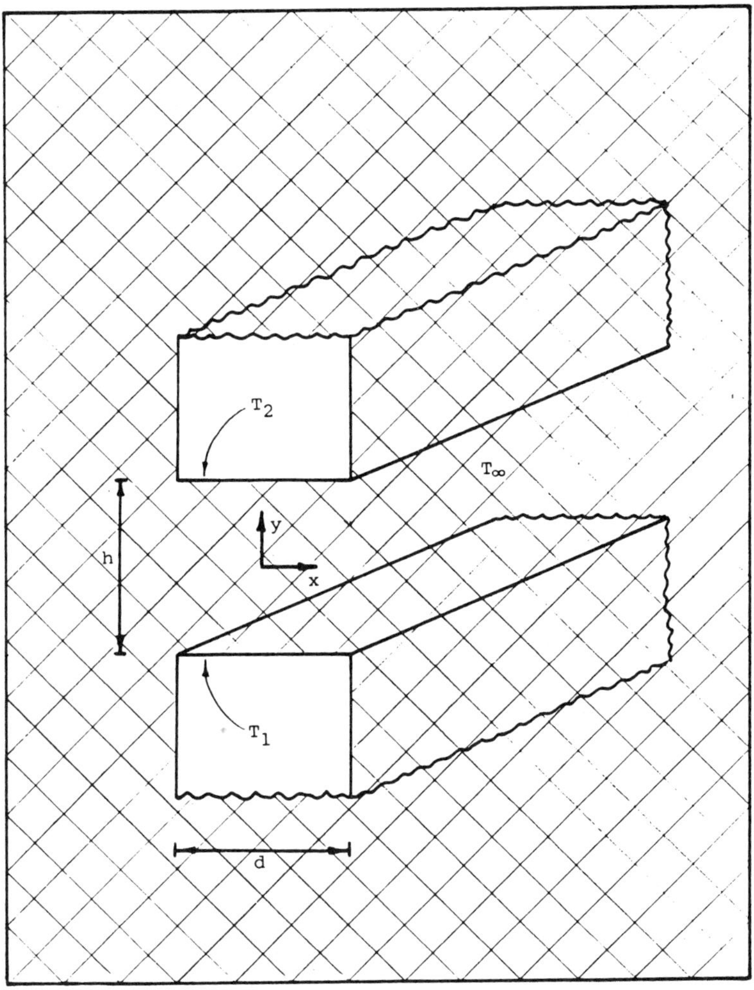

(j)

FIG. 3. Geometric configuration for natural convection heat transfer (confined flow).
(a) Rectangular porous cavity heated from the side.
(b) Rectangular cavity, partially filled with a porous medium, heated from the side.
(c1) Horizontal cylinder heated from the side.
(c2) Horizontal Annulus heated from the side.
(d) Vertical porous annulus heated from the side.
(e) Horizontal concentric cylinders or spheres with radial temperature gradient.
(f) Horizontal porous layers heated from below.
(g) Horizontal porous layers with localized heating from below.
(h) Inclined porous layer heated from below.
(i) Wedge-shaped porous layer heated from below.
(j) Open-ended cavity with a porous obstructing medium.

reported significant contributions of the viscous diffusion term at high Rayleigh and Darcy numbers. Poulikakos and Bejan (1985) used the Forschheimer-extended Darcy model to obtain boundary layer solutions and numerical results. Chan, Ivey, and Barry (1970), Lauriat and Prasad (1987), and Tong and Orangi (1986) performed numerical analyses of a rectangular cavity heated from the side for various aspect ratios, Darcy numbers and Rayleigh numbers, using the Brinkman-extended Darcy model.

In the limit of low aspect ratios the problem becomes one of natural convection in shallow cavities with different end temperatures. Walker and Homsy (1978) studied such a problem by using matched asymptotic expansions and found the following expression for the Nusselt number based on a Darcy model

$$\mathrm{Nu} = 1 + \frac{1}{120}\left(\mathrm{Ra}_H \frac{H}{L}\right)^2. \tag{3.13}$$

The same result was also reported by Bejan and Tien (1978). Hickox and Gartling (1981) also performed a numerical investigation of this problem using the Galerkin form of the finite element method. To account for the momentum boundary layer, recently Sen (1987) used the Brinkman-extended Darcy model to obtain the boundary layer solutions for shallow rectangular cavities. In order to investigate the effect of infiltration into a shallow cavity Haajizadeh and Tien (1983) performed a numerical and experimental study of natural convection in a shallow cavity with one permeable vertical wall.

In a series of investigations, Blythe, Simpkins, and Daniels (1982, 1983) discussed the large Rayleigh number limit for a fixed aspect ratio and presented some new scaling laws for the problem of natural convection in a rectangular cavity. Assuming a constant buoyant term, Philip (1982) found some exact solutions, using the separation of variables techniques, in the small Rayleigh number limit in cavities of rectangular as well as of other geometric cross-sections. Prasad and Kulacki (1984a) studied numerically the effect of aspect ratio on fluid flow and heat transfer for the rectangular cavity problem. It was shown that for a tall cavity, an increase in the width decreases heat transfer except when the flow exhibits boundary layer behavior on the vertical walls.

Bejan (1983a) considered the case where the uniform heating and cooling effects are prescribed along the vertical side walls. His solution was based on the Darcy flow model, the thermal boundary layer approximation and the use of the modified Oseen method. For this case the Nusselt number in the high Rayleigh number limit was found to be

$$\mathrm{Nu} = \frac{1}{2}\left(\frac{H}{L}\right)^{1/5} \mathrm{Ra}_H^{*2/5}, \tag{3.14}$$

where $\mathrm{Ra}_H^* = Kg\beta H^2 q/\alpha\nu k$. To account for the various non-Darcian effects on fluid flow and heat transfer, Poulikakos (1985) used the Forschheimer-extended Darcy model to obtain the boundary layer solution, and Vasseur and Robillard (1987) used the Brinkman-extended Darcy model to study the boundary effects, which were shown to have a significant influence on the flow field and heat transfer. Also Prasad and Kulacki (1984b) studied numerically two-dimensional, steady, free convection for a rectangular cavity with constant heat flux on one vertical wall while the other vertical wall as isothermally cooled.

Furthermore, Bejan (1983b) studied numerically the effect of internal flow obstructions, created by the addition of a variable size horizontal or vertical plate to the corresponding vertical or horizontal boundary, on heat transfer through a two-dimensional porous layer heated from the side. Finally, in an attempt to observe the effect of non-uniform permeability and thermal diffusivity on natural convection through a porous layer heated from the side, Poulikakos and Bejan (1983a) numerically investigated the flow field in a rectangular enclosure with vertical (or horizontal) layers of various permeability porous media.

In building insulation, it is common to fill the enclosure entirely with a porous material. To study the effect of partial filling of an enclosure with a porous medium, Tong and Subramanian (1983) have numerically analyzed natural convection in a rectangular enclosure that is vertically divided into an open region and a region filled with a porous medium. By assuming an impermeable interface, the open and the porous regions were essentially decoupled. In essence their results could also be obtained by a combination of available literature on open and porous cavities by using an iteration procedure. An experimental investigation of the problem by Tong, Faruque, Orangi, and Sathe (1986) led the authors to conclude that the heat transfer could be minimized by filling the enclosure partially with a porous medium rather than filling it completely. A more practical case where there is no impermeable partition between the fluid and the porous medium was considered both numerically and experimentally by Beckermann, Ramadhyani, and Viskanta (1987).

For a horizontal cylinder and annulus with ends at different temperatures Bejan and Tien (1978) presented a parametric solution. It was based on representation of the velocity components and the temperature field by what amounts to a perturbation in the core region and integral analyses in the end regions. Bejan and Tien (1979) also studied natural convection in a porous medium bounded by two concentric cylinders with different end temperatures.

Hickox and Gartling (1982) studied the case of a vertical porous annulus heated from the side, in which the inner and outer walls are isothermally heated and cooled, respectively, while the top and bottom walls are insulated. They used a finite element method for obtaining heat transfer results for Rayleigh numbers up to 10^2. This type of geometry is typically encountered in the storage of canisters containing high-level nuclear waste material. However, in recent years the emphasis has been placed in the horizontal emplacement of these canisters, thereby making a modified version of the previously cited case more appropriate for this type of application. Havstad and Burns (1982) used a finite difference method, a perturbation technique, and an approximate analysis to analyze the same problem. Prasad and Kulacki (1984c, 1985) also considered the same problem for a range of Rayleigh numbers up to 10^4 and for the height-to-gap width ratio of $1 \leq A \leq 20$ and $0.3 \leq A \leq 0.9$ respectively. The work, based on a finite difference formulation, concentrated on an investigation of the boundary layer and curvature effects on temperature and flow-field structures. Furthermore, in their latter work Prasad and Kulacki (1985), as well as Prasad *et al.* (1985), performed experimental investigations of the same problem.

Studies of natural convective flow and heat transfer in porous concentric cylinders and spheres with radial temperature gradients have applications to buried pipelines, storage of solar energy in underground containers, and insulating material for some types of nuclear reactors and LNG facilities. Caltagirone (1976) examined free convection in porous media when the cylindrical surfaces of an annulus are kept at different temperatures. He obtained the steady, two-dimensional solutions numerically and also observed the three-dimensional flows and associated fluctuations experimentally. Facas and Farouk (1983) studied numerically two-dimensional transient convection for the same problem. They were able to show that the transient state finally does converge to a steady state with no significant intermediate oscillatory behavior. Recently Muralidhar and Kulacki (1986) numerically investigated the non-Darcian effects of inertia, boundary friction, and variable porosity for the same problem.

Burns and Tien (1979) analyzed two-dimensional steady natural convection in a porous medium bounded by concentric spheres and horizontal cylinders. Their solutions were based on the finite difference formulation and the use of regular perturbation method for the case where the inner boundary was isothermal and the outer boundary was allowed to interact with the surroundings. Masuoka, Ishizaka, and Katsuhara (1980) also performed a numerical and experimental investigation of natural convection in porous media bounded by two concentric spheres.

For horizontal porous layers heated from below, there has been a considerable amount of analytical, numerical, and experimental work concerning the Benard convection. Lapwood (1948) examined the onset of convection in an infinitely long horizontal layer heated from below by using linearized stability analysis and found that the critical Rayleigh number for the onset of convection is given by

$$\mathrm{Ra}_H = 4\pi^2, \tag{3.15}$$

where $\mathrm{Ra}_H = Kg\beta H(T_H - T_C)/\alpha\nu$. Later, Katto and Masuoka (1967) performed an experimental investigation for the same problem.

Nusselt number dependence on Rayleigh numbers higher than the critical value has been reviewed by Cheng (1978a), who compiled results from nine sources for work that has been done over the past three decades. These sources were Combarnous and Bories (1975), Schneider (1963), Buretta and Berman (1976), Yen (1974), Elder (1967a), Kaneko, Mohtadi, and Aziz (1974), Gupta and Joseph (1973), Strauss (1974), and Combarnous and Bia (1971).

Robinson and O'Sullivan (1976), Somerton (1983), and Combarnous and Boris (1975) tried to account for the wide spread in the data from the sources that were reviewed by Cheng (1978a). Recently, Georgiadis and Catton (1986) and Jonsson and Catton (1987) have shown through numerical and experimental studies, respectively, that this scatter can be attributed to the Prandtl number effect on Benard convection in porous media heated from below. Georgiadis and Catton (1986) used the Brinkman–Forschheimer-extended Darcy equation to account for the inertial and boundary effects.

In a related configuration, Poulikakos (1986) studied numerically the problem of buoyancy-driven flow instability in a horizontal composite (fluid–porous) system heated from below. The composite system consists of a fluid layer overlaying a porous substrate and is bounded by two isothermal horizontal walls and two adiabatic vertical walls. This configuration has important applications to fibrous and granular thermal insulation where the insulation occupies only part of the space separating the hot boundary from the cold boundary. A Forschheimer–Brinkman-extended Darcy equation was used to describe the flow inside the porous bed. Also, the dependence of the flow and temperature field characteristics on the pertinent dimensionless groups was discussed.

Natural convection through horizontal porous layers, where only a portion of the bottom surface is heated while the rest of it is either adiabatic or isothermally cooled, has some applications with respect to a geologic repository for the storage of nuclear waste and also for some geothermal operations. Elder (1967a) performed a numerical study of the steady-state natural convection in a horizontal porous cavity with walls that are isothermally cooled except for a centrally heated portion of the bottom surface. Later, Elder (1967b) studied numerically and experimentally the transient natural convection for the same geometry and boundary conditions.

Horne and O'Sullivan (1974) studied numerically and experimentally the oscillatory flow behavior in a horizontal porous cavity with insulated vertical walls and isothermally cooled top and bottom surfaces, except for a centrally heated portion of the bottom surface. Also Horne and O'Sullivan (1978) studied the effect of temperature dependent viscosity and thermal expansion coefficient on flow and temperature fields.

Recently Prasad and Kulacki (1986, 1987) analyzed natural convection in a horizontal porous cavity with insulated vertical walls, an isothermally cooled top surface, and an insulated bottom surface with a centrally heated portion. They used the finite difference formulation to characterize the effects of the size and strength of the heat source as well as the aspect ratio on heat transfer and fluid flow. They observed a plume-like behavior above the heated region as the Rayleigh number was increased.

Bories and Combarnous (1973) have performed a theoretical and experimental investigation of convective heat transfer in an inclined porous layer heated isothermally from below. The three main structures of free convection movements have been observed depending on the values of the Rayleigh number, the inclination angle, and the longitudinal extension of the layer. It was found that if $\mathrm{Ra}_H < 4\pi^2/\cos\theta$, where θ is the angle between the heated wall and the horizontal plate, the convective motion takes the form of a two-dimensional unicellular flow. For the Rayleigh numbers higher than this critical value, a transition from unicellular flow to stable three-dimensional flow is observed to occur in such a way that the convective motion takes the form of either polyhedral cells or longitudinal coils, depending on the value of θ itself. The investigations of Bories and Combarnous (1973) were based on a finite-sized porous layer. For an infinite layer, Weber (1975b) analyzed natural convection in a tilted porous layer using a perturbation technique. Caltagirone and Bories (1985) obtained solutions and stability criteria for natural convection in an inclined porous layer using a linear stability theory and Galerkin's spectral method.

Recently, Moya, Ramos, and Sen (1987) studied numerically two-dimensional natural convective flow in a tilted rectangular porous cavity where two opposing walls are kept at constant but different temperatures while the other two walls are thermally insulated. Also, Vasseur, Satish, and Robillard (1987) studied analytically and numerically natural convection in an inclined thin layer of porous medium when a constant heat flux is applied on the two opposing walls while the other two walls are insulated.

A somewhat related configuration to the horizontal layer is a wedge-shaped porous layer. This type of geometry has direct use in estimating the performance of the thermal insulation used in an attic-shaped application. Bejan and Poulikakos (1982) studied natural convection in an attic-shaped horizontal space heated from below by using an asymptotic analysis that was also used by Walker and Homsy (1978). Also Poulikakos and Bejan (1983b) performed scaling analysis and numerical simulation on transient and steady-state convection in the limit of high Rayleigh numbers.

The final configuration considered in this section is natural convection in open-ended porous cavities. The engineering applications of open cavities are important. Examples include fire spread in rooms, solar thermal central receiver systems, connections between reservoirs, nuclear waste repositories, etc. This very important and fundamental geometry can be modified to analyze a variety of complex geometries. It should be mentioned that while that there has been many investigations on the two-dimensional natural convection in porous enclosures, the related but more complex problem of open cavities has received very little attention. One of the reasons for this imbalance in the above-mentioned investigations can be traced to the difficulty in specifying the appropriate boundary conditions for an open cavity configuration due to its significantly more complex geometry. Furthermore, in this configuration, the physical attributes of both confined and external natural convection are present, thus leading to a complex chain of interactions. Ettefagh and Vafai (1988) have analyzed such a configuration and their results clearly show the influence of the external corners in an open-ended cavity in augmenting of the flow instabilities. They also discussed in detail the role of the open boundaries, different temperature levels, and different aspect ratios.

IV. Multiphase Transport in Porous Media

The subject of multiphase transport and phase change in porous media has gained considerable attention during the past two decades. Among the wide variety of the current applications of this subject the following examples could be cited: drying of different porous materials (Berger and Pei, 1973; Plumb *et al.*, 1985), condensation in porous materials (Ogniewicz and Tien, 1981; Vafai and Whitaker, 1986); heat pipe applications (Udell, 1985); and geothermal applications (Faust and Mercer, 1979; Cheng, 1978b).

A. Theoretical Basis

The earliest studies of fluid transport including some aspects of heat transport in porous materials were reported in the 1920s. Some early examples are the works by Gardner and Widtsoe (1920), Lewis (1921), Sherwood (1931), Newman (1931), and Richards (1931). In the pioneering works of chemical engineers in drying phenomena, it was assumed that fluid motion in the porous media occurred by diffusion. Lewis (1921), Newman (1931), and Sherwood (1929a,b, 1930) all used diffusion equations for representing fluid transport. On the other hand, researchers from other disciplines, such as hydrologists, ceramicists, and colloid chemists, found out that diffusion equations were not adequate to explain the liquid motion in unsaturated porous media and that capillary forces (surface tension) were also important. Studies by Gardner and Widtsoe (1920), Richards (1931), and Westman (1929) demonstrated this influence.

While the emphasis in the pioneering works on drying was given to momentum transport, the influence of heat transport was ignored. The first attempts to include the effect of heat transport were made by Krischer (1938, 1940a,b). This, afterwards, led to more serious consideration of heat transport in related problems.

Relatively complete models of multiphase transport processes considering phase change and capillary action in porous media were first established in the 1950s. Some of the related and more commonly used works include: the studies reported by Philip and DeVries (1957) and DeVries (1958) in which the multiphase transport processes in soil were formulated, the model for drying of hygroscopic capillary porous solids developed by Berger and Pei (1973) as an improved version of Krischer's model, the model of simultaneous transport processes in capillary-porous media developed by Luikov (1964, 1975), the studies by Eckert and Pfender (1978) and Eckert and Faghri (1980) in which moisture migration due to temperature gradient and transport processes in unsaturated porous media are examined, and the drying model developed by Whitaker (1977) using local volume-averaging technique.

One of the main problems encountered in the modeling of multiphase tranport processes in porous media is the lack of information about the numerous transport coefficients required for formulating the problem. These coefficients may be functions of temperature as well as the composition of the fluids flowing through the porous medium. In the case of anisotropic media, some of these may even be tensors with nine components. For simplifying the formulation of the multiphase transport processes in porous media, a number of assumptions are commonly made.

The differences between various models of multiphase transport processes in porous media usually arise from the different level of details of the physical phenomena considered and the different simplifying assumptions made according to the prevailing physical conditions. Moreover, the use of different empirical correlations for transport coefficients and different thermodynamic relations also gives rise to differences in formulations.

Numerous studies dealing with specific problems related to multiphase transport processes with change of phase in porous materials have been reported. Some of the more widely used models and the pertinent references on which they are based include the models by Whitaker (1977), Luikov (1964, 1975), Philip and DeVries (1957), Berger and Pei (1973), Eckert and Pfender (1978), and Eckert and Faghri (1980).

Although some of these models have been developed for specific problems, their different versions have been used quite extensively. The different references cited for each of DeVries' and Eckert's models are complementary to each other for the model concerned. In Luikov's case, there are a few differences in the two references cited. In what follows, these models are analyzed and compared qualitatively, the assumptions employed in each model are evaluated, and the advantages, disadvantages, and general applicability of each model are discussed.

In Whitaker's model the local volume-averaging technique is applied to the governing balance equations resulting in volume-averaged equations. With the help of constitutive equations such as Fourier's law of heat conduction and Darcy's law for fluid flow, volume constraint relations, and thermodynamic relations such as Kelvin equation and Clausius–Clapeyron equation, a rigorous model of governing equations is formed. Details of transport phenomena such as diffusion, dispersion, and convective flow as well as capillary action in liquid phase are taken into account.

Luikov, too, has established a very comprehensive system of governing equations for multiphase transport processes in capillary-porous systems. The primitive form of the equations for the liquid, vapor, and inert gas phases is quite appealing. However, the simplifying assumptions used for reducing the entire system of governing equations to three coupled differential equations with temperature, pressure, and total moisture content as system variables are not realistic.

The model developed by DeVries for simultaneous heat and mass transport with phase change in soils is also widely used. It contains, to a certain extent, information on the physics of the transport of liquid and vapor in the porous medium. The fluid motion is represented by Darcy's law, and it is shown how the liquid flux term can be split into temperature gradient, liquid moisture gradient, and gravity terms. This model is suitable for applications with no noncondensibles or for applications in which the effects of noncondensibles may be neglected. One drawback of this model is that the correlation given for the liquid diffusivity is valid only for the capillary condensation region. Analytical solutions for some simplified steady-state cases have been presented in DeVries' work.

The model presented by Berger and Pei (1973) for drying of hygroscopic porous materials is a simplified model in which the motion of both liquid and vapor is represented by Fick's law. The vapor flow is taken to be by diffusion, and the liquid motion is considered to be by capillary flow. The simplified nature of this model arises from the assumption of constant transport coefficients, namely the vapor diffusivity and the liquid conductivity. Further simplification is carried out by assuming that heat transport occurs only by conduction through the solid matrix and by transfer of latent heat due to phase change. This model is applicable to cases with small temperature changes. For the different stages of drying, two different coupling relations are used for the governing equations.

Eckert, in his works with Pfender (1978) and Faghri (1980), developed a model for simultaneous heat and mass transfer in an unsaturated porous medium. In his work with Faghri, no phase change is considered. Rather, the motion of moisture due to temperature gradients is analyzed. Vapor motion is represented by Fick's law, and the problem is formulated and solved for the one-dimensional case where the porous medium is bounded by two impermeable walls. In this model, too, no noncondensibles are taken into account. Some analytical solutions are obtained for simplified cases with constant transport properties. Eckert, in his work with Pfender, considers multiphase transport in porous media in the presence of phase change. In this study, the liquid and vapor mass fluxes are formulated by Darcy-like and diffusive terms respectively. Although the energy transport equation is not developed, the heat flux per unit area is defined and a description of the experimental work performed to determine the thermal conductivity of the soil is given.

The models developed for simultaneous heat and mass transfer processes in multiphase porous systems may have differences due to the various assumptions made in developing each model. These assumptions may change to a great extent depending on the physical conditions of the particular problem considered. Moreover, there is lack of universal methods for defining the transport coefficients and transport potentials. This is a great drawback in establishing a generalizable model.

B. Application Areas

The application of multiphase transport in porous media cover a wide variety of problems. The simplest problem that has been considered is the effect of a temperature gradient on the transport processes in porous media. Dybbs and Schweitzer (1973) formulated the problem of non-isothermal flow in porous media using the volume-averaging technique for low Reynolds number flows in which inertia terms are negligible.

Dinulescu and Eckert (1980) studied the problem of moisture migration due to temperature gradient in porous media. They performed a one-dimensional analysis of this problem and came up with an analytical solution. The first set of boundary conditions used were constant temperature and impermeable wall boundaries at the two sides. Another solution was obtained for constant heat flux boundary conditions. Huang (1979) established a model of governing equations for the multiphase transport processes in porous media. In Huang's work the diffusive and convective transport were described carefully and the effect of the temperature gradient was emphasized. The governing equations for drying processes in both funicular and pendular stages were developed. In the funicular stage, the gaseous phases were assumed to be stationary, while in the pendular stage, the liquid phase was taken to be immobile. Huang's model looks like a blend of Whitaker's (1977) and Luikov's (1964) models. An account of the thermodynamic relations to be used in conjunction with the transport equations was also presented in Huang's work. Huang *et al.* (1978) obtained a solution for the problem of moisture migration in a concrete slab for natural drying conditions. Baladi *et al.* (1981) studied the transport processes in soil around a buried heat source. The phenomena of moisture migration and energy transport were analyzed by a one-dimensional model. Numerical and experimental works were carried out. In addition, closed-form solutions were obtained by using approximate models in this study. In modeling the porous system, two distinct wet and dry regions were assumed to exist.

Drying phenomena form an important part of the applications of multiphase transport in porous media. One of the earlier works in this area was carried out by Harmathy (1969) who investigated the drying problem in a porous medium. An important assumption used in this work was to consider all moisture transfer to take place in the gaseous phase only. More recent studies have been reported by Lyczkowski and Chao (1984). In their work, the authors give a comparison of the two-phase drying model developed by Lyczkowski with the Stefan model for coal drying. In the two-phase model discussed, noncondensibles are not taken into account. Plumb *et al.* (1985) studied the problem of drying of softwood. They formed a model for multiphase transport processes taking into account both diffusion and capillary action in the liquid phase. By the use of a mechanistic model they expressed the capillary porous properties (permeability and capillary pressure) in terms of geometric properties and measured permeability parameters. Numerical and experimental studies were carried out for obtaining moisture profiles. Another example is the study reported by Haber *et al.* (1984) on drying in a semi-infinite porous system with emphasis on heat convection. A one-dimensional model was developed in which two separate regions with a moving interface were considered to form the system.

The problem of condensation in porous materials is one of the recent problems studied. Ogniewicz and Tien (1981) reported an analytical work on condensation in porous wall insulation. The model developed is one-dimensional and mainly steady state, and it partially accounts for convective and diffusive transport. The problem studied consists of a porous slab subjected to two different environments on the two sides. Vafai and Sarkar (1986, 1987) reported a rigorous transient analysis of moisture migration and condensation in porous and partially porous enclosures. The analysis considered variation of properties as well as noncondensible effects. The effect of Lewis and Prandtl numbers was investigated by a model that utilized the volume-averaging technique. In this problem, impermeable and adiabatic horizontal wall and permeable vertical wall boundary conditions were employed. Properties such as temperature, pressure, liquid fraction, and vapor density were solved for separately, and the interface location was determined as part of the analysis and not on a trial basis.

In another study, Vafai and Whitaker (1986) analyzed a different problem using a two-dimensional transient model. In this study, the multiphase transport processes with phase change in a porous slab were also modeled by using the volume-averaging technique. Permeable wall boundaries with constant temperature and liquid moisture content boundary conditions were employed. The results of a numerical investigation solving for a large number of parameters were presented, and the identification of significant transport mechanisms was performed. Based on these findings, a simplified model for multiphase transport processes in the presence of phase change in a porous medium was suggested.

Nilson and Romero (1980) investigated another condensation problem in porous media. They formulated a one-dimensional problem of a porous matrix with an inflow of saturated vapor at the boundary. Hyperbolic/parabolic equations were obtained by Darcy formulation of two-phase flow, and similarity solutions were obtained for the propagation of the condensation wave. In this study, correlations for relative permeabilities of liquid and vapor as functions of liquid saturation have been used. In a recent investigation by White and Tien (1987a), laminar film condensation in a porous medium was analyzed. In their analysis, the no-slip condition was used for the velocity at the wall, and an exponential model for the porosity and modified boundary layer equations was employed for the region near the wall. Experimental studies on the same topic were also reported by White and Tien (1987b).

Chuah and Carey (1985) investigated the transport processes in a medium with variable porosity. In this study the porosity of the medium is expressed in exponential form. Darcy's law is used for liquid and vapor transport. In this study, empirical correlations are used extensively for some variables based on previous experimental data. Permeability is expressed by Kozeny–Carman formula in terms of porosity. Relative permeabilities of liquid and vapor are expressed in terms of liquid saturation based on experimental data. Capillary pressure is also based on Leverett's correlation from experimental data, which expresses capillary pressure in terms of porosity, permeability, and surface tension. Besides local thermodynamic equilibrium, the assumption of equal chemical potentials in the liquid and vapor phases is employed. This yields an additional thermodynamic relation that relates temperature, saturation pressure, and liquid and vapor pressures.

Another area of application of multiphase transport in porous media is the heat pipe technology. A comprehensive review of the physical phenomena and applications of heat pipes is given by Tien (1975, 1985a) and Tien and Rohani (1972). Udell (1985) reported analytical and experimental studies for a one-dimensional problem of heat pipe effect in porous media. In the model developed, gravity forces, capillarity, and change of phase were considered. The problem studied was a steady state one. Leverett correlation relating capillary pressure to permeability, porosity, and surface tension was also used in this study. Relative permeability correlations found by Flatt and Klikoff (1959) were also employed. Numerical solution and experimental work were carried out for model verification in Udell's work. Udell and Fitch (1985) reported a further study considering the effects of noncondensible gases. In this work, the effect of noncondensible gases on Kelvin's equation is explained and utilized. Some additional empirical correlations are introduced from previous studies for the thermal conductivity of a partially saturated granular medium and for the effective molecular diffusivity.

Almost all the models developed for analyzing problems of multiphase transport processes and phase change utilize the assumption of local thermodynamic equilibrium between the phases in the porous system.

While this is a very good assumption for many practical applications, it might not be a realistic assumption for cases in which the fluid velocity is high and/or where the system is subjected to an intense hat flux. Under such situations, it may not be possible to have local thermal equilibrium between the solid and fluid phases. One of the few examples of studies that do not utilize the local thermal equilibrium assumption was carried out by Wong and Dybbs (1978). In this work, the volume-averaged energy equations for a single phase fluid flow are obtained for a saturated porous medium. The heat transfer between the solid and fluid phases is formulated by a convection term. An approximate means of accounting for the radiative heat transfer between these phases by a convective-type heat transfer coefficient is explained. A relationship is also established for the effective thermal conductivity of a saturated porous medium in terms of the effective thermal conductivities of the individual phases.

Limited experimental work has been performed for obtaining transport coefficients for flow in porous materials. One investigation was reported by Singh and Dybbs (1979). Besides forming a model for multiphase transport processes, they have carried out experimental work for determining the permeability of the porous matrix and the effective thermal conductivity of the fluid saturated porous system. The permeability measurement was performed in an isothermal condition. The effective thermal conductivity was measured, for the case where the porous matrix was fully saturated with stagnant fluid, by the steady-state method of comparison. The sintered metallic porous media used for forming the solid matrix in this study were copper and nickel.

Other experimental works considering phase change in packed beds have been reported by Tsai *et al.* (1984). Their study investigated the dryout of an indutively heated packed bed made of metal particles. Water was used as the working fluid. The variation of the dryout heat flux with mass flux was determined. Similar work studying the effect of pressure on the dryout of a packed bed of steel particles was carried out by Catton and Jakobsson (1987).

For experimental investigations related to transport processes and phase change in porous media, measurement of different variables such as temperature, pressure, and relative humidity may be performed by thermocouples and pressure and relative humidity measuring devices, which are available in relatively small sizes. However, to our knowledge, a very reliable commercial device for detection of the liquid moisture and for measurement of the liquid moisture content in a porous medium does not seem to be available. This is crucial for experimental work aiming at investigating condensation problems in porous media. The design and construction of a device for *in situ* measurement of the liquid content in a fibrous insulation material was reported by Motakef and El-Masri (1985). The device utilizes a small electrical resistance probe.

The subject of multiphase transport processes and phase change in porous media is now a very active research area. Besides the applications considered to date, new frontiers are being explored for the use of these results in technological applications. Among the recent proposed systems for spacecraft applications related to this subject are the packed-bed type heat exchangers utilizing phase change materials for forming the packed bed and heat pipes with porous wicks. Both of these require a thorough understanding of the fundamentals and analysis of the transport processes in porous media. From the research carried out to date, it may be seen that these physical phenomena are relatively well understood. But the models established to date lack general applicability. This is partly because of the lack of universally accepted methods of definition for transport coefficients and transport potentials and partly because of the different assumptions employed in forming the model in each particular problem considered. Further theoretical work is required for obtaining a generalizable model. Experimental work for determining empirical correlations for the transport coefficients in different porous materials is also an important necessity.

V. Radiative Heat Transfer in Porous Beds

A. Thermal Radiation Characteristics of Porous Beds

The porous beds in which thermal radiation can become important are usually multiphase systems consisting of solid particulates and gases. The role of thermal radiation in these porous beds is of major importance in the design of fluidized beds, packed beds, catalytic reactors, and many other advanced energy conversion systems, especially when the beds operate at high temperatures. Radiation in these porous beds brings up additional difficulties and challenging aspects in the analysis. These difficulties arise in two major areas: the development of a radiation transfer model, which prescribes the propagation of radiation within an absorbing, emitting, and scattering medium, and the corresponding radiative properties for the model, and the development of a heat conduction and/or convection model with properly determined thermophysical properties.

Previous studies of radiative heat transfer through porous beds have employed a variety of analytical and experimental techniques. Vortmeyer (1978) summarized some earlier radiation models that use unit cell representations for analyzing packed beds. Such cell and layer models, in conjunction with Monte Carlo methods, are also used by Chan and Tien (1974a) to evaluate radiative characteristics of packed beds of fixed porosity and regular structure, by Yang *et al.* (1983) for randomly packed beds of uniform spheres, and by Kudo *et al.* (1985), who examined different types of packing and variation in volume fraction. Also Borodulya and Kovensky (1983) used the cell approach by evaluating exact view factors in the unit cell by assuming the surfaces to be isothermal and diffuse.

Thermal radiation within these beds usually is the result of emission by the hot walls and the gas–particle mixture. This radiation undergoes complex interactions with the bed, primarily due to absorption and scattering processes. The three primary radiative properties that characterize the interactions of radiation with the particulate bed are the scattering coefficient, the extinction coefficient (i.e., sum of scattering and absorption coefficients), and the scattering phase function. These radiative coefficients are defined as the fractions of the corresponding energy losses from the propagating wave per length of travel. The units for the absorbing, emitting, and scattering coefficients are inverse of length, whereas the phase function, which specifies the scattering distribution, is dimensionless. The radiative coefficients are functions of the optical constants of the bed materials and of the particle size, shape, and packing, while the optical constants are functions of the wavelength. The phase function is a strong function of shape and varies from predominantly forward scattering for large particles to semi-diffuse for small ones.

In homogeneous media such as gases, absorption and emission are the major radiative mechanisms. If the medium contains inhomogeneities, such as the particles in packed or fluidized beds, the additional mechanism of scattering is introduced. These absorption and scattering processes are governed by electromagnetic field equations and their associated boundary conditions at all interfaces. The resulting analytical problem is formidable and is usually solved by using simplifying assumptions: idealized geometry of the scatterers, independent scattering and absorption, and homogeneous distribution of particles.

Scattering and absorption characteristics of a particle are governed by three factors: the particle shape, the particle size relative to the wavelength of the incident radiation, and the optical properties of the particle and the background medium (Tien, 1985b). Three characteristics of a single particle are described by the solution of the electromagnetic field equations. Physically, they can be explained by the processes of reflection, refraction, and diffraction. When an electromagnetic wave strikes the particle surface, a portion of it is reflected while the remainder penetrates the particle. The beam within the particle may experience some absorption and multiple internal reflections before it escapes out of the particle in different directions, giving rise to scattering. This scattering is the contribution by refraction. The diffraction scattering process originates from the bending of the incident beams near the edge of the particle. Consequently, even a completely absorbing particle scatters radiation.

Scattering and absorption characteristics of many particles in close packing or fluidization can be obtained from the single-particle characteristics. The procedure depends on the scattering regime of the system to which the system of particles belongs. Based on the size parameter and the particle volume fraction f_v the regime map shown in Figure 4 (Brewster and Tien, 1982a; Yamada *et al.*, 1986; Tien and Drolen, 1987; Drolen *et al.*, 1987) is divided into two parts: one where independent theory is an adequate representation and the other where interparticle interactions must be accounted for. The independent theory is based on the assumption that each particle in the assembly scatters and absorbs radiation unaffected by the presence of other particles. Thus, the extinction and scattering of energy by the system is expressed by a simple algebraic addition of the energy extinguished and scattered by each parimary particle. The cross-section for the system of N particles is the sum of the cross-sections of these particles, and the individual particles are assumed to scatter and absorb radiation independently of the others.

In essence, the independent theory assumes no electromagnetic interaction between the various particles in the system and that each particle has the same incidence. Interference of the waves scattered from different particles is also neglected. Thus, the resultant properties of the system are simply an algebraic sum of the corresponding properties of the

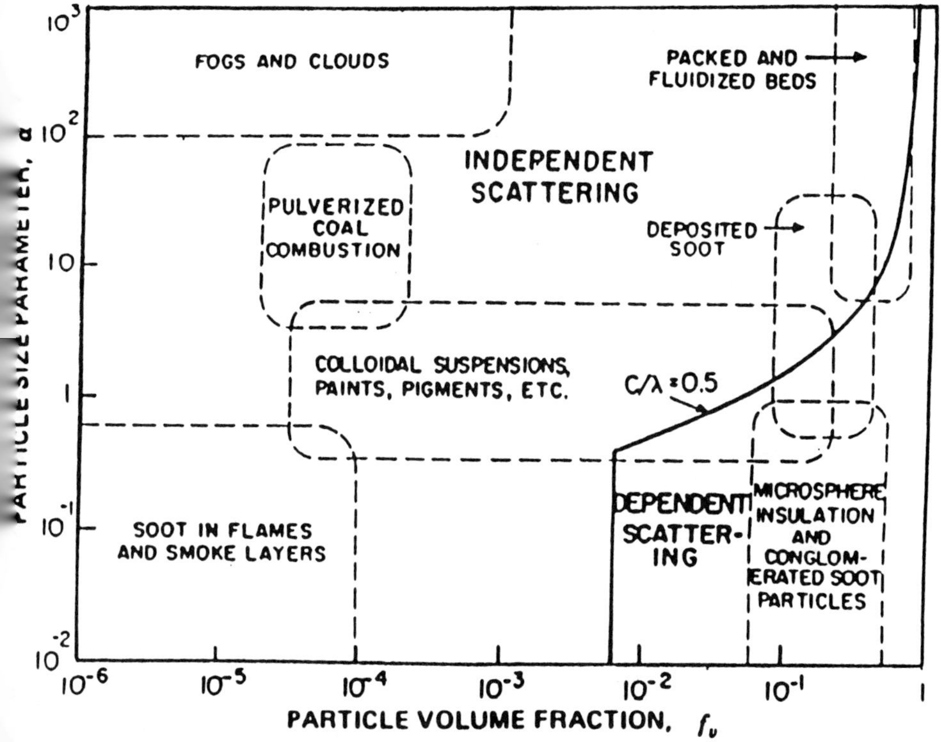

FIG. 4. Independent and dependent scattering regime map.

individual particles that constitute the system, where each particle is assumed to be alone in the imposed incident field.

Intuitively the assumptions associated with the independent theory seem to be valid when the clearance between particles is significantly larger than the particle diameter, as well as when the particles are randomly distributed in space and time. Furthermore, the ratio of interparticle clearance to the diameter of the particles is not of significance for randomly distributed homogeneous particles, whereas the ratio of clearance to the wavelength is. Results using the independent theory are far simpler, and wherever justified, it is advantageous to use them over the exact expressions that accounts for interparticle effects.

Departure from independent theory occurs in a densely packed system, where the close spacing of the particles renders invalid the assumption that each particle acts independent of the others. Dependent effects are introduced into the radiative characteristics by two mechanisms. The first is the near-field interparticle effect by which the net incidence on the particles as well as the internal fields is modified and consequently both the extinction and scattering characteristics of the system are changed. The second is due to coherent addition of scattered radiation in the far-field that is manifested by a change in the scattering characteristics only (Tien and Drolen, 1987; Kumar and Tien, 1987). In reality, dependent effects are always present but may be neglected under certain conditions. It is thus important to quantify demarcation criteria that separate the regions where the independent assumption is a good approximation from those where dependent effects cannot be ignored (Tien and Drolen, 1987).

Experimental meaurements of radiant transmission through packed and fluidized media have been reported by various researchers. Experimentally, the radiative characteristics of packed beds are determined by first determining the extinction characteristics through direct transmission measurements. Next, the scattering cross-section and phase function are obtained by measuring the angular distribution of scattered thermal radiation. Finally, the absorption characteristics are inferred from the difference between the extinction and scattering of the incident radiation. The radiation source in the experiment is either a laser or glow-bar (Drolen *et al.*, 1987; Brewster and Tien, 1982a; Hottel *et al.*, 1971).

The experimental results consist of transmittance and reflectance data for a planar slab containing the scatterers from the particulate bed. Data reduction to yield the desired radiative properties is achieved by adopting a suitable radiative transfer model. For example, Brewster and Tien (1982a) used the two flux model (Brewster and Tien, 1982b) for the data inversion. Chen and Churchill (1963) used an open-ended tubular electric furnace as a blackbody source. The test section comprised of packed spheres supported on a screen and the source was placed underneath. Using thermocouples as detectors, the characteristics of glass, aluminum oxide, steel, and silicon carbide spheres, cylinders, and irregular grains were found for various source temperatures. Two-flux models, with the inclusion of an emission term, were used for the data inversion.

Whereas characteristics of packed beds can be measured in *ex situ* situations, the thermal radiation characteristics of fluidized beds have to be measured *in situ* in order to preserve the operating conditions. This is achieved by laser scattering techniques. Alavizadeh *et al.* (1984) have designed an instrument for the measurement of the radiative component of the heat transfer in a high-temperature gas fluidized bed. The design uses a silicon window to transmit the radiative flux to a thermopile detector located at the base of a cavity.

B. Modeling of Radiative Transfer in Porous Beds

Computation of the transport of thermal radiation in a particulate system requires an accurate knowledge of the primary radiative properties. These properties are adopted primarily due to the following considerations: (i) the theory of electromagnetic interaction with particles, (ii) direct experimental measurements, and (iii) inferred measurements. For example, the absorption coefficient cannot be directly measured from light scattering experiments. It is obtained indirectly by measuring the scattering and extinction losses from the incident beam and evaluating the difference between the corresponding extinction and the scattering coefficients.

The propagation of radiation within an absorbing, emitting, and scattering medium is governed by the equation of transfer (Kerker, 1961; Siegel and Howell, 1981; Ozisik, 1973):

$$\mathbf{e}_\Omega \cdot \nabla I_\lambda(\mathbf{r}, \mathbf{e}_\Omega) = -(\sigma_{a\lambda} + \sigma_{s\lambda}) I_\lambda(\mathbf{r}, \mathbf{e}_\Omega) + \sigma_{a\lambda} I_{b\lambda}(T(\mathbf{r})) + \frac{\sigma_{s\lambda}}{4\pi} \int_{4\pi} I_\lambda(\mathbf{r}, \mathbf{e}_{\Omega'}) \Phi(\mathbf{e}_{\Omega'} \rightarrow \mathbf{e}_\Omega)\, d\Omega', \tag{5.1}$$

where I_λ is the monochromatic radiation intensity, T the medium local temperature, $\mathbf{r}$ the position vector, $\mathbf{e}_\Omega$ the unit vector in the direction of consideration, and Ω the solid angle centered around $\mathbf{e}_\Omega$. The coefficients are denoted by σ and the subscripts a and s refer to absorption and scattering, respectively. The first term on the right hand side of the equation of transfer represents the attenuation of intensity due to absorption and scattering, the second term represents the gain due to emission, and the last term is the gain due to the in-scattering into the direction $\mathbf{e}_\Omega$ from all other directions. The intensity I_λ is defined as the energy per unit area per unit solid angle per unit wavelength and the scattering phase function $\Phi(\mathbf{e}_{\Omega'} \rightarrow \mathbf{e}_\Omega)$ is a specification of the radiation intensity scattered from the direction $\mathbf{e}_{\Omega'}$ into the direction under consideration, normalized by the isotropic scattered radiation intensity, i.e., $\Phi(\mathbf{e}_{\Omega'} \rightarrow \mathbf{e}_\Omega) = 1$ for isotropic scattering. It should be noted that the equation of transfer treats the medium as a continuum where each volume element absorbs, emits, and scatters radiation. The exact positions of the different particles in the volume are not considered; only volume-averaged values of the radiative properties are used. Other methods, which do not treat the medium as a continuum, but instead take into account the position of particles and the boundaries between the solid and the gas phase, are termed the discrete models of radiative transfer.

Neglecting the viscous dissipation, while including radiation, convection, and conduction modes of heat transfer (Siegel and Howell, 1981; Ozisik, 1973), will lead to the following energy equation

$$\rho C_p\left[\frac{\partial T}{\partial t} + (\mathbf{v}\cdot\nabla)T\right] = \nabla\cdot k\nabla T + \nabla\cdot\mathbf{q}_r, \tag{5.2}$$

where $\nabla\cdot\mathbf{q}_r$ is given by

$$\nabla\cdot\mathbf{q}_r(\mathbf{r}) = 4\pi\int_\lambda \sigma_{a\lambda}I_{b\lambda}(T(\mathbf{r}))\,d\lambda - \int_\lambda \sigma_{a\lambda}\int_{4\pi} I_\lambda(\mathbf{r},\mathbf{e}_\Omega)\,d\Omega\,d\lambda. \tag{5.3}$$

Here, T is the temperature, $\mathbf{v}$ the velocity vector, ρ the density, C_p the specific heat, $\mathbf{k}$ the thermal conductivity, and t time. The energy equation (5.2) is an integro-differential equation which in general does not lend itself to straight-forward solutions even for simple cases. Furthermore, due to the nature of the equation, numerical computations tend to be on the heavy side.

Several methods have been used for modeling radiative transfer in packed and fluidized beds. These approaches can be classified, according to the modeling of the medium itself, into two major groups: discontinuous or discrete models and continuous or pseudo-continuous models. In discontinuous or discrete models, the medium is considered as a regular assembly of units or cells of idealized geometry, resulting in a simple algebraic formulation of the problem. In this case, volume-averaged radiative properties of the medium, determined from the radiative characteristics of particles, are used in the equation of transfer. The second approach visualizes the medium as a random collection of particles with some number density N/V, a reasonable approximation when the characteristic dimension of the system is much larger than the characteristic size of the particles in the system. In this case, the radiative transfer modeling involves either the integro-differential equation of radiative transfer or a simplified version of this equation, or Monte Carlo and/or ray tracing methods.

Discontinuous models characteristically treat a particulate system as a regular assembly of cells or units of idealized geometry such as parallel flat plates, close-packed spheres, and cubic-packed spheres. In these models, the scattering diagram of a unit or cell is first determined. Standard resistance network or layer theory approaches are then applied to calculate radiation transfer through the system. Vortmeyer (1978) reviewed many models that fall into this category for predicting radiant conductivity. Chan and Tien (1974a), Kudo *et al.* (1985), and Borodulya and Kovensky (1983) used different unit cells for calculating the optical properties. In these works, Chan and Tien (1974a) examined a single cubic cell that is representative of a cubic packed geometry. Using ray tracing techniques and assuming specular surfaces, they calculated the optical properties of a single layer based on the fraction of rays transmitted, reflected and absorbed by a single cell. In the work of Borodulya and Kovensky (1983), rather than using the ray tracing, diffuse and isothermal surfaces were assumed and then the transmitted, absorbed, and reflected fluxes were computed. In contrast to Borodulya and Kovensky (1983), Kudo *et al.* (1985) used Monte Carlo techniques to calculate the reflected, transmitted, and absorbed energy.

Predictions of radiative transfer in packed/fluidized beds based on discontinuous or discrete models are expected to be very sensitive to the packing geometry assumed. Nevertheless, the discontinuous approach is the natural one to apply in order to predict radiative transfer in many discrete systems such as nuclear fuel rods, solid-matrix beds, screens, and packed beds of low tube–particle diameter ratio, such as the packed-bed tubular reactors widely used in the chemical industry. Careful consideration of the solid particle arrangement and recourse to a ray tracing/Monte Carlo method or to other more complicated methods is then required to accurately predict radiative transfer in these systems. Furthermore, the modification of the ray tracing technique to account for energy diffracted, especially for intermediate particle size parameters, is then required. The application of the theory of geometric diffraction, originally developed by Keller (1962) to radiative transfer problems deserves special attention.

In the continuous model approach, the particulate medium is modeled as a random assembly of particles with some number density. The medium extinction characteristics can be calculated by a Monte Carlo and/or ray tracing approach or by the solution of an integro-differential transport equation such as Equation (5.1) or a simplified version of this equation such as the two-flux model and the diffusion approximation.

The Monte Carlo and/or ray tracing models are best represented by the work of Yang *et al.* (1983), who presented a novel approach for modeling packed-bed radiative transfer. In their model, a randomly packed bed of spheres is created mathematically, using a model that describes the slow sequential settling of individual rigid spheres of equal diameter, where each sphere must be supported by at least three others, and no sphere can overlap.

For this packing, rays are most likely to have their first interaction at about one quarter of a sphere diameter. It was also found that the mean penetration depth is about 0.66 sphere diameter, and that virtually all rays have hit a sphere surface after traveling a distance of six diameters. The above-mentioned probability information was then used to perform a ray tracing analysis on thick beds. A ray was projected in a random direction across a randomly chosen distance based upon the cumulative distribution function. The process was then continued by randomly choosing the position of a sphere around the endpoint of this ray and assuming specular reflections at the sphere surface. Each of the models that were discussed neglects the diffraction contribution by the particles. This is a reasonable assumption for large α ($\alpha > 100$), where α is the size parameter, since the diffracted energy is concentrated in the direction of propagation. However, for intermediate and small α the diffracted energy can no longer be considered as part of the propagating beam. Thus, reflection at the particle surface is not the only contributor to the scattered energy; diffraction must be considered as well.

The integro-differential approach involves the solution of an equation of transfer as if the medium were a continuum. The extinction characteristics of this continuum are based on the properties of the discrete particles. Chen and Churchill (1963) used a two-flux approximation of the equation of transfer to correlate their experimental data for transmission of radiation through a packed bed of spheres.

The diffusion approximation method greatly simplifies the equation of transfer in the case of optically thick media by assuming the radiative transfer to be a diffusion process. Wang and Tien (1983) and Brewster and Tien (1982b) have presented a diffusion based solution of the radiative flux assuming a two-flux model for the radiative transfer in a one-dimensional planar geometry.

The two-flux model was extended by Chandrasekhar (1960) by approximating the radiation field in terms of a number N of discrete streams. This reduced the problem to the solution of N coupled differential equations. The direction of these streams, $\mu_i = \cos \omega_i$, are typically chosen so as to correspond to the zeros of the Legendre polynomials $P_N(\mu)$. The integral in Equation (5.1) can then be approximated by Gaussian quadrature. The model just discussed is called the method of discrete ordinates and can be thought of as the exact solution to the problem given a large enough N, say $N \geq 20$. It is best applied to problems involving anisotropic scattering and/or requiring directional results. A simple algorithm for solving the general equation of transfer using discrete ordinates is presented by Kumar *et al.* (1988). Other methods for analyzing the radiative transfer equation have been reviewed by Viskanta (1982, 1984) and Menguc and Viskanta (1983).

A few general comments are in order regarding the models that have been proposed to predict the radiative transfer in particulate systems. Models have been developed to predict primarily the transmission of radiation and/or the effective radiant conductivity of the particulate medium. Corresponding experiments have been performed to test these two different approaches. Of these two major experimental methods, the measurement of radiation transmittance through an evacuated and isothermal bed of spheres is the fundamental and by far more reliable method for evaluating theoretical models since these experiments completely isolate radiation from other modes of heat transfer. An excellent review of many models developed to predict radiant conductivity can be found in the literature (Vortmeyer, 1978). By far the most quoted experimental data relating to radiant transport in packed beds are those of Chen and Churchill (1963). Many of the authors previously referenced have compared results of their respective models to these experimental data. This experiment was the first to isolate the radiant mode of heat transfer from the convective and conductive modes. This was achieved by illuminating a bed of spheres (2–16 diameters deep) with a modulated, high-temperature (700–1366 K) blackbody source and measuring the transmitted energy via a spectrally independent detector.

Based on experimental comparisons it is concluded that the two-flux model predicts consistently high values of the exchange factor, a quantity which is used in the definition of the radiative conductivity. Particularly, as the emittance is increased the experimental data indicate an increase in exchange factor E, whereas the two-flux model predicts a decrease. In the two-flux model, the individual spheres are assumed to be isothermal and scattering independently. These may be poor assumptions for welded-steel spheres. A model proposed by Vortmeyer (1978), in which he modified the plane-layer model to incorporate transmission of energy through void spaces in the layer, appears to give the best predictions. A simple model proposed by Schotte (1960), in which the exchange factor is approximated by the particle emissivity, also gives surprisingly good results.

C. Multimode Heat Transfer in Porous Beds

Combinations of radiation with the other modes of heat transfer were first studied for cases where only radiation and conduction were present. These types of multimode heat transfer processes in porous beds have been the subject of many studies (Chan and Tien, 1974b; Bergquam and Seban, 1971). Simultaneous radiative–convective transfer in packed and fluidized beds has also been studied (Echigo *et al.*, 1974; Tabanfar and Modest, 1987). In fluidized bed systems, convection and radiation are the important mechanisms of energy transfer as indicated by experimental studies (Goshayeshi *et al.*, 1986). Solutions have been obtained by incorporating simplifying approximations such as an isotropically scattering gray medium (Yener and Ozisik, 1986), and a linearly anisotropic scattering medium (Azad and Modest, 1981). A review of some of these topics in heat transfer in fluidized beds was presented by Saxena *et al.* (1978).

Equations (5.1)–(5.3) are all formulated in terms of volumetric-averaged variables and are coupled through temperature. This set of equations is a highly nonlinear system of differential and integro-differential equations and it is a formidable task even to attempt a numerical solution. The general equation of radiative transfer alone has been the subject of many investigations, with several approximate methods being proposed for its solution (Chandrasekhar, 1960; Siegel and Howell, 1981; Viskanta, 1984; Menguc and Viskanta, 1983). An iterative method or a variant of this method has been traditionally used to solve the coupled energy and radiative transfer equation. In this method, a temperature profile is assumed and used in Equation (5.3) to calculate the divergence of the radiative heat flux, this being then substituted in Equation (5.2) for recalculating temperature distributions. The solution of these complex problems has been substantially facilitated with the recent development of the Differential-Discrete-Ordinate (DDO) method (Kumar *et al.*, 1988) for solving the general equation of radiative transfer. This simple but powerful method allows the simultaneous solution of Equations (5.1)–(5.3) in a direct and computationally efficient way. The method uses a discrete ordinate technique to reduce the integro-differential equation, Equation (5.1), to a system of ordinary differential equations. The resulting set of coupled differential equations are then solved by utilizing the existing software routines.

Heat transfer by simultaneous conduction and radiation between two reflecting surfaces with an intervening medium that absorbs, emits, and scatters thermal radiation is a problem of considerable practical importance in stagnant packed beds, such as microsphere insulation and nuclear fuel rods. There has been a significant number of investigations concentrating on the conduction radiation modeling and interactions (e.g., Viskanta, 1982; Yuen and Wong, 1980; Bergquam and Seban, 1971; Chan and Tien, 1974b; Vortmeyer, 1974; Vafai and Ettefagh, 1988).

The solution of combined convection and radiation in a particulate medium is much more complicated than the solution of conduction radiation problems because of the difficulties in the modeling of the fluid in these systems. The understanding of convection heat transfer in packed beds has developed substantially in recent years. Theoretical models that include wall porosity variation effects and inertia effects (non-Darcy effects) have yielded results in excellent agreement with experiments (Hunt and Tien, 1987). The problem of combined convection and radiation in these systems has not been fully investigated.

Acknowledgments

The authors wish to thank Dr. S. J. Kim, Dr. M. Sozen, and Dr. H. C. Tien for their careful reading of this manuscript.

References

Alavizadeh, N., Adams, R. L., Welty, J. R., and Goshayeshi, A. (1984). An instrument for local radiative heat transfer measurement in a gas-fluidized bed at elevated temperatures. *ASME HTD* **31**, 1–8.

Azad, F. H., and Modest, M. F. (1981). Combined radiation and convection in absorbing, emitting and anisotropically scattering gas-particulate tube flow. *Int. J. Heat Mass Transfer* **24**, 1681–1698.

Baladi, J. Y., Ayers, D. L., and Schoenhals, R. J. (1981). Transient heat and mass transfer in soils. *Int. J. Heat Mass Transfer* **24**, 449–458.

Balakrishnan, A. R., and Pei, D. C. T. (1974). Heat transfer in fixed bed. *Industrial and Engineering Chemical Process Design Development* **13**, 441–446.

Bankvall, C. G. (1974). Natural convection in vertical permeable space. *Warme und Stoffubertragung* **7**, 22–30.

Beavers, G. S., and Joseph, D. D. (1967). Boundary conditions at a naturally permeable wall. *J. Fluid Mech.* **30**, 197–207.

Beckermann, C., Ramadhyani, S., and Viskanta, R. (1987). Natural convection flow and heat transfer between a fluid layer and a porous layer inside a rectangular enclosure. *ASME J. Heat Transfer* **109**, 363–370.

Bejan, A. (1978). Natural Convection in an infinite porous medium with a concentrated heat source. *J. Fluid Mech.* **89**, 97–107.

Bejan, A. (1979). On the boundary layer regime in a vertical enclosure filled with a porous medium. *Lett. Heat Mass Transfer* **6**, 93–102.

Bejan, A. (1983a). The boundary layer regime in a porous layer with uniform heat flux from the side. *Int. J. Heat Mass Transfer* **26**, 1339–1346.

Bejan, A. (1983b). Natural convection heat transfer in a porous layer with internal flow obstructions. *Int. J. Heat Mass Transfer* **26**, 815–822.

Bejan, A. (1984). "Convection Heat Transfer." Wiley, New York.

Bejan, A., and Anderson, R. (1981). Heat transfer across a vertical impermeable partition imbedded in a porous medium. *Int. J. Heat Mass Transfer* **24**, 1237–1245.

Bejan, A., and Anderson, R. (1983). Natural convection at the interface between a vertical porous layer and an open space. *ASME J. Heat Transfer* **105**, 124–129.

Bejan, A., and Poulikakos, D. (1982). Natural convection in an attic-shaped space filled with porous material. *ASME J. Heat Transfer* **104**, 241–247.

Bejan, A., and Poulikakos, D. (1984). The non-Darcy regime for vertical boundary layer natural convection in a porous medium. *Int. J. Heat Mass Transfer* **27**, 717–722.

Bejan, A., and Tien, C. L. (1978). Natural convection in a horizontal porous medium subject to an end-to-end temperature difference. *ASME J. Heat Transfer* **100**, 191–198.

Bejan, A., and Tien, C. L. (1979). Natural convection in horizontal space bounded by two concentric cylinders with different end temperatures. *Int. J. Heat Mass Transfer* **22**, 919–927.

Berger, D., and Pei, D. C. T. (1973). Drying of hygroscopic capillary porous solids—A theoretical approach. *Int. J. Heat Mass Transfer* **16**, 293–301.

Bergquam, J. B., and Seban, R. A. (1971). Heat transfer by conduction and radiation in absorbing and scattering materials. *ASME J. Heat Transfer* **93**, 236–239.

Blythe, P. A., Daniels, P. G., and Simpkins, P. G. (1982). Thermally driven cavity flows in porous media I. The vertical boundary layer structure near the corners. *Proc. Roy. Soc. London Ser. A* **380**, 119–136.

Blythe, P. A., Simpkins, P. G., and Daniels, P. G. (1983). Thermal convection in a cavity filled with a porous medium: A classification of limiting behaviors. *Int. J. Heat Mass Transfer* **26**, 701–708.

Bories, S. A., and Combarnous, M. A. (1973). Natural convection in a sloping porous layer. *J. Fluid Mech.* **57**, 63–79.

Borodulya, V. A., and Kovensky, V. I. (1983). Radiative heat transfer between a fluidized bed and a surface. *Int. J. Heat and Mass Transfer* **26**, 277–287.

Brewster, M. Q., and Tien, C. L. (1982a). Radiative transfer in packed fluidized-beds: Dependent versus independent scattering. *ASME J. Heat Transfer* **104**, 573–579.

Brewster, M. Q., and Tien, C. L. (1982b). Examination of the two-flux model for radiative transfer in particulate systems. *Int. J. Heat Mass Transfer* **25**, 1905–1907.

Buretta, R. J., and Berman, A. S. (1976). Convective heat transfer in a liquid saturated porous layer. *J. Appl. Mech.* **43**, 249–253.

Burns, P. J., and Tien, C. L. (1979). Natural convection in porous media bounded by concentric spheres and horizontal cylinders. *Int. J. Heat Mass Transfer* **22**, 929–939.

Burns, P. J., Chow, L. C., and Tien, C. L. (1977). Convection in a vertical slot filled with porous insulation. *Int. J. Heat Mass Transfer* **20**, 919–926.

Caltagirone, J. P. (1976). Thermoconvective instabilities in a porous medium bounded by two concentric horizontal cylinders. *J. Fluid Mech.* **76**, 337–362.

Caltagirone, J. P., and Bories, S. A. (1985). Solutions and stability criteria of natural convective flow in an inclined layer. *J. Fluid Mech.* **155**, 267–287.

Catton, I., and Jakobsson, J. O. (1987). The effect of pressure on dryout of a saturated bed of heat-generating particles. *ASME J. Heat Transfer* **109**, 185–195.

Chan, C. K., and Tien, C. L. (1974a). Radiative transfer in packed spheres. *ASME J. Heat Transfer* **96**, 52-58.

Chan, C. K., and Tien, C. L. (1974b). Combined radiation and conduction in packed spheres. *In* "Proceedings of the 5th International Heat Transfer Conference," Vol. 1, Tokyo, Japan. pp. 72–74.

Chan, B. K. C., Ivey, C. M., and Barry, J. M. (1970). Natural convection in enclosed porous media with rectangular boundaries. *ASME J. Heat Transfer* **2**, 21–27.

Chandrasekhar, S. (1960). "Radiative transfer." Dover, New York.

Chang, I. D., and Cheng, P. (1983). Matched asymptotic expansions for free convection about an impermeable horizontal surface in a porous medium. *Int. J. Heat Mass Transfer* **26**, 163–173.

Chen, J. C., and Churchill, S. W. (1963). Radiant heat transfer in packed beds. *AIChE Journal* **9**, 35–41.

Chen, C. K., Hung, C. I., and Cleaver, J. W. (1987). NonDarcian effects on vertical plate transient natural convection in porous media with high-porosities. In "Proceedings of the ASME–JSME Thermal Engineering Joint Conference", Vol. 2, pp. 313–318. Honolulu, Hawaii.

Cheng, P. (1977). Combined free and forced convection flow about inclined surfaces in porous media. *Int. J. Heat Mass Transfer* **20**, 807–814.

Cheng, P. (1978a). Heat Transfer in geothermal systems. *Adv. Heat Transfer* **14**, 1–105.

Cheng, P. (1978b). Convective heat transfer in porous layers by integral methods. *Lett. Heat Mass Transfer* **5**, 243–252.

Cheng, P. (1981). Thermal dispersion effects in nonDarcian convective flows in a saturated porous medium. *Lett. Heat Mass Transfer* **8**, 267–270.

Cheng, P. (1982). Mixed convection about a horizontal cylinder and a sphere in a saturated porous medium. *Int. J. Heat Mass Transfer* **25**, 1245–1247.

Cheng, P., and Ali, C. L. (1981). An experimental investigation of free convection about an inclined surface in a porous medium. ASME Paper No. 81-HT-85.

Cheng, P., and Chang, I. D. (1976). Buoyancy induced flows in a saturated porous medium adjacent to impermeable horizontal surfaces. *Int. J. Heat Mass Transfer* **19**, 1267–1272.

Cheng, P., and Chang, I. D. (1979). Convection in a porous medium as a singular perturbation problem. *Lett. Heat Mass Transfer* **6**, 253–258.

Cheng, P., and Hsu, C. T. (1984). Higher-order approximations for Darcian free convective flow about a semi-infinite vertical flat plate. *ASME J. Heat Transfer* **106**, 143–151.

Cheng, P., and Hsu, C. T. (1986). Fully-developed, forced convective flow through an annular packed-bed with wall effects. *Int. J. Heat Mass Transfer* **29**, 1843–1853.

Cheng, P., and Minkowycz, W. J. (1977). Free convection about a vertical plate embedded in a saturated porous medium with application to heat transfer from a dike. *J. Geophys. Res.* **82**, 2040–2044.

Cheng, P., and Pop, I. (1984). Transient free convection about a vertical flat plate embedded in a porous medium. *Int. J. Engng. Sci.* **22**, 253–264.

Cheng, P., and Vortmeyer, D. (1988). Transverse thermal dispersion and velocity profile in a packed bed with forced convective flow. *Chem. Engng. Sci.* **43**, 2523–2532.

Cheng, P., and Zheng, T. M. (1985). Thermal dispersion effects in forced convection plume above a horizontal line source of heat in a packed bed. *Int. Comm. Heat Mass Transfer.*

Cheng, P., Ali, C. L., and Verma, A. K. (1981). An experimental study of nonDarcian effects in free convection in a saturated porous medium. *Lett. Heat Mass Transfer* **8**, 261–265.

Chuah, Y. K., and Carey, V. P. (1985). Analysis of boiling heat transfer and two-phase flow in porous media with nonuniform porosity. *Int. J. Heat Mass Transfer* **28**, 147–154.

Combarnous, M. A., and Bia, P. (1971). Combined free and forced convection in porous media. *Soc. Pet. Eng. J.* **11**, 399–405.

Combarnous, M. A., and Bories, S. A. (1975). Hydrothermal convection in saturated porous media. *Adv. Hydrosci.* **10**, 231–307.

DeVries, D. A. (1958). Simultaneous transfer of heat and moisture in porous media. *Trans. Amer. Geophys. Union* **39**, 909–916.

De Wasch, A. P., and Froment, G. F. (1972). Heat transfer in packed beds. *Chem. Engng. Sci.* **27**, 567–576.

Dinulescu, H. A., and Eckert, E. R. G. (1980). Analysis of the one-dimensional moisture migration caused by temperature gradients in a porous medium. *Int. J. Heat Mass Transfer* **23**, 1069–1078.

Drolen, B. L., Kumar, S., and Tien, C. L. (1987). Experiments on dependent scattering of radiation. *In* "Proceedings of the AIAA 22nd Thermophysics Conference", Honolulu, Hawaii.

Dybbs, A., and Schweitzer, S. (1973). Conservation equations for nonisothermal flow in porous media. *J. Hydrology* **20**, 171–180.

Echigo, R., Hasegawa, S., and Nakano, S. (1974). Simultaneous radiative and convective heat transfer in a packed bed with high porosity. *Trans. JSME* **40**, 479.

Eckert, E. R. G., and Faghri, M. (1980). A general analysis of moisture migration caused by temperature differences in an unsaturated porous medium. *Int. J. Heat Mass Transfer* **23**, 1613–1623.

Eckert, E. R. G., and Pfender, E. (1978). Heat and mass transfer in porous media with phase change. In "Proceedings of the Sixth International Heat Transfer Conference," Vol. 6, pp. 1–12. Hemisphere Publishing Corp., Washington, D.C. Toronto, Canada.

Elder, J. W. (1967a). Steady free convection in a porous medium heated from below. *J. Fluid Mech.* **27**, 29–48.

Elder, J. W. (1967b). Transient convection in a porous medium. *J. Fluid Mech.* **27**, 609–623.

Ettefagh, J., and Vafai, K. (1988). Natural convection in open-ended cavities with a porous obstructing medium. *Int. J. Heat Mass Transfer* **31**, 673–693.

Evans, G. H., and Plumb, O. A. (1978). Natural convection from a vertical isothermal surfaces embedded in a saturated porous medium. ASME Paper No. 78-HT-85.

Facas, G. N., and Farouk, B. (1983). Transient and steady-state natural convection in a porous medium between two concentric cylinders. *ASME J. Heat Transfer* **105**, 660–663.

Fand, R. M., Steinberger, T. E., and Cheng, P. (1986). Natural convection heat transfer from a horizontal cylinder embedded in a porous medium. *Int. J. Heat Mass Transfer* **29**, 119–133.

Farouk, B., and Shayer, H. (1985). Natural convection around a heated cylinder buried in a saturated porous medium. *In* "Natural Convection in Porous Media," ASME HTD—Vol. 56, pp. 181–189.

Faust, C. R., and Mercer, J. W. (1979). Geothermal reservoir simulation: 1. Mathematical models for liquid- and vapor-dominated hydrothermal systems. *Water Resources Res.* **15**, 23–30; Geothermal reservoir simulation: 2. Mathematical models for liquid- and vapor-dominated hydrothermal systems. *Water Resources Res.* **15**, 31–46.

Fernandez, R. T., and Schrock, V. E. (1982). Natural convection from cylinders buried in a liquid-saturated porous medium. *In* "Proc. 7th Int. Heat Transfer Conference," München, Germany. Vol. 2, pp. 335–340.

Flamant, G., and Arnaud, G. (1984). Analysis and theoretical study of high-temperature heat transfer between a wall and a fluidized bed. *Int. J. Heat Mass Transfer* **27**, 1725–1735.

Flatt, I., and Klikoff, W. A. (1959). Effect of functional wettability on multiphase flow through porous media. AIME Technical Note #2043. *AIME Transactions* **216**, 256.

Gardner, W., and Widtsoe, J. A. (1920). The movement of soil moisture. *Soil Science* **11**, 215–232.

Georgiadis, J., and Catton, I. (1986). Prandtl number effect on Benard convection in porous media. *ASME J. Heat Transfer* **108**, 284–290.

Goshayeshi, A., Welty, J. R., Adams, R. L., and Alavizadeh, N. (1986). Local heat transfer coefficients for horizontal tube arrays in high-temperature large-particle fluidized beds: An experimental study. *ASME J. Heat Transfer* **108**, 907–912.

Gupta, V. P., and Joseph, D. D. (1973). Bounds for heat transport in a porous layer. *J. Fluid Mech.* **57**, 491–514.

Haajizadeh, M., and Tien, C. L. (1983). Natural convection in a rectangular porous cavity with one permeable endwall. *ASME J. Heat Transfer* **105**, 803–808.

Haber, S., Shavit, A., and Dayan, J. (1984). The effect of heat convection on drying of porous semi-infinite space. *Int. J. Heat Mass Transfer* **27**, 2347–2353.

Harmathy, T. Z. (1969). Simultaneous moisture and heat transfer in porous systems with particular reference to drying. *Ind. Eng. Chem. Fundamentals* **8**, 92–103.

Havstad, M. A., and Burns, P. J. (1982). Convective heat transfer in vertical cylindrical annuli filled with a porous medium. *Int. J. Heat Mass Transfer* **25**, 1755–1766.

Hickox, C. E. (1981). Thermal convection at low Rayleigh number from concentrated sources in porous media. *ASME J. Heat Transfer* **103**, 232–236.

Hickox, C. E., and Gartling, D. K. (1981). A numerical study of natural convection in a horizontal porous layer subjected to an end-to-end temperature difference. *ASME J. Heat Transfer* **103**, 797–802.

Hickox, C. E., and Gartling, D. K. (1982). A numerical study of natural convection in a vertical annular porous layer. ASME Paper No. 82-HT-68.

Hickox, C. E., and Watts, H. A. (1980). Steady thermal convection from a concentrated source in a porous medium. *ASME J. Heat Transfer* **102**, 248–253.

Hong, J. T., and Tien, C. L. (1987). Analysis of thermal dispersion effect on vertical-plate natural convection in porous media. *Int. J. Heat Mass Transfer* **30**, 143–150.

Hong, J. T., Tien, C. L., and Kaviany, M. (1985). NonDarcian effects on vertical-plate natural convection in porous media with high porosities. *Int. J. Heat Mass Transfer* **28**, 2149–2157.

Horne, R. N., and O'Sullivan, M. J. (1974). Oscillatory convection in a porous medium heated from below. *J. Fluid Mech.* **66**, 339–352.

Horne, R. N., and O'Sullivan, M. J. (1978). Convection in a porous medium heated from below: The effect of temperature dependent viscosity and thermal expansion coefficient. *ASME J. Heat Transfer* **100**, 448–452.

Hottel, H. C., Sarotim, A. F., Dalzell, W. H., and Vasalos, I. A. (1971). Optical properties of coatings, effect of pigment concentration. *AIAA Journal* **9**, 1895–1898.

Hsu, C. T., and Cheng, P. (1985). The Brinkman model for natural convection about a semi-infinite vertical flat plate in a porous medium. *Int. J. Heat Mass Transfer* **28**, 683–697.

Huang, C. L. D. (1979). Multi-phase moisture transfer in porous media subjected to temperature gradients. *Int. J. Heat Mass Transfer* **22**, 1295–1307.

Huang, C. L. D., Siang, H. H., and Best, C. H. (1978). Heat and moisture transfer in concrete slabs. *In* "Proc. 8th U.S. National Congress of Applied Mechanics," Los Angeles, CA.

Huenefeld, J. S., and Plumb, O. A. (1981). A study of nonDarcy natural convection from a vertical heated surface in a saturated porous medium. ASME Paper No. 81-HT-45.

Hunt, M. L., and Tien, C. L. (1987). NonDarcian convection in cylindrical packed beds. *In* "Proceedings of the ASME–JSME Thermal Engineers Joint Conference," Vol. 2, pp. 433–438. Honolulu, Hawaii.

Jonsson, T., and Catton, I. (1987). Prandtl number dependence of natural convection in porous media. *ASME J. Heat Transfer* **109**, 371–377.

Joshi, Y., and Gebhart, B. (1984). Vertical natural convection flows in porous media: Calculations of improved accuracy. *Int. J. Heat Mass Transfer* **27**, 69–75.

Kalthoff, O., and Vortmeyer, D. (1980). Ignition/extinction phenomena in a wall cooled fixed bed reactor, *Chem. Engng Sci.* **35**, 1637–1643.

Kaneko, T., Mohtadi, M. F., and Aziz, K. (1974). An experimental study of natural convection in inclined porous media. *Int. J. Heat Mass Transfer* **17**, 485–496.

Katto, Y., and Masuoka, T. (1967). Criterion for the onset of convective flow in a fluid in a porous medium. *Int. J. Heat Mass Transfer* **10**, 297–309.

Kaviany, M. (1985a). Laminar flow through a porous channel bounded by isothermal parallel plates. *Int. J. Heat Mass Transfer* **28**, 851–858.

Kaviany, M. (1985b). An integral solution to vertical plate natural convection in porous media. ASME Paper No. 85-WA/HT-1.

Kaviany, M. (1987). Boundary-layer treatment of forced convection heat transfer from a semi-infinite flat plate embedded in porous media. *ASME J. Heat Transfer* **109**, 345–349.

Kaviany, M., and Mittal, M. (1987). Natural convection heat transfer from a vertical plate to high permeability porous media; an experiment and an approximate solution. *Int. J. heat Mass Transfer* **30**, 967–977.

Keller, J. B. (1962). Geometrical theory of diffraction. *J. Optical Society of America* **52**, 116–130.

Kerker, M. (1961). "The Scattering of Light and Other Electromagnetic Radiation." Academic Press, New York.

Kim, S. J., and Vafai, K. (1989). Analysis of natural convection about a vertical plate embedded in a porous medium. *Int. J. Heat Mass Transfer* **32,** 665–677.

Kimura, S., Bejan, A., and Pop, I. (1985). Natural convection near a cold plate facing upward in a porous medium. *ASME J. Heat Transfer* **107**, 819–825.

Klarsfeld, S. (1963). Champs de temperature associes aux mouvements de convection naturelle dons un milieu poreux limite. *Rev. Gen. Thermique* **108**, 1403–1424.

Koh, J. C. Y., and Colony, R. (1974). Analysis of cooling effectiveness for porous material in a coolant passage. *ASME J. Heat Transfer* **96**, 324–330.

Koh, J. C. Y., and Stevens, R. L. (1975). Enhancement of cooling effectiveness by porous material in coolant passage. *ASME J. Heat Transfer* **97**, 309–311.

Krischer, O. (1938). Fundamental law of moisture movement in drying by capillary flow and vapor diffusion. *VDIZ* **82**, 373–378.

Krischer, O. (1940a). The heat, moisture and vapor movement during drying porous materials. *VDIZ, Beih.* **1**, 17–24.

Krischer, O. (1940b). Heat and mass transfer in drying. *VDI-Forschungsh.* 415.

Kudo, K., Yang, W. J., Taniguchi, H., and Hayasaka, H. (1985). Radiative heat transfer in packed spheres by Monte Carlo method. *In* "Proceedings of the U.S.–Japan Heat Transfer Joint Seminar," *J*-4, pp. 1–12.

Kumar, S., and Tien, C. L. (1987). Dependent scattering and absorption of radiation by small particles. *In* "24th National Heat Transfer Conference," Vol. 72, pp. 1–7. Pittsburgh, Pennsylvania.

Kumar, S., Majumdar, A., and Tien, C. L. (1988). The differential-discrete-ordinate method for solving the general equation of radiative transfer. *In* "25th National Heat Transfer Conference," pp. 24–27. Houston, Texas.

Lapwood, E. R. (1948). Convection of a fluid in a porous medium. *Proc. Camb. Phil. Soc.* **44**, 508–521.

Lauriat, G., and Prasad, V. (1987). Natural convection in a vertical porous cavity: A numerical study for Brinkman-extended Darcy formulation. *ASME J. Heat Transfer* **109**, 688–696.

Lerou, J. J., and Froment, G. F. (1977). Velocity, temperature and conversion profile in fixed bed catalytic reactors. *Chem. Engng. Sci.* **32**, 853–861.

Lewis, W. K. (1921). The rate of drying of solid materials. *Ind. Eng. Chem.* **13**, 427–432.

Li, C., and Finlayson, B. A. (1977). Heat transfer in packed beds—A reevaluation. *Chem. Engng. Sci.* **32**, 1055–1066.

Luikov, A. V. (1964). Heat and mass transfer in capillary-porous bodies. *Adv. Heat Transfer* **1**, 123–184.

Luikov, A. V. (1975). System of differential equations of heat and mass transfer in capillary-porous bodies (review). *Int. J. Heat Mass Transfer* **18**, 11–4.

Lyczkowski, R. W., and Chao, Y.-T. (1984). Comparison of Stefan model with two-phase model of coal drying. *Int. J. Heat Mass Transfer* **27**, 1157–1169.

Masuoka, T., Ishizaka, K., and Katsuhara, T. (1980). Heat transfer by natural convection in porous media between two concentric spheres. *In* "Natural Convection in Enclosures," ASME HTD—Vol. 8, pp. 115–120.

Menguc, M. P., and Viskanta, R. (1983). Comparison of radiative transfer approximations for highly forward scattering planar medium. *J. Quantitative Spectroscopy and Radiative Transfer* **29**, 381–394.

Merkin, J. H. (1979). Free convection boundary layers on axi-symmetric and two-dimensional bodies of arbitrary shape in a situated porous media. *Int. J. Heat Mass Transfer* **22**, 1461–1462.

Minkowycz, W. J., and Cheng, P. (1976). Free convection about a vertical cylinder embedded in a porous medium. *Int. J. Heat Mass Transfer* **19**, 805–813.

Motakef, S., and El-Masri, M. A. (1985). Liquid diffusion in fibrous insulation. *ASME J. Heat Transfer* **107**, 299–306.

Moya, S. L., Ramos, E., and Sen, M. (1987). Numerical study of natural convection in a tilted rectangular porous material. *Int. J. Heat Mass Transfer* **30**, 741–756.

Muralidhar, K., and Kulacki, F. A. (1986). NonDarcian natural convection in a saturated horizontal porous annulus. In "Natural Convection in Porous Media," ASME HTD—Vol. 56, pp. 23–32.

Newman, A. B. (1931). The drying of porous solids: Diffusion calculations. *AIChE Journal* **27**, 310–333.

Nilson, R. H., and Romero, L. A. (1980). Self-similar condensing flows in porous media. *Int. J. Heat Mass Transfer* **23**, 1461–1470.

Ogniewicz, Y., and Tien, C. L. (1981). Analysis of condensation in porous insulation. *Int. J. Heat Mass Transfer* **24**, 421–429.

Ozisik, M. N. (1973). "Radiative transfer and interactions with conduction and convection." Wiley, New York.

Philip, J. R. (1982). Free convection at small Rayleigh number in porous cavities of rectangular, elliptical, triangular and other cross-sections. *Int. J. Heat Mass Transfer* **25**, 1503–1510.

Philip, J. R., and DeVries, D. A. (1957). Moisture movement in porous materials under temperature gradients. *Trans. Amer. Geophs. Union* **38**, 222–232.

Plumb, O. A. (1983). The effect of thermal dispersion on heat transfer in packed bed boundary layers. *In* "Proc. ASME–JSME Thermal Engineering Conference," Vol. 2, pp. 17–22. Honolulu.

Plumb, O. A., and Huenefeld, J. A. (1981). NonDarcy natural convection from heated surfaces in saturated porous media. *Int. J. Heat Mass Transfer* **24**, 765–768.

Plumb, O. A., Spolek, G. A., and Olmstead, B. A. (1985). Heat and mass transfer in wood during drying. *Int. J. Heat Mass Transfer* **28**, 1669–1678.

Pop, I., and Cheng, P. (1983). The growth of thermal layer in a porous medium adjacent to a suddenly heated semi-infinite horizontal surface. *Int. J. Heat Mass Transfer* **26**, 1574–1576.

Poulikakos, D. (1985). A departure from the Darcy model in boundary layer natural convection in a vertical porous layer with uniform heat flux from the side. *ASME J. Heat Transfer* **107**, 716–720.

Poulikakos, D. (1986). Buoyancy-driven convection in a horizontal fluid layer extending over a porous substrate. *Phys. Fluids* **29**, 3949–3957.

Poulikakos, D., and Bejan, A. (1983a). Natural convection in vertically and horizontally layered porous media heated from the side. *Int. J. Heat Mass Transfer* **26**, 1805–1814.

Poulikakos, D., and Bejan, A. (1983b). Numerical study of transient high Rayleigh number convection in an attic-shaped porous layer. *ASME J. Heat Transfer* **105**, 476–484.

Poulikakos, D., and Bejan, A. (1985). The departure from Darcy flow in natural convection in a vertical porous layer. *Phys. Fluids* **28**, 3477–3484.

Poulikakos, D., and Renken, K. (1987). Analysis of forced convection in a duct filled with porous media. *In* "Heat Transfer in Geophysical and Geothermal Systems," ASME HTD—Vol. 76, pp. 9–20.

Prasad, V., and Kulacki, F. A. (1984a). Convective heat transfer in a rectangular porous cavity: Effect of aspect ratio on flow structure and heat transfer. *ASME J. Heat Transfer* **106**, 158–165.

Prasad, V., and Kulacki, F. A. (1984b). Natural convection in a rectangular porous cavity with constant heat flux on one vertical wall. *ASME J. Heat Transfer* **106**, 152–157.

Prasad, V., and Kulacki, F. A. (1984c). Natural convection in a vertical porous annulus. *Int. J. Heat Mass Transfer* **27**, 207–219.

Prasad, V., and Kulacki, F. A. (1985). Natural Convection in horizontal porous layers with localized heating from below. *In* "Heat Transfer in Porous Media and Particulate Flows," ASME HTD—Vol. 46, 199–208.

Prasad, V., and Kulacki, F. A. (1986). Effects of the size of heat source on natural convection in horizontal porous layers heated from below. *In* "Proceedings 8th International Heat Transfer Conference," Vol. 5, pp. 2677–2682. San Francisco, California.

Prasad, V., and Kulacki, F. A. (1987). Natural convection in horizontal porous layers with localized heating from below. *ASME J. Heat Transfer* **109**, 795–798.

Prasad, V., Kulacki, F. A., and Keyhani, M. (1985). Natural convection in porous media. *J. Fluid Mech.* **150**, 89–119.

Richards, L. A. (1931). Capillary conduction of liquids through porous media. *J. Appl. Phys.* **1**, 318–333.

Robinson, J. L., and O'Sullivan, M. J. (1976). A boundary layer model of flow in a porous medium at high Rayleigh number. *J. Fluid Mech.* **75**, 459–467.

Rohsenow, W. M., and Hartnett, J. P. (1973). "Handbook of Heat Transfer." McGraw–Hill, New York.

Saxena, S. C., Grewal, N. S., Gabor, J. D., Zabrodsky, S. S., and Galershtein, D. M. (1978). Heat transfer between a gas fluidized bed and immersed tubes. *Adv. Heat Transfer* **14**, 150–248.

Schneider, K. J. (1963). Investigation of the influence of free thermal convection in heat transfer through granular material. Paper 11-4. *In* "Proc. 11th Int. Congr. Refrig.," pp. 247–254.

Schotte, W. (1960). Thermal conductivity of packed beds. *AIChE Journal* **6**, 63–67.

Schrock, V. E., Fernandez, R. T., and Kesavan, K. (1970). Heat transfer from cylinders embedded in a liquid filled porous medium. *In* "Proc. 4th Int. Heat Transfer Conference," Vol. 7, p. CT-3.6. Paris-Versailles.

Seki, N., Fukusako, S., and Inaba, H. (1978). Heat transfer in a confined rectangular cavity packed with porous media. *Int. J. Heat Mass Transfer* **21**, 985–989.

Sen, A. K. (1987). Natural convection in a shallow porous cavity—The Brinkman model. *Int. J. Heat Mass Transfer* **30**, 855–868.

Sherwood, T. K. (1929a). The drying of solids. I. *Ind. Eng. Chem.* **21**, 12–16.

Sherwood, T. K. (1929b). The drying of solids. II. *ind. Eng. Chem.* **21**, 976–980.

Sherwood, T. K. (1930). The drying of solids. III. *Ind. Eng. Chem.* **22**, 132–136.

Sherwood, T. K. (1931). Application of the theoretical diffusion equations to the drying of solids. *AIChE Journal* **27**, 190–202.

Shiralkar, G. S., Haajizadeh, M., and Tien, C. L. (1983). Numerical study of high Rayleigh number convection in a vertical porous enclosure. *Numerical Heat Transfer* **6**, 223–234.

Siegel, R., and Howell, J. R. (1981). "Thermal Radiation Heat Transfer," Second Edition. Hemisphere, New York.

Simpkins, D. G., and Blythe, P. A. (1980). Convection in a porous layer. *Int. J. Heat Mass Transfer* **23**, 881–887.

Singh, B. S., and Dybbs, A. (1979). Nonisothermal liquid flow and heat transfer in sintered metallic porous media. *Int. J. Heat Mass Transfer* **22**, 1049–1057.

Somerton, C. W. (1983). The Prandtl number effect in porous layer convection. *Appl. Sci. Res.* **40**, 333–344.

Strauss, J. M. (1974). Large amplitude convection in porous media. *J. Fluid Mech.* **64**, 51–63.

Tabanfar, S., and Modest, M. F. (1987). Combined radiation and convection in absorbing, emitting, nongray gas-particulate tube flow. *ASME J. Heat Transfer* **109**, 478–484.

Tien, C. L. (1975). Fluid mechanics of heat pipes. *Ann. Rev. Fluid Mech.* **7**, 167–185.

Tien, C. L. (1985a). Heat Pipes. *In* "Handbook of Heat Transfer Applications," Second Edition, Chapter 5. Editors: Rohsenow, W. M., Hartnett, J. P., and Ganic, E. N. Hemisphere, New York.

Tien, C. L. (1985b). Radiation properties of particulates. *In* "Handbook of Heat Transfer Fundamentals," Second Edition, pp. 14.83–14.91. Hemisphere, New York.

Tien, C. L., and Cunnington, G. R. (1973). Cryogenic insulation heat transfer. *Adv. Heat Transfer* **9**, 349–417.

Tien, C. L., and Drolen, B. L. (1987). Thermal radiation in particulate media with dependent and independent scattering. *Ann. Rev. Num. Fluid Mech. Heat Transfer* **1**, 1–32.

Tien, C. L., and Rohani, A. R. (1972). Theory of two component heat pipes. *J. Heat Transfer* **94**, 479–484.

Tong, T. W., and Orangi, S. (1986). A numerical analysis for high modified Rayleigh number natural convection in enclosures containing a porous medium. *In* Proceedings 8th International Heat Transfer Conference," Vol. 5, pp. 2647–2652. San Francisco, California.

Tong, T. W., and Subramanian, E. (1983). Natural convection in rectangular enclosures partially filled with a porous medium. *In* "Proceedings of the ASME–JSME Thermal Engineering Joint Conference," Vol. 1, pp. 331–338. Honolulu, Hawaii.

Tong, T. W., and Subramanian, E. (1985). A boundary-layer analysis for natural convection in vertical porous enclosures—Use of the Brinkman-extended Darcy model. *Int. J. Heat Mass Transfer* **28**, 563–571.

Tong, F. W., Faruque, M. A., Orangi, S., and Sathe, S. B. (1986). Experimental results for natural convection in vertical enclosures partially filled with a porous medium. *In* "Natural Convection in Porous Media," ASME HTD—Vol. 56, pp. 85–94.

Tsai, F. P., Jakobsson, J., Catton, I., and Dhir, V. K. (1984). Dryout of an inductively heated bed of steel particles with subcooled flow from beneath the bed. *Nuclear Technology* **65**, 10–15.

Udell, K. S. (1985). Heat transfer in porous media considering phase change and capillarity—The heat pipe effect. *Int. J. Heat Mass Transfer* **28**, 485–495.

Udell, K. S., and Fitch, J. S. (1985). Heat and mass transfer in capillary porous media considering evaporation, condensation and noncondensible gas effects. *In* "Heat Transfer in Porous Media and Particulate Flows," pp. 103–110.

Vafai, K. (1984). Convective flow and heat transfer in variable-porosity media. *J. Fluid Mech.* **147**, 233–259.

Vafai, K. (1986). Analysis of the channeling effect in variable porosity media. *ASME J. Energy Resources Technology* **108**, 131–139.

Vafai, K., and Ettefagh, J. (1988). Analysis of the radiative and conductive heat transfer characteristics of a waste package canister. *ASME J. Heat Transfer* **110**, 1011–1014.

Vafai, K., and Sarkar, S. (1986). Condensation effects in a fibrous insulation slab. *ASME J. Heat Transfer* **108**, 667–675.

Vafai, K., and Sarkar, S. (1987). Heat and mass transfer in partial enclosures. *AIAA J. Thermophysics Heat Transfer* **1.3**, 253–259.

Vafai, K., and Thiyagaraja, R. (1987). Analysis of flow and heat transfer at the interface region of a porous medium. *Int. J. Heat Mass Transfer* **30**, 1391–1405.

Vafai, K., and Tien, C. L. (1981). Boundary and inertia effects on flow and heat transfer in porous media. *Int. J. Heat Mass Transfer* **24,** 195–203.

Vafai, K., and Tien, C. L. (1982). Boundary and inertia effects on convective mass transfer in porous media. *Int. J. Heat Mass Transfer* **25**, 1183–1190.

Vafai, K., and Whitaker, S. (1986). Simultaneous heat and mass transfer accompanied by phase change in porous insulation. *ASME J. Heat Transfer* **108**, 132–140.

Vafai, K., Alkire, R. L., and Tien, C. L. (1985). An experimental investigation of heat transfer in variable porosity media. *ASME J. Heat Transfer* **107**, 642–647.

Vasseur, P., and Robillard, L. (1987). The Brinkman model for boundary layer regime in a rectangular cavity with uniform heat flux from the side. *Int. J. Heat Mass Transfer* **30**, 717–727.

Vasseur, P., Satish, M. G., and Robillard, L. (1987). Natural convection in a thin, inclined, porous layer exposed to a constant heat flux. *Int. J. Heat Mass Transfer* **30**, 537–549.

Viskanta, R. (1982). Radiation heat transfer: Interaction with conduction and convection and approximate methods in radiation. *In* "Proceedings of the Seventh International Heat Transfer Conference," Vol. 1, pp. 103–121. Munchen, Germany.

Viskanta, R. (1984). Radiative heat transfer. *Progress Chemical Engineering* (*Fortschritte der Verfahrenstechniks*) *Section A* **22**, 51–81.

Vortmeyer, D. (1974). Radiation interaction with conduction and convection. *In* "Proceedings of the 5th International Heat Transfer Conference," Vol. 6, pp. 129–135. Tokyo, Japan.

Vortmeyer, D. (1978). Radiation in packed solids. *In* "Proceedings of the 6th International Heat Transfer Conference," Vol. 6, pp. 525–539. Toronto, Canada.

Vortmeyer, D., and Schuster, J. (1983). Evaluation of steady flow profiles in rectangular and circular packed beds by a variational method. *Chem. Eng. Sci.* **38**, 1691–1699.

Walker, K. L., and Homsy, G. M. (1978). Convection in a porous cavity. *J. Fluid Mech.* **87**, 449–474.

Wang, K. Y., and Tien, C. L. (1983). Radiative heat transfer through opacified fibers and powders. *J. Quantitative Spectroscopy Radiative Transfer* **30**, 213–223.

Weber, J. E. (1975a). The boundary layer regime for convection in a vertical porous layer. *Int. J. Heat Mass Transfer* **18**, 569–573.

Weber, J. E. (1975b). Thermal convection in a tilted porous layer. *Int. J. Heat Mass Transfer* **18**, 474–475.

Westman, A. E. R. (1929). The capillary suction of some ceramic materials. *J. Amer. Ceram. Soc.* **12**, 585–595.

Whitaker, S. (1977). Simultaneous heat, mass, and momentum transfer in porous media: A theory of drying. *Adv. Heat Transfer* **13**, 119–203.

White, S. M., and Tien, C. L. (1987a). Analysis of laminar condensation in a porous medium. *In* "Proc. of the ASME–JSME Thermal Engineering Joint Conference," Vol. 2, pp. 401–406. Honolulu, Hawaii.

White, S. M., and Tien, C. L. (1987b). Experimental investigation of film condensation in porous structures. *In* "Proc. of the 6th International Heat Pipe Conference, Fundamental and Basic Research: Theoretical and Experimental Studies," pp. 148–153.

Wong, K. F. and Dybbs, A. (1978). Energy equations for convection in saturated porous media. ASME Paper No. 78-HT-54.

Wooding, R. A. (1963). Convection in a saturated porous medium at large Rayleigh number or Peclet number. *J. Fluid Mech.* **15**, 527–544.

Yamada, Y., Cartigny, J. D., and Tien, C. L. (1986). Radiative transfer with dependent scattering by particles: Part 2—experimental investigation. *ASME J. Heat Transfer* **108**, 614–618.

Yang, Y. S., Howell, J. R., and Klein, D. E. (1983). Radiative heat transfer through a randomly packed bed of spheres by the Monte Carlo method. *ASME J. Heat Transfer* **105**, 325–332.

Yen, Y. C. (1974). Effects of density inversion on free convective heat transfer in porous layer heated from below. *Int. J. Heat Mass Transfer* **17**, 1349–1356.

Yener, Y., and Ozisik, M. N. (1986). Simultaneous radiation and forced convection in thermally developing turbulent flow through a parallel-plate channel. *ASME J. Heat Transfer* **108**, 985–988.

Yuen, W. W., and Wong, L. W. (1980). Heat transfer by conduction and radiation in a one-dimensional absorbing, emitting, and anisotropically scattering medium. *ASME J. Heat Transfer* **102**, 303–307.

ADVANCES IN APPLIED MECHANICS, VOLUME 27

Solitons Induced by Boundary Conditions

C. K. CHU

Department of Applied Physics
Columbia University
New York, New York

RU LING CHOU

Lamont Doherty Geological Observatory
Columbia University
New York, New York

and

NASA Goddard Institute for Space Studies
New York, New York

I. Introduction

Since the sighting of the solitary wave by Russell (1844) over a century ago, the discovery of the remarkable properties of solitons by Zabusky and Kruskal (1965) over two decades ago, and the invention of the inverse scattering transform by Gardner, Greene, Kruskal, and Miura

ISBN 0-12-002027-0

(1967) shortly afterwards, the study of solitons has mushroomed into an active field of research. Considerable interest has also been shown in the generation of solitons.

In water, solitary waves can be generated by three different methods: an initial profile evolving into one or many solitary waves, a moving ship or equivalent pressure source on the surface of water, or boundary excitation, such as a sluice opening or a wall pushing. In other media, solitons can be produced by analogous situations.

In theoretical work, the first method corresponds to a pure initial value problem (for some governing differential equation or system of equations), the second to a driving term or an inhomogeneous differential equation (or equations), and the third to a mixed initial boundary value problem. Indeed, the most exciting advances in soliton theory, such as soliton interaction and collision and the inverse scattering transform, have all come from pure initial value problems. Only recently have steps been taken in the theoretical and numerical studies of the mixed initial boundary value problems.

The differential equations used in the theoretical and numerical works vary. The simplest and best studied is the Korteweg–deVries equation (KdV equation), either in the ordinary form or in the regularized form (the latter replaces the third x-derivative in the former by a mixed xxt-derivative). For water, the Boussinesq systems (also in different forms), and for nonlinear optics, the nonlinear Schrödinger equations have been used extensively.

In this paper, we summarize briefly experiments and then describe numerical results, mainly for water waves. We then describe some theoretical results obtained to date for mixed initial boundary value problems for the KdV equation and for the nonlinear Schrödinger equation.

In the preparation of this review, the authors have benefited enormously from valuable and insightful discussions with many colleagues, particularly the following: Drs. M. Ablowitz, Y. Baransky, T. Fokas, M. Tabor, and T. Y. Wu. We thank them deeply for their help.

II. Experiments

There have been numerous experiments done on solitons in water, as well as in other media, with the express purpose of seeing the nonlinear interaction that preserves wave identities, as predicted by numerical experiments and by theory. The main results of these experiments need not concern us in this paper, but the methods of generation of the solitons are of significant interest. Indeed, in most experiments, the solitons are boundary generated.

Maxworthy (1976) used a uniform bottom trough to study soliton reflection and collision in water. In studying reflection, the method he used to generate the solitons was to push a movable partition along the length of the trough. The waves thus generated then reflect off a fixed end wall. In studying collisions of solitons of different strengths, he used partitions near both ends of the trough, and filled water behind the partitions to different heights. The solitons are generated by withdrawing the partitions, and solitons of different strengths then run toward each other. Clearly, all the excitation processes involved here—pushing a wall (or piston), reflecting waves from a fixed wall, and producing waves from water at different heights—all correspond to boundary generation. The last case can be argued as generation from initial conditions, but if either the water level is maintained constant behind the partition after withdrawal, or if the end wall is close to the partition, then the excitation is, strictly speaking, still by boundary action.

Hammack and Segur (1974) also studied solitons in a trough by raising or lowering a piston in the vertical direction, i.e., perpendicular to the trough. This could approximate generation by initial conditions. Again, however, the piston is small in comparison with the trough length, and it is placed right next to a fixed wall. Thus, strictly speaking, the solitons are affected by boundary action as well.

The ingenious laboratory experiments of J. Scott Russell (1844), which followed his famous sighting of solitary waves in the Union Canal in 1834, were similar to the Hammack and Segur experiments, except that instead of raising or lowering a piston, he dropped one end of a solid weight into the water to produce positive waves (bumps), and lifted the weight out of the water to produce negative waves (depressions). This too is boundary generation of solitary waves.

Bona *et al.* (1981) studied nonlinear periodic waves in a trough. The waves are generated by a paddle wheel at one end of the trough. The generation is again obviously by boundary action.

These examples suffice to indicate the practical importance of studying boundary generation of solitons, despite the theoretical convenience of considering generation by initial conditions. The latter is rarely achievable experimentally, and the same situation holds in other media (e.g., solitons in ion-acoustic plasmas, nonlinear optics, etc.).

III. Numerical Results

A. Soliton Formation

The work of Chu, Xiang, and Baransky (1983) was one of the earliest papers explicitly aimed at studying soliton formation in the quarterplane solutions of the KdV equation, in its ordinary form

$$u_t + uu_x + \varepsilon u_{xxx} = 0. \tag{3.1}$$

The initial condition and boundary conditions are the simplest possible:

$$\begin{aligned} &u(x, 0) = 0, \\ &u(0, t) = UH(t) \quad \text{or} \quad U[H(t) - H(t - t_0)], \end{aligned} \tag{3.2}$$

where U is a constant and $H(t)$ is the Heavyside function. The results they found were striking and are summarized concisely below:

1. For a constant boundary excitation, $u(0, t) = UH(t)$, a continuous stream of identical solitons are produced. As measured from the computed results, the amplitude of each soliton is $2U$, and the speed is $2U/3$ in agreement with classical theory (Figure 1). Each soliton "matures" in a certain formation time before the next soliton is "born".

2. For a pulse excitation of length t_0, $u(0, t) = U[H(t) - H(t - t_0)]$, the first solitons reach maturity, but the last one in general does not. Its amplitude reaches a value determined by the cut-off time t_0 (Figure 2).

3. Subsequntly, Camassa and Wu (1988) confirmed these findings using more accurate numerical calculations. In addition, Chou (Chou, 1987 and Chou and Chu, 1988) showed in her doctoral thesis that solitons calculated with the Boussinesq system

$$\begin{aligned} &\frac{\partial h}{\partial t} + \frac{\partial}{\partial x}(hu) = 0 \\ &\frac{\partial u}{\partial t} + u\frac{\partial u}{\partial x} + \frac{\partial h}{\partial x} + \tfrac{1}{3}\beta\frac{\partial^3 h}{\partial x^3} = 0 \end{aligned} \tag{3.3}$$

Here h and u are the dimensionless height and velocity respectively, and

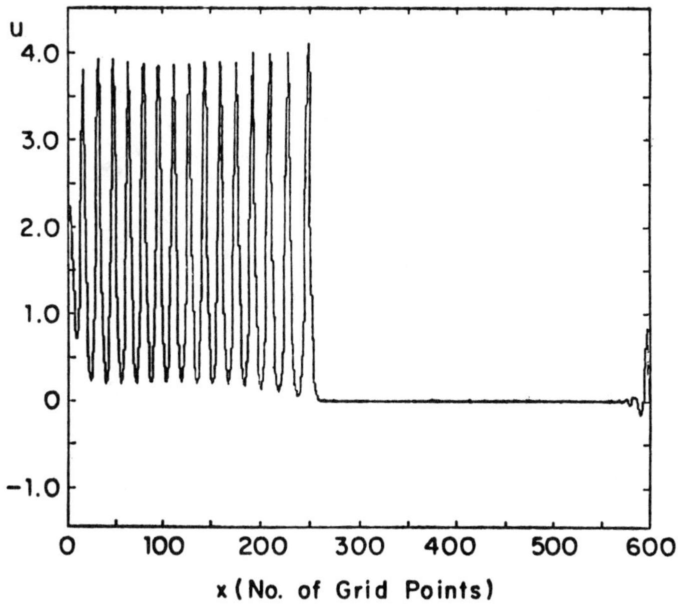

FIG. 1. Solitons generated from KdV equation by the constant boundary condition $u(0, t) = 2$ and the zero initial condition $u(x, 0) = 0$ (Chu *et al.*, 1983. Copyright © 1983 by John Wiley & Sons, Inc. Reprinted by permission of John Wiley & Sons, Inc.).

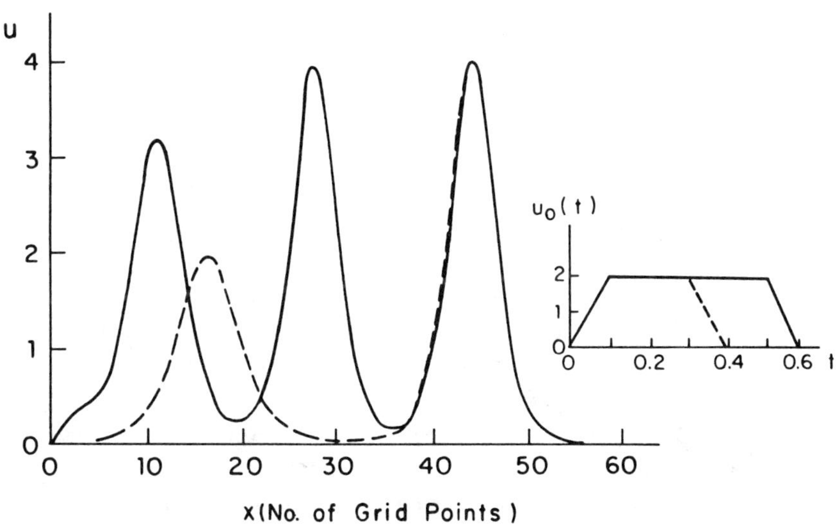

FIG. 2. Solitons produced from two pulses $u(0, t)$ of different durations (0.4—dashed line and 0.6—solid line) showing maturing (Chu *et al.*, 1983. Copyright © 1983 by John Wiley & Sons, Inc. Reprinted by permission of John Wiley & Sons, Inc.).

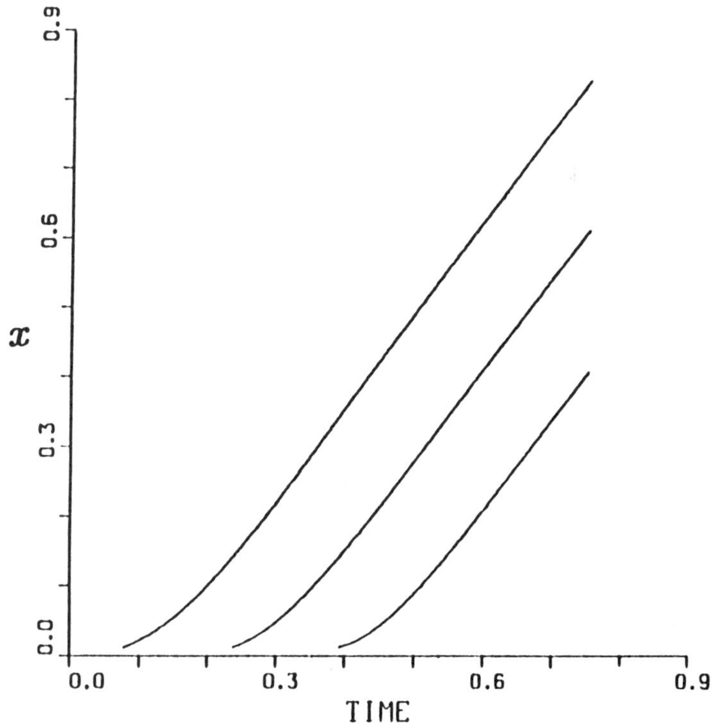

FIG. 3. Trajectories in x–t space of soliton peaks, showing acceleration during maturing, for a finite duration pulse (Camassa and Wu, 1988).

β is related to the ε in the Korteweg–deVries equation: $\varepsilon = \frac{1}{6}\beta$, for these initial and boundary conditions are virtually identical to these solitons, thus confirming that the two nonlinear equations indeed approximate each other very well.

4. Solitons accelerate as they grow, until they reach the speed appropriate for the mature solitons. Figure 3, taken from Camassa and Wu (1988), shows the x–t curves of the soliton crests as they form, accelerate, and reach steady speed.

5. Chou (Chou, 1987 and Chou and Chu, 1988) also considered more general boundary conditions than square pulses. One aim of her work was to see whether soliton "turbulence" would result if randomly fluctuating boundary conditions were applied. This did not occur. In fact, the random boundary data produced perfectly uniform solitons, with amplitude equal to twice the expected value of the random values (Figure

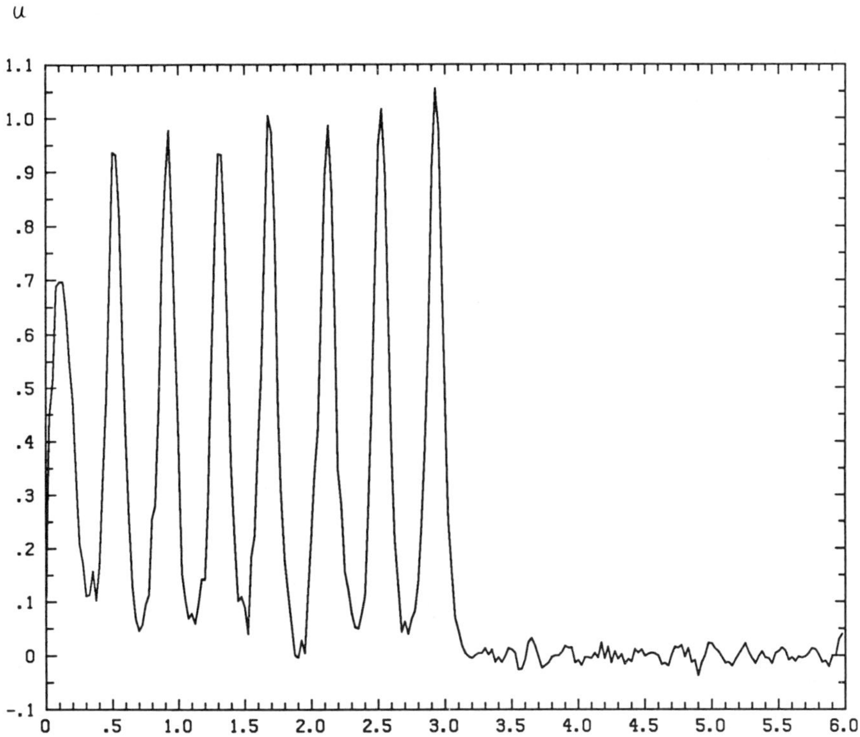

FIG. 4. Solitons generated from random boundary values of very short duration—1 time step (Chou, 1987 and Chou and Chu, 1988).

4). However, more slowly varying boundary values did produce modulated amplitudes in the solitons (Figure 5). From this, and from slowly varying sinusoidal boundary data, Chou concluded that the time variation in the boundary data has a strong influence on the modulated soliton amplitudes only if the characteristic time of the variation is comparable to or greater than the soliton formation time; if it were much smaller, than the solitons formed sense only averaged boundary values.

6. In Chu *et al.* (1983), they also considered the KdV–Burgers equation, i.e.

$$u_t + uu_x - \nu u_{xx} + \varepsilon u_{xxx} = 0. \tag{3.4}$$

This equation has both dissipation and dispersion, but the interesting case

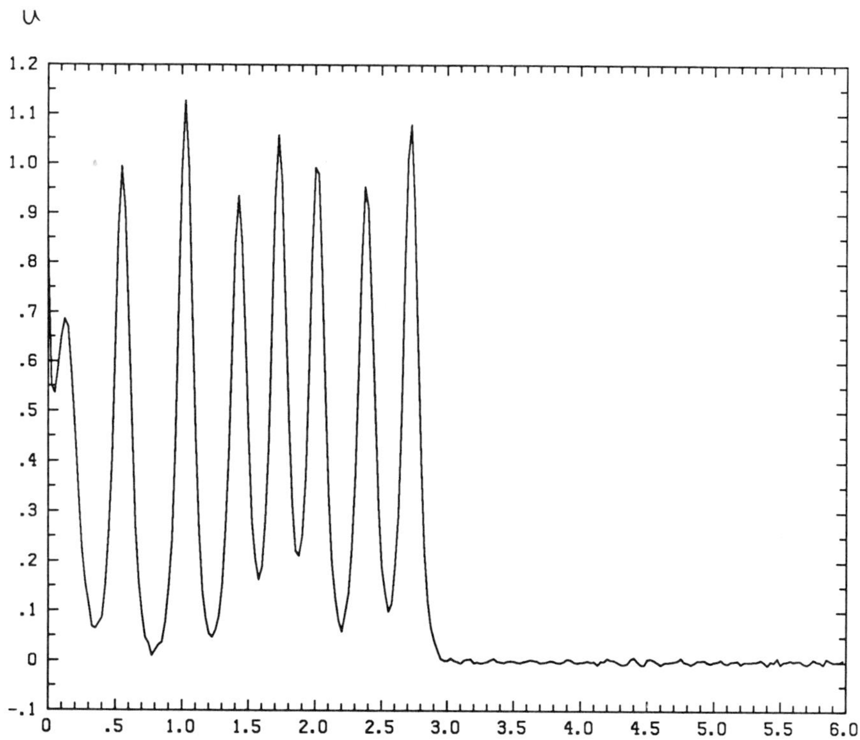

FIG. 5. Solitons generated from random boundary values of longer duration—10 time steps (Chou, 1987 and Chou and Chu, 1988).

is when the effect of dissipation is small. For the same initial boundary value problem as described above the pure KdV equation (i.e., zero initial data and a pure Heavyside function for the boundary data), the resulting wave pattern now appears as in Figures 6a and 6b. The pattern given in Figure 6b is a weak bore or collisionless shock, and is significantly different from the solitons in Figure 1. While the former is a continuously lengthening train of waves, the latter is a steady structure that connects two uniform states, in this case, $u = 0$ and $u = U$. This structure can be readily calculated from a set of ordinary differential equations when we perform a Galilean transformation at the speed of the wave in the usual way for obtaining shock wave structure. The length of the train depends on the dissipation ν, and the speed is $U/2$, (as derived

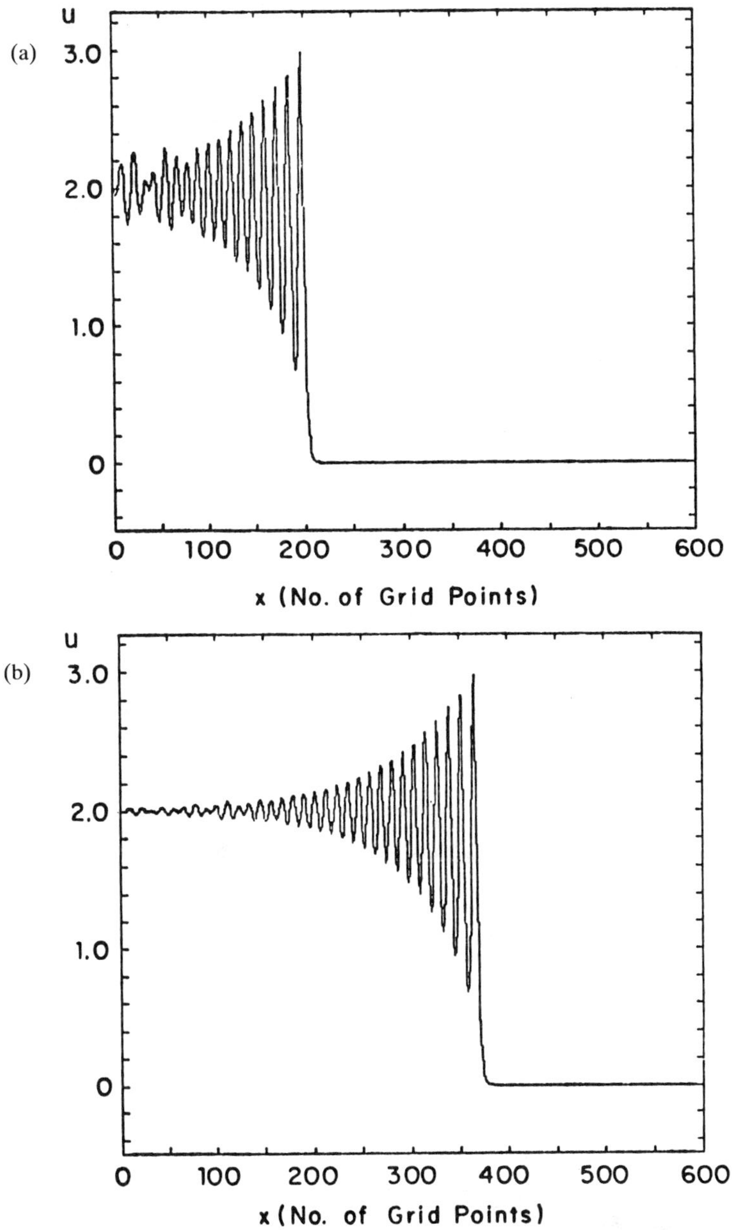

FIG. 6. "Collisionless" shock from KdV–Burgers equation ($\varepsilon = 0.022^2$, $\nu = 2\varepsilon$). (a) Partially formed $t = 2.4$ (189 steps). (b) Fully formed $t = 4.4$ (34.9 steps). (Chu *et al.*, 1983. Copyright © 1983 by John Wiley & Sons, Inc. Reprinted by permission of John Wiley & Sons, Inc.).

from the Hugonot relation and as measured from the computed results), in contrast to the speed of the leading soliton in Figure 1, which is $2U/3$.

7. In Chou (Chou, 1987 and Chou and Chu, 1988), she extended the class of initial conditions to $u(x, 0) = u_0$, a constant, from the simple case $u(x, 0) = 0$ of Figure 1. Rather surprisingly, waves of the form of Figure 7 now appear. The nonzero initial state $u_0 > 0$ has the effect of increasing the soliton maturing time, so that now there are many solitons awaiting maturation, and only the few near the front have attained full maturity. In the case of Figure 1, we recall, each soliton matures almost fully before the next one is born. This wave pattern of solitons growing is not to be confused with the collisionless shocks of the previous paragraph, as the shapes are completely different, and the former is an elongating wave train while the latter is a steady structure.

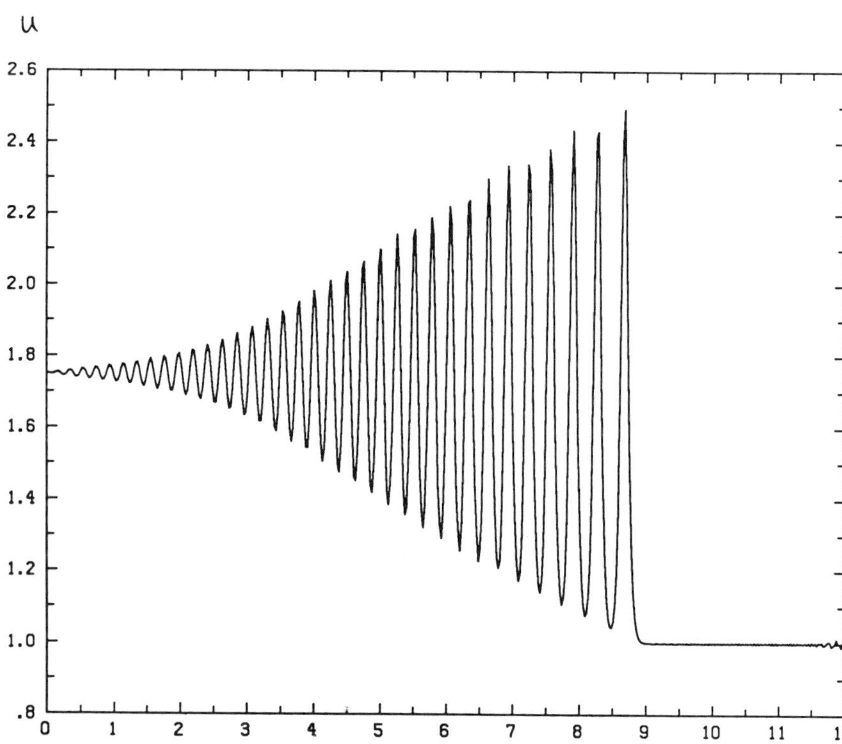

FIG. 7. KdV solitons with $u(x, 0) = u_0 = 1$, $u(0, t) = 1.75$, showing much slower maturing process (Chou, 1987 and Chou and Chu, 1988).

This situation can be seen more clearly if we make a transformation. Thus, let

$$v = u - u_0,$$

$$y = x - u_0 t.$$

The original problem with $u(x, 0) = u_0$ and $u(0, t) = UH(t)$ then transforms into

$$v_t + vv_y + \varepsilon v_{yyy} = 0$$

with

$$v(x, 0) = 0, \qquad v = U - u_0 \quad \text{on} \quad y = -u_0 t.$$

Thus, the problem is equivalent to the original problem with $u = U$ given not on a fixed wall $x = 0$, but on a receding wall with velocity u_0. The later solitons will take a longer time to catch up with the earlier ones, and thus the wave train will have the shape of Figure 7.

8. The equations used throughout these studies are the KdV equation in its ordinary form (3.1), or the Boussinesq system (3.3). The regularized form, where u_{xxx} is replaced by u_{xxt}, was not used. The motivation for not regularizing is that the authors wanted to see the explicit behavior of the KdV equation in the quarterplane, and hoped to benefit from and extend the vast theoretical literature on inverse scattering. Regularization is necessary if one is particularly interested in the behavior of short waves, which is not the aim of these authors. The numerical schemes used were the same as given by Zabusky and Kruskal, i.e., leap frog for the convective term and 5-point centered difference for the third derivative terms; in Chu (1983), the leap-frog scheme used is the standard second-order, while in Chou (1987, 1988), both second-order and fourth-order schemes were used. A matter of concern is that when the wave lengths are short, "instability" would occur. Figure 8, taken from Chou *et al.* (1988), however, insures that this does not occur. Shown in Figure 8 are the dispersion relations of the KdV equation, the regularized KdV equation, and the numerical schemes using second- and fourth-order leap-frog schemes for the convective operator. These equations are first linearized with respect to a constant $u = a \neq 0$. We remark that a is not the velocity at infinity, which would be zero, but the local fluid velocity u; the dispersion relation, as always, is local. The KdV equation and the regularized KdV equation differ in the dispersive terms: εu_{xxx} and $-(\varepsilon/a)u_{xxt}$ respectively, and the dispersion relations are found

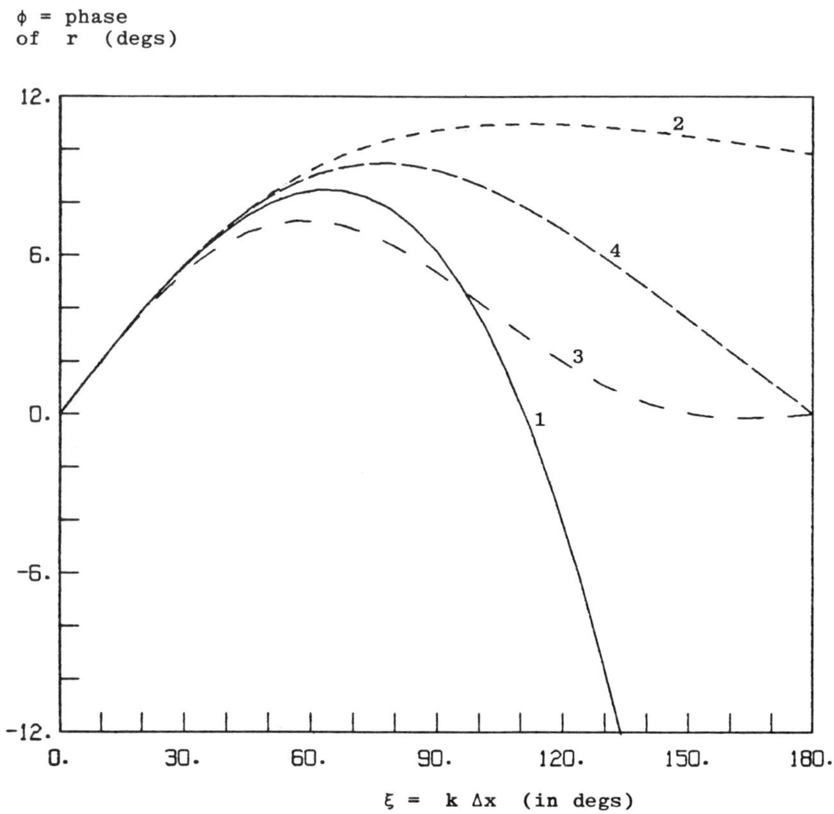

FIG. 8. Dispersion relation for (linearized) KdV equation (Curve 1), regularized KdV equation (Curve 2), and leap-frog schemes (second order—Curve 3, fourth order—Curve 4). Abscissa is wave number k multiplied by grid size Δx. Ordinate is phase ϕ of amplification factor ($= \omega k \, \Delta t$, the frequency multiplied by k and the time step). (Chou *et al.*, 1988.)

in the standard manner. For the numerical schemes, the dispersion relations are found from the phase of the amplification factor, a complex number, obtained from the usual von Neumann analysis. In contrast to the exact KdV equation, which has negative ω for k large, these numerical approximations to the KdV equation have ω always nonnegative, thus having a regularizing effect on the equation. In any case, the equation is not intended to be applied at small wave lengths, and this regularizing effect protects against any short wave length instabilities.

B. Nonlinear Periodic Waves

Unknown to the authors of Chu *et al.* (1983), the KdV equation in the quarterplane had been considered in great detail by Bona and coauthors. A theory paper, Bona and Winther (1983), on the proper posing of the problem will be discussed later, and an earlier paper by Bona, Pritchard, and Scott (1981) solves the KdV equation numerically in the quarter-plane. The authors were interested mainly in studying nonlinear periodic waves and in obtaining the accurate dissipation and dispersion coefficients to match experimental results, and not in soliton generation. Nevertheless, the problems are closely related, and their numerical scheme is much more sophisticated, so that describing their work is in order.

The authors used the regularized KdV equation with dissipation included, thus

$$\eta_t + (1 + \tfrac{3}{2}\eta)\eta_x - \mu\eta_{xx} - \tfrac{1}{6}\eta_{xxt} = 0. \tag{3.5}$$

The regularizing, as mentioned earlier, kills off short-wave instabilities. In contrast to the previously described works, here this regularization is desirable, as the more sophisticated numerical scheme may preserve the nature of the dispersion relation to large values of k. The initial value of $\eta(x, 0)$ is taken as 0, while the boundary value $\eta(0, t)$ is taken to be a given function $h(t)$.

The problems studied by Bona *et al.* (1981) are for initial conditions and boundary conditions: $\eta(x, 0) = 0$, $\eta(0, t) =$ small-amplitude sine function of t or more general periodic function of t. If we transform Equation (3.5) into (3.4), or into (3.1) with μ set equal to zero, we have $\varepsilon = \frac{1}{6}$, while the initial condition corresponds to $u(x, 0) = 1$, and the boundary condition corresponds to a periodic oscillation about $u = 1$.

The numerical scheme is based on an integral equation equivalent to (3.5), which is also used for proving existence and uniqueness of the solution:

$$\begin{aligned}\eta_t(x, t) = h'(t)\mathrm{e}^{-\sqrt{6}x} &+ \int_0^\infty K(x, y)(\eta + \tfrac{3}{4}\eta^2)(y, t)\,\mathrm{d}t \\ &+ 6\mu[\eta(t)\mathrm{e}^{-\sqrt{6}x} - \eta(x, t)] - \mu\int_0^\infty H(x, y)\eta(y, t)\,\mathrm{d}t,\end{aligned} \tag{3.6}$$

where

$$K(x, y) = 3[\mathrm{e}^{-\sqrt{6}(x+y)} + \mathrm{sgn}(x - y)\mathrm{e}^{-\sqrt{6}|x-y|}]$$

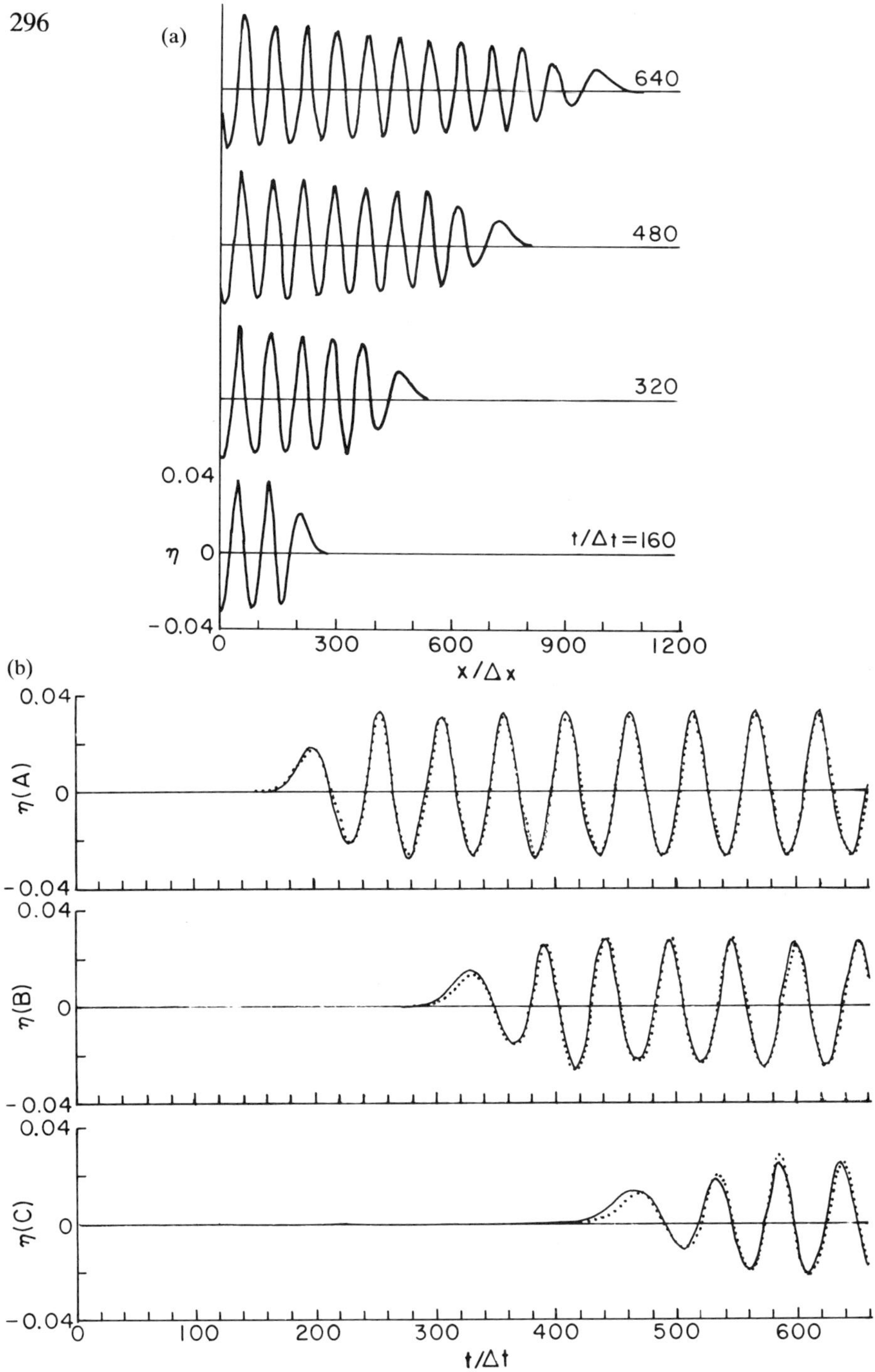

FIG. 9. (a) Waves computed from KdV–Burgers equation at different instants. (b) Time histories of computed waves compared with experiments at three different locations. (Bona *et al.* 1981.)

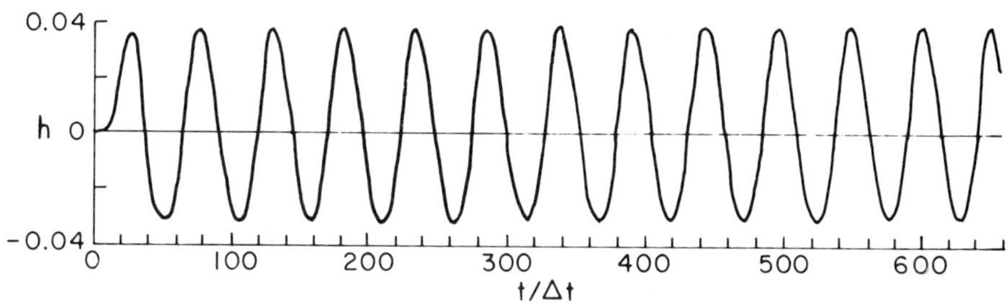

FIG. 10. Typical boundary data used to generate the waves of Figure 9 (Bona *et al.*, 1981).

and

$$H(x, y) = 3\sqrt{6}\,[e^{-\sqrt{6}(x+y)} - e^{-\sqrt{6}|x-y|}.$$

The numerical scheme uses a quadrature in x for the integrals on the right hand side, and uses a fourth-order prediction–correction scheme in t for the time integration. The method is considerably more accurate than the scheme used in Chu *et al.* (1983). To test convergence, it was used on a single soliton generated by boundary excitation, which compared favorably with the exact solution.

The periodic waves were compared with experiment to determine the correct parameter of μ for water waves. Typical results are given in Figure 9, where Figure 9a shows the shape of the excited waves at a fixed time (for $\mu = 0.014$), and Figure 9b shows the time history of the wave height at a fixed location (for $\mu = 0.00168$, and other coefficients all altered), compared with experimental observations. A typical boundary condition used for these studies is shown in Figure 10.

Based on these cited results, we can summarize boundary generated waves to the KdV equation (3.1) and the KdV–Burgers equation (3.4) (plus variants) in Table 1. Essentially, all the wave patterns that could be generated by these two equations have been covered, and a reasonably consistent picture, though still qualitative, has emerged from these works seen together.

IV. Theory

There is a small amount of theoretical work for the KdV equation, and somewhat more for the nonlinear Schrödinger equation, in the quarterplane. No such work has appeared in the literature on the Boussinesq equation.

TABLE 1

WAVE PATTERNS OF KdV AND KdV–BURGERS EQUATION

Initial state	Boundary condition	Wave patterns
No dissipation ($\nu = 0$):		
$u = 0$	$u =$ constant	Continuous stream of identical solitons, Figure 1
	$u =$ random	Similar to above, Figure 4
	$u =$ periodic and >0	Stream of solitons, modulated, Chou (1987)
$u = u_0 > 0$	$u =$ constant	Continuous stream of solitons maturing slowly, Figure 7
	$u =$ periodic and >0	Figure 9a and Bona *et al.* (1981)*
Dissipation present ($\nu > 0$):		
$u = 0$	$u =$ constant	Collisionless shock, Figure 4
$u = u_0 > 0$	$u =$ periodic and >0	Figure 9a and Bona *et al.* (1981)*

* Recall these results are for the perturbed level η. The total height $u = 1 + \eta$ is always positive.

The theoretical results that have appeared in the literature so far can be divided into two main groups. The first is the study of the existence and uniqueness of the solution to the KdV equation in the quarterplane, i.e., the well-posed nature of the problem. The second consists of the attempts at extending the inverse scattering transform method from pure initial value problems to those in the quarterplane.

A. WELL-POSED PROBLEM

The problem of existence and uniqueness has been addressed in detail by Bona and his coworkers (Bona and Winther, 1983). They showed that prescribing initial conditions $u(x, 0)$ and a single boundary condition $u(0, t)$ with suitable smoothness and compatibility conditions results in a unique solution of the KdV equation in the quarterplane, with appropriate smoothness. With minimal assumptions on smoothness, the existence of a unique distribution solution is proved; with more stringent smoothness assumptions, the unique existence of a classical solution is proved. The method of proof uses the integral equation described earlier, from which *a priori* estimates were obtained. The smoothness requirements and the detailed proofs are technical, and we refer the interested readers to their original papers.

B. Inverse Scattering

Extension of the inverse scattering method to the quarterplane is a direction of research still in progress, and it is being studied by many authors. There are two different paths of attack in this case. The first, started by Kaup and Hansen (1986) for the nonlinear Schrödinger equation and adopted by Camassa and Wu (1988) for the KdV equation, is intended to model the numerical results, and necessitates some ad hoc assumptions. The other, pursued by Fokas, Ablowitz, and colleagues (Fokas, 1987 and Fokas and Ablowitz, 1988) uses Fourier sine transforms to extend the initial data on the semi-infinite line to the entire line. The procedure is thus self-consistent and rigorous, but up to the present, results have only been obtained for the nonlinear Schrödinger equation, and it is by no means clear how to extend the method for the KdV equation.

We first summarize the inverse scattering transform method for pure initial value problems. Consider the eigenvalue problem for the steady state Schrödinger equation

$$\psi_{xx} - u\psi = -\lambda\psi. \tag{4.1}$$

The scattering problem is, given $u(x)$, to determine the discrete eigenvalues λ and the eigenfunctions ψ, assuming square integral eigenfunctions entering from say, $+\infty$. With this assumption, the eigenvalues and eigenfunctions can be represented as

$$\kappa = \kappa_1, \kappa_2, \ldots, \kappa_n, \qquad \psi = \psi_1, \psi_2, \ldots, \psi_n,$$

$$\lambda = -\kappa^2,$$

$$\int_{-\infty}^{\infty} \psi^2(x)\,\mathrm{d}x = 1 \qquad \text{and} \qquad c(\kappa) = \lim_{x\to\infty} \mathrm{e}^{\kappa x}\psi(x),$$

For the continuous spectrum, $\lambda > 0$,

$$\psi(x) = \begin{cases} \mathrm{e}^{-\mathrm{i}\kappa x} + b(\kappa)\mathrm{e}^{-i\kappa x} & \text{as} \quad x \to \infty, \\ a(\kappa)\mathrm{e}^{-i\kappa x} & \text{as} \quad x \to -\infty. \end{cases}$$

For the continuous spectrum, $a(\kappa)$ and $b(\kappa)$ are the transmission and reflection coefficients respectively; for the discrete eigenvalues $\lambda = -\kappa^2$, $c(\kappa)$ are the normalizing coefficients. The inverse scattering problem is to give these coefficients and determine $u(x)$, using the well-known Gelfand–Levitan or Marchenko integral equation. These facts and procedures are standard in quantum mechanics.

The remarkable part of the inverse scattering transform method is the theorem (Gardner *et al.*, 1967) that if u is regarded as a function of x and a parameter t in equation (4.1), and u satisfies the KdV equation (3.1) with uu_x replaced by $-6uu_x$, with x and t now regarded as independent variables, then the λ's are independent of t, the a's are independent of t, while the b's and c's evolve in t in a simple manner. This exciting result permits the exact solution of the initial value problem of the KdV equation using the following procedure: From the initial data $u(x, 0)$, we calculate by *direct scattering* the various values of λ, a, b, and c at $t = 0$. Then these values are known at all t. Hence by *inverse scattering* (now considering t as a parameter), we get $u(x, t)$ for all t.

In Kaup and Hansen (1986) and in Camassa and Wu (1988), the different authors attempted to extend this procedure to the quarterplane problem. We reinterpret their procedures in a simple manner. Two linearly independent eigenfunctions of the steady-state Schrödinger equation (4.1) can be given as $\psi(x)$ and $\bar{\psi}(x)$, which behave as

$$\psi(x) \sim \mathrm{e}^{i\kappa x} \quad \text{and} \quad \bar{\psi}(x) \sim \mathrm{e}^{-i\kappa x} \quad \text{as} \quad x \to \infty.$$

Two other linearly independent eigenfunctions can be given as $\phi(x)$ and $\bar{\phi}(x)$, which behave as

$$\phi(x) \sim \mathrm{e}^{i\kappa x} \quad \text{and} \quad \bar{\phi}(x) \sim \mathrm{e}^{-i\kappa x} \quad \text{as} \quad x \to - >.$$

From the definition of the scattering coefficients a and b, we have

$$a\phi(x) = \psi(x) - b\bar{\psi}(x). \tag{4.2}$$

Moreover, ψ satisfies the integral equation

$$\psi(x) = \mathrm{e}^{i\kappa x} + \int_x^\infty \frac{\sin \kappa(x - \xi)}{\kappa} u(\xi; t)\psi(\xi)\, \mathrm{d}x. \tag{4.3}$$

In addition, $\bar{\psi}$ satisfies the same equation with $e^{-i\kappa x}$ for the first term, while ϕ and $\bar{\phi}$ satisfy similar equations except with the integral extended from $-\infty$ to x.

The authors assume that the potential $u(x; t)$, which satisfies the KdV equation, to be $u(x;t)=0$ for all $x<0$ and all t. Then, the integral equations satisfied by ϕ and $\bar{\phi}$ all have integrals extended from 0 to x. From (4.2) and (4.3), one easily deduces that $\psi(0,t)$ is a linear combination of the scattering coefficients a and b. The t-evolution of a and b thus depends on the evolution of $\psi(0,t)$.

The time evolution of ψ cannot be described by a simple equation for ψ alone, and is best seen in the form given by Zakharaov and Shabat, cf. Ablowitz *et al.* (1974). Let $v_1=\psi$, and let $v_2=i\kappa\psi-\psi x$. Then the eigenvalue equation (4.1) can be written for the column vector $\boldsymbol{v}=(v_1,v_2)$ as a 2×2 system, and the time evolution equations can now be written as a 2×2 system. A compact integral equation now replaces the many equations corresponding to (4.3). From the time evolution equations, the evolution for $\psi(0,t)$ can be determined, and hence the evolution of the scattering coefficients are also found. Nevertheless, the evolution equation for $\psi(0,t)$ requires the derivatives $u_x(0,t)$ and $u_{xx}(0,t)$ for the KdV equation, and the derivative $u_x(0,t)$ for the nonlinear Schrödinger equation. These are not prescribable data for the initial boundary value problems in question. The authors thus make the ad hoc assumptions of setting them to zero, and obtained results that are somewhat close to the numerical results. Hence, they describe the procedure as a model and not yet a theory as such

Fokas (1987) used the Fourier sine transform (in x) as a starting point, and he did succeed in obtaining a theory for the nonlinear Schrödinger equation. The sine transform, however, works for even derivatives in x, but not for odd derivatives. Hence the method cannot be extended directly to the KdV equation, and future work is still needed.

V. Conclusion

We have summarized scattered work on boundary generation of solitons. Various authors doing numerical experiments have expressed the hope that theory would develop, just as the inverse scattering transform method blossomed following the early numerical experiments of Zabusky and Kruskal on initial value problems. It is gratifying to see that some beginnings in theoretical work have taken place, but considerable work is needed before mixed initial boundary problems reach the same degree of understanding as pure initial value problems with the inverse scattering transform.

The works described here merely scratch the skin of a vast field of interesting problems. First, theory should also develop for the forced KdV equation, relevant for ship generated solitons, for example. For linear equations, an inhomogeneous forcing function can always be replaced by a homogeneous equation and inhomogeneous boundary data, but for nonlinear equations, this is quite different. Second, there appears to have been no work yet done on boundary generated solitons in more than one dimension, even numerically. This seems to be a very fruitful and exciting direction waiting to be pursued. The physical problems that can be modelled are numerous, and undoubtedly, many new physical phenomena will be discovered from such numerical experiments.

References

Ablowitz, M. J., Kaup, D. J., Newell, A. C., and Segur, H. (1974). The inverse scattering transform—Fourier analysis for nonlinear problems. *Stud. Appl. Math.* **53,** 249–315.

Bona, J. L., Pritchard, W. G., and Scott, L. R. (1981). An evaluation of a model equation for water waves. *Philos. Trans. Roy. Soc. London Ser. A* **302,** 457–510.

Bona, J. L., and Winther, R. (1983). The Korteweg–deVries equation posed in a quarterplane. *SIAM J. Math. Anal.* **14,** 1056–1106.

Camassa, R., and Wu, T. Y. (1988). The KdV model with boundary forcing. To appear.

Chou, R. L. (1987). Solitons induced by boundary conditions. Ph.D. dissertation, City University of New York.

Chou, R. L., and Chu, C. K. (1988). Solitons induced by boundary conditions from Boussinesq equation. *Phys. Fluids.,* submitted.

Chu, C. K., Xiang, L. W., and Baransky, Y. (1983). Solitary waves induced by boundary action. *Comm. Pure Appl. Math.* **36,** 495–504.

Fokes, T. (1987). An initial-boundary value problem for the nonlinear Schrodinger equation. Clarkson Univ. Rep. INS *81. (Accepted *Phys. D.*)

Fokas, T., and Ablowitz, M. J. (1988). Forced nonlinear evolution equations and the inverse scattering transform. Clarkson Univ. Rep. INS *99.

Gardner, C. S., Greene, J. M., Kruskal, M. D., and Miura, R. M. (1967). A method for solving the Korteweg–deVries equation. *Phys. Rev. Lett.* **19,** 1095–1097.

Hammack, J. L. and Segur, H. (1974). The Korteweg–deVries equation and water waves: Part 2, Comparison with experiments. *J. Fluid Mech.* **65,** 289–314.

Kaup, D. J., and Hansen, P. J. (1986). The forced nonlinear Schrödinger equation. *Phys. D* **18,** 77–84.

Maxworthy, T. (1976). Experiments on collisions between solitary waves. *J. Fluid Mech.* **76,** 177–185.

Russell, J. S. (1844). Report on waves. "*In* Reports from the 14th Meeting of the British Association for the Advancement of Science", pp. 331–390. John Murray, London.

Zabusky, N. J., and Kruskal, M. D. (1965). Interaction of solitons in a collisionless plasma and the recurrence of initial states. *Phys. Rev. Lett.* **15,** 240–243.

ADVANCES IN APPLIED MECHANICS, VOLUME 27

Physical Limnology

JÖRG IMBERGER

and

JOHN C. PATTERSON

Centre for Water Research
The University of Western Australia
Nedlands, Western Australia

I. Introduction

Our understanding of the hydrodynamics of lakes has developed rapidly over the last few years. With new, sophisticated instruments, our powers of observation are greater, giving us deeper insight into the physical principles governing the motion and mixing of water in a lake.

ISBN 0-12-002027-0

The hydrodynamics of lakes is intimately connected with the motion of a stratified fluid affected by external forcing, the instability of these motions, and the dynamics of turbulence in a stratified environment.

Most lakes become stratified as solar energy is input at the surface. Through the expansion of the water, this thermal energy is converted to a mechanical stability, which arrests motion in the lake. The stronger the thermal insolation, the stronger the stratification and, in general, the more quiescent the water body. Wind, stream inflow, and the withdrawal of water all modify the stratification. However, as explored in this review, there are mechanisms induced by the stratification that cause active horizontal water transport and that become more intense as stratification increases.

The dynamical balance within a lake may be likened to an engine; the disturbing forces work against the potential energy gradient set up by radiation. The internal dynamics of this engine are extremely inefficient. A basin scale disturbance must begin large scale basin seiching, which then degenerates into smaller scale internal waves and intrusions, ultimately used by a host of instability mechanisms to produce turbulence. This turbulence then induces a buoyancy flux upwards, raising the overall center of gravity of the lake. Up to 90% of this mechanical energy is lost to dissipation.

Lake hydrodynamics appears to have attracted research in two main areas. First, there is a large body of work in the complicated patterns of seiching in basins of arbitrary shapes, with and without rotation (Hutter, 1984, 1986; Stocker and Hutter 1986). The second focuses more on the consequences that follow these internal wave motions, particularly stress, small scale motions, and mixing processes (Fischer *et al.*, 1979; Imberger and Hamblin, 1982; Imboden *et al.*, 1983; Imberger, 1985; Imberger, 1987).

Interest in small scale mean motions and turbulence as important components of lake hydrodynamics is recent: reviews of lake hydrodynamics by Mortimer (1974) and Csanady (1982) mention almost nothing about these aspects, concentrating instead on internal and surface seiching introduced by wind stresses.

The similarities between physical oceanography and physical limnology are many, especially in the second category of work, and with this in mind, the reader is referred to the following excellent reviews: Garrett and Munk (1979), Gregg and Briscoe (1979), Caldwell (1983), Fritts (1984), Holloway (1986), Gregg (1987), Hopfinger (1987), Thorpe

(1987), and Muller and Garwood (1988), which discuss the motion of a stratified fluid and mixing due to turbulence. There also have been two recent major conferences dealing with the oceanic internal gravity waves (Muller and Pujalet, 1984) and the dynamics of the oceanic surface mixed layer (Muller and Henderson, 1987). Much of this material is relevant to an understanding of the hydrodynamics of reservoirs and lakes. To avoid duplication, this review concentrates on smaller scale motions and the resulting mixing. A description of the action of rotation has therefore been almost completely omitted since it concerns the dynamics of internal waves, and it is adequately described in Hutter (1984, 1986) and Stocker and Hutter (1986). Circulation and mixing in ice-covered waters is an extremely important subject for a large number of lakes, but this has also been omitted because of the existence of the review by Carmack (1986).

Because of the rapid development of the subject, much of the description in this review is speculative and concentrates on findings at the Center for Water Research. Data from the Wellington, Canning, and Harding Reservoirs in Western Australia are used as illustrations of processes. These data sets have been used to set the stage for the review and literature on the dynamics displayed by these data has been referenced extensively.

The review begins with a discussion of seasonal behavior and the role of the geographic location of a lake. Through a new, dimensionless parameter the dynamics of a lake in the tropics at the time of peak stratification can be compared with the dynamics of a lake in the temperate zones at the beginning and end of the stratification cycle. This section is followed by four sections on the dynamics of the surface layer (formerly called the mixed layer) beginning with a discussion of the surface fluxes, followed by the vertical dynamics of the surface layer. Upwelling, differential deepening, and differential heating and cooling are presented in separate sections as deviations from the vertical picture. Our knowledge on outflow has increased considerably since the last review by Imberger (1980) and this material is updated here. There are also a number of significant developments in our understanding of the inflow dynamics, particularly where the inflow separates from the river channel.

Mixing below the surface layer is also discussed. Because of recent reviews by Gregg (1987), Hopfinger (1987), and Thorpe (1987), individual mechanisms are not discussed, and the discussion concentrates on the underlying dynamics contributing towards mixing in the hypolimnion and the metalimnion, on the classification of turbulent activity, and on the parametrization of mixing internal to the hypolimnion and boundary mixing. These sections set the stage for the review in Section XI of modeling of lakes where stratification is the dominant force. Lastly, because of eutrophication problems in lakes throughout the world and the attention they are receiving, we have included a discussion of some recent advances in the use of bubble aerators in reservoir destratification.

In this review, the words "lake" and "reservoir" are synonomous, with "lake" defined as an impoundment that stores water from a catchment. A reservoir is a man-made lake, which differs from a natural lake in some ways. In a reservoir, the outflow can be controlled, and, because the throughflow is more rapid, the water residence time is shorter. The hydrodynamics of the two are not greatly different, unless the throughflow is so strong that it affects stratification.

II. Seasonal Behavior

A lake or reservoir is a body of water contained in a river valley, insulated by the bottom but with heat transfers at the surface. Short wave solar radiation (of wavelength 300 nm to 1000 nm) and long wave radiation (of wavelength greater than 1000 nm) emitted by clouds and atmospheric water vapor heat the water, while evaporation, sensible heat transfer, and radiation from the surface most often cool the water. The net heat input from these sources depends on the season and the meteorological conditions at the time and site; the balance may change dramatically not only from day to day but from hour to hour.

The net daily heat flux at the surface of the Canning Reservoir, 50 km southeast of Perth, Western Australia, is shown in Figure 1. The reservoir is at a latitude of 32.5°S with a Mediterranean climate of cool, wet winters from June to August, and hot, dry summers from December to March. The sensible and latent heat fluxes used in Figure 1 are calculated from meteorological data gathered 2.0 m above the water surface using bulk aerodynamic flux equations (see Section III) with coefficient values applicable to a neutral atmosphere (Tennessee Valley Authority, 1972; Fischer *et al.*, 1979).

The variance of the daily net heat flux was larger in spring and summer (September to March) than in autumn and winter (April to August). This was mainly due to variable cloud cover and higher water surface

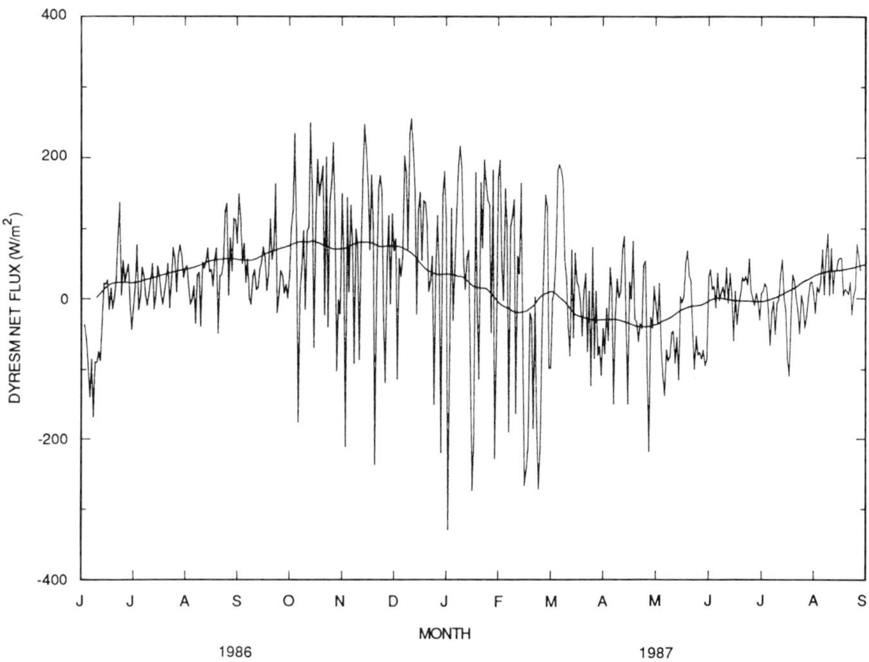

FIG. 1. Net heat flux at the surface of the Canning Reservoir. The data were collected with a meteorological station mounted on a specially designed buoy kept in position with a taut mooring. The sensors were 2.0 m above the water level at all times. The smooth line represents the averaged flux, and the tick marks represent the start of the month.

temperature, which led to large back radiation and strong evaporation on cloudy, windy summer days. Also, the smoothed net heat flux was negative only for a brief period in autumn (March to May). During winter, the net flux was approximately zero. In spring and summer, the smoothed net heat flux was positive with a mean value of approximately 80 W m^{-2}. The total net heat exchange for the year was positive, but the average lake temperature did not increase between June 1986 and June 1987.

This implies that the sum of the heat flux due to the river inflow, which tends to be cold, and the loss of heat due to the outflow, which is somewhat warmer, was comparable in magnitude to the fluxes through the surface. Inflow and outflow patterns, which differentiate lakes from reservoirs, must therefore have a significant influence on the heat budget of a lake (see also Myrup *et al.* (1979) for a detailed heat budget of Lake Tahoe).

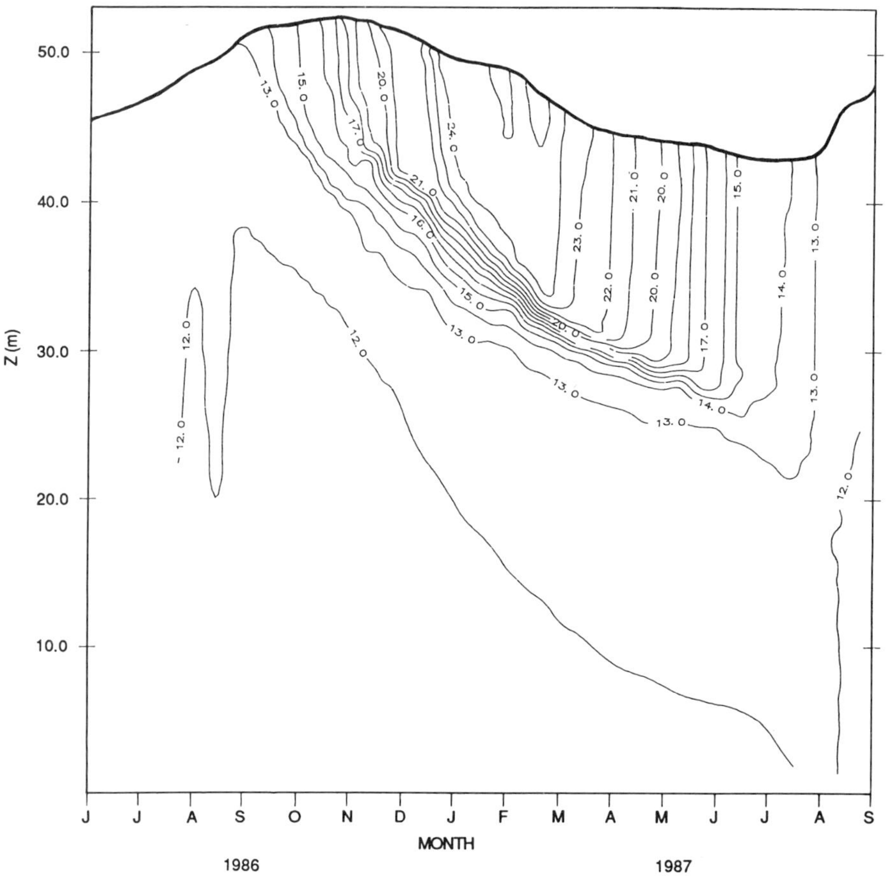

FIG. 2. The thermal characteristics of the water in the Canning Reservoir corresponding to the net heat flux shown in Figure 1. Drops were collected at fortnightly intervals throughout the 15 months.

The Canning Reservoir thermal characteristics corresponding to the period described in Figure 1, are shown in Figure 2. During winter, cold weather mixed the lake over the whole depth, leading to an isothermal water body of approximately 12°C. The inflow during the winter had much the same temperature, as seen in Figure 2, and the temperature at the bottom of the lake remained at 12°C. Just before the minimum temperature was reached at the bottom of the lake, the surface layers began to develop a weak thermal gradient in direct response to the positive net heat flux (see Figure 1). This heating intensified until, at the end of December, the temperature at the surface had reached 24°C over a depth of nearly 10 m. The stratification during this period appeared to be unable to suppress the turbulence in the deep part of the lake between 20 and 30 m depth, where the temperature rose from a low just above 11°C in September to 12°C in December. After December, the warmer surface temperatures increased the back radiation, the evaporation, and the sensible heat loss, and the surface temperatures rose only slightly, reaching a maximum of 25°C in late February.

After this time, penetrative convection, introduced by a net surface cooling, led to a characteristic decrease of the surface water temperature of the top 15 m, which continued from late February to June. Also noticeable was the gradual deepening of the isotherms at the base of the surface layer, induced by the erosion at the base by both natural convection and wind-induced turbulence. Surface cooling has two effects. First, the kinetic energy of the penetrative convection thermals erodes the stable structure underlying the surface layer. Second, the cooling decreases the stability (the density difference between the surface layer and the underlying water), thus making the base more susceptible to erosion by turbulence from within the surface layer. By June, the lake again had become isothermal.

In limnology textbooks (Hutchinson, 1957; Wetzel, 1975; Henderson-Sellers, 1984b), the upper warm layer is called the epilimnion, the strong temperature gradient region below is called the metalimnion, and the colder water in the lowest part of the lake the hypolimnion. These terms are useful for description, but as Imberger (1985) pointed out, the understanding of the properties of water behavior in these regions has recently been revised.

As the following sections will show, the surface layer is not as well mixed as is generally believed, nor is it in a state of uniform and constant turbulence. The surface layer, or epilimnion, responds to diurnal changes in surface fluxes. The metalimnion contains the major temperature changes between the surface and bottom water. It is made up of many temperature steps, and it is unrealistic to define the thermocline as that depth at which the temperature gradient is maximum, as has been suggested by previous authors (see Wetzel (1975) and Straskraba (1980)). Typical profiles from the data set of Figure 2 are shown in Figure 3. The temperature gradient is a strong function of the resolution of the instrument. For the smoothed profile with a resolution roughly matching the vertical extent of the interface (Figure 3b), the gradient reaches only

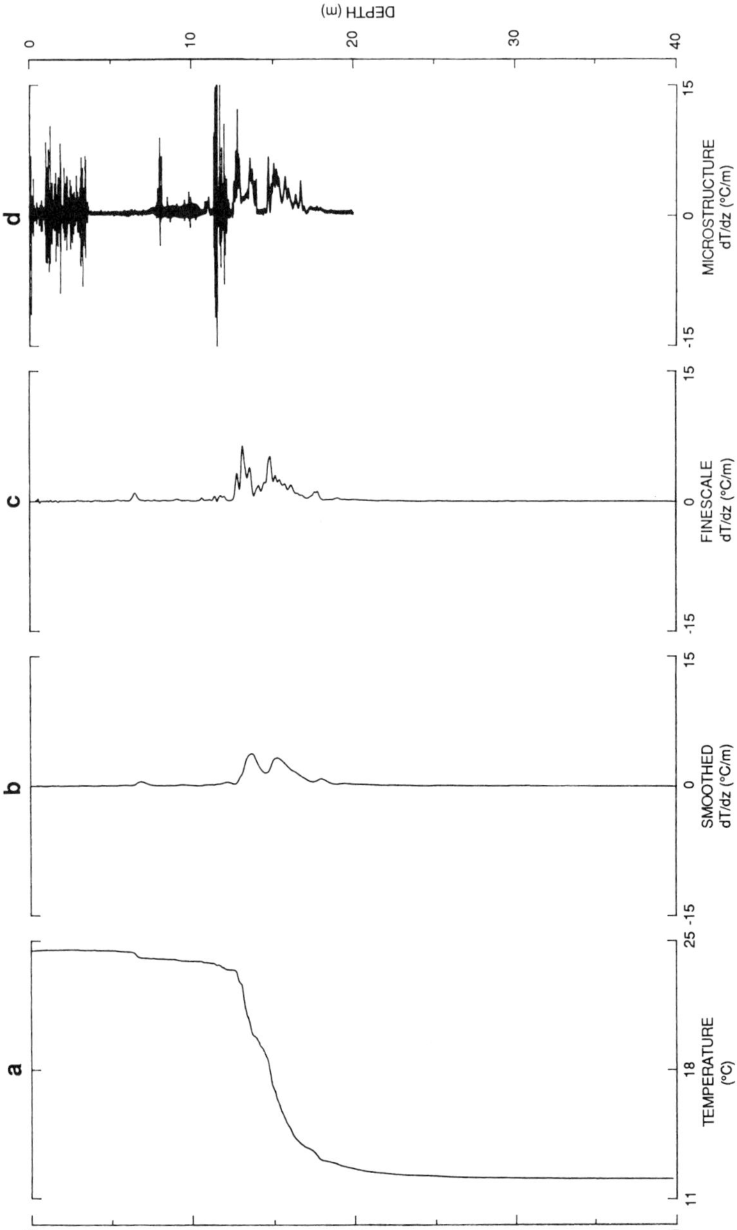

FIG. 3. Typical profile data obtained in the Canning Reservoir on 26 January 1987 at a central location. (a) Temperature obtained with a standard conductivity, temperature, depth profiling instrument. (b) Smoothed temperature gradient. (c) Enhanced temperature gradient with an effective resolution of 0.02 m. (d) Temperature gradient obtained with a microstructure profiler which had a resolution of 1 mm. This profile was taken a little earlier than that shown in (a), (b), and (c) and only extended to a depth of 20 m.

3°C per meter, whereas at the finest resolution obtained with a temperature microstructure instrument (which resolved the profile to 1 mm intervals) a maximum temperature gradient of 15°C per meter was observed (Figure 3d). The coarse definitions find their origin in the early measurements of Wedderburn (1906, 1911, 1912) where temperatures were measured at every one or two meters.

The metalimnion is also poorly defined, for example, given the data in Figure 3a, it is not clear whether the temperature step at 6 m depth should be included. For this reason Imberger (1985) and Imberger (1987) introduced the classification of a diurnal surface layer, a parent thermocline, a metalimnion, and hypolimnion. The parent thermocline is the temperature step caused by the most severe mixing or deepening event in the immediate past. The metalimnion is the water body from the parent thermocline to a depth where the smoothed temperature has reached a certain percentage of the coldest temperature in the hypolimnion.

The parent thermocline is illustrated in Figure 4 with data collected during a hot, cloud-free day at the Wellington Reservoir. At 0830 hours, there is a clear surface layer, uniform in temperature from the surface to approximately 7.0 m. This layer was mixing (Monismith and Imberger, 1988) by penetrative convection; the parent thermocline is the gradient region found at 6.5 m. Following this, the weather became extremely hot (temperature at 1200 hours was 42°C), there was no wind, and the water surface was glassy all day. By 1500 hours (Figure 4), considerable surface heating had formed a diurnal thermocline to a depth of approximately 1 m, with a temperature rise in excess of 4°C. By 1700 hours a brisk wind of 6 m s^{-1} caused the surface to mix (Figure 4), causing the diurnal thermocline to deepen toward the parent thermocline. By definition, once the diurnal thermocline has reached its maximum depth, it becomes the new parent thermocline. In some cases, this may be deeper than the previous parent thermocline (see Imberger (1987), Schertzer *et al.* (1987), and Strub (1983)). Generally during spring, lakes stratify and the location of the parent thermocline becomes shallower with time; in autumn, during the lake's cooling phase, the parent thermocline lies progressively deeper.

The temperature distribution of a lake depends on the cumulative effect of the surface fluxes, the inflows, the outflows, and the rate of vertical mixing. There has, however, been a great effort made (see Straskraba (1980)) to provide empirical formulae to predict the thermal characteristics of a lake according to its location. Three parameters are

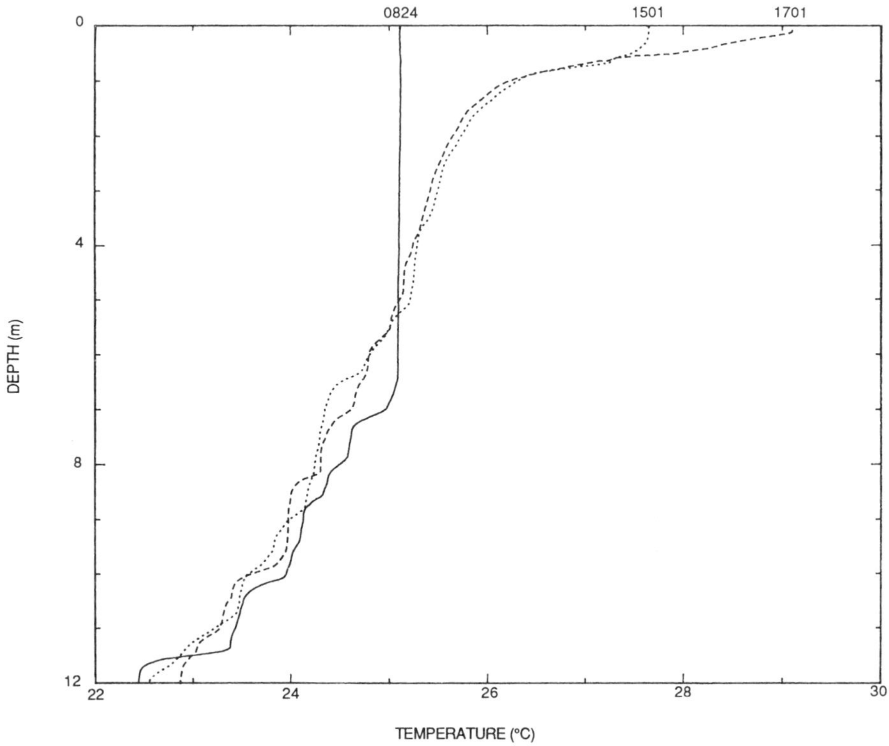

FIG. 4. Evolution of a diurnal surface layer. Data taken in the Wellington Reservoir on 25 February 1985. Solid line: data collected at 0824 hours during calm conditions. Dashed line: data collected at 1501 hours at time of maximum surface temperature gradient. Dotted line: data collected at 1701 hours; cooling had commenced.

usually characterized: the surface temperature, the bottom temperature, and the depth of the seasonal thermocline. Given that the solar incident radiation is determined, at a particular time, by the latitude of the lake ϕ, Straskraba (1980) suggested the following relationship for the surface temperature:

$$\theta_s = 28.1 - 0.34\,\phi' + (0.54 - 0.045\,\phi' + 0.0146\,\phi'^2 - 1.97 \times 10^4\,\phi'^3)\sin(\tau + \gamma), \tag{2.1}$$

where

$$\phi' = \phi - 3.4,$$

$$\gamma = \begin{cases} 240 & \text{for } \phi > 0 \text{ (northern hemisphere)}, \\ 60 & \text{for } \phi < 0 \text{ (southern hemisphere)}, \end{cases}$$

and τ is the Julian day.

This formula was derived by curve fitting to the measured surface temperatures of over 50 lakes, both in the southern and northern hemispheres. Straskraba (1980) mentions that the expected temperatures may be higher for shallower lakes and lower for deeper lakes than that predicted by (2.1). Equation (2.1) was tested for two lakes in Western Australia: Lake Argyle is very large, with a volume of $5.6 \times 10^9\,\text{m}^3$ and a depth of 45 m, located at a latitude of 16.5°S; Wellington Reservoir (Patterson *et al.*, 1984) has a volume of $186 \times 10^6\,\text{m}^2$, a crest level depth of 30 m, and is located at 33.5°S. Data from these lakes are shown in Figure 5. The fit for the Wellington Reservoir is reasonable, but for Lake Argyle the departure both in the mean annual surface temperature and the annual variation is severe, so such empirical generalizations must be treated with caution. However, (2.1) illustrates the general trend with latitude: tropical lakes are warmer and have smaller annual variability, while lakes at higher latitudes are colder and have larger annual variations.

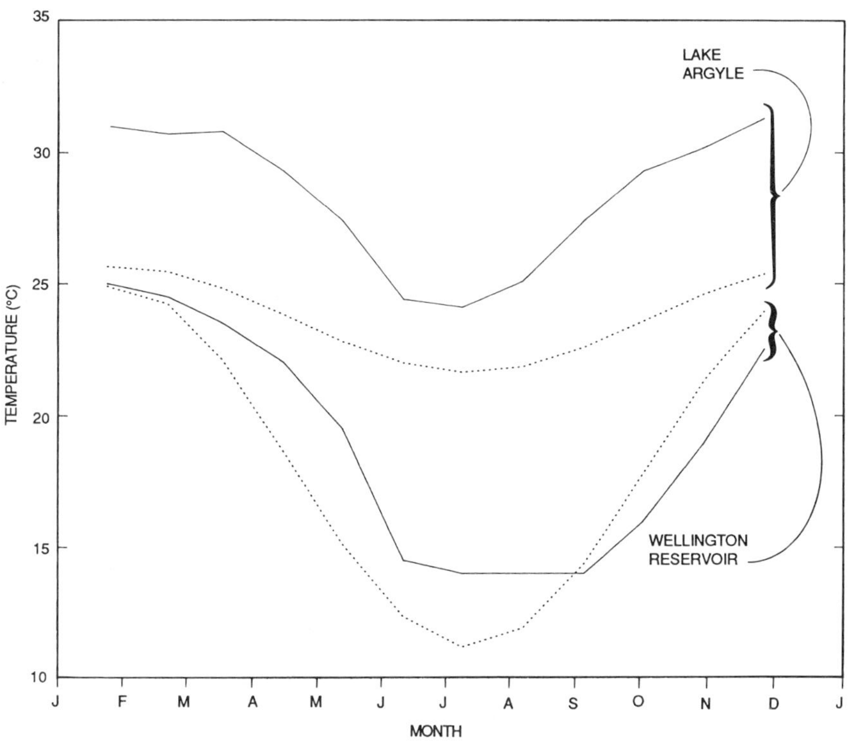

FIG. 5. Mean surface water temperature for the Wellington Reservoir and Lake Argyle. Solid line: data. Dashed line: empirical relationship (2.1) suggested by Straskraba (1980).

The temperature of water at the bottom of a lake is determined either by the temperature of the coldest inflow (Hebbert *et al.*, 1979) or the temperature at the time of lake overturn (Robarts *et al.*, 1982). Straskraba (1980) assumes that the bottom temperature remains approximately constant, which is true for deep lakes such as Lake Tahoe (Henderson-Sellers, 1984b), but is not true for lakes with active mixing. Even deep lakes such as Lake Baldegg (maximum depth 70 m) show definite signs of warming during the summer months (Imboden *et al.*, 1983).

Ignoring the warming due to mixing, Straskraba (1980) suggested that the bottom temperatures are given by

$$\theta_B = \begin{cases} 28.1 - 0.6\,\phi & \text{for } \phi \le 40°, \\ 4 & \text{for } \phi > 40°. \end{cases} \tag{2.2}$$

This does not encompass lakes at extreme latitudes such as Lake Erie (Schertzer *et al.*, 1987), Lake Ontario (Boyce *et al.*, 1983), or Lake Babine (Farmer and Carmack, 1982) where temperatures fall below 4°C in winter. The coldest temperatures given by (2.1) should match the temperatures given by (2.2), since during the coldest period, most lakes are isothermal. Again, agreement between (2.1) and (2.2) is only marginal.

Such empirical generalizations ignore many factors that have a profound influence on the temperature distribution and heat budget. As mentioned above, inflows and outflows often contribute significantly to the thermal budget, and this is illustrated in Figure 6 with data from Patterson *et al.* (1984). In 1976, water was withdrawn from Wellington Dam only from an offtake at mid-depth. In 1977, however, most of the water was withdrawn from a scour valve at the base of the dam wall. The mid-depth draw policy intensified the metalimnion, which in turn increased the stability of the lake and thus reduced mixing throughout (Fischer *et al.*, 1979). By contrast, the bottom withdrawal reduced the volume of the cold, salty bottom water, which reduced stability and allowed surface fluxes to mix the lake much earlier in the year.

These variations cannot be incorporated into a simple empirical formula, but require an estimate of stability of the lake water. Such stability measures have long been advocated. Hutchinson (1957) introduced the definition for stability:

$$S_t = \int_0^{z_m} (z - z_g) A(z) \rho(z)\, dz, \tag{2.3}$$

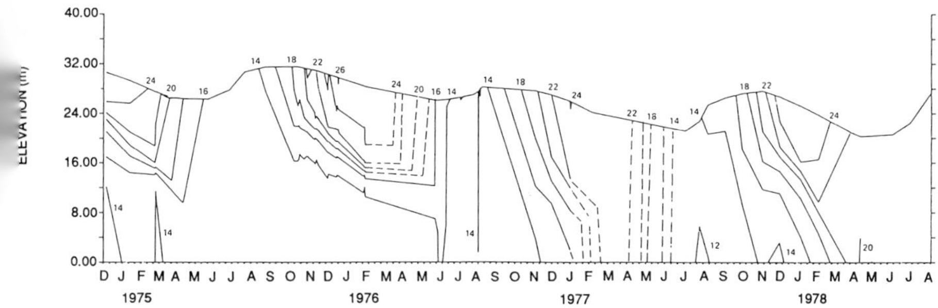

FIG. 6. Field measurements of the thermal characteristics of the Wellington Reservoir from 1975–1978. The metalimnion present in 1976 was absent in 1977 due to the bottom scour policy instituted in 1977 (Patterson *et al.*, 1984).

where z is the vertical coordinate from the bottom of the lake, $A(z)$ is the area of the lake at height z, $\rho(z)$ is the water density at a height z, z_{m} is the water depth, and z_{g} is the height to the center of volume of the lake, defined by

$$z_{\mathrm{g}} = \frac{\displaystyle\int_0^{z_{\mathrm{m}}} zA(z)\,\mathrm{d}z}{\displaystyle\int_0^{z_{\mathrm{m}}} A(z)\,\mathrm{d}z}. \tag{2.4}$$

Straskraba (1980) suggests that the depth of the center of the metalimnion and the rate of turbulent mixing depend on the magnitude of S_t. He also observed that for a particular lake basin undergoing temperature stratification given by (2.1) and (2.2), S_t should be small near the equator, rise to a maximum at middle latitudes, and again reach zero at latitudes of approximately 70°; the actual stability being dependent on the density of the water which is a highly nonlinear function of the temperature (UNESCO, 1981). Idso (1973) reintroduced this concept of stability, but multiplied S_t by gravity g and divided (2.3) by $A(z_{\mathrm{m}})$, removing the absolute size of the lake from the definition of the stability. However, even with these modifications, S_t is still dimensional. Further, the concept of stability expressed by (2.3) is deficient since it only accounts for the stabilizing influence of the stratification and not the destabilizing influence of the disturbances of wind, inflows, and outflows.

The problem can be addressed by generalizing the surface layer stability criterion first introduced by Sverdrup (1945), developed further by Hurley Octavio *et al.* (1977), Spigel and Imberger (1980), Thompson and Imberger (1980), and formalized by Imberger and Hamblin (1982) as the Wedderburn number

$$W = \frac{g'h^2}{u_*^2 L}, \tag{2.5}$$

where g' is the modified acceleration due to gravity across the base of the surface layer, h is the thickness of the surface layer, u_* is the water friction velocity due to wind stress, and L is the fetch length. Here $g' = g\,\Delta\rho/\rho_0$, where $\Delta\rho$ is the density jump across the base of the surface layer and ρ_0 is the hypolimnion density. The friction velocity u_* is defined by (3.1) below. This number is dimensionless and represents, as explained in Imberger (1985) and Imberger (1987), the ratio of the baroclinic pressure force $g'h^2$ at the point of upwelling and the surface force $u_*^2 L$ imposed by the wind stress; both forces being calculated per unit width of the lake. This number will be elaborated upon in Section IV, but a simple generalization is suggested to describe the behavior of the lake as a whole.

Consider a general lake with an arbitrary stratification $\rho(z)$ being acted upon by a general wind field with a surface friction velocity $u_*(x, y)$, where x and y are the horizontal coordinates embedded in the lake's surface. As this wind stress is imposed on the surface layer, there will be a net force acting to overturn the density structure of the water column. Taking moments about the center of volume located at z_g, we obtain for equilibrium

$$\int_{A(z_m)} \rho_0 u_*^2 \, \mathrm{d}A (z_m - z_g) = (z_g - z_0) M g \beta, \tag{2.6}$$

where z_0 is the center of gravity of the actual water mass with a density stratification $\rho(z)$, M is the total mass of water, and β is the angle subtended to the vertical by the center of mass at the center of volume. This leads to the ratio we shall call the Lake number

$$L_N = \frac{(z_g - z_0) M g \beta}{\displaystyle\int_{A(z_m)} \rho_0 u_*^2 \, \mathrm{d}A (z_m - z_g)}. \tag{2.7}$$

Now noting that z_0 is defined by

$$z_0 = \frac{\displaystyle\int_0^{z_m} \rho(z) z A(z)\, \mathrm{d}z}{\displaystyle\int_0^{z_m} \rho(z) A(z)\, \mathrm{d}z}, \tag{2.8}$$

the Lake number may be rewritten as

$$L_N = \frac{S_t g \beta}{\displaystyle\int_{A(z_m)} \rho_0 u_*^2 \, \mathrm{d}A(z_m - z_g)}. \tag{2.9}$$

Following the work of Spigel and Imberger (1980), the angle β may be fixed so that L_N is calculated at the point where upwelling commences, that is when the metalimnion intersects the surface, so that

$$\beta = \frac{z_m - z_T}{A^{1/2}(z_m)}, \tag{2.10}$$

where z_T is the height to the center of the metalimnion, and the fetch is scaled with $A^{1/2}(z_m)$, yielding the relationship

$$L_N = \frac{g S_t (1 - z_T/z_m)}{A_0^{1/2}(1 - z_g/z_m) \displaystyle\int_{A(z_m)} \rho_0 u_*^2 \, \mathrm{d}A}. \tag{2.11}$$

Assuming that the wind stress is constant over the surface, then L_N reduces to

$$L_N = \frac{g S_t (1 - z_T/z_m)}{\rho_0 u_*^2 A_0^{3/2}(1 - z_g/z_m)}, \tag{2.12}$$

where A_0 is the surface area of the lake $A(z_m)$. For large Lake numbers, the stratification will be severe and dominate the forces introduced by the surface wind stress. Under these circumstances, stratification is expected to be horizontal, with little or no seiching and little turbulent mixing in the metalimnion or the hypolimnion. Horizontality of the isotherms at large L_N is illustrated by the data set taken at Canning Reservoir on the 18 January 1987, which is depicted in Figure 7. A weak, warm lense of water resident in the sheltered upstream part of the basin and a very weak slope of $10^{-4}\,°\mathrm{C\,m^{-1}}$ towards the dam wall, attributable to a first mode seiche, are the only departures from this horizontality. As discussed in Shay and Imberger (1988), the turbulence levels in the epilimnion and the hypolimnion were extremely low. At the time, the wind mean shear velocity (averaged over 24 hours) was $0.0033\,\mathrm{m\,s^{-1}}$, leading to a value of $L_N = 86$. The full implication of L_N with respect to deep mixing will be discussed in Section X.

The interplay between the temperature distribution variation with latitude and the changing water density due to the elevated temperatures at lower latitudes may be captured effectively by taking the data displayed in Figure 2, transposing this through the various latitudes via (2.1) and (2.2), and then calculating L_N from (2.12). This was done by scaling the actual measured surface temperature by the ratio $\theta_s(t, \phi_2)/\theta_s(t, 32.5)$. The bottom temperatures were adjusted to vary linearly so that at the equator the bottom temperature equalled the surface temperature, and at 32.5°S the bottom temperature was equal to that measured in the Canning Reservoir. Here, t is the time of a particular profile and ϕ_2 is the second latitude. The scaled temperature profiles were then used to compute L_N from (2.12), yielding the two-dimensional variation of L_N shown in Figure 8. The diagram shows

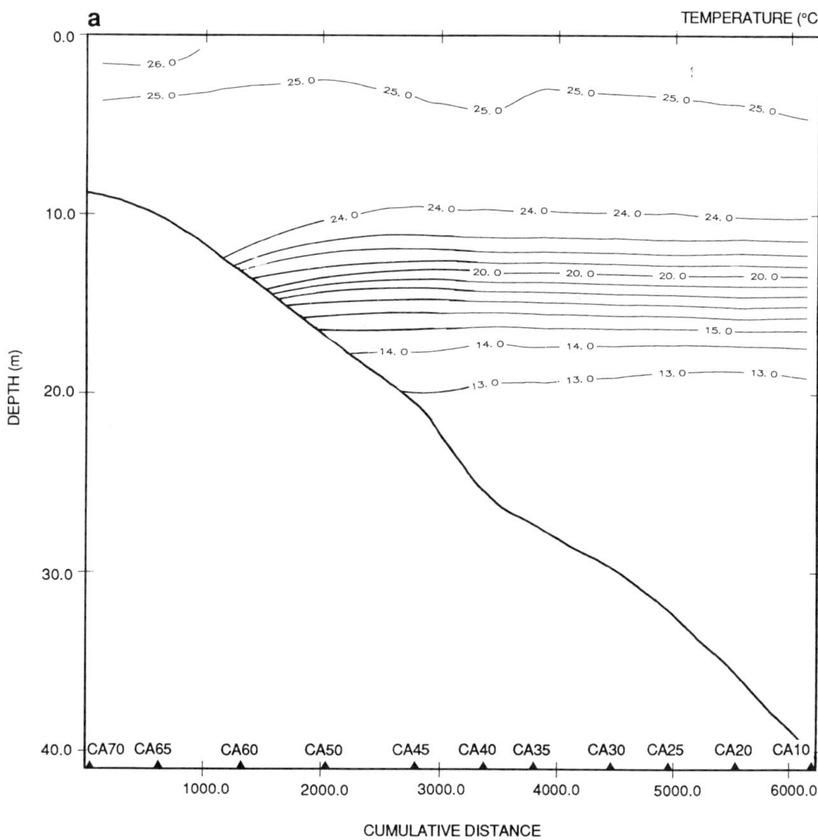

FIG. 7. (a) Longitudinal transect of isotherms in the Canning Reservoir taken on 18 January 1987. (b) Map of Canning Reservoir showing location of stations.

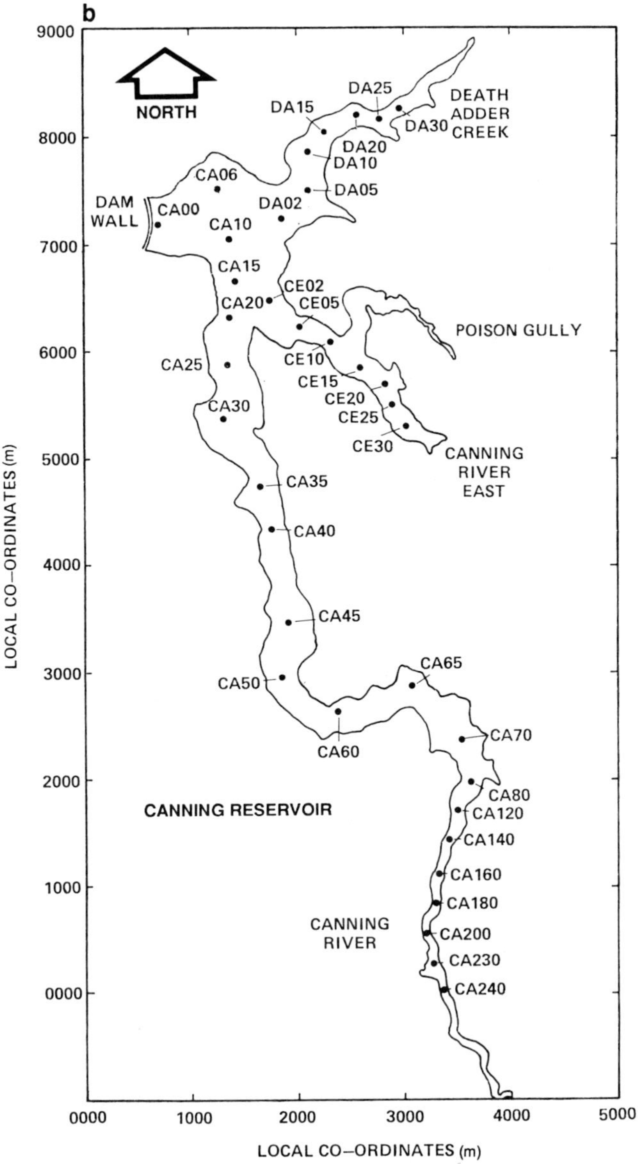
b
NORTH
DAM
WALL
CA00
CA06
CA10
CA15
CA20
CA25
CA30
CA35
CA40
CA45
CA50
CA60
CA65
CA70
CA80
CA120
CA140
CA160
CA180
CA200
CA230
CA240
DA02
DA05
DA10
DA15
DA20
DA25
DA30
DEATH
ADDER
CREEK
CE02
CE05
CE10
CE15
CE20
CE25
CE30
POISON GULLY
CANNING
RIVER
EAST
CANNING RESERVOIR
CANNING
RIVER
LOCAL CO-ORDINATES (m)
LOCAL CO-ORDINATES (m)

FIG. 7. (*contd.*)

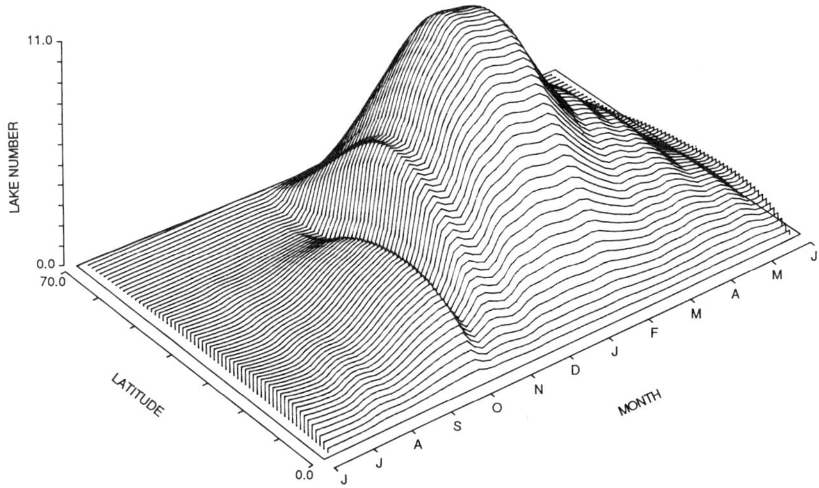

FIG. 8. An example of the variation of the Lake number as a function of season and latitude. The surface was constructed by computing the Lake number from data gathered in the Canning Reservoir, but which was transposed to different latitudes using the empirical relationship (2.1) suggested by Straskraba (1980). A constant wind speed of $8.0\,\mathrm{m\,s^{-1}}$ was applied throughout.

that the maximum stability is obtained at mid-latitude, around the summer period. Further, it illustrates that it is possible to trade latitude and season when analyzing the dynamics of a lake. This would mean that the Canning Reservoir in August, leaving aside the influence of the earth's rotation, should be dynamically similar (in the sense of dynamic similitude) to Lake Argyle in February. More generally, it may be said that lakes in the tropics behave dynamically similar to lakes at mid-latitudes at the beginning or end of the stratification cycle.

The seasonal pattern described above is often strongly influenced by chemicals dissolved in the water, which also influence the density of the water, thus contributing to the stability of the lake. The origin of these dissolved salts are threefold (Wetzel, 1975). First, in coastal lakes, sea water may enter the lake periodically. Second, ground or river salt water intrusions may be present, and third, salts may be accumulated from the decomposition of sediments and organic matter. If the dissolved salts influence the density of water enough to prevent the normal winter overturn, the lakes are called meromictic. Examples in the literature are too numerous to list here, and it suffices to mention that they range from very large lakes, such as the Dead Sea (Steinhorn and Assaf, 1980;

Steinhorn and Gat, 1983); to small shallow lakes (MacIntyre and Melack, 1982; Bunn and Edward, 1984). The Wellington Reservoir is an example (see Imberger, 1987) where a strong variability of the salinity of the inflow led in some years to meromictic behavior.

In general, the dissolved salts influence the dynamics of the lake only through the equation of state, and provided the Lake number and other parameters allow for salinity variation, their dynamics are similar to single species stratification. In the section on deep mixing, we discuss an exception to this statement, and under certain circumstances, such as in Lake Kivu (Newman, 1976), double diffusive instabilities may arise in salt-stratified lakes due to different rates of molecular diffusion between temperature and salt (see Turner (1985) for a complete review on multi-components).

Many attempts exist to empirically correlate the dynamics of a lake to variables such as the surface area (Ward, 1977) and geographical trends such as altitude (Hutchinson, 1957; Wright, 1961; Loffler, 1968; Khomskis, 1969; Lerman and Stiller, 1969; Bella, 1970; Lerman, 1971; Darbyshire and Colclough, 1972; Khomskis and Filatova, 1972; Blanton, 1973; Idso and Cole, 1973; Lewis, 1973; Sundaram and Rehm, 1973; Tzur, 1973) and length of the lake (Yoshimura, 1936; Patalas, 1960 and 1961; Arai, 1964; Ventz, 1973). These correlations are useful only in that they are special cases of the lake number L_N formulation.

III. Surface Fluxes

The interaction between the atmosphere and a lake occurs at the lake's surface. A great deal of work has been done to determine the fluxes of momentum, energy and mass across the air–water interface and many numerical formulae have been developed. These methods are reviewed by Tennessee Valley Authority (1972), Straskraba (1980), and Henderson-Sellers (1986).

Short wave radiation (300 nm to 1000 nm) is usually measured directly. Long wave radiation (greater than 1000 nm) emitted from clouds and atmospheric water vapor can be measured directly or calculated from cloud cover, air temperature, and humidity (Tennessee Valley Authority, 1972). The reflection coefficient, or albedo, of the short wave and long wave radiation varies from lake to lake and depends on the angle of the sun, the color of the water, and the surface wave state. Back radiation from the warm water surface may be calculated from the black body radiation law, it also may be measured by pointing a radiation instrument towards the water surface.

It was shown by Strub and Powell (1987), Marti and Imboden (1986), Keijman (1974), Sadhuram *et al.*, (1988), and many other authors that simple aerodynamic bulk formulae, with constant transfer coefficients, can be used to calculate the momentum, the sensible heat, and the latent heat fluxes on a seasonal basis:

$$-\frac{\tau}{\rho} = \overline{u'w'} = -u_* u_* = -C_D U_z^2, \tag{3.1}$$

$$\frac{H}{\rho C_p} = \overline{\theta' w'} = -u_* \theta_* = -C_H U_z(\theta_z - \theta_s), \tag{3.2}$$

$$\frac{E}{\rho L_v} = \overline{q'w'} = -u_* q_* = -C_W U_z(q_z - q_s), \tag{3.3}$$

where τ is the surface stress, ρ is the density of the air, u' and w' are the horizontal and vertical fluctuation of velocity, the overbar is a time average long enough to average out the short term energy but short enough to allow for synoptic variability (Busch, 1977; Smith, 1980; Geernaert *et al.*, 1987), u_* is the shear velocity in the air, C_D is the momentum or drag coefficient, U_z is the air velocity at a certain height z above the water surface, H is the sensible heat transfer, C_p is the specific heat of water, θ' is the temperature fluctuation, θ_* is the temperature scale, C_H is the heat transfer coefficient, often called the Stanton number (Geernaert *et al.*, 1987), θ_s is the water surface temperature, θ_z is the temperature of the air at the height z above the water, E is the latent heat flux, L_v is the latent heat of vaporization, q' is the specific humidity, q_* is the specific humidity scale, C_W is the latent heat transfer coefficient or Dalton number (Geernaert *et al.*, 1987), q_z is the specific humidity at a height of z above the water surface, and q_s is the specific humidity at saturation pressure at the water surface temperature.

Equations (3.1), (3.2), and (3.3) are only valid for a stationary surface. Since the water moves, U_z should be replaced by $U_z - U_s$, where U_s is the water surface mean velocity (see Businger (1973) for a full discussion), although for most wind speeds, this is a small correction.

The value of the transfer coefficients C_H and C_W for such bulk modeling has received a great deal of attention and, in the case of lakes, may be estimated by carrying out a heat and water budget over seasonal time scales (for example, Strub and Powell (1987) for Castle Lake, Marti and Imboden (1986) for Lake Sempach, Myrup *et al.* (1979) for Lake Tahoe, and Taylor and Aquise (1984) for Lake Titicaca). These references suggest a value of $C_H = C_W = 1.9 \times 10^{-3}$ (referenced to 10 m) yields an adequate description of the total stored energy over a seasonal time scale. This may be compared with a lower value of 1.45×10^{-3} suggested by Hicks (1972). Constant values are sufficient for seasonal lake modeling since the thermal budget is self-regulating; an underestimate of the heat loss will cause the water surface to heat, thus increasing heat loss, while an overestimate of heat loss will cause the water surface to cool and, by (3.2) and (3.3), this will lead to the heat loss being decreased.

At smaller time scales (from hours to days), this negative feedback mechanism no longer suffices because the thermal inertia of the water in the surface layer is too great. The influence of the air column stability and the water surface roughness must therefore be accounted for, and these may introduce considerable variability to all three transfer coefficients. Excellent reviews exist on this topic: Businger (1973), Dyer (1974), Stewart (1974), Garratt (1977), Wu (1980), Donelan (1982), and Blanc (1985). There are three main measurement techniques to determine the instantaneous surface fluxes at the water surface: eddy correlation, profile, and dissipation. These have been used to calibrate (3.1) to (3.3) thus giving values for C_D, C_H, and C_W. A description of each technique is given by Dobson *et al.* (1980) and Blanc (1983). The eddy correlation method is direct; the profile method uses mean velocity, temperature, and humidity profile data; and the dissipation method uses the spectral characteristics of the high wave number turbulence. Results from the eddy correlation technique usually agree with each other within 10% (Blanc 1987); profile techniques usually agree with eddy correlation methods to within 25% (Miyake *et al.*, 1970; Wucknitz, 1976). Large and Pond (1981) and Smith and Anderson (1984), in comparisons between the dissipation method and the eddy correlation technique, found that the two agreed to within 40%. Within similar schemes, there can be even greater variability (see Lo and McBean (1978) for interprofile comparisons).

The reliability of the bulk aerodynamic equations (3.1) to (3.3) depends on the sensors being located in the internal boundary layer of the lake. Air coming from land will meet a different roughness, humidity and temperature over the lake and the internal boundary layer will be established with new equilibrium profiles (Taylor, 1970; Venkatram, 1977; Mulhearn, 1981; Claussen, 1987; Garratt, 1987). Peterson (1969) recommends that the fetch be at least 100 times the height of the sensor, but Hicks (1975) advocates a more cautious ratio of 1000. The recent simulations of Garratt (1987) suggest that the growth of the internal thermal internal boundary layer for stable conditions is given by

$$h = 0.014\, x^{1/2} U_\infty \left(\frac{g\, \Delta\theta}{\theta} \right)^{-1/2}, \tag{3.4}$$

where h is the height of the internal boundary layer, x is the distance downstream, U_∞ is the free stream air velocity, g is the acceleration due to gravity, $\Delta\theta$ is the potential temperature difference between the continental air (constant throughout the approaching boundary layer and equal to the land surface temperature) and the water surface temperature, and θ is the absolute temperature of the air. For a typical $\Delta\theta = 10°\text{C}$ and $U_\infty = 5\ \text{m s}^{-1}$, this would indicate a layer thickness at 1000 m of 3.8 m, which lies between the recommendations of Hicks (1975) and Peterson (1969). Equation (3.4) also agrees well with the results presented by Mulhearn (1981) and Garratt and Ryan (1988) for boundary layers close to the shore, even though (3.4) was derived for considerably larger length scales. The momentum and water vapor internal boundary layers may be expected to have similar thickness (Garratt, personal communication).

The unstable case, where the water is warmer than the air, does not appear to have been solved, but the case of cool ocean air overflowing a warm land mass was recently reviewed by Venkatram (1986). The roughness was fixed and not dependent on the shear stress, as in the case of the model by Garratt (1987), but the internal boundary layer was still observed to grow as $x^{1/2}$.

For very small boundary layer thicknesses, the application of equilibrium profiles severely constrains the position of the instruments, as they must be placed in that area of the boundary layer described by the law of the wall (Tennekes and Lumley, 1972) a region that extends only to about 10% of the layer thickness (Sorbjan, 1986). In the example above, where the layer thickness was 3.8 m, the instruments must be placed approximately 0.4 m from the water's surface. From Donelan (1980), the root mean square surface roughness due to wave action may be estimated, for a steady wind, to be 0.02 m, which implies a value for $H_{1/3}$ of 0.1m, where $H_{1/3}$ is the significant wave height. The region available for positioning of the sensors is thus limited to between 0.1 m and 0.4 m. To prevent spray damage, the sensors would probably have to be placed close to 0.4 m from the water surface.

If the meteorological sensors are embedded in this law-of-the-wall region of the internal boundary layer, and the horizontal advection effects within the boundary layer can be ignored, the normal turbulence relationships can be used. Monin and Obukhov (1954) parameterized the non-dimensional gradients of wind, potential energy, and specific humidity in terms of the stability parameter z/L, where

$$L = \frac{-\rho u_*^3 \theta_v}{kg\left[\dfrac{H}{C_p} + 0.61\dfrac{\theta E}{L_v}\right]}, \tag{3.5}$$

$\theta_v = \theta(1 + 0.61q)$ is the virtual temperature (°K), k is the von Karman constant and L is the estimate of the Monin and Obukhov length.

The similarity relationships introduced by Monin and Obukhov (1954) and Businger (1955), amongst others, are assumed to hold in the law-of-the-wall region:

$$\frac{kz}{u_*}\frac{dU}{dz} = \phi_M\left(\frac{z}{L}\right), \tag{3.6}$$

$$\frac{kz}{\theta_*}\frac{d\theta}{dz} = \phi_H\left(\frac{z}{L}\right), \tag{3.7}$$

$$\frac{kz}{q_*}\frac{dq}{dz} = \phi_W\left(\frac{z}{L}\right), \tag{3.8}$$

Dyer (1974) summarized the various proposed forms for the functions ϕ_M, ϕ_H, and ϕ_W. For a convecting boundary layer ($z/L < 0$),

$$\phi_M = \left[1 - 16\left(\frac{z}{L}\right)\right]^{-1/4}, \tag{3.9}$$

$$\phi_H = \phi_W = \left[1 - 16\left(\frac{z}{L}\right)\right]^{-1/2}, \tag{3.10}$$

and for a stratified boundary layer with $z/L > 0$,

$$\phi_M = \phi_H = \phi_W = 1 + 5\left(\frac{z}{L}\right). \tag{3.11}$$

Hicks (1976) reanalyzed the Wangara data (Clarke *et al.*, 1971) and found (3.11) to hold for only weakly stable ($0 < z/L < 0.5$) boundary layers. He suggested that for $z/L > 10$,

$$\phi_M = 0.8\frac{z}{L}, \tag{3.12}$$

whereas Carson and Richards (1978) postulated that for the in-between region $(0.5 < z/L < 10)$,

$$\phi_M = \left[8 - 4.25\left(\frac{z}{L}\right)^{-1} + \left(\frac{z}{L}\right)^{-2}\right]. \tag{3.13}$$

Paulson (1970) substituted (3.9) to (3.13) into (3.6) to (3.8) and carried out the integration to obtain

$$U = \frac{u_*}{k}\left(\ln\frac{z}{z_M} - \psi_M\right), \tag{3.14}$$

$$\theta - \theta_s = \frac{\theta_*}{k}\left(\ln\frac{z}{z_H} - \psi_H\right), \tag{3.15}$$

$$q - q_s = \frac{q_*}{k}\left(\ln\frac{z}{z_W} - \psi_W\right), \tag{3.16}$$

where z_M, z_H, and z_W are roughness lengths, and defining

$$x = \left[1 - 16\left(\frac{z}{L}\right)\right]^{1/4}, \tag{3.17}$$

for the unstable case,

$$\psi_M = 2\ln\left(\frac{1+x}{2}\right) + \ln\left(\frac{1+x^2}{2}\right) - 2\tan^{-1}(x) + \frac{\pi}{2}, \tag{3.18}$$

$$\psi_H = \psi_W = 2\ln\left(\frac{1+x^2}{2}\right), \tag{3.19}$$

and for the stable case,

$$\psi_M = \psi_H = \psi_W = \begin{cases} -5\left(\frac{z}{L}\right), & 0 < \frac{z}{L} < 0.5, \quad (3.20) \\ 0.5\left(\frac{z}{L}\right)^{-2} - 4.25\left(\frac{z}{L}\right)^{-1} - 7\ln\left(\frac{z}{L}\right) - 0.852, & 0.5 < \frac{z}{L} < 10.0, \quad (3.21) \\ \ln\left(\frac{z}{L}\right) - 0.76\left(\frac{z}{L}\right) - 12.093, & \frac{z}{L} > 10.0. \quad (3.22) \end{cases}$$

These relationships appear to have been tested over the range from $-15 < z/L < 15$, but for the values larger than 15, the similarity functions must be extended as suggested by Sorbjan (1986).

Equations (3.18) to (3.22) can now be used to relate the neutral to the actual transfer coefficients, leaving only the surface roughness to be estimated. The roughness lengths are related directly to the drag coefficients, as can be seen by substituting (3.14) to (3.16) into (3.1) to (3.4). Under neutral conditions,

$$C_{\mathrm{DN}} = k^2\left[\ln\left(\frac{z}{z_{\mathrm{M}}}\right)\right]^{-2}, \tag{3.23}$$

$$C_{\mathrm{HN}} = k^2\left[\ln\left(\frac{z}{z_{\mathrm{M}}}\right)\right]\ln\left(\frac{z}{z_{\mathrm{H}}}\right)\Bigg]^{-1}, \tag{3.24}$$

$$C_{\mathrm{WN}} = k^2\left[\ln\left(\frac{z}{z_{\mathrm{M}}}\right)\ln\left(\frac{z}{z_{\mathrm{W}}}\right)\right]^{-1}. \tag{3.25}$$

For nonneutral conditions, the procedure yields

$$\frac{C_a}{C_{a\mathrm{N}}} = \left[1 + \frac{C_{a\mathrm{N}}}{k^2}\left(\psi_{\mathrm{M}}\psi_a - \frac{k\psi_a}{C_{\mathrm{DN}}^{1/2}} - \frac{k\psi_{\mathrm{M}}C_{\mathrm{DN}}^{1/2}}{C_{a\mathrm{N}}}\right)\right]^{-1}, \tag{3.26}$$

where a stands for D, H, or W. Geernaert and Katsaros (1986) point out that the roughness, which is a function of u_*, will be influenced by the stability of the air and the boundary layer so that (3.23) becomes

$$C_{\mathrm{DN}} = k^2\left[\ln\left(\frac{z}{z_{\mathrm{MN}}}\right)\right]^{-2}. \tag{3.27}$$

Now eliminating z and U from (3.1), (3.27), and (3.14) yields

$$C_{\mathrm{D}} = \left[C_{\mathrm{DN}}^{-1/2} - \frac{\psi_{\mathrm{M}}}{k} + \frac{1}{k}\ln\left(\frac{z_{\mathrm{MN}}}{z_{\mathrm{M}}}\right)\right]^{-2}, \tag{3.28}$$

rather than (3.26). The correction $(1/k)\ln(z_{\mathrm{MN}}/z_{\mathrm{M}})$ is small, seen by noting that the roughness lengths may be given by the Charnock (1955) relationship

$$z_{\mathrm{M}} = \frac{\alpha u_*^2}{g}, \tag{3.29}$$

$$z_{\mathrm{MN}} = \frac{\alpha u_{*\mathrm{N}}^2}{g}, \tag{3.30}$$

so that (3.28) becomes

$$C_D = \left[C_{DN}^{-1/2} - \frac{\psi_M}{k} + \frac{1}{k}\ln\left(\frac{C_{DN}}{C_D}\right)\right]^{-2}. \tag{3.31}$$

The correction at $z/L = 1$ is about 10%; a reduction for stable conditions and an increase for unstable conditions. However, Keller *et al.* (1985) also investigated the influence of stability on the surface roughness and found an even greater correction was necessary.

The stability correction can now be implemented in two ways. First, given the relationships (3.14) to (3.16), these may be substituted into the definition of the bulk Richardson number (Panofsky, 1963; Deardorff, 1968; Rayner, 1981),

$$\mathrm{Ri_B} = \frac{gz}{\theta_v}\left(\frac{\Delta\theta + 0.61\theta_v\Delta q}{U^2}\right), \tag{3.32}$$

to yield a relationship between z/L and $\mathrm{Ri_B}$ that reduces to

$$\mathrm{Ri_B} = \frac{z}{L}\frac{(kC_{DN}^{1/2}/C_{HWN} - \psi_{HW})}{(k/C_{DN}^{1/2} - \psi_M)^2}, \tag{3.33}$$

where it was assumed that $C_{HN} = H_{WN} = C_{HWN}$. Deardorff (1968), however, did not make this assumption, and inverted the equilibrium form of (3.32) to produce the following relationship (see Strub and Powell 1987):

$$\frac{C_H}{C_{HN}} = e^{-18\,\mathrm{Ri_B}}, \tag{3.34}$$

$$\frac{C_D}{C_{DN}} = e^{-14\,\mathrm{Ri_B}}, \tag{3.35}$$

$$\frac{C_W}{C_{WN}} = e^{-24\,\mathrm{Ri_B}}, \tag{3.36}$$

for stable conditions, and

$$\frac{C_H}{C_{HN}} = \frac{C_W}{C_{WN}} = 1 + \left(\frac{11}{b}\right)\ln(1 - b\,\mathrm{Ri_B}), \tag{3.37}$$

$$\frac{C_D}{C_{DN}} = 1 + \left(\frac{7}{a}\right)\ln(1 - a\,\mathrm{Ri_B}), \tag{3.38}$$

for unstable conditions, where

$$a = 0.83C_{\mathrm{DN}}^{-0.62}, \tag{3.39}$$

$$b = 0.25C_{\mathrm{DN}}^{-0.8}. \tag{3.40}$$

These relationships were successfully used by Strub and Powell (1987) in their heat budget calculations at Castle Lake.

The second approach was suggested by Hicks (1975) and involves an iterative procedure. Neutral values of C_{DN}, C_{HN} and C_{WN} are chosen and neutral fluxes are calculated by (3.1) to (3.3). These values are substituted into (3.5) to calculate the Monin and Obukov length L and thus z/L. New values of the transfer coefficients, now partially corrected for stability, may be obtained from (3.26). This procedure may be repeated and the scheme is said to converge when the estimates for L converge.

This scheme was implemented by Rayner (1981), and some typical results from the data set presented in Figure 1 are shown in Figures 9a–i. The air column, judging by the difference in the air and water temperature shown in Figure 9a was unstable during the nights 29 January and 2 February particularly, but was stable for most of the day period. The evaporative flux (Figure 9d) is a major contributor to the heat budget, so the value of C_{W} is extremely important for an overall heat budget calculation. The variation of the stability corrected transfer coefficient for the anemometer height of 2.0 m (Figures 9g and h) show large variations throughout 24-hour periods. The algorithm was bounded so that $-15 < z/L < 15$; larger values were not allowed. Dependence on stability is now generally accepted to be given by (3.26), although direct verification in terms of C_{D}, C_{H}, and C_{W} is scarce. For lakes, the reader is referred to Strub and Powell (1987) for verification under stable conditions, and to Graf *et al.* (1984) for both stable and unstable boundary layers. In general, it appears that the stability corrections implied by (3.26) describe a diurnal heat flux variation and are necessary to obtain thermal budget closures in diurnal surface layer modeling (Imberger 1985).

The remaining question concerns the choice of a neutral drag coefficient, or as indicated by (3.23) to (3.25), the choice of the surface roughness z_{M}, z_{H}, and z_{W}. Charnock (1955) put forward, from purely dimensional grounds, the relationship (3.29) and Stewart (1974) suggested in his review, by fitting data available at the time, a value of α

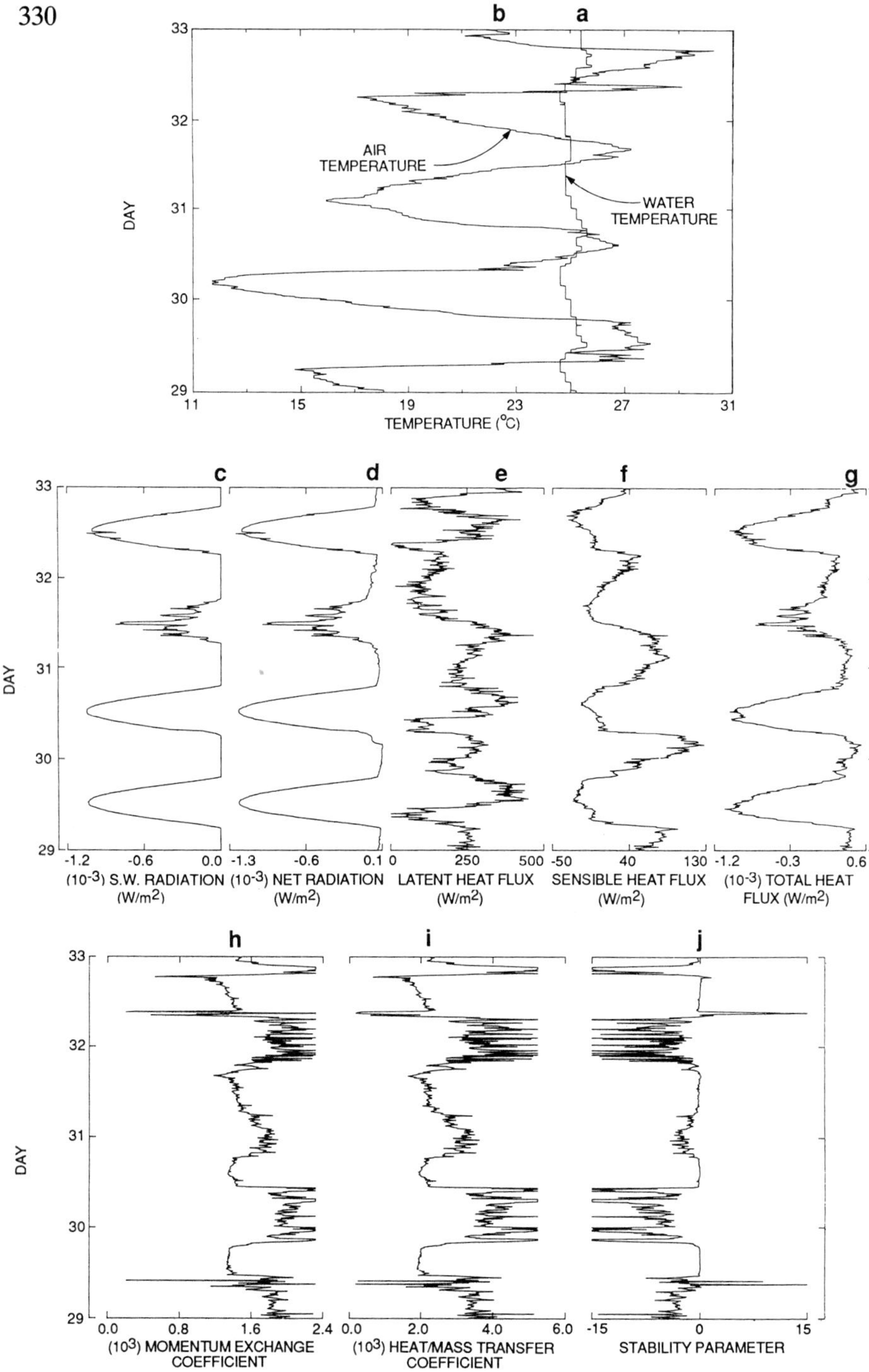

a
b
AIR TEMPERATURE
WATER TEMPERATURE
DAY
TEMPERATURE (°C)
c
d
e
f
g
(10⁻³) S.W. RADIATION (W/m2)
(10⁻³) NET RADIATION (W/m2)
LATENT HEAT FLUX (W/m2)
SENSIBLE HEAT FLUX (W/m2)
(10⁻³) TOTAL HEAT FLUX (W/m2)
h
i
j
(10³) MOMENTUM EXCHANGE COEFFICIENT
(10³) HEAT/MASS TRANSFER COEFFICIENT
STABILITY PARAMETER

between 0.0123 and 0.351. Garratt (1977) added the data from Kitaigorodski *et al.* (1973), Wieringa (1974), Kondo (1975), and Smith and Banke (1975) to obtain a value of α equal to 0.0144; a value that also seemed to fit inferred drag coefficients applicable to wind speeds up to $50\,\mathrm{m\,s^{-1}}$. More recently, Wu (1980) arrived at a value of 0.0185. Large and Pond (1981) derived a value of $C_{DN} = 0.012$ for wind speeds between 4 and $11\,\mathrm{m\,s^{-1}}$, but increasing with wind speed beyond $11\,\mathrm{m\,s^{-1}}$. Equating the two formulations at a wind speed of $25\,\mathrm{m\,s^{-1}}$ by using (3.27) leads to a value of $\alpha = 0.0097$, considerably smaller than suggested by most other investigators. At low wind speeds (less than $5\,\mathrm{m\,s^{-1}}$), the situation is unclear, but in general, the results from the above investigators suggest a constant value of 1×10^{-3} for C_{DN} (see Hicks (1972)). In general, the inter-technique variability and data scatter (Blanc, 1985, 1987) outweigh the above differences, but the most recent values of α, estimated by Wu (1980), (0.0185); Geernaert *et al.* (1986), (0.0178); and Geernaert *et al.* (1987), (0.0165) appear to suggest a mean value of α equal to 0.0175 for a fully developed wave field.

The large variability in the surface roughness has recently received considerable attention. Kitaigorodskii (1968) and Kitaigorodskii and Zaslavskii (1974) already indicated that Charnock's (1955) relationship (3.29) should be modified to include the wave age C_0/u_*:

$$\frac{z_M g}{u_*^2} = \alpha\left(\frac{C_0}{u_*}\right), \tag{3.41}$$

where C_0 is the phase velocity of the dominant wave (wave number at maximum energy) and large values of C_0/u_* represent decaying waves or large swell and small values indicate a growing sea. This can be rewritten in terms of the non-dimensional fetch gL/U^2 (Wu, 1985). Many authors (see Donelan, 1982) have pointed out that for a young sea (short fetch or small duration), the roughness is greater than under the equivalent equilibrium conditions and so the neutral drag coefficients may be expected to be larger. This explains why storm surge modelers (short

FIG. 9. Example of typical heat flux variations computed from data shown in Figure 1. Day 29 is 29 January 1987. (a) Surface water temperature. (b) Air temperature. (c) Short-wave incoming radiation. (d) Net total radiation. (e) Latent heat flux. (f) Sensible heat flux. (g) Total heat flux at the water surface. (h) Drag coefficient C_D. (i) Heat and mass transfer coefficients $C_H = C_W$. (j) The bulk Richardson number.

duration winds) generally require higher values for the neutral drag coefficient.

Hsu (1974, 1986) and Geernaert *et al.* (1987) derived a relationship between the wave age and wave steepness:

$$\alpha = \frac{H_{1/3}}{L_W}, \tag{3.42}$$

where $H_{1/3}$ is defined above and L_W is the wave length at the peak of the energy spectrum. By using both statistics from the JONSWAP formulation and from Chapter 3, Section 4, of the U.S. Army Corps of Engineers (1977), the wave steepness for equilibrium seas becomes 0.023, yielding a value of α slightly higher than that proposed by Wu (1980) (α is equal to 0.0185). Graf *et al.* (1984) investigated drag coefficients for shoaling waves and large wave steepness and showed that water roughness using (3.42) yielded good results. Geernaert *et al.* (1987) carried out a direct correlation between the neutral drag coefficient C_{DN} and the wave age:

$$C_{DN} = 0.0119\left(\frac{C_0}{u_*}\right)^{-0.661}, \qquad 10 < \frac{C_D}{u_*} < 40, \tag{3.43}$$

indicating again an increased coefficient for small values of the wave age (a developing wave surface). The relationships (3.41), (3.42), and (3.43) also explain the reduction of drag coefficient in shallow water where C_0/u_* tends to be larger (Hicks *et al.*, 1974; Emmanuel, 1975; Geernaert *et al.*, 1987).

Donelan (1982) partitioned the water surface roughness into two parts:

$$z_e = \beta\left[\int_0^{2\omega_p} E(\omega)\, d\omega\right]^{1/2}, \tag{3.44}$$

and

$$z_s = \beta\left[\int_{2\omega_p}^{\infty} E(\omega)\, d\omega\right]^{1/2}, \tag{3.45}$$

where ω_p is the frequency at the peak of the spectrum and β is a constant equal to 0.0125. He then defined a neutral drag coefficient

$$C_{DN} = \left\{[\hat{C}_D]_e\, |\cos\theta| \left|u - \frac{C_e}{\cos\theta}\right| \left(U - \frac{C_e}{\cos\theta}\right)\Big/ U_N^2 + [\hat{C}_D]_s (U_N - C_S)^2 / U_N^2\right\}(0.07 + 0.2\log_{10} R_s), \tag{3.46}$$

where $\hat{C}$ is the drag coefficient for an immobile plate with roughness either z_e or z_s given by (3.27). Here U_N is the neutral wind velocity at 10 m, $C_e = 0.83\,g/\omega_p$, $C_s = 0.83\,g/2\omega_p$, ϕ is the angle between the wind and the waves at the spectral peak, $R_s = U_N\sigma/\nu$, σ is the root mean square surface deviation, and ν is the kinematic viscosity of water. Once again, the parameters C_e, C_s, σ, z_e, and z_s can all be estimated from an understanding of the wave climate (Melville, 1977; Donelan, 1980; Donelan and Peirson, 1987) and depend on properties such as wind duration, fetch length, and depth of water. Thus three major new procedures for estimating neutral drag coefficients over a lake water surface exist: from wave steepness, equations (3.41) and (3.42); from the age of the wave field, equation (3.43); and from a knowledge of the wave frequency spectrum, equations (3.44), (3.45), and (3.46).

The dependence on water surface state is extremely important for the study of lakes as the fetch is usually limited, the winds are often diurnal and the water depth is highly variable. However, there does not appear to be available a study comparing the accuracy of any of the above techniques and the choice remains mainly one of personal preference.

The scatter of data in measurements of C_{HN} and C_{WN} so far exceeds the trends with changing wind speeds or sea state. Rayner (1981) suggested using the constant value of $C_{HN} = C_{WN} = 1.35 \times 10^{-3}$, based on the works of Hicks (1972 and 1975) and Pond *et al.* (1974). Friehe and Schmitt (1976) analyzed data from nine different experimental programs for various stabilities. They found a value of 0.86×10^{-3} for C_H under stable conditions and 0.97×10^{-3} for unstable conditions. However, Friehe and Schmitt (1976) also pointed out that if the large flux data from Smith and Banke (1975) are added, the unstable coefficients become 1.46×10^{-3}; the value of C_{WN} was 1.32×10^{-3}, independent of stability. By contrast, Large and Pond (1981) found good agreement with similarity formulations for C_{DN} and C_{WN} but found that C_{HN} was equal to 1.13×10^{-3} for unstable conditions and 0.66 for 10^{-3} for stable layers. The data analyzed by Smith (1980), when corrected for stability (see his Figure 13), indicated a constant value of $C_{HN} = 1 \times 10^{-3}$ with perhaps a slightly larger value for the unstable profiles. Anderson and Smith (1981) found a mean value of C_{WN} of 1.3×10^{-3}, but again with a large amount of scatter. So far, the dependence on the wave climate parameters has not been clearly demonstrated (Geernaert *et al.*, 1987).

In closing, it is important to mention that McBean and Paterson (1975), Bean *et al.* (1975), and Parker and Imberger (1986) have shown that the wind field over a lake can be extremely variable: this has enormous consequences for the behavior of the surface layer, as will be seen in Section VI.

IV. The Surface Layer

The diurnal surface layer is the water counterpart to the atmospheric internal boundary layer. However, it should be remembered from the outset that the characteristic time scale h/u_* with which the internal boundary layer responds is of order 30 seconds, whereas the same time scale for the water surface layer is in the vicinity of 900 seconds for a layer thickness of 5 m. The energy spectrum of the wind record used to calculate the fluxes depicted in Figure 9 is shown in Figure 10; it is seen that there is a thousand fold decrease in wind energy from $10^{-6}\,s^{-1}$ to

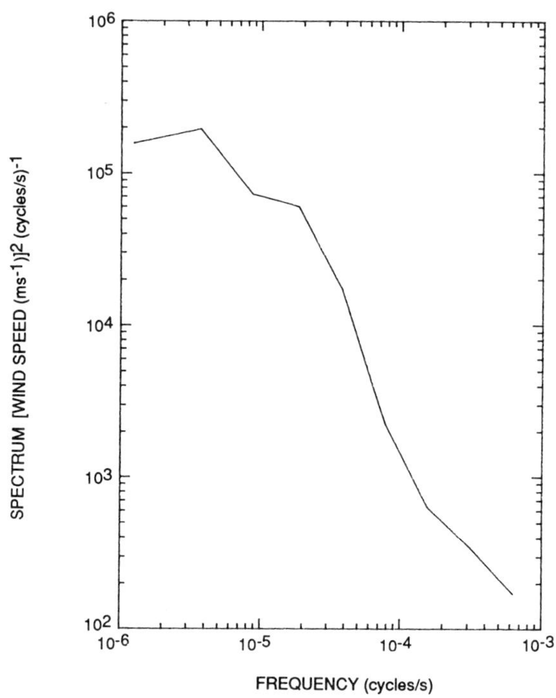

FIG. 10. Power spectral density of the wind speed during the period shown in Figure 9. Seventeen days of data were used to compute the spectrum.

$10^{-3}\,s^{-1}$. However, the energy at the higher frequencies is sufficient to introduce considerable turbulent kinetic energy to the surface mixed layer; since the period is comparable to the time scale of adjustment of the water surface layer, it is unlikely that one would find, except perhaps in the very near surface region, equilibrium profiles of the form given by (3.14) to (3.16) (Dillon *et al.*, 1981; Imberger, 1985; Muller and Garwood, 1988). Even in the very surface region, it is unlikely that equilibrium profiles exist because of the distortions introduced by wave induced velocities. The only major exception would be in the case of a strongly convecting surface layer, where the wind activity is small and the time scale for the heat loss from the surface is set by the diurnal cycle or by an even longer synoptic weather pattern. Such conditions have recently been documented in lakes by Imberger (1985) and Brubaker (1987) and in the ocean by Shay and Gregg (1984, 1986).

Before proceeding with the discussion of the dynamics of the surface layer and the associated vertical energy budget, let us consider some actual examples of the evolution of the temperature field in a typical lake situation. Recent microstructure measurements taken in the Wellington Reservoir reveal the evolution of the diurnal surface layer. The data were taken in February 1985 during an extremely hot period, when the morning heating was strong and where, in the early morning, a southerly wind with speeds up to $4\,m\,s^{-1}$ prevailed which turned to a northwesterly with a peak wind speed of $6\,m\,s^{-1}$ (at a height of 2 m) by 1300 hours. The westerly remained active until 1900 hours (see Monismith and Imberger (1988) for a full description of meteorological conditions).

The initial profile from a region sheltered from the southerly wind is shown in Figure 11a-I. The surface (top 1 m) had become strongly stratified with a temperature differential of almost 1°C; the gradient signal (Figure 11b-I) shows little activity in the surface region except for some isolated high gradient points (Imberger, 1988). The water below this was turbulent, sustained by a weak gravitational underflow that had been set up by the early morning differential cooling (Monismith and Imberger, 1988). Ten minutes after the wind had moved north, causing a gentle roughening of the surface, the temperature had changed to that shown in the second profile. The temperature signal (Figure 11a-II) and the temperature gradient signal (Figure 11b-II) show the formation of a small mixed layer at the surface and also that the profile had become active to a depth of 1.3 m, even though the time since the surface had become exposed to the wind was less than the time scale h/u_*. The

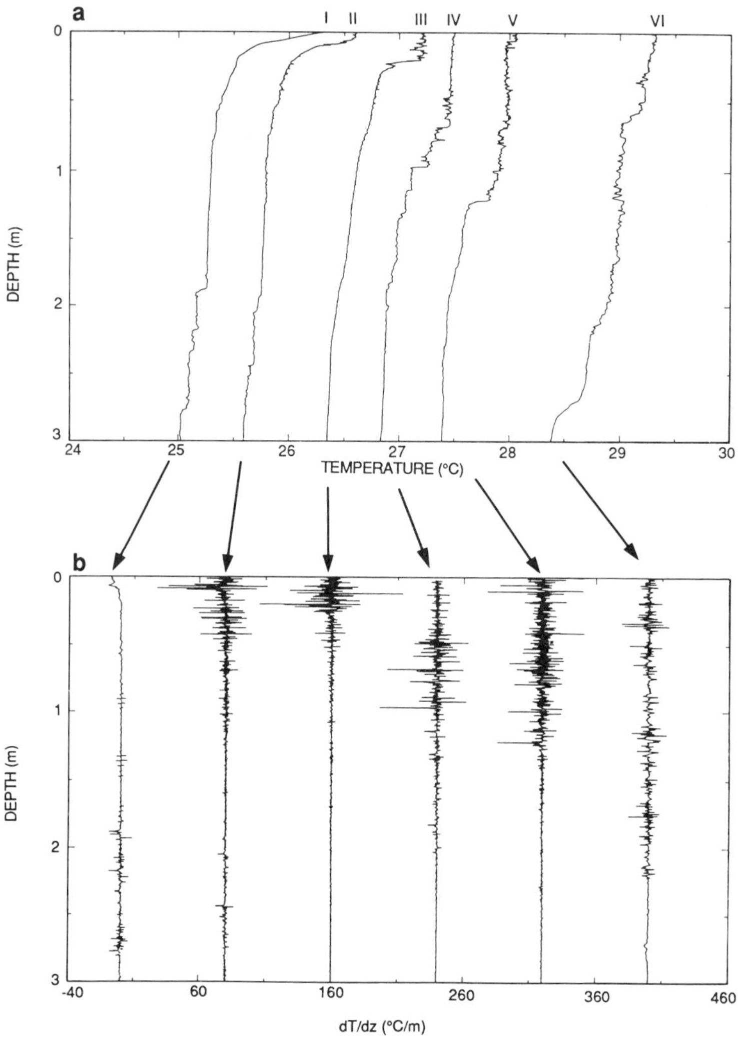

FIG. 11. (a) Examples of microstructure profiles collected at the entrance of Salmon Brook, Wellington Reservoir on 26 February 1985. See Figure 26a for station locations. Temperature:

I.	0928	SB20
II.	0939	SB20
III.	1002	C10
IV.	1021	C10
V.	1057	C10
VI.	1726	SB10

(b) Microstructure temperature gradient signals corresponding to data in Figure 11a.

turbulence was established at small scales and was not the result of a large overturn as has been normally associated with turbulence in a stratified fluid (Gregg, 1987). Imberger (1988) attributed this rapid spread of the energy to the leakage by internal wave action from the base of the small surface mixed layer. Twenty minutes later, and at a somewhat more exposed site further away from the shore, the surface layer had deepened to about 0.3 m and the temperature distribution within the lake was more uniform (Figure 11a-III). There is clear evidence of active turbulence extending below the base of the surface layer to almost 2 m (Figure 11b-III). This deepening process continued and the profile taken another twenty minutes later reveals a "uniform" surface layer extending to nearly 0.7 m. However, as described by de Szoeke and Rhines (1976), Imberger (1985), and Spigel *et al.* (1986), the interface at the base of the surface layer, initially sharp and sustained by the surface-introduced turbulence, is now smeared by billow activity energized by the momentum of the surface layer, which is clearly illustrated in Figures 11a-IV and 11b-IV. The leakage through the base of the mixed layer remained and the turbulent activity extended to nearly 2 m.

During the measuring period, the lake was exposed to a very strong net heating flux of around 750 W m^{-2}. At approximately 1030 hours, the wind intensified, reaching a peak of nearly 6 m s^{-1} at about 1100 hours. A well-defined diurnal thermocline formed at 1.2 m, reflected both in the temperature (Figure 11a-V) and the temperature gradient signal (Figure 11b-V), but two features are noticeable. First, the heat flux introduced sufficient buoyancy at the surface to prevent a completely mixed surface layer from forming, but the temperature gradient signal shows strong turbulence in this surface layer. Second, the temperature gradient signal continued to show turbulence below the base of the surface layer, extending to about 2.0 m.

The wind strength remained at about 4–6 m s^{-1} for the remainder of the day and only began to decrease at 1700 hours. At the time the last profile was collected, the wind had decreased to about 3 m s^{-1}. As seen from Figure 11a-VI, the surface layer was not well mixed, even though the turbulence obviously extended down to 2.3 m (see Figure 11b-VI). The gradient signal shows an intermittent, patchy turbulence within the surface layer, although there was a well-defined wind-stirred near-surface region extending down to 0.06 m. The water column below 2.2 m appeared to have become quiescent.

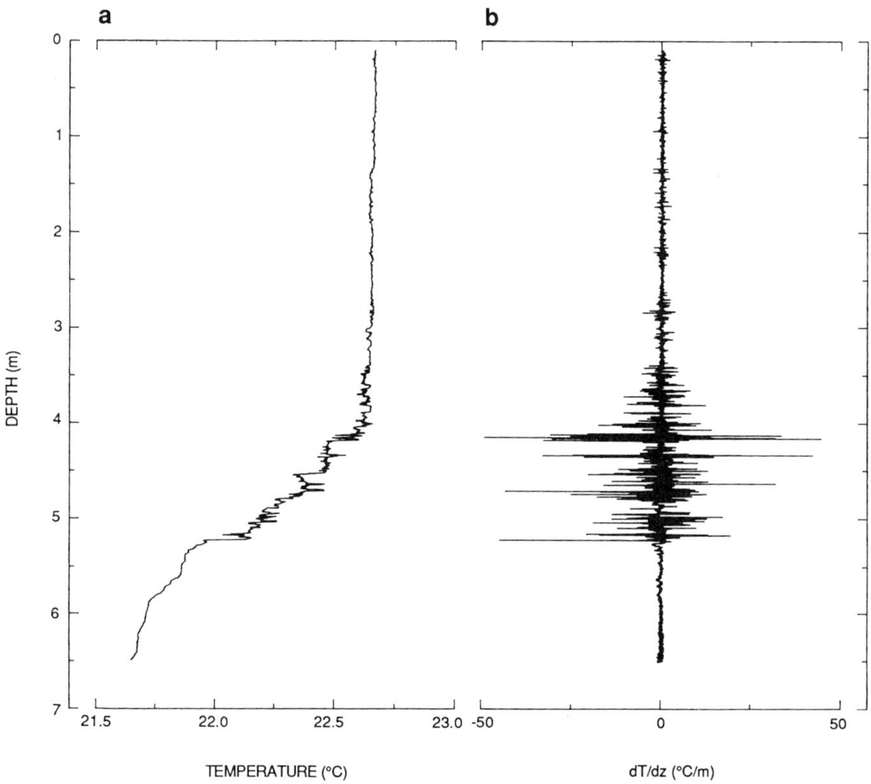

FIG. 12. Temperature microstructure data from the Wellington Reservoir (13 March 1982) at the central basin after a very strong sea breeze had deepened and tilted the surface layer. (a) Temperature. (b) Temperature gradient. (After Imberger (1985).)

Now consider an example (Figures 12a and 12b) from the diurnal sequence documented in Imberger (1985). The profile was taken at a time when there was a net heat loss of about $250\,\mathrm{W\,m^{-2}}$, a strong wind-induced surface layer velocity of about $0.12\,\mathrm{m\,s^{-1}}$, and an active wind stress. As seen in Figure 12, the surface layer was extremely well-defined (variation less than 0.03°C), the base of the mixed layer was diffuse due to billow activity (Imberger, 1985), and the water below the base of the surface layer was relatively quiescent.

The third example, shown in Figure 13, also comes from the diurnal data set presented in Imberger (1985) but was collected when natural convection completely dominated the surface layer dynamics. The profile is characterized by a well-mixed surface layer with obvious thermals

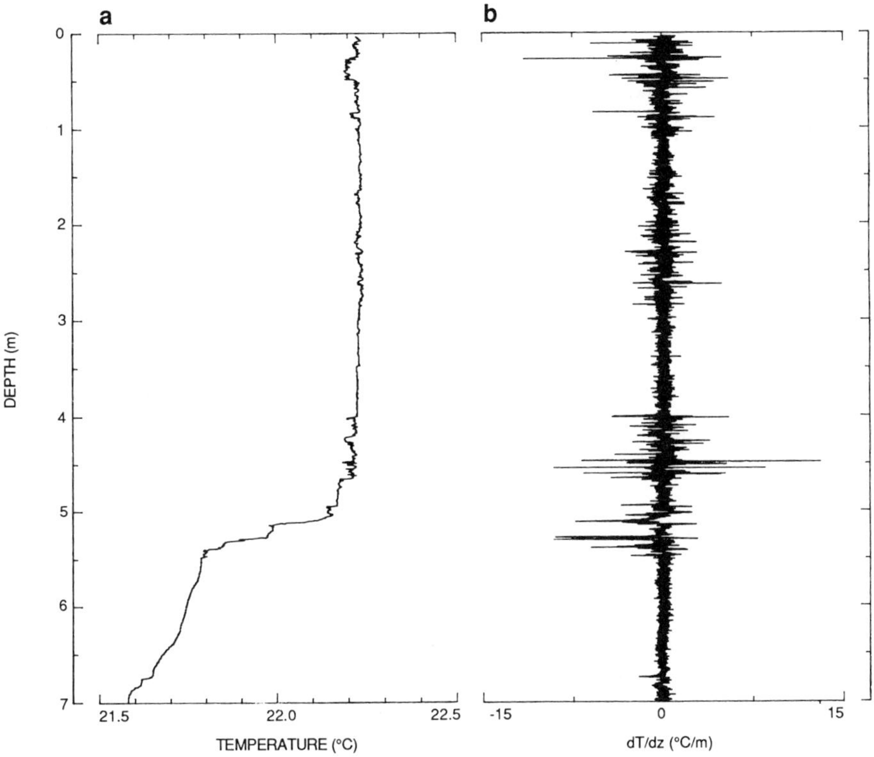

FIG. 13. Temperature microstructure data from a penetrative convective period, collected in the Wellington Reservoir on 14 March 1982. (a) Temperature. (b) Temperature gradient. (After Imberger (1985).)

falling through the full extent of the layer and impinging on a sharp interface.

In summary, these examples show that it is difficult to define the surface layer in terms of the temperature profile. It is better to define the layer as that depth of water directly energized by the surface fluxes and the mean shear of the surface layer. The data also show that the temperature, and thus the density, is rarely absolutely uniform in the surface layer, especially when there is strong surface heating or when the layer is retreating (wind stress reducing with time). Lastly, the surface fluxes not only introduce turbulence into the surface layer but also, via leakage through the base of the surface layer, into the deeper waters making the above definition of the surface layer somewhat ambiguous. Some of this surface-introduced energy energizes the water column as a whole, but the exact fraction has not been documented in field measurements.

The overall dynamics of the surface layer is determined by the magnitude of the surface Wedderburn number (Imberger and Hamblin, 1982) given by (2.5). For $W \gg 1$, tilting of the isotherms due to the applied wind stress will be small and horizontal variations are negligible. This corresponds to strong stratification, light winds, and slow deepening of the mixed layer dominated by surface-introduced turbulence reaching the base of the mixed layer and eroding the interface. For $W \ll 1$, deepening is dominated by internal shear production and occurs on a time scale much shorter than horizontal convection in the surface layer. This leads to a sharp interface downwind and a broad upwelling at the upwind end of the lake (Monismith, 1986). In the initial classification (see Spigel *et al.* (1986)), this regime was misinterpreted as having a horizontal interface due to the envisaged rapid deepening. This question is addressed in the next section. For intermediate values of the Wedderburn number, $W \sim 1$, upwelling and horizontal mixing become important (Spigel and Imberger, 1980). This was confirmed by the model of Imberger and Monismith (1986), the laboratory finding of Monismith (1986) and by the field data of Imberger (1985) and Strub and Powell (1986).

In the case where $W \gg 1$, the processes are essentially one-dimensional, and as shown by Spigel (1980), the layer deepening can be uncoupled from any internal seiching that may exist. There is a long history of interest in such a one-dimensional surface layer. Munk and Anderson (1948) discussed earlier observations and presented a theory based on an eddy diffusion coefficient, dependent on stability, which enabled the prediction of the surface layer behavior. The central task in understanding the surface mixed layer is to quantify the rate at which the kinetic energy of the turbulent velocity fluctuations is converted to potential energy within and at the base of the surface layer as denser water from beneath the layer is entrained and then mixed into the surface layer. The turbulence available for this mixing may be generated at the surface through pressure work, shear production, and wave breaking (Kraus and Turner, 1967; Kraus, 1977; Cavaleri and Zecchetto, 1987; Muller and Garwood, 1988) or in the surface layer and the base by shear production (Pollard *et al.*, 1973; Imberger, 1985). The transport and redistribution of the turbulent kinetic energy within the surface layer is dominated by large scale or secondary motions if present. The most popular concept is of a coherent, well-ordered Langmuir circulation (Langmuir, 1938; Scott *et al.*, 1969; Leibovich and Radhakrishnan, 1977; Pollard 1977; Leibovich, 1977a, 1977b, and 1980; Leibovich and Paolucci, 1980; Leibovich, 1983). From the examples of surface layer development discussed above, the presence of such organized cellular motions are apparently not ubiquitous (see also Muller and Garwood (1988)) and only few direct observations exist (Thorpe and Hall, 1982; Weller *et al.*, 1985; Smith *et al.*, 1987).

The turbulent kinetic energy budget in the surface layer has so far remained elusive. Because direct observations of the vertical turbulent fluxes of momentum, mass, or energy are still lacking, one must rely completely on indirect evidence from measurement of dissipation of turbulent kinetic energy. Such measurements can be used to validate integral models (Imberger, 1985), scaling laws (Kitaigorodskii *et al.*, 1983; Dillon *et al.*, 1981; Shay and Gregg, 1986; Brubaker, 1987; Gregg, 1987) and estimates of the Reynolds stress by equating dissipation to production (Dillon *et al.*, 1981). Detailed comparisons of this kind have led to better agreement in the lake cases than those in the deep ocean, suggesting that better estimation of surface fluxes, the smaller variance of the turbulent fluxes (smaller waves), and the weaker influence of horizontal advection all contribute to this better match. Further, it suggests that the problem lies in capturing the full variability with the measurements rather than with turbulent kinetic energy budget assumptions. An example of the mismatch encountered in the ocean is given in Figure 14 (from Muller and Garwood (1988)), where the observed dissipation rates are considerably larger than the estimated surface input of energy.

A greater understanding of mixed layer energetics has come from one-dimensional models that simulate vertical processes but neglect all horizontal variations. These models are in two categories. First, a number of attempts have been made to obtain solutions to the fundamental equations of motions using some type of closure assumption. Munk and Anderson (1948) may be viewed as a zeroth order model (Andre and Lecarrere, 1985), whereas the models by Mellor and Durbin (1975), Mellor and Yamada (1982), Findikakis and Street (1982a, 1982b, 1983), Edinger and Buchak (1983), Franke *et al.* (1987), Shih *et al.* (1987), and Murota *et al.* (1988) are higher order schemes. Excellent summaries of the applicability of such closure schemes to mixed layer problems is given in the review by Rodi (1987). The range of length

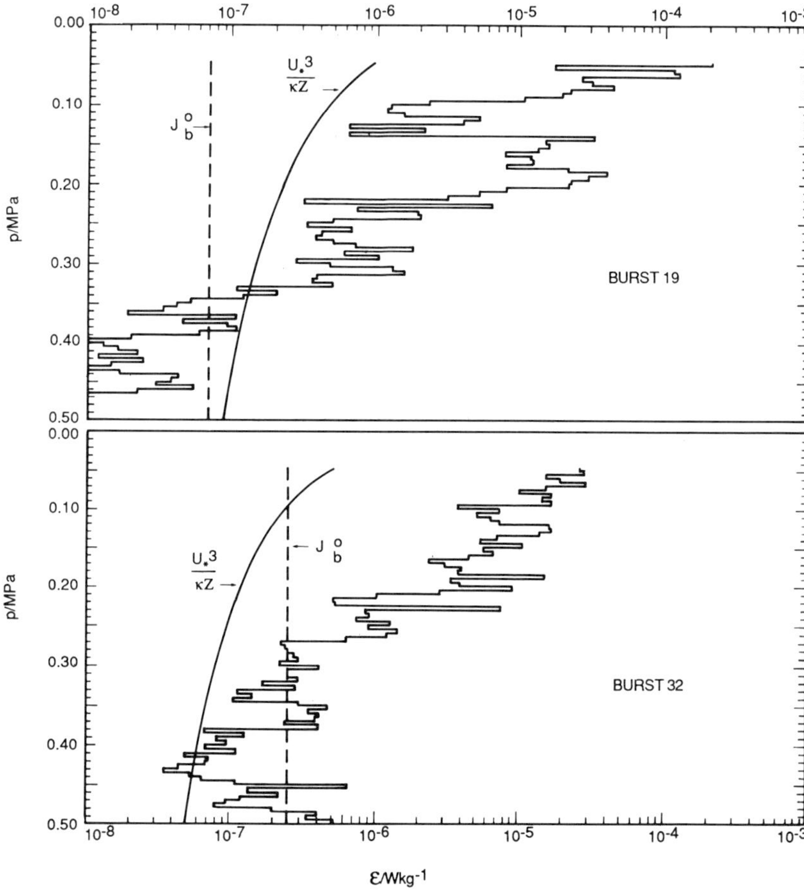

FIG. 14. Two examples of dissipation as a function of depth taken in the ocean and compared with the similarity scaling for turbulence produced by both wind stress and by convection. (From Muller and Garwood (1987); data originally from Gregg, unpublished.)

scales characterizing the surface layer dynamics and the unsteadiness of the turbulence makes the application of closure schemes difficult to justify. For instance, it is now well known that in a penetrative convection surface layer, the heat flux in the lower half of the layer is against the mean gradient, a process that requires the application of a high order closure model.

Large eddy simulation schemes fall into this same class of models, but are even more computer intensive. The method has been applied successfully to a homogeneous wind-driven reservoir by Ivetic *et al.*

(1986), but once again, in the presence of stratification, the diversity of scales and the non-stationarity of the turbulence has prevented the use of the method.

Second, integral formulations pioneered by Niiler and Kraus (1977) continue to be developed (Atkinson and Harleman, 1983; Spigel *et al.*, 1986; Heathershaw and Martin, 1987). As these authors point out, the simplicity of physical insight afforded by the integral approach is adequate justification for pursuing it.

Shear production of turbulence at the base of the surface layer was recognized by Pollard *et al.* (1973) as an important source of energy for the deepening process. Niiler (1975), Zeman and Tennekes (1977), Sherman *et al.* (1978), and others have included shear production, together with surface stirring as originally advocated by Kraus and Turner (1967), in their mixed-layer models. Observations by Price *et al.* (1978) verified the importance of shear production for oceanic mixed-layer deepening, while the measurements of Thorpe (1978) in Loch Ness provided evidence of billowing accompanying strong velocity shear at the base of the mixed layer. Spigel and Imberger (1980) introduced a classification for mixed-layer deepening that predicted the conditions under which shear production would become important in small to medium-sized lakes. Their mixed-layer-deepening algorithm was based on the model of Sherman *et al.* (1978) but did include a separate calculation to simulate billowing. However, Spigel and Imberger (1980) did not explicitly account for the effects of billowing in the turbulent energy budget, and their parametrization of the turbulent energy budget assumed (in common with Sherman *et al.*) that mixed-layer turbulent kinetic energy was a fixed proportion of external energy input by wind and surface cooling. The assumption of fixed proportionality implies that mixed-layer turbulence adjusts rapidly to changing external inputs as was shown above; this may not be valid for diurnal simulations in which meteorological forcing is varying rapidly. In addition, in deepening by convective overturn, Denton and Wood's (1981) experiments have shown that the assumption of fixed proportionality is valid only under certain boundary conditions.

The role of billowing is not so straightforward. Consider two parallel streams separated by a sharp interface across which a velocity jump ΔU and density jump $\Delta\rho$ (with the heavier layer on the bottom) occur. Theoretical considerations (cf. Drazin and Reid (1981), Nishida and Yoshida (1984), and Lawrence *et al.* (1987)) show that the interface is unstable, and in practice, Kelvin–Helmholtz billows are always observed along such an interface (Thorpe, 1969). The time for billows to form, grow, and then collapse into small-scale turbulence is relatively short, of order $T_b = 20\,\Delta U/g'$ (Thorpe, 1973; Chu and Baddour, 1984), where $g' = \Delta\rho g/\rho_0$ (ρ_0 is a reference density). Values of T_b in lakes are of the order of minutes or less (Spigel, 1978). The ultimate mixing that occurs after the breakdown of billows is of more relevance here than the details of billow formation, and we use "billowing" broadly to include this final mixing. The net result of billowing is that the sharp interface is replaced by a shear layer of thickness δ across which density and velocity vary continuously. Simple energy arguments (Sherman *et al.*, 1978) show that δ is proportional to $\Delta U^2/g'$; on the basis of numerical and laboratory experiments, Sherman *et al.* suggest that

$$\delta = 0.3\,\Delta U^2/g'. \tag{4.1}$$

Such a configuration is stable, and if nothing occurs to increase ΔU or decrease the thickness δ, no further billowing will occur. If ΔU grows (possibly due to an increasing wind stress) or δ decreases (possibly due to mixed-layer deepening—see Figure 11a-V), then billowing will occur until a new stable configuration is achieved.

There is thus a fundamental distinction between mixing caused by billowing and mixing caused by thermocline erosion. Billowing broadens an interface more or less symmetrically about the point of maximum velocity gradient and thereby reduces the density gradient between layers. Billowing by itself does not cause thermocline erosion or produce any net mixed-layer deepening. Thermocline erosion is a one-way process: an upper turbulent layer grows thicker at the expense of a lower nonturbulent, or less turbulent, layer by entrainment of quiescent fluid into the upper mixed layer. There is no tendency for density gradients to be weakened during thermocline erosion and in some cases they may be sharpened. Sharpening of density gradients is associated with convective penetration or other stirring processes.

In lakes, billowing accompanies mixed-layer deepening (Thorpe, 1978; Imberger, 1985) yet both processes drain kinetic energy from the mean flow shear ΔU to produce turbulent kinetic energy. In this sense both processes compete for a given supply of mean flow kinetic energy proportional to $\frac{1}{2}\rho_0\,\Delta U^2$ per unit volume. The processes complement each other in that a fraction of the turbulent kinetic energy produced in both is used in working against gravity to increase the potential energy of the water column; i.e., both processes cause mixing. Billowing utilizes mean flow kinetic energy that would otherwise be available to produce mixed-layer turbulence for entrainment, but at the same time weakens the density gradient at the base of the mixed layer. Billowing reduces the energy required for further mixed-layer deepening, since the energy required to entrain heavier fluid is roughly proportional to the strength of the density gradient across the base of the mixed layer. There thus arises a rather complex and unsteady interaction whereby mixed-layer deepening sharpens a gradient, making it unstable to shear so that billowing occurs. Billowing weakens the gradient, which is then more easily eroded by further deepening, which leads to further billowing.

One of the principal goals of the model by Rayner (1981), and later Spigel *et al.* (1986), was to demonstrate that the interaction between billowing and deepening can be successfully modeled by accounting simultaneously for the energetics of surface stirring, shear production, billowing, and variability in the store of turbulent kinetic energy.

In these simple one-dimensional integral models, it is assumed that the surface layer may be approximated in the integral sense by three distinct zones (Niiler and Kraus, 1977; Spigel *et al.*, 1986): a comparatively thin constant stress surface layer of thickness γ (see Figure 15) where turbulent kinetic energy is produced and then exported to the fluid below; a uniform central layer in which part of the surface energy is used to mix the fluid and is lost to dissipation; and a thin front of thickness

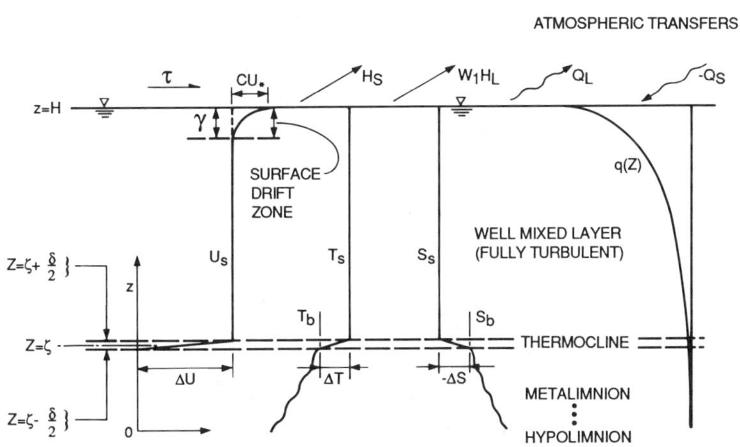

FIG. 15. Definition sketch of the surface layer model. (After Spigel *et al.* (1986).)

δ at the base of the surface layer. This transition marks the pronounced density jump described above as the diurnal thermocline. Here, the remainder of the turbulent kinetic energy generated at the surface plus that generated internally by shear and less that which is locally dissipated or radiated downwards by internal waves (Linden, 1975; E and Hopfinger, 1986; Imberger, 1988) is used to entrain underlying fluid into the central layer above. Using the symbols defined in Figure 15, the governing differential equations (Niiler and Kraus, 1977; Spigel *et al.*, 1986) may be written as

$$\frac{\partial \theta}{\partial t} = -\frac{\partial}{\partial z}(\overline{\theta' w'}) - \frac{1}{\rho_0 C_P}\frac{\partial Q}{\partial z}, \tag{4.2}$$

$$\frac{\partial S}{\partial t} = -\frac{\partial}{\partial z}(\overline{s' w'}), \tag{4.3}$$

$$\frac{\partial \rho}{\partial t} = -\frac{\partial}{\partial z}(\overline{\rho' w'}) + \frac{\alpha}{C_P}\frac{\partial Q}{\partial z}, \tag{4.4}$$

$$\frac{\partial U}{\partial t} = -\frac{\partial}{\partial z}(\overline{u' w'}), \tag{4.5}$$

$$\frac{1}{2}\frac{\partial E}{\partial t} = -\overline{u' w'}\frac{\partial U}{\partial z} - \frac{\partial}{\partial z}\left[\overline{w'\left(\frac{p'}{\rho_0} + \frac{E}{2}\right)}\right] - \frac{g\overline{\rho' w'}}{\rho_0} - \varepsilon, \tag{4.6}$$

where $E = \overline{u'^2 + v'^2 + w'^2}$, Q is the short-wave radiation, and the simplest form of the momentum equation (4.5) is used. Niiler and Kraus (1977), de Szoeke (1980), Ivey and Patterson (1984), Janowitz (1986), and others have included the influence of longitudinal pressure gradient and Coriolos force. The rationale for using such a simple balance is discussed in Spigel and Imberger (1980), Kranenburg (1984), Monismith (1985), and Strub and Powell (1986). Monismith (1985) shows that (4.5) can be used until the base of the mixed layer develops a baroclinic tilt due to the second mode seiche; the time to the first deceleration is $T_i/4$, where T_i is the period of the second mode seiche (the seiche associated with the upper surface layer).

The boundary conditions at the free surface assume continuity of fluxes across the interface:

$$\text{Momentum:} \quad -\overline{u' w'}(H) = u_*^2, \tag{4.7}$$

$$\text{Heat:} \quad \rho_0 C_P \overline{\theta' w'}(H) = Q_L + H_L + H_S, \tag{4.8}$$

$$\text{Salt:} \quad -\overline{s' w'}(H) = W S_s, \tag{4.9}$$

$$\text{Mass:} \quad \overline{\rho' w'}(H) = \rho_0[-\alpha \overline{\theta' w'}(H) + \beta \overline{s' w'}(H)], \tag{4.10}$$

where W is the water loss at the surface, Q_L is the net long-wave radiation, H_L is the latent heat loss, H_S is the sensible heat loss, α is the thermal coefficient of expansion, and β is the salt coefficient of volume.

Spigel *et al.* (1986) assumed that the turbulent heat and momentum fluxes at $z = \zeta - \delta/2$ (the very bottom of the base of the surface layer) are zero, so that

$$\overline{\theta' w'} = \overline{s' w'} = \overline{\rho' w'} = \overline{u' w'} = 0. \tag{4.11}$$

The leakage or transfer of turbulent kinetic energy through the base may be written

$$\Delta_L = \overline{w' \left[\frac{p'}{\rho_0} + \frac{E}{2} \right]} \Bigg|_{\zeta - \delta/2}. \tag{4.12}$$

As seen from Figures 11, 12, and 13, this is usually nonzero. Imberger (1988) postulated that the mechanism for turbulent energy transport was by internal wave propagation, away from the base of the surface layer along rays. Linden (1975) estimated from laboratory experiments that up to 50% of the turbulent kinetic energy budget arriving at the base of the mixed layer seeps through the base. By constrast, E and Hopfinger (1986) suggested that while internal waves were generated, energy radiation generally did not affect the entrainment rate. Scaling suggests that the leakage Δ_L be parameterized by

$$\Delta_L = C_L A^3 N^3 \tag{4.13}$$

where C_L is a coefficient, A is the amplitude of the internal waves generated, N is the buoyancy frequency below the base, and h is the depth of the mixed layer. At present, there are no conclusive data that allows the determination of the correctness of (4.13) or the magnitude of C_L, and for this reason, Spigel *et al.* (1986) assumed $C_L = 0$. It is worth remarking that since their model gave very good comparisons with observed field data, the leakage term must be proportional to the flux of turbulent kinetic energy arriving at the interface, which is seen by noting that $A \sim E_s^{1/2}/N$.

Spigel *et al.* (1986) followed Niiler and Kraus (1977) and integrated (4.2) to (4.6) from the bottom to the surface of the lake using the boundary conditions given by (4.7) to (4.12). Further, closure was achieved by introducing a set of efficiencies for each of the source and sink terms. Shear production in the base layer minus the dissipation was approximated by

$$\frac{U_s^2}{2}\frac{dh}{dt}+\frac{1}{12}\frac{d}{dt}(U_s^2\delta)-\varepsilon_b\delta=C_S\left[\frac{U_s^2}{2}\frac{dh}{dt}+\frac{1}{12}\frac{d}{dt}(U_s^2\delta)\right], \tag{4.14}$$

where ε_b is the average dissipation in the base layer and C_S is the efficiency of production of turbulent kinetic energy. Experimental evidence that allows the evaluation of C_S is sparse. Sherman *et al.* (1978) reanalyzed the experiments of Kato and Phillips (1969) and Kantha *et al.* (1977) to arrive at a value of between 0.2 and 0.5. Imberger (1985) presented field data giving a value of 0.24 and Spigel *et al.* (1986) obtained best simulation results with a value of 0.2.

The turbulence introduced at the surface constant stress layer may be parameterized (Kraus and Turner, 1967; Niiler and Kraus, 1977) as

$$cu_*^3-\overline{w'\left(\frac{p'}{\rho_0}+\frac{E}{2}\right)}\bigg|_H=\frac{C_N^3}{2}u_*^3, \tag{4.15}$$

where c is the velocity scale in the surface drift layer (see Figure 15), C_N is the efficiency of energy production at the water surface, and u_* is the water shear velocity.

The source of buoyancy flux due to surface heating may be written (Deardorff, 1970; Zeman and Tennekes, 1977; Rayner, 1981) in the form

$$\begin{aligned} w_*^3 = {} & g\left(h+\frac{\delta}{2}\right)\left\{\frac{\alpha}{\rho_0 C_P}\left[Q(H)-Q\left(\zeta-\frac{\delta}{2}\right)\right]+\alpha\overline{\theta' w'}(H)\right\} \\ & +\beta W S_s-\frac{2\alpha g}{\rho_0 C_P}\int_{\zeta-\delta/2}^{H}Q\,dz. \end{aligned} \tag{4.16}$$

The dissipation in the mixed layer may be written (Spigel *et al.*, 1986)

$$\varepsilon_s h=\frac{C_E}{2}E_s^{3/2}, \tag{4.17}$$

where ε_s is the average dissipation in the mixed layer. This form was introduced by Mahrt and Lenschow (1976); similar forms were suggested by Garwood (1977) and Zeman and Tennekes (1977), as well as by the results of Willis and Deardorff (1974). The alternative parameterization used by Niiler and Kraus (1977), Sherman *et al.* (1978), and others consists of introducing a set of efficiencies to reduce each energy source by a certain fraction; an approach that does not allow for the storage of turbulent kinetic energy. Rayner (1981) proposed a necessary final closure assumption which states that a fraction of the turbulent kinetic energy $\frac{1}{2}C_F E_s^{2/3}$ in the mixed layer together with the available shear production (4.14) is utilized at the base to spin up the turbulence and to energize the buoyancy flux (neglecting any leakage). This may be expressed as

$$\frac{C_F}{2}E_s^{3/2} + \frac{C_s}{2}\left[U_s^2\frac{dh}{dt} + \frac{1}{6}\frac{d}{dt}(U_s^2\delta)\right]$$
$$= \frac{E_s}{2}\frac{dh}{dt} + \frac{g}{2\rho_0}\left[\Delta\rho h\frac{dh}{dt} - \frac{\delta}{2}\frac{d}{dt}(\Delta\rho h) + \frac{1}{12}\frac{d}{dt}(\Delta\rho\delta^2)\right]. \qquad (4.18)$$

Given these assumptions and carrying out the integration yields (Spigel *et al.*, 1986) the necessary set of equations for the unknowns U_s, h, θ_s, S_s, δ, and E_s:

$$h\frac{dE_s}{dt} = -(C_F + C_E)E_s^{3/2} + w_*^3 + C_N^3 u_*^3, \qquad (4.19)$$

$$h\frac{d\theta_s}{dt} = -\Delta\theta\frac{dh}{dt} - \frac{\delta}{2}\frac{d\theta_b}{dt} - \frac{1}{\rho_0 C_P}[H_s + H_L + Q_L + Q_s - Q(\zeta)], \qquad (4.20)$$

$$h\frac{dS_s}{dt} = -\Delta s\frac{dh}{dt} - \frac{\delta}{2}\frac{dS_b}{dt} + WS_s, \qquad (4.21)$$

$$\frac{d}{dt}(U_s h) = u_*^2, \qquad t < \frac{T_i}{4}, \qquad (4.22)$$

and (4.1) and (4.18).

The evaluation (Spigel *et al.*, 1986) of the remaining three coefficients $C_N = 1.33$, $C_F = 0.25$, and $C_E = 1.15$ follows directly from a careful analysis of all available data (Kato and Phillips, 1969; Tennekes and Lumley, 1972; Wu, 1973; Deardorff, 1974; Kamail *et al.*, 1976; Mahrt and Lenschow, 1976; Stull, 1976; Willis and Deardorff, 1974).

Recently, there has been a renewed interest in the entrainment problem described by (4.1) and (4.18)–(4.22), and there is greater awareness that there are two distinct sources of energy for entrainment at the base of the surface layer as expressed by (4.18),

$$\tfrac{1}{2}C_F E_s^{3/2} \quad \text{and} \quad \tfrac{1}{2}C_s\left[U_s^2\frac{dh}{dt} + \frac{1}{6}\frac{d}{dt}(U_s^2\delta)\right].$$

Atkinson (1988) reviewed the literature for shear driven experiments (see also Christodoulou (1986)) and concluded that the data support the empirical entrainment law

$$\frac{1}{U}\frac{dh}{dt}=\frac{\alpha-\beta\,\mathrm{Ri}\,P_e^{-1/2}}{\gamma+\mathrm{Ri}}, \tag{4.23}$$

where α, β, and γ are constants, $P_e = UH/\kappa$, κ is the coefficient of species diffusion, and $\mathrm{Ri} = g'h/U^2$. Given that shear production dominated the deepening process, (4.18) reduces to (assuming δ to be small)

$$h=\frac{C_s U_s^2}{g'}. \tag{4.24}$$

Comparison with (4.1) indicates that the depth of the mixed layer scales the same as δ, which means that for a particular U_s the depth is fixed and cannot continue to deepen, contrary to (4.23). This was the basis of the original Pollard *et al.* (1973) model, and may be explained by noting that once a particular surface layer has a certain shear across its base, both h and δ will adjust until (4.1) and (4.24) are satisfied, after which time there will be no further adjustment (Sherman *et al.*, 1978; Lawrence *et al.*, 1987; Thorpe, 1987). The experiments of Narimousa *et al.* (1986) and Narimousa and Fernando (1987) rely on a surface energy source u_*^3 that will continually sharpen the interface and reduce δ, and the entrainment law (4.23) therefore depends on the type of experiment and must be used with caution.

This is best seen by substituting (4.22) into (4.24) and noting that $g'h$ is constant for no net heat flux, which then yields

$$\frac{1}{u_*}\frac{dh}{dt}=\frac{C_s^{1/2}}{\mathrm{Ri}^{1/2}}, \qquad \text{where } \mathrm{Ri}=g'h/u_*^2. \tag{4.25a}$$

Simple grid experiments contain no shear so that (4.18) reduces to the relationship

$$\frac{1}{E_s^{1/2}}\frac{dh}{dt}=\frac{C_F}{1+g'h/E_s}. \tag{4.25b}$$

Hannoun *et al.* (1988) have shown that the energy flux that arrives at the interface is the same with and without the interface being present, and is given by u^3, where u is the root mean square velocity, which was also observed to decay inversely proportional to the distance from the grid. In grid experiments, we should therefore expect C_F, the fraction of the mean kinetic energy flux that arrives at the interface, to be a function of both the depth of the surface layer and the turbulent intensities. Ignoring the spin-up term in (4.25) and defining $\bar{u}$ as the mean (over the depth of the mixed layer) root mean square velocity allows (4.25) to be rewritten

$$\frac{1}{\bar{u}}\frac{dh}{dt}=\frac{C_F}{\overline{\mathrm{Ri}}}, \tag{4.26}$$

where

$$\overline{\mathrm{Ri}}=\frac{g'h}{\bar{u}^2}. \tag{4.27}$$

Nokes (1988) has done a thorough review of data from all available grid experiments and concluded that the rate of non-dimensional deepening ranges in proportionality from Ri^{-1} to $\mathrm{Ri}^{-1.75}$. In this context, the results from Kranenburg (1984) and Murota and Michioku (1986b) give convincing evidence that the non-dimensional deepening is proportional to Ri^{-1}. In all recent work, investigators have defined the Richardson number $\mathrm{Ri}=g'l/u^2$, where l is the integral scale of the turbulence at the position of the interface (but in the absence of the interface) and u is the root mean square velocity of the turbulence arriving at the position of the interface. Unfortunately, there are no data available on the relationship between the depth of the mixed layer and the integral scale for surface layers so we cannot take this comparison any further. It would be constructive to carry out an integral analysis of the grid configuration to determine the relationship between $\bar{u}$ and u, and to establish if (4.26) would lead to the relationship suggested by Nokes (1988):

$$\frac{1}{u}\frac{dh}{dt}=\frac{0.15}{\mathrm{Ri}^{-1.2}}. \tag{4.28}$$

This is now possible as all the necessary spatial variations can be derived from Hopfinger and Toly (1976), Hannoun and List (1988), and Hannoun *et al.* (1988). Last, it is important to reconcile, as discussed by Nokes (1988), the difference in the measurement of the entrainment rate dh/dt: indirectly from a measure of the change of buoyancy in the surface layer and directly from the propagation rate of the interface.

Murota and Michioku (1986b) also carried out an experiment where they used grid stirring together with a convective energy source and successfully compared their results with the model put forward by Sherman *et al.* (1978) and embodied in (4.19), although their value for C_N was 2.9, not 1.33.

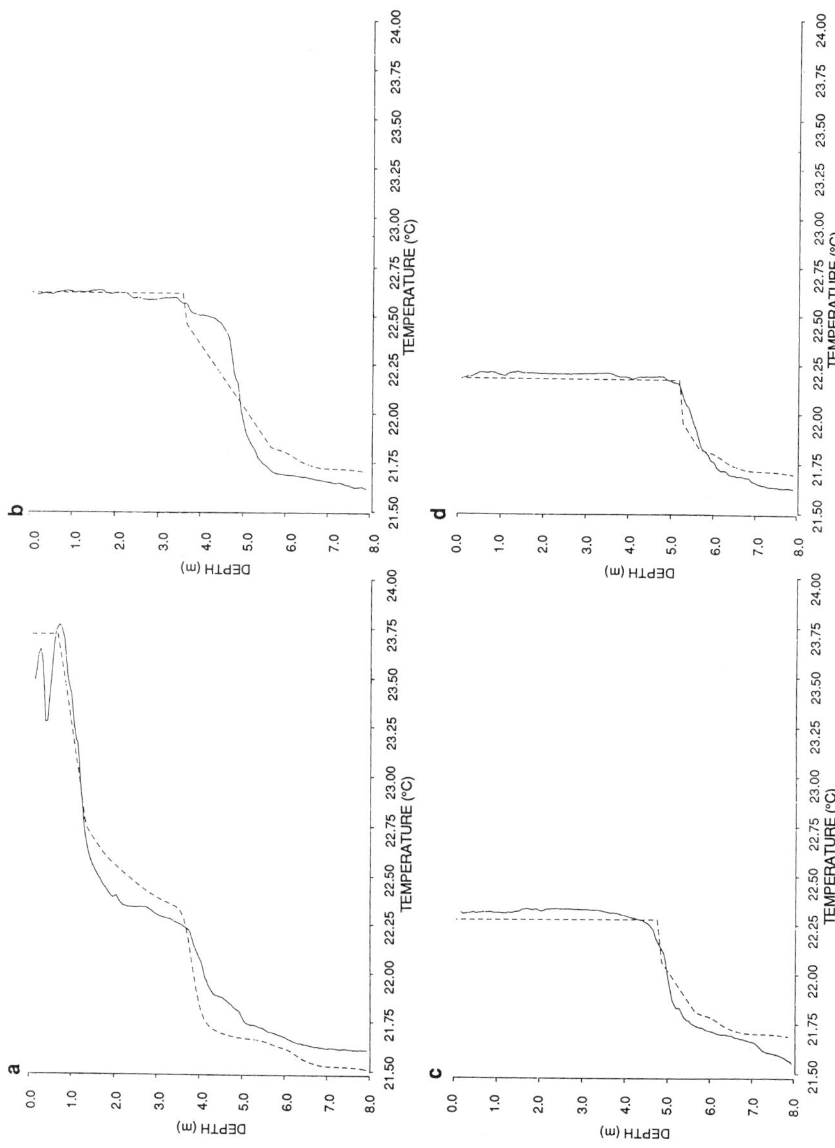

FIG. 16. Comparison of measured (solid line) and predicted (dashed line) temperature profiles. The profiles b, c, and d have been corrected for effects of advection. (After Spigel *et al.* (1986).)

The model described by (4.11) and (4.18)–(4.22) was used by Spigel *et al.* (1986) to simulate the diurnal cycle documented in Imberger (1985) for a surface layer in the Wellington Reservoir in March 1982. The 24-hour period included strong heating in the early morning, severe wind in the afternoon when the Wedderburn number decreased to below 0.02, and intense penetrative convection from midnight to early morning. Comparison from these simulations is shown in Figure 16. Given that none of the coefficients C_S, C_N, C_F, or C_E were adjusted for the data and also given that a period of heating, stirring, shear production, and penetrative convection all contributed to the surface layer deepening, the comparisons are remarkably good.

Recently it was pointed out (Dickey and Simpson, 1983; Stefan *et al.*, 1983) that the exact distribution of solar radiation $q(z)$ is most important to the dynamical response of the surface layer. From (4.16), the departure from the linear absorption curve determines the mechanical buoyancy flux, and it is this departure that contributes to the direct influence on the dynamical response.

In summary, we have shown that the identification of the energy sources in the surface layer has allowed, over the last few years, the construction of successful simulation algorithms. The success of these models, however, depends on an accurate description of the surface fluxes, a realistic model for the momentum in the surface layer, and the removal of changes due to horizontal advection. Last, the model allows intercomparisons with the full range of laboratory experiments such as natural convection, shear-generated turbulence, and grid-generated turbulence, but as yet this comparison does not seem to have been carried out. The leakage of turbulent kinetic energy through the base of the surface layer remains to be parameterized.

V. Upwelling

When the surface wind stress increases, the isopycnals surface at the upwind end and deepen at the downwind end (Wedderburn, 1912; Keulegan and Brame, 1960; Blanton, 1973; Stefan and Ford, 1975; Spigel and Imberger, 1980; Monismith, 1986; and many others). This phenomenon is called upwelling.

A demonstration of upwelling is given by the longitudinal transects taken in the Wellington Reservoir (Figure 17a) in August 1988. The isopycnals prior to the start of a wind event are shown in Figure 17d; the wind event started at about 1000 hours and lasted for about 6 hours (Figure 17b), with a direction of 320° (Figure 17c), the direction of the transect. The isopycnals after 4 hours are shown in Figure 17e, indicating a high degree of disturbance. The hypolimnion was also disturbed but retained a degree of horizontality. The upwind surface layer showed strong upwelling with a well-defined horizontal gradient supported by cold water brought up from the metalimnion.

The idealized density profile assumed by Heaps and Ramsbottom (1966) and later used by Spigel and Imberger (1980) in their lake classification scheme consisted of only two layers. This two-layer approximation was used by Thompson and Imberger (1980) and Imberger and Hamblin (1982) to formulate the Wedderburn number (2.5) as the nondimensional parameter determining the response of the surface layer to an impulsive wind stress.

The initial value problem for the two-layer situation was solved by Spigel (1980). For the case with rotation, the reader is referred to Csanady (1982), Heaps (1984), Kielman and Simons (1984), and Horn *et al.* (1986); the influence of rotation will not be discussed here. From Spigel (1980), in the first stages of development of the surface layer following the startup of the wind stress, the surface layer moves according to (4.22). This surface layer velocity was shown by Imberger (1985) to be applicable until the interface tilt reaches the center of the lake and induces a baroclinic pressure gradient, which retards the motion. The period of the surface layer seiche was given by ($h_1 < H$)

$$T_1 = \frac{2L}{\left[\dfrac{\Delta\rho}{\rho} g \dfrac{h_1 h_2}{H}\right]^{1/2}}, \tag{5.1}$$

where L is the length of the lake, $\Delta\rho$ is the density difference across the

FIG. 17. Upwelling as recorded along the central channel of the Wellington Reservoir in response to a strong northwesterly wind. Data collected on 24 August 1988. (a) Location map showing plane of projection and measuring stations. (b) Wind speed as recorded in the central valley of the Wellington at 1.4 m above the water surface. (c) Wind direction recorded as above. (d) Isopycnals at 0930, before wind had commenced. (e) Isopycnals at 1330 extending from the Gervase River to the main basin. (f) Example of a temperature microstructure record, collected during the strong upwelling event shown in Figure 17e. (g) Corresponding temperature gradient. (h) Corresponding dissipation as computed from the Batchelor spectra fitting technique (Bar graph) and the Wigner–Ville technique (sticks).

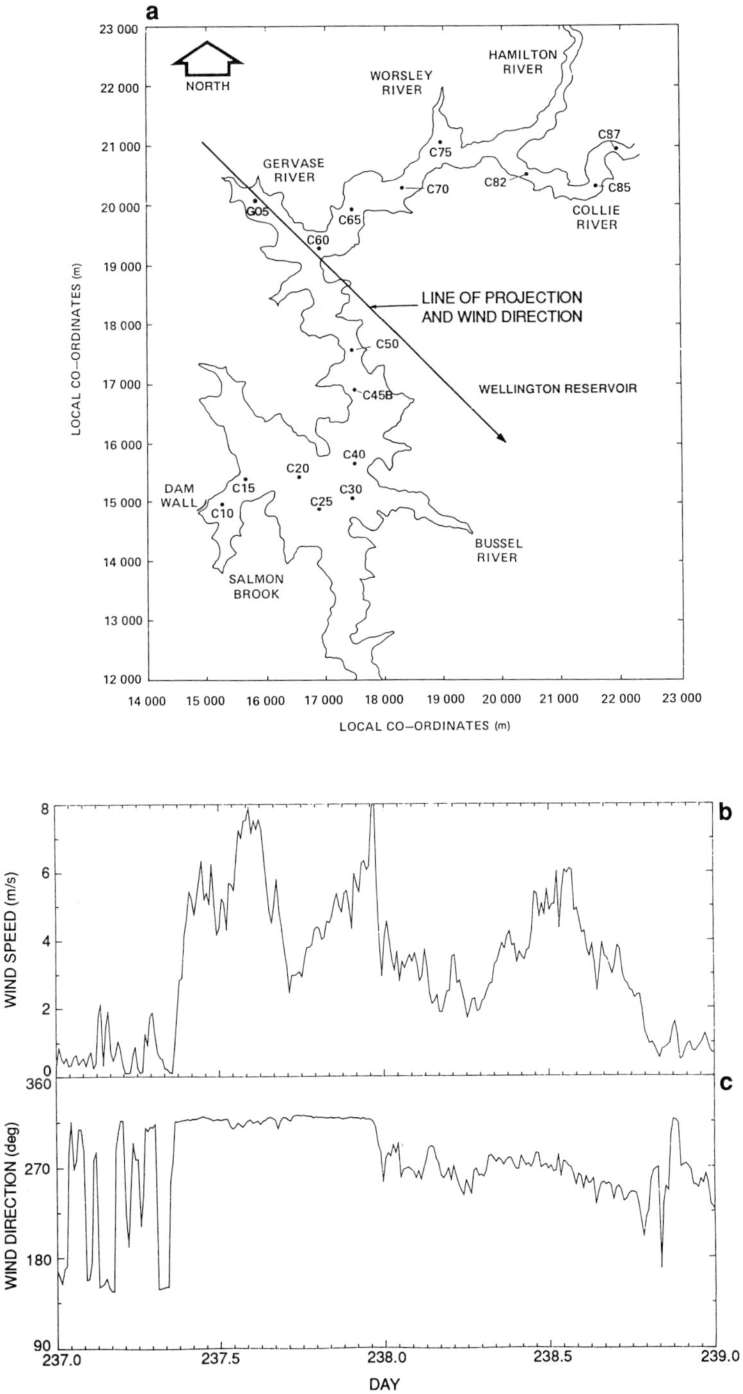
a
NORTH
HAMILTON RIVER
WORSLEY RIVER
GERVASE RIVER
COLLIE RIVER
C87
C75
C70
C82
C85
G05
C65
C60
LINE OF PROJECTION AND WIND DIRECTION
C50
WELLINGTON RESERVOIR
C45B
C40
C20
C15
C30
C25
C10
DAM WALL
BUSSEL RIVER
SALMON BROOK
LOCAL CO-ORDINATES (m)
LOCAL CO-ORDINATES (m)
b
WIND SPEED (m/s)
c
WIND DIRECTION (deg)
DAY

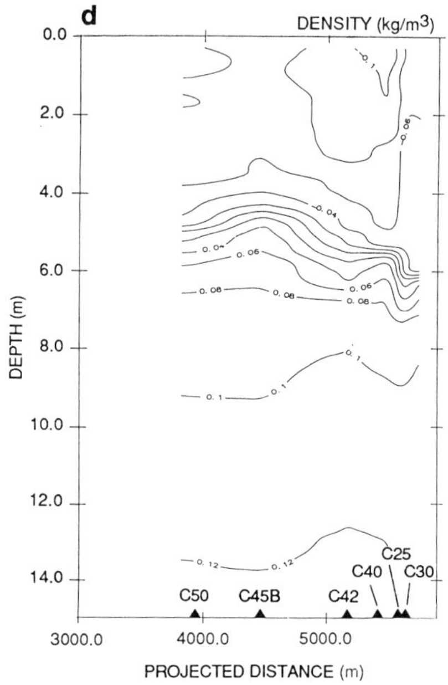
d
DENSITY (kg/m3)
DEPTH (m)
PROJECTED DISTANCE (m)
C50
C45B
C42
C40
C25
C30

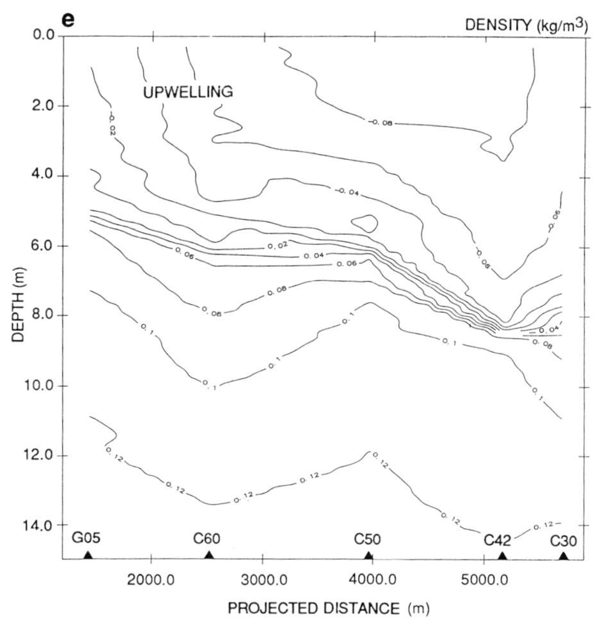
e
DENSITY (kg/m3)
UPWELLING
DEPTH (m)
PROJECTED DISTANCE (m)
G05
C60
C50
C42
C30

FIG. 17. (*contd.*)

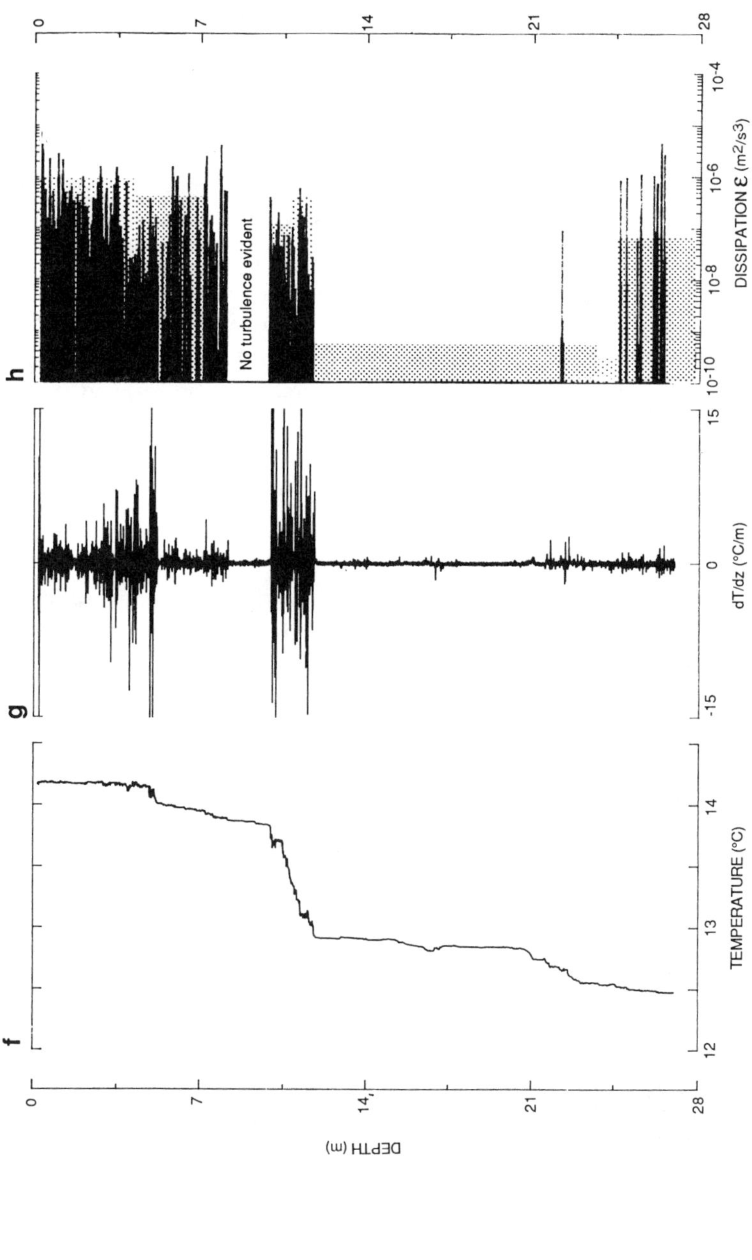
f
g
h
DEPTH (m)
TEMPERATURE (°C)
dT/dz (°C/m)
DISSIPATION ε (m2/s3)
No turbulence evident

FIG. 17. *(contd.)*

two-layer idealized density structure, h_1 is the depth of the surface layer, h_2 is the depth and ρ the density of the lower layer, and H is the depth of the lake. Throughout it was assumed that the theory was applicable for times up to $T_1/4$ when the base of the surface layer begins to tilt. The theory developed by Spigel and Imberger (1980), which provided the background for the Wedderburn number concept, relied completely on the two-layer model. However, it is apparent from the work of Monismith (1986) that this model does not explain the upwelling that already occurs at Wedderburn numbers considerably larger than one.

Monismith (1985), using the modal decomposition technique developed by Lighthill (1969) (see also Gill and Clarke (1974) and Csanady (1972)), solved the n-layer case for wind stress varying arbitrarily in time. He applied this general theory in three-layer form to the Wellington field data reported by Imberger (1985). From the analysis, he showed that the shear across the surface layer base ΔU_{12} was given by

$$\frac{\partial}{\partial t}(\Delta U_{12}) = \frac{g\Delta\rho_{12}}{\rho_1}\frac{\partial\zeta_{12}}{\partial x} + \frac{u_*^2}{h_1}, \tag{5.2}$$

which is the generalization of (4.22) to the case of an arbitrary number of layers and where ζ_{12} is the displacement of the surface layer base (between layer 1 and 2), $\Delta\rho_{12}$ is the density difference between layer 1 and 2, and h_1 is the thickness of layer 1 (the surface layer). The first term on the right side of (5.2) is the pressure gradient induced by the slope of the interface ζ_{12}.

The importance of Monismith's (1985) analysis is that when h_1 and $\Delta\rho_{12}$ are small (that is, W is small), mode 2 is excited, leading to a large shear across the base of the surface layer with an associated strong tilt of the interface ζ_{12}; the second interface ζ_{23} remains essentially horizontal (Figure 18a). Conversely, where $\Delta\rho_{12}$ is of the same order as $\Delta\rho_{23}$ and the top and bottom layers are of comparable size or larger than the middle layer, the response is strongly influenced by the first mode and the second mode with a comparable tilt in ζ_{12} and ζ_{23} and the introduction of a large velocity in the bottom layer. By taking a range of surface layer depths as measured in the Wellington field experiment, Monismith (1985) obtained excellent comparisons between his theory and experimental results for the thermocline tilt and induced shear; the three-layer stratification example is a satisfactory model of the majority of lake stratification patterns.

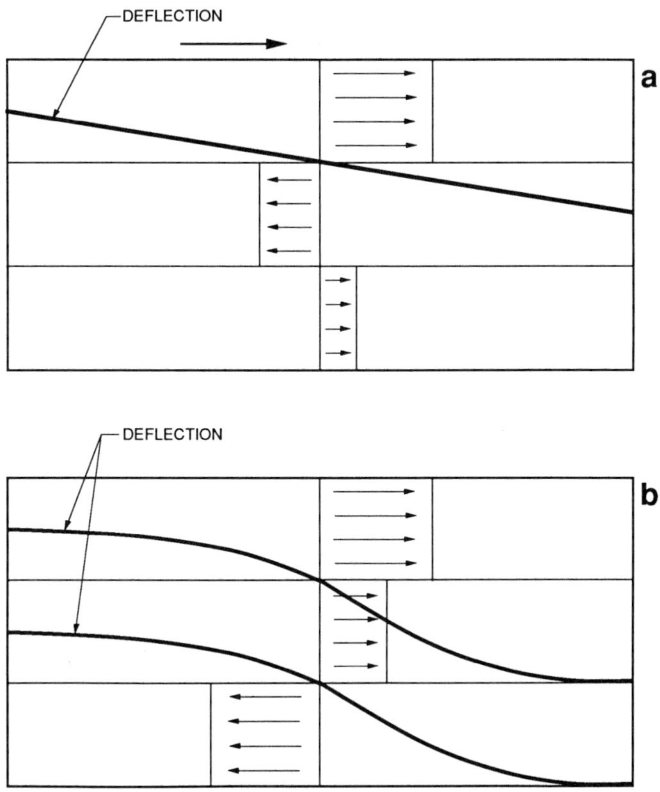

FIG. 18. Schematic of the modal response for a three layer system. (a) Second mode response: W small, L_N large. (b) First mode response: W small, L_N small.

The results from this model have two profound implications:

(a) A mode 2 response (Figure 18a) concentrates the shear, and thus induced turbulence, in the upper surface layer. The strong tilt induced at the base of the surface layer introduces active upwelling at even quite moderate Wedderburn numbers (Monismith, 1986). Further, the lower layer remains essentially stationary with ζ_{23} remaining horizontal.

(b) A mode 1 response (Figure 18b) introduces motion throughout the lake, with tilting isopycnals in both the metalimnion and the hypolimnion. This motion leads to active turbulence throughout the water column.

It is important to categorize a lake's response as either mode 1 or 2. It is clear from the theory underlying the Wedderburn number W that a mode 2 response is induced when W becomes small (less than 10); the tilting of the base of the surface layer becomes stronger with decreasing W (Monismith, 1986). The three-layer model has been used successfully to explain why upwelling is induced for moderately small Wedderburn numbers. With increasing wind stress relative to the buoyancy influence, W decreases and the amplitude of the mode 1 response increases, causing the metalimnion (layer 2) to tilt, which induces upwelling not only from the metalimnion but also from the hypolimnion. However, it is not the Wedderburn number that determines the mode 1 response, but the Lake number L_N (2.12); L_N is the mode 1 counterpart to W in mode 2. This may be illustrated by examining in more detail the experiments of Monismith (1986). The stratification used in these experiments is well approximated by a three-layer stratification with a constant density in layers 1 and 3 and a linear transition between the two. For a rectangular basin of length L, width B and total density difference between the top and bottom layers of $\Delta\rho$, the Lake number takes the form

$$L_N = W\left[\left(\frac{h_3}{h_1}\right)^2 - \frac{h_3 H}{h_1^2} + \frac{h_2 h_3}{h_1^2} + \frac{2h_2^2}{3h_1^2} - \frac{Hh_2}{h_1^2}\right]\left(\frac{h_3}{H} + \frac{h_2}{2H}\right), \tag{5.3}$$

where

$$W = \frac{\Delta\rho g h_1^2}{\rho_1 u_*^2 (BL)^{1/2}}, \tag{5.4}$$

and h_1, h_2, and h_3 have the same definition as above.

The results from Monismith's (1986) Experiment 8 (Figure 19a–d) show that a mode 2 response was induced with the base of the metalimnion remaining essentially horizontal, but with the surface of the metalimnion developing distinct upwelling very similar to that in Figure 17c for the Wellington Reservoir field data. The value of W was 3.9 and that of L_N was 6.4, indicating that the base of the surface layer was relatively susceptible to the mode 2 response, but that the overall stratification contained a large amount of potential energy. By contrast, the values for Experiment 22 were $W = 0.4$ and $L_N = 0.15$; the value of W had decreased by a factor of 10 but L_N had decreased by as much as a factor of 50, indicating that in this experiment both the base of the surface layer and stratification were relatively weak and could not resist the overturning moment supplied at the surface by the belt drive (Figure 19e–h).

These experiments must thus be reinterpreted as both W and L_N were allowed to vary. The low Lake number value in Experiment 22 explains why the boundaries of the metalimnion remained almost parallel at even

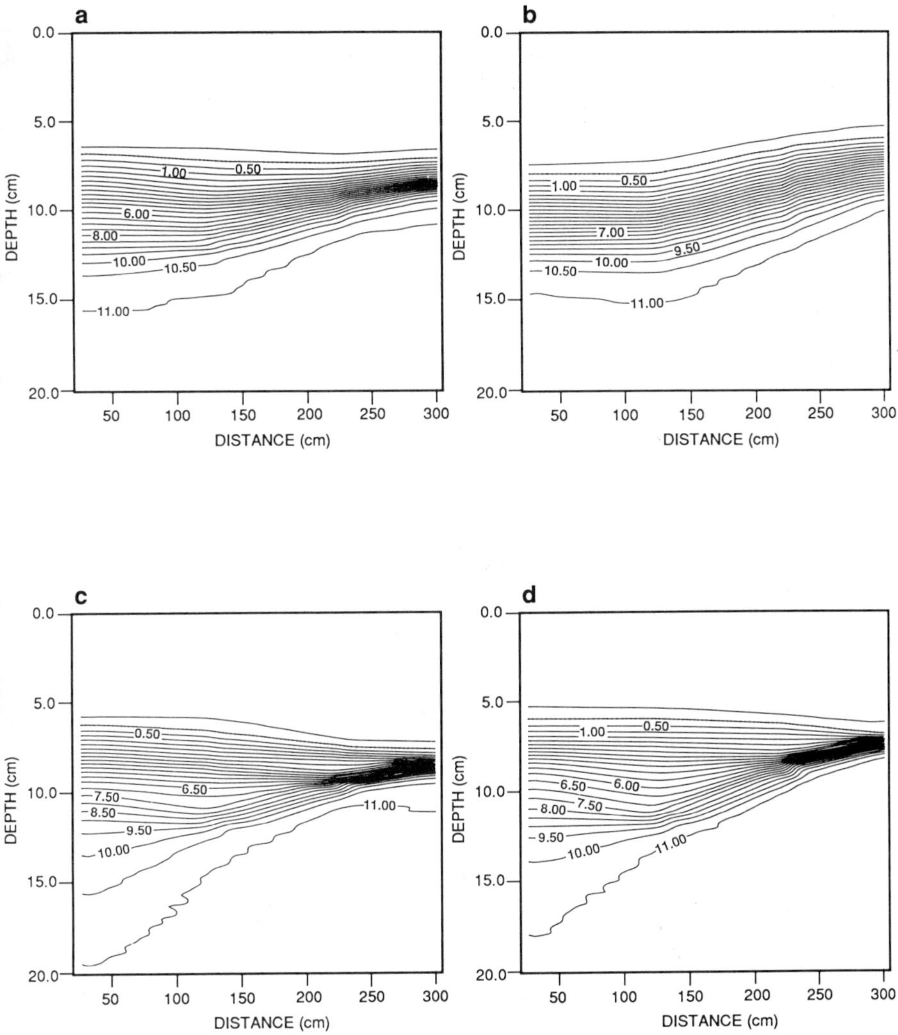

FIG. 19. Contours of density field measured by Monismith (1986) in a laboratory tank. The stress was introduced at the bottom with a belt moving from left to right.
(a) Experiment 8: $W = 3.9$, $L_N = 6.4$, $t = 0.32\ T_1$.
(b) Experiment 8: $W = 3.9$, $L_N = 6.4$, $t = 0.64\ T_1$.
(c) Experiment 8: $W = 3.9$, $L_N = 6.4$, $t = 0.96\ T_1$.
(d) Experiment 8: $W = 3.9$, $L_N = 6.4$, $t = 8.0\ T_1$.
(e) Experiment 22: $W = 0.4$, $L_N = 0.15$, $t = 0.35\ T_1$.
(f) Experiment 22: $W = 0.4$, $L_N = 0.15$, $t = 0.53\ T_1$.
(g) Experiment 22: $W = 0.4$, $L_N = 0.15$, $t = 0.70\ T_1$.
(h) Experiment 22: $W = 0.4$, $L_N = 0.15$, $t = 1.05\ T_1$.

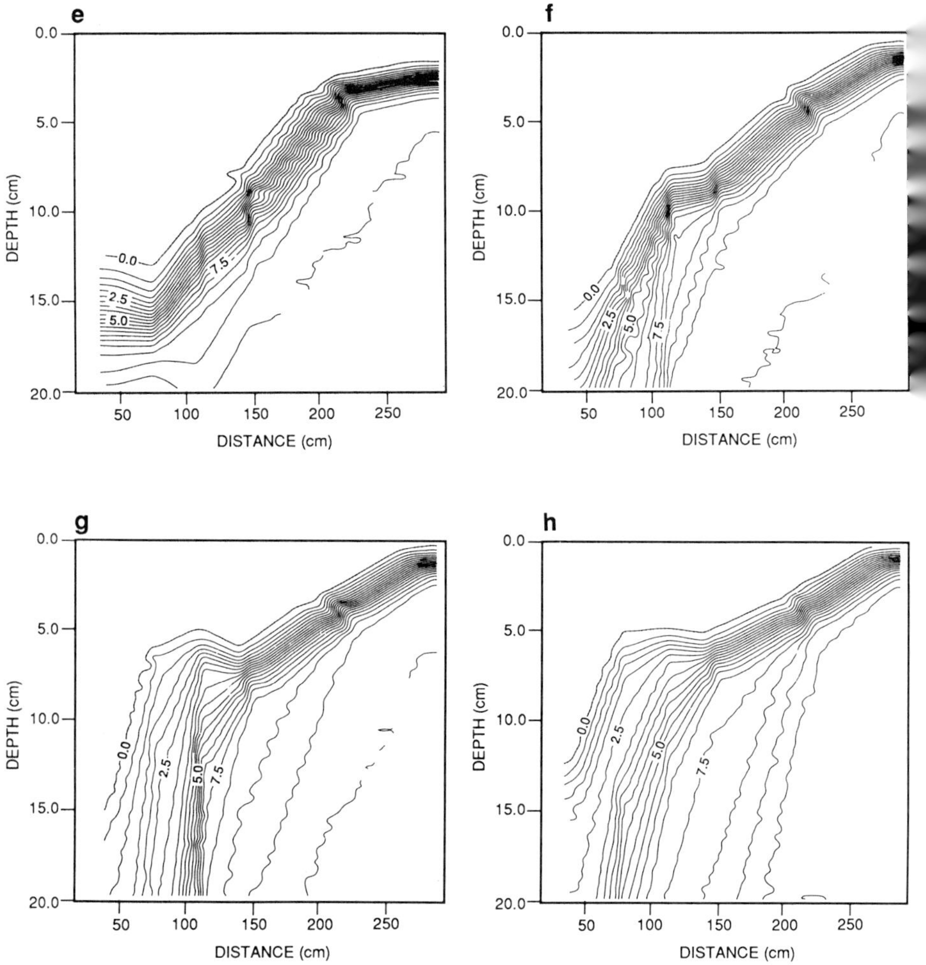

FIG. 19. (*contd.*)

extreme deflections (Figure 19f). Indeed, the whole metalimnion surfaced, leading to a direct mixing of water from the bottom and surface layers. Experiments in which L_N is kept large while W is reduced have not been carried out, but field data from Imberger (1985), in the initial phases of the wind build-up, had a Wedderburn number of 0.02 and a Lake number of 12. The isotherms (isopycnals) below the base showed essentially no tilt throughout the wind period. The Lake number decreased to a value of 1.4 at the time of peak wind speed, and the data indicated some tilting of the isotherms and some mixing below the surface layer. By contrast, the Lake number for the data shown in Figure 17 was considerably smaller (0.24), explaining the greater mixing observed at depth (see Section X).

W and L_N are separate indicators, and for the case where W is small but L_N is large, only the surface layers respond to wind stress. Where W and L_N are small, the lake as a whole responds and we may expect that vertical mixing will greatly increase throughout the lake.

The response in the case of continuous stratification (Monismith, 1986) can now be interpreted in terms of these two nondimensional numbers. For a linear stratification that reaches the surface, the mixed layer depth is zero, as is the density jump across the base of the surface layer (continuous stratification may be viewed as the limit where both approach zero). This would mean that W is zero but L_N is finite. The expected response is a strong tilt of the surface isopycnals with active upwelling, but at depth, the isopycnals remain horizontal. This case was recently solved by Monismith (1987), using the same decomposition technique mentioned above and assuming that the horizontal velocity profile $u(x, z, t)$ and the force field $f(x, z, t)$ could be decomposed as follows:

$$(u, f) = \sum_{n=1}^{\infty} (u_n, f_n)(x, t)\phi_n'(z), \tag{5.5}$$

where $\phi_n(z)$ are solutions of the long wave form of the internal eigenvalue problem (Gill, 1966) such that

$$\zeta = \sum_{n=1}^{\infty} \zeta_n(x, t)\phi_n(z), \tag{5.6}$$

and u_n, f_n, and ζ_n are the amplitude functions. Comparison with experimental results from Monismith (1986) indicate that the fundamental mode was predicted well, but higher orders, which depend strongly on the exact nature of the assumed stress distribution within the water column were not as well predicted, although qualitatively, the agreement was good.

So far the discussion has concentrated completely on the response of times up to the quarter internal wave period for either mode 1 or 2. Beyond this time, the internal displacement degenerates into seiches (W large, L_N large) or damped motions (W and L_N small). The damping of the oscillations is caused by internal and boundary turbulence and we shall postpone the discussion of these processes to Section X, where turbulence is reviewed.

For times longer than the damping time, it is possible to estimate the deepening rate due to entrainment at the surface layer base for the case where L_N is large and W is small. Imberger and Monismith (1986) used the results from Keulegan and Brame (1960), Kranenburg (1985), and Monismith (1986) to formulate a model of a steady upwelling resulting from a mode 2 response. Under such conditions, the experimental data and the initial value problems discussed above lead to the following conclusions:

(a) The interface tilts and opens at the upwind end over a period of time approximately equal to the internal seiche period, T_1 (given by (5.1)). The data suggest that the isopycnals are almost horizontal at the bottom of the interface but slope considerably upwards at the top of the interface, establishing a weak horizontal density gradient in the mixed layer by a combination of upwelling and horizontal variation in turbulent entrainment. The internal circulation within the surface layer causes a general upwelling and divergence of the flow at the upwind end, leading to a diffuse interface. At the downwind end, there is downwelling, causing a convergence at the interface, leading to an extremely sharp, well-defined entraining interface.

(b) The velocity profile in the surface layer is such that the velocity at the stress surface is approximately $20u_*$ downwind (Kranenburg, 1985; Monismith, 1986) and about $2u_*$ downwind at the interface. The flow at the interface was observed by Monismith (1986) to be jet-like, concentrated a very short distance above the interface and contained by a weak density gradient there. This differs from the assumption made by Spigel and Imberger (1980) that the recirculation flow in the mixed layer could be neglected for $W > 1$.

(c) The upwelling region is confined to the upwind end of the basin. The density in the surface layer increases with time; for $T > T_n$, where T_n is the characteristic longitudinal mixing time, the density profile in the epilimnion varies linearly with distance along the length of the surface layer.

(d) The net entrainment rule which best fits the experimental data from Keulegan and Brame (1960), Kranenburg (1985), and Monismith (1986) may be written

$$\frac{dh_1}{dt} = C_1 u_* \,\mathrm{Ri}^{-1}, \tag{5.7}$$

where C_1 is a constant between 0.07 (Kranenburg, 1985; Monismith, 1986) and 0.23 (Wu, 1973).

These conclusions permit the construction of a simple model of upwelling and mixed layer deepening for reservoirs where $W \sim 1$ and $L_N \gg 1$ (Imberger and Monismith, 1986, Figure 20).

For this case, as the stress is applied to the free surface, the interface tilt is set up by the mode 2 response so that the upper isopycnals take on an angle given by the dynamic balance developed by Wu (1973), and the boundary layer thickness is proportional to the displacement of the upper isopycnals

$$\delta \sim \zeta \sim \frac{Lu_*^2}{g_0' h_1}, \tag{5.8}$$

where δ here is shown in Figure 20, $g_0' = (\rho_1 - \rho_0)g/\rho_0$, $\rho_1 - \rho_0$ is the initial value of the density difference at the base of the surface layer, h_1 is the initial surface layer depth, and ζ is the mean interfacial deflection.

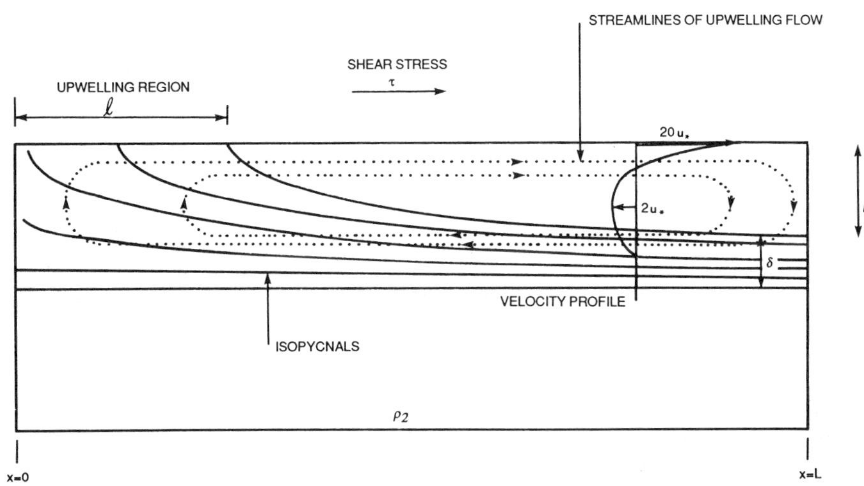

FIG. 20. Definition sketch for model of mixed-layer deepening due to upwelling. (After Imberger and Monismith (1986).)

The model assumes that the return flow of $2u_*$ drags interfacial fluid with it from the downwind end to the upwelling region. It is assumed that this fluid, which can be visualized as being planed off the interface, enters the upwelling region and then is distributed longitudinally by shear flow dispersion. A simple mass balance would indicate that

$$g'_{\mathrm{a}}\,\delta u_* \sim L g'_{\mathrm{a}} \frac{dh}{dt}, \tag{5.9}$$

where ρ_{a} is the average density in the surface layer, $g'_{\mathrm{a}} = g(\rho_1 - \rho_{\mathrm{a}})/\rho_0$, and h is the mixed layer depth at any time. It follows that

$$\frac{dh}{dt} = \frac{\delta u_*}{L}. \tag{5.10}$$

If δ satisfies (5.8) then (5.10) produces a result consistent with (5.7). Thus, the Ri^{-1} entrainment law can be arrived at without invoking interfacial shear entrainment or entrainment energized by turbulence imported from the surface; it is instead the consequence of upwelling.

The buoyancy flux (given by (5.9)) introduced by upwelling into the main part of the surface layer at the upwind end is mixed downwind by shear flow dispersion. Thus, the perturbation buoyancy field in the surface layer $g'_{\mathrm{m}}(x, t) = [\rho(x, t) - \rho_0]g/\rho_0$ satisfies the diffusion equation (Fischer *et al.*, 1979)

$$\frac{\partial g'_{\mathrm{m}}}{\partial t} = \varepsilon_x \frac{\partial^2 g'_{\mathrm{m}}}{\partial x^2}, \tag{5.11}$$

where

$$\varepsilon_x = C_2 h u_* \tag{5.12}$$

is the effective diffusion coefficient. The coefficient C_2 depends on the shape of the velocity profile and on the distribution and magnitude of the vertical diffusivity ε_z. For the log profile, characteristic of two-dimensional open channel flow, Elder (1959) derived a value of 6. However, C_2 will remain unspecified for the moment. The width l of the end region is approximately h_1 and can be neglected (Cormack *et al.*, 1974).

A simple quantitative model of how upwelling leads to mixed-layer deepening can be formulated by assuming that the mixed layer is initially homogeneous and by modeling the effect of upwelling with an impulsively started, and later constant, buoyancy flux of strength β at $x = 0$. Thus, the temporal and spatial variations of g'_{m} can be found by solving

(5.11) subject to the following boundary and initial conditions.

$$\frac{\partial g'_{\mathrm{m}}}{\partial x}(0, t) = \frac{\beta}{\varepsilon_x} H(t), \tag{5.13}$$

$$\frac{\partial g'_{\mathrm{m}}}{\partial x}(L, t) = 0, \tag{5.14}$$

$$g'_{\mathrm{m}}(x, 0) = 0. \tag{5.15}$$

The required solution can be found in Carslaw and Jaeger (1978) as

$$g'_{\mathrm{m}}(x, t) = 2\beta(t/\varepsilon_x)^{1/2} \sum_{n=0}^{\infty} [\mathrm{ierfc}\{(2nL + L - x)(2(\varepsilon_x t)^{1/2})^{-1}\} + \mathrm{ierfc}\{(2nL + x)(2(\varepsilon_x t)^{1/2})^{-1}\}], \tag{5.16}$$

where $\mathrm{ierfc}(x) = \{\exp(-x^2)/\pi^{1/2}\} - x\,\mathrm{erfc}(x)$ and $\mathrm{erfc}(x)$ is the complementary error function. Equation (5.16) is plotted in Figure 21 in terms of the quantity $\Gamma = g'_{\mathrm{m}}(x, t)\varepsilon_x^{1/2}\beta^{-1}t^{-1/2}$ for different values of

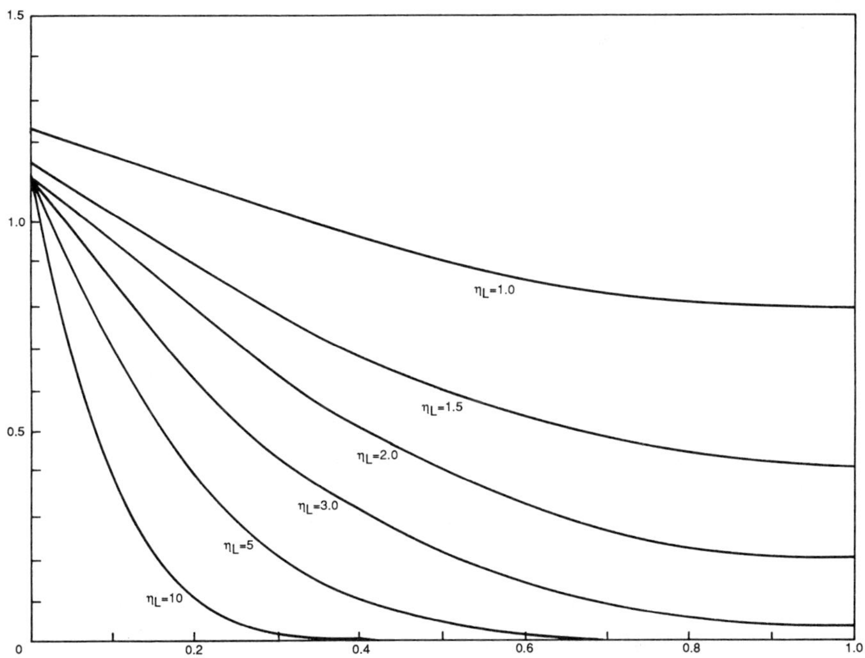

FIG. 21. Curves of the solution (5.16) for $\Gamma = g'_{\mathrm{m}}(x, t)\varepsilon_x^{1/2}\beta^{-1}t^{-1/2}$ as functions of x at different values of $\eta_L = L/(\varepsilon_x t)^{1/2}$. (After Imberger and Monismith (1986).)

$\eta_L = L/(\varepsilon_x t)^{1/2}$. When comparing experimental data with this solution, an allowance must be made for the fact that horizontal density gradients are created during the initial period of set-up by upwelling and by shear-driven entrainment, i.e. $g'_m(x, 0) = f(x)$. This can be done by adjusting the time origin either through addition or subtraction of an offset. Ideally, this offset should be much less than the characteristic timescale of the dispersion process

$$T_0 = \frac{L^2}{\varepsilon_x}. \tag{5.17}$$

In addition, since β depends on h, it is not really constant throughout the experiment. The solution will only be valid for times somewhat less than

$$T_v = \frac{h_1 \mathrm{Ri}}{C_1 u_*}, \tag{5.18}$$

the time required for the mixed-layer depth to double.

From the solution plotted in Figure 21, it can be seen that the effect of the upwelling flux is not felt at $x = L$ until $t \approx 0.11\, T_0$ at which time $g'_m(L) \approx 0.04 g'_m(0)$. The total density difference in the mixed layer reaches a maximum at $t \approx 0.25\, T_0$, remains constant until $t \approx T_0$ and then begins to drop as the mixed layer "heats up". The solution for $g'_m(0, t)$ can be shown to be

$$g'_m(0, t) = 2\beta t^{1/2} \varepsilon_x^{-1/2} \pi^{1/2}, \tag{5.19}$$

for times less than $0.44\, T_0$. Finally, $g'_m(0, t)$ will equal to g'_0 when

$$t = T_f = \frac{\pi}{4}\left(\frac{C_2}{C_1^2}\right) W^2 \left(\frac{h}{u_*}\right). \tag{5.20}$$

If $t > 0.44\, T_0$, the factor appearing in (5.20) will not be $\pi/4$.

Figure 22 shows the data from three experiments reported in Keulegan and Brame (1960), from two experiments reported in Kranenburg (1985), and from Experiment 8 of Monismith (1986). In reducing these data, Imberger and Monismith (1986) carried out the following operations:

(a) C_1 was first chosen to be 0.07 and, for all except Experiment 8, C_2 was chosen to be 10.
(b) The resulting "raw" curves were plotted.
(c) So as to best fit the data to the theoretical curves, the dimensionless time was offset by an amount $\Delta\tau$ and C_1 was altered.

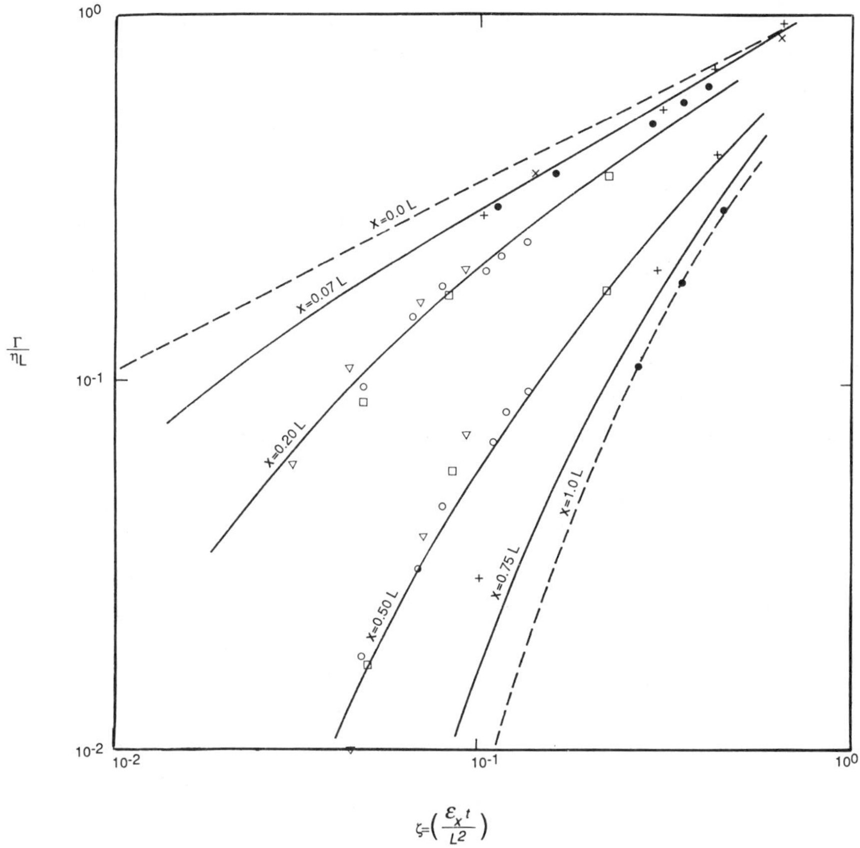

FIG. 22. Plot of values of (Γ/η_L) calculated from densities measured with correction for the initial conditions and adjusted for a best-fit of data at two values of ζ. The solid lines are the theoretical solutions while the symbols represent:
• Monismith (1986), $W = 3.9$, $\Delta\tau' = 0.060$, $C_1 = 0.07$, $C_2 = 6$;
△ Keulegan and Brame (1960), $W = 3.0$, $\Delta\tau' = 0.015$, $C_1 = 0.11$, $C_2 = 10$;
○ Keulegan and Brame (1960), $W = 2.4$, $\Delta\tau' = 0.025$, $C_1 = 0.11$, $C_2 = 10$;
□ Keulegan and Brame (1960), $W = 1.5$, $\Delta\tau' = 0.030$, $C_1 = 0.05$, $C_2 = 10$;
+ Kranenburg (1985), $W = 3.7$, $\Delta\tau' = -0.015$, $C_1 = 0.11$, $C_2 = 10$;
× Kranenburg (1985), $W = 2.3$, $\Delta\tau' = +0.010$, $C_1 = 0.11$, $C_2 = 10$.
(After Imberger and Monismith (1986).)

The last step was interactive and performed first for the data from the ζ position, say ζ_1, closest to $\zeta = 0$. Once a satisfactory fit was obtained for ζ_1, the data at a second position, $\zeta_2 > 0.5$, was reduced using the final values of C_1 and $\Delta\tau$. From trial and error, it became apparent that for each experiment analyzed, a best set of values of $\Delta\tau$, C_1, and C_2 existed that appeared to minimize the differences between theory and measurements for both values of ζ. In all cases, reasonable values of all three parameters were obtained: the time offsets were always less than $0.04\ T_0$, and the entrainment coefficient C_1 was between 0.05 and 0.133, well within the range reported in the literature and cited above. The difference between values of C_2 chosen for the belt-driven flow and the wind-driven flow may reflect differences in the structure of those flows, especially in the way the shear stress and surface velocity vary with x.

Recently Imberger and Spigel (1987) have carried out measurements in Lake Rotognaio in New Zealand during conditions of strong heating, weak winds, and very turbid water (active algal bloom reduced the light extinction depth to 0.5 m). The general upwelling circulation remained, but a strong stratification persisted within the whole surface layer, inhibiting turbulence and eliminating the entrainment surface. The surface buoyancy flux outpaced the mixing caused by wind-induced circulation. This type of upwelling is characterized by W being very small but L_N very large, and the circulation appeared as that described by Monismith (1986) for the continuously stratified case. The major difference observed was that with the buoyancy flux present a steady state circulation appeared to form quite rapidly.

The other important variation involves the case of a variable wind field. If the wind changes direction (Dickey and Simpson, 1983) or speed, gravitational adjustments lead to large intrusions that may degenerate into solitons (Kao *et al.*, 1978; Maxworthy, 1980) or undular bores (Thorpe, 1974; Farmer, 1978; Smyth and Holloway, 1988).

In summary, the upwelling response of the epilimnion (diurnal thermocline) is determined by the magnitude of W, and that of the metalimnion by the magnitude of L_N. Where W is small, but L_N large, it is possible to model the entrainment process at steady state as simple recirculation, where the entrained fluid is swept along the interface into the upwind upwelling region, then is mixed in the surface layer downwind by shear dispersion. If the stratification is complicated and admits numerous eigenfunctions of equal importance, then an equivalent number of nondimensional numbers become important.

VI. Differential Deepening

The wind field over a lake surface is rarely uniform. Mesoscale wind variability (McBean and Paterson, 1975; Bean *et al.*, 1975) causes variation over large lakes, and wind sheltering by the surrounding terrain is the major factor affecting smaller lakes (Parker and Imberger, 1986). Wind variability has two major effects. First, in areas with relatively high wind speed, evaporation, and thus latent heat transfer, will be greater, introducing a horizontal gradient in the surface heat flux. This phenomenon, called differential cooling (Imberger, 1982), is discussed in Section VII. Second, higher wind speeds on the exposed parts of the lake cause both a more rapid local deepening of the surface layer and a greater introduction of momentum. Both must ultimately be averaged across the lake surface by gravitational adjustments of the isopycnals. The variability in the surface layer deepening is called differential deepening (Imberger and Parker, 1985).

Data from a field investigation near the headland of Salmon Brook in the Wellington Reservoir illustrate these three phenomena extremely well (Parker and Imberger, 1986). The data were collected along a transect extending from a station in a sheltered embayment, station W4, to the center of the basin exposed to the wind, Station C10 (Figure 23a). The morning was warm and there was no wind. Shortly thereafter, a southerly wind began to blow (wind at C10 was $2\ \mathrm{m\ s^{-1}}$ and at W4 was $1\ \mathrm{m\ s^{-1}}$); the surface layer was warm in the sheltered part of the transect with some noticeable cooling out towards C10 (Figure 23b). This difference was due to greater evaporative cooling in the open parts of the lake. The wind intensified sharply at around 1540 hours (rose to $4\ \mathrm{m\ s^{-1}}$ at C10, but remained less than $0.6\ \mathrm{m\ s^{-1}}$ at W4) leading to noticeable deepening at C10 (Figure 23c). The wind continued strongly until about 2000 hours, when it decreased markedly to less than $1\ \mathrm{m\ s^{-1}}$. The horizontal density gradient associated with the isotherms is shown in Figure 23d, implying a further deepening than that at 1805 hours. After this time, the isotherms progressively relaxed (Figure 23e) until at 2300 hours (Figure 23f), the isotherms were once again virtually horizontal.

Parker and Imberger (1986) investigated the evolution of the water profiles by using the surface layer models described in the previous section. They showed that the local deepening laws could be used to describe the observations if local wind conditions were used and the effects of horizontal advection were accounted for, although this appeared to be a relatively minor correction. The results from Parker and Imberger (1986) suggest that to obtain an average deepening law, careful averaging must be done over the surface of the lake.

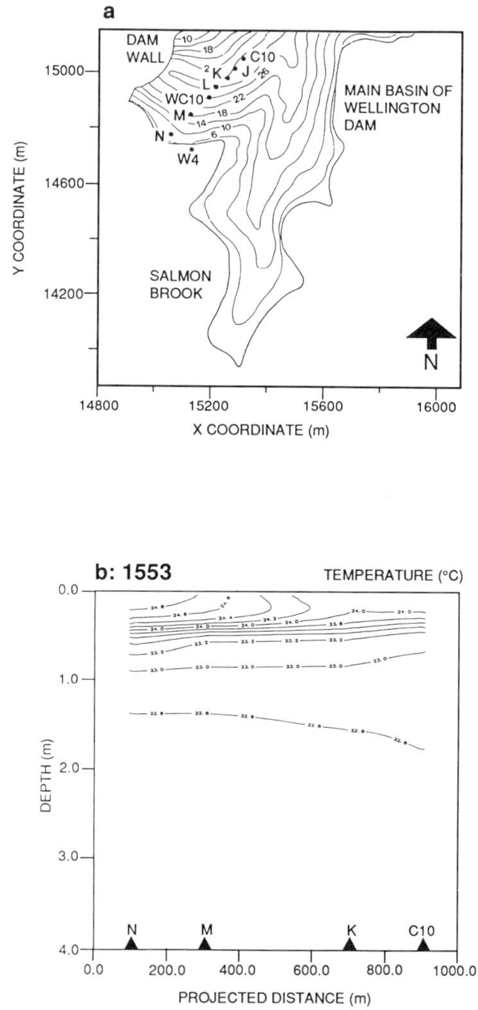

FIG. 23. An example of differential deepening recorded in the Wellington Reservoir in March 1984. (a) Location map showing the isotherms. (b) Isotherms before the wind had started at either the central or the sheltered station. (c) Data recorded at 1800 hours—the wind speed in the central station had risen to 4 m s^{-}1, but in the sheltered corner, the wind was still less than 0.6 m s^{-1}. (d) Data recorded at 1948 hours—wind speed had dropped everywhere to less than 1.0 m s^{-1} immediately prior to taking this data set. (e) Data recorded at 2137 hours—wind at both stations was less than 0.5 m s^{-1}, thermal structure was relaxing and intruding out into the central part of the lake. (f) Data recorded at 2310 hours—the wind remained calm during the whole period, allowing the thermal structure to adjust. The heat flux was beginning to go negative, causing slight cooling at the surface. After Parker and Imberger (1986).

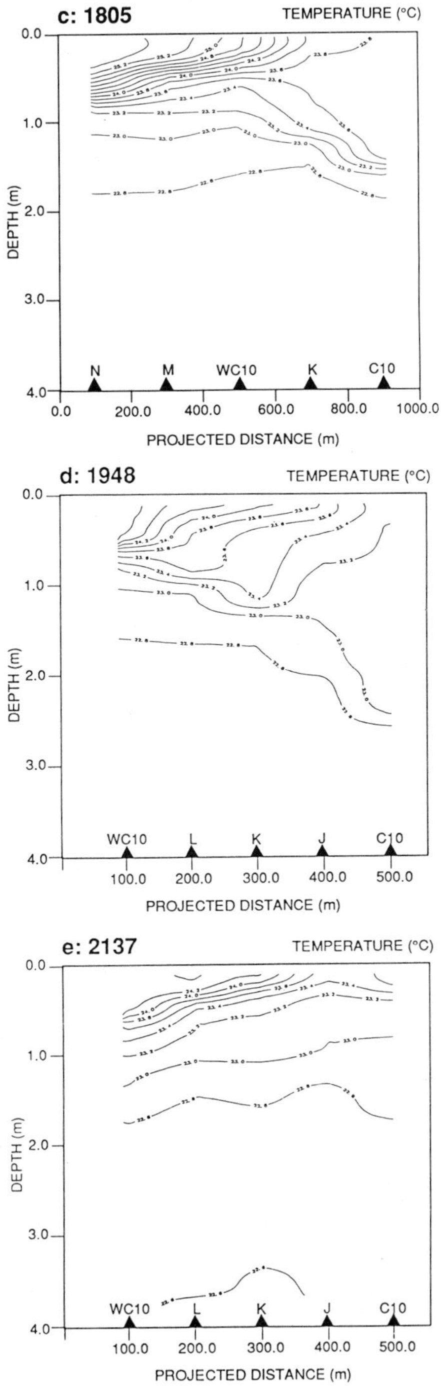
c: 1805
TEMPERATURE (°C)
DEPTH (m)
0.0
1.0
2.0
3.0
4.0
N
M
WC10
K
C10
0.0
200.0
400.0
600.0
800.0
1000.0
PROJECTED DISTANCE (m)
d: 1948
TEMPERATURE (°C)
DEPTH (m)
WC10
L
K
J
C10
100.0
200.0
300.0
400.0
500.0
PROJECTED DISTANCE (m)
e: 2137
TEMPERATURE (°C)
DEPTH (m)
WC10
L
K
J
C10
100.0
200.0
300.0
400.0
500.0
PROJECTED DISTANCE (m)

FIG. 23. (*contd.*)

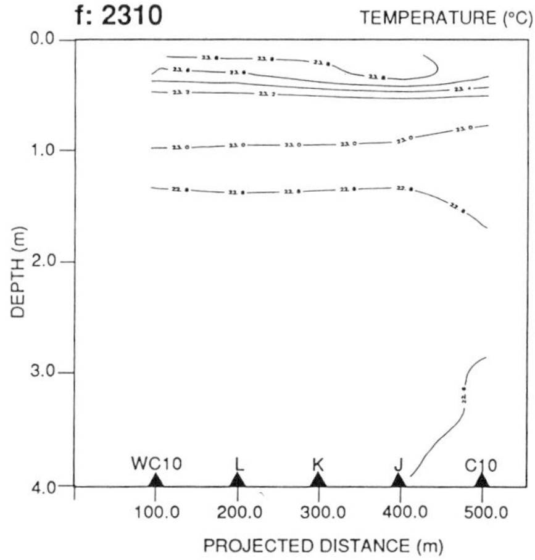

FIG. 23. (*contd.*)

The case just discussed was for conditions of moderate W and large L_N. In Figure 24a, we show a much larger transect extending along the length of the Wellington Reservoir approximately six hours after a westerly wind had begun. The speed of the wind was $2\,\mathrm{m\,s^{-1}}$ at 0600 hours, then rose steadily and peaked at $6\,\mathrm{m\,s^{-1}}$ at 1430 hours. The direction remained quite steady at due west (see Figures 17b and 17c). The dramatic two-dimensionality of the structure is observed in the isopycnals in Figure 24a with severe deepening between Stations C10 and C45 and again between Stations C80 and C90; both sections were more exposed to the strong westerly. The deepening of the isotherms between C10 and C45 was due to the setdown associated with upwelling at C10, whereas the slope of the isopycnals between C40 and C60 was due to the more sheltered nature of this part of the valley. Confirmation of the general tilt west–east due to upwelling is given in Figure 24b, which shows data from a west–east transect at Station C40. The water in both exposed areas was warmer than its surroundings, which is distinct from what is observed in Figures 23b to 23f. This indicates that advection was dominating the deepening process (Imberger, 1985). Further, the average Wedderburn number for the central basin surface flow at 0927 hours was 0.25, and L_N was approximately 1, implying a much less stable lake as a whole, with mixing expected throughout; the shear associated with upwelling (4.20) was sufficient even at Station C40, close to the end of

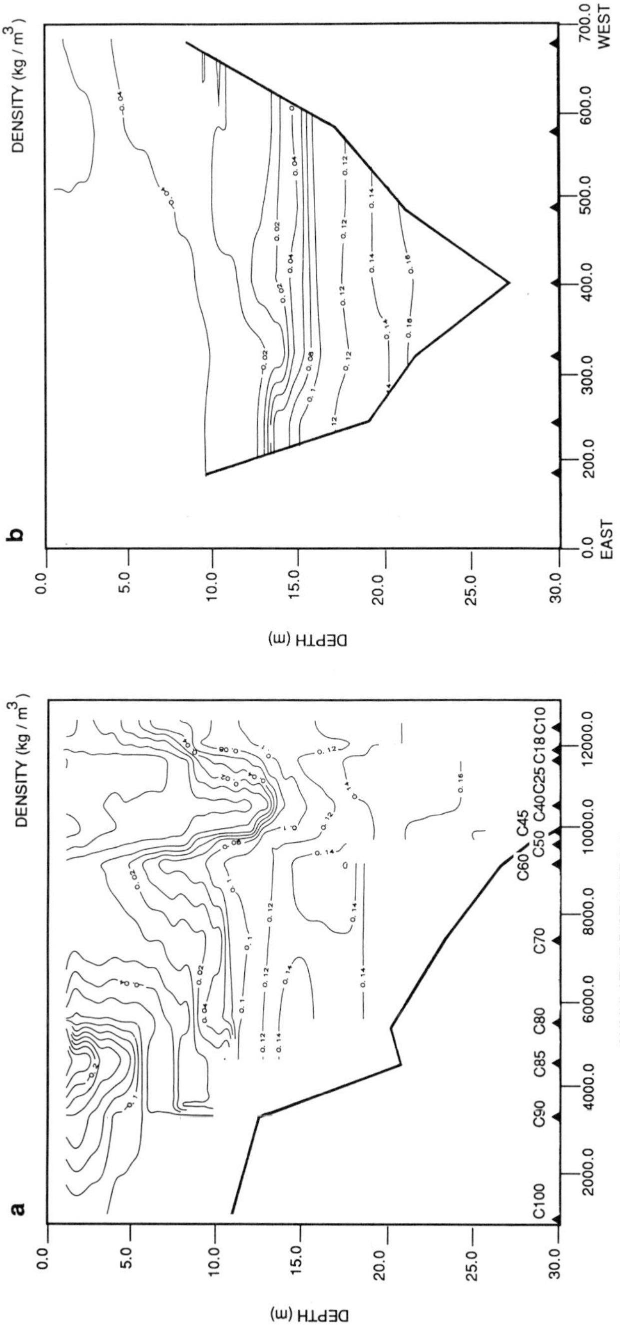

FIG. 24. (a) Transect along the Wellington Valley (see Figure 17a). The wind was coming from the west (as shown in Figures 17b and 17c) and blowing across the lake between the 2000 and 5000 m mark as well as between the 12000–16000 m mark. The isopycnals show extreme differential deepening, the wind strength at the time was $6\,\mathrm{m\,s^{-1}}$. (b) Cross section at Station C40, from east to west, the differential deepening, shown in Figure 24a was accompanied by strong upwelling. (c) Temperature. (d) Gradient temperature. (e) Dissipation as recorded with microstructure instrument at Station C40 at 0947 hours. Dissipation as measured by both the Wigner–Ville technique (sticks) and the Batchelor spectrum fitting technique (bar chart).

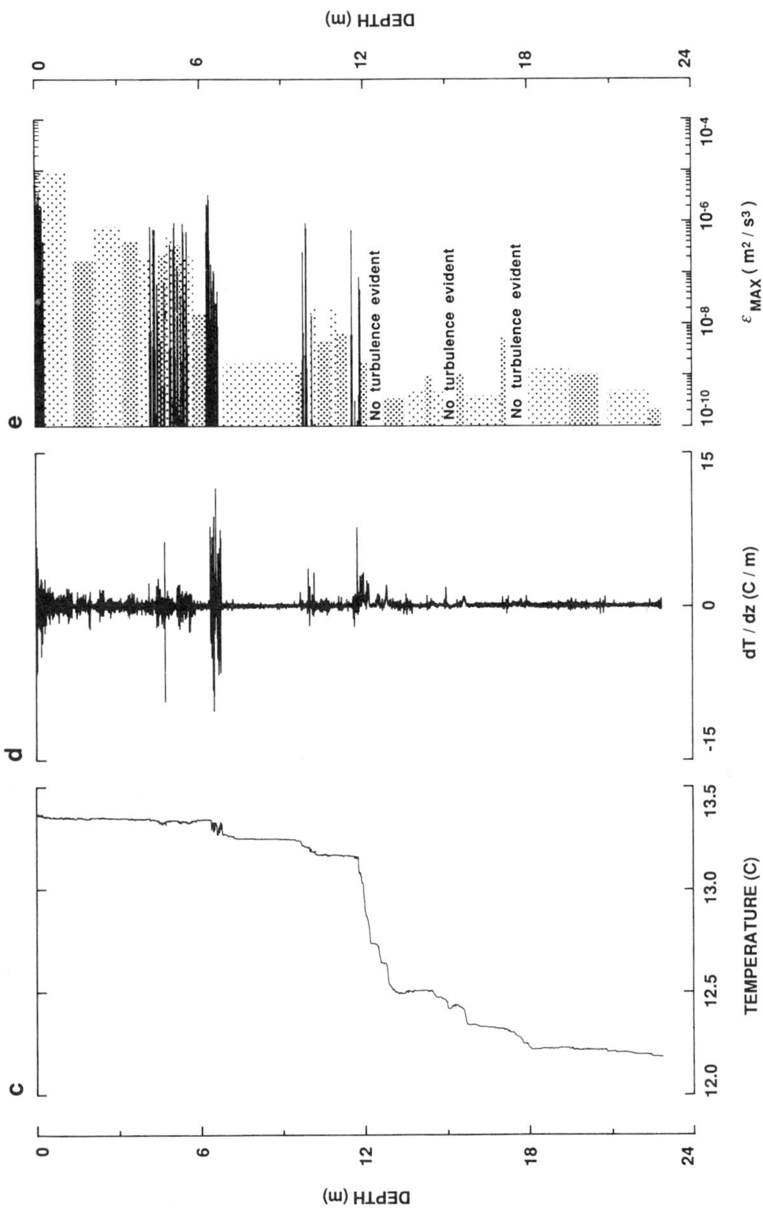
c
d
e
DEPTH (m)
TEMPERATURE (C)
dT / dz (C / m)
ε_{MAX} (m²/ s³)
No turbulence evident
No turbulence evident
No turbulence evident

FIG. 24. *(contd.)*

the basin, to induce very active mixing throughout the water column (see Figures 24c, 24d, and 24e).

De Szoeke (1980) addressed the problem of a variable wind stress over the ocean. However, he aimed to determine the variability induced by the Ekman pumping driven by the curl of the surface wind stress and not by the differential rates of vertical entrainment. Maxworthy and Monismith (1988) were motivated, in part, by the observations in the Wellington (Parker and Imberger, 1986) and conducted a grid stirring experiment where the grid extended over only part of a long, stratified tank. Their results illustrated the sequence of events immediately after the commencement of the grid oscillations. First, the turbulence introduced by the grid mixed the underlying fluid and set up a downward propagating entraining front immediately below the grid. Second, this front continued to propagate downwards until either the turbulence became weaker due to the greater distance from the grid (Hopfinger and Toly, 1976), or the mixed fluid collapsed and intruded horizontally into the neighboring quiescent ambient fluid. Third, as the intrusion propagated out into the main tank, it reached a stage where the end of the tank began to slow its progress, once again influencing the rate of descent of the entrainment front immediately below the grid.

Using the various relationships for intrusion speed and entrainment rates, Maxworthy and Monismith (1988) derived a relationship for the speed of propagation of the entrainment front under the grid and satisfactorily verified these relationships with the experiments. The grid mechanism, while convenient, is a poor analogue for wind-induced turbulence. Thus, even though the flow scenario will be the same, it is more useful, as Maxworthy and Monismith (1988) did, to write down the relationships for the case where the mixing is carried out by a wind stress.

Consider a strongly stratified surface layer over part of which a strong wind stress is suddenly initiated with a shear velocity of u_*. Under the wind stress, the initial deepening will be given by (4.25b)

$$\frac{1}{u_*}\frac{dh}{dt}=\frac{C_1}{\mathrm{Ri}^*}, \tag{6.1}$$

which is the relationship relevant for a linearly stratified fluid and where $C_1 = 0.1$ (Kranenburg, 1985). As shown by de Szoeke and Rhines (1976), surface-induced turbulence and deepening dominates initially, until the shear at the base of the mixed layer (5.2) builds up. For longer times, shear production (4.25a) dominates and the entrainment rate is described by:

$$\frac{1}{u_*}\frac{dh}{dt} = (2C_s)^{1/2}\left(\frac{u_*}{Nh}\right), \tag{6.2}$$

where $C_s = 0.2$ to 0.3 (Spigel *et al.*, 1986).

These deepening processes will continue (Maxworthy and Monismith, 1988) until the outflows/inflows become appreciable (see Figure 25) and dilute the effort expended by the vertical mixing processes.

It is well established (Maxworthy, 1972; Imberger *et al.*, 1976; Manins, 1976) that the intrusion can either be governed by an inertial buoyancy balance or a viscous inertial buoyancy balance; in the reservoir, however, the scales are such that invariably the inertial buoyancy force balance prevails. This means that the intrusion thickness is given by (for a two-dimensional intrusion)

$$\delta = \frac{4\,q^{1/2}}{N^{1/2}}, \tag{6.3}$$

and the speed by

$$c = 0.44(Nq)^{1/2}, \tag{6.4}$$

where q is the two-dimensional volume flux and N the buoyancy frequency. However, as observed by Maxworthy and Monismith (1988), δ is not necessarily equal to the depth H of the entraining front under the

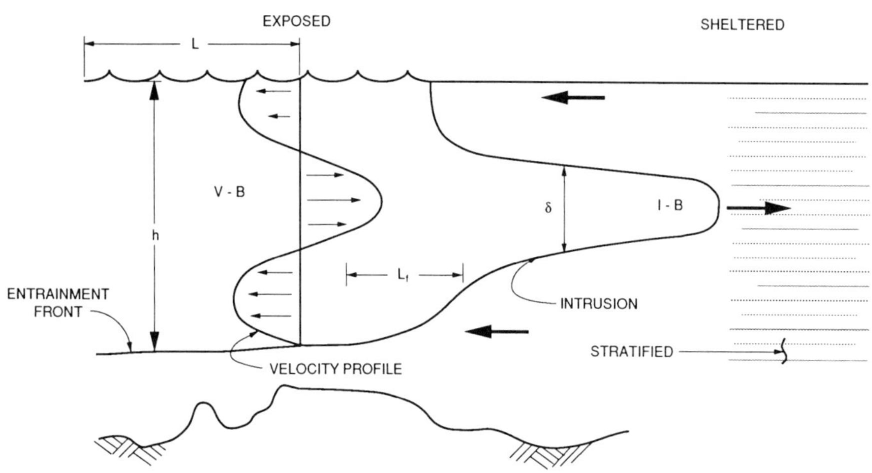

FIG. 25. Schematic of differential deepening.

grid. The flow out of the turbulent area is governed by a gravitationally driven shear flow, which is retarded by the vertical turbulent transport. Such flows have received extensive treatment in estuarine situations; detailed solutions are described in Scott and Imberger (1988). The flow itself was recently modeled in the laboratory by Linden and Simpson (1986), who, however, do not appear to have connected the dynamics of their flow with the pioneering work of Taylor (1954) on shear flow dispersion (see Fischer *et al.* (1979) for a general reference). It is therefore better to re-analyze the Linden and Simpson (1986) experiments in light of the estuarine work and the analysis presented by Maxworthy and Monismith (1988).

Suppose the wind introduces a vertical mixing coefficient given by (Fischer *et al.*, 1979)

$$\varepsilon_z = 0.06\, u_* h. \tag{6.5}$$

The flow under the grid then may be derived, if it is assumed that the grid is much larger than the mixed region is deep ($h/L < 1$), from the long box convection problem, where L is the length of the box.

A simple balance between baroclinic vorticity generation and vertical diffusion leads to the scale of the velocity under the grid (Imberger, 1974)

$$u = \frac{8 \times 10^{-3} g' h^3}{\varepsilon_z L_f}, \tag{6.6}$$

where g' is the reduced gravity across the turbulent front and L_f is the length of the density variation at the outflow end; this length increases with time as the intrusion increases to bring outer fluid into the turbulent zone. This balance may be expected to be valid provided $L_f > h^2 u/\varepsilon_z$ (Fischer *et al.*, 1979).

The effective longitudinal dispersion coefficient is given by Fischer *et al.* (1979) for such a velocity profile:

$$K_x = 5.3 \times 10^{-4} \frac{h^2 u^2}{\varepsilon_z}, \tag{6.7}$$

and a good approximation to L_f is given by

$$L_f = (K_x t)^{1/2} \tag{6.8}$$

for times which are short enough so that $L > L_f$. Combining (6.5) and (6.8) yields the relationships for both the dispersion coefficient and the peak velocity:

$$K_x = 1.2 \times 10^{-2}\left(\frac{g'h^{5/2}}{u_*^{3/2}t^{1/2}}\right), \tag{6.9}$$

$$u = 1.2\left(\frac{g'^2h^3}{u_*t}\right)^{1/4}, \tag{6.10}$$

where it was assumed that the depth h adjusts slowly compared to the time it takes for the flow to adjust to the changing L_f.

Unfortunately, the data presented in Linden and Simpson (1986) are not sufficiently detailed to allow comparison. Maxworthy and Monismith (1988) balanced the deepening given by (4.23) with the vertical advection induced by the outflowing intrusional such that

$$L_f \frac{dh}{dt} = uh. \tag{6.11}$$

However, if we consider the example shown in Fig. 23c, then $K_x = 4 \times 10^{-3}\ \mathrm{m\,s^{-1}}$, so that in the time deepening has taken place, L_f has only grown to 6 m, far too small to influence the process; it is therefore not surprising that Parker and Imberger (1986) obtained good comparisons between the one-dimensional model and the field data once they used local wind speeds. The oversight in the paper by Maxworthy and Monismith (1988) was that no distinction was made between the length scale L_f and L.

In summary, the concept of differential deepening appears to be well established and documented in the field. The laboratory experiments by Maxworthy and Monismith (1988) have given an explanation for the connection between the deepening under the exposed areas and the intrusional flows set up by the buoyancy imbalance introduced by the wind. However, it appears that for diurnal cycles, intrusional flow is of secondary importance to the deepening. However, the intrusional flows, which can reach velocities of up to 5 cm per second (Parker and Imberger, 1986), are of primary importance in redistributing water horizontally once the wind event is over.

VII. Differential Heating and Cooling

Field data collected in Salmon Brook, a side arm of the Wellington Reservoir (Monismith and Imberger, 1988), provides a good illustration of differential heating and cooling. Salmon Brook (Figure 26a) runs north–south with a central uniform valley. The surrounding terrain is hilly and the upper reaches of the side arm are sheltered from the wind.

Data were collected in late February 1985 during a hot (39°C), clear day, where the peak net heat flux at the surface reached 960 W m^{-2}. The wind was less than 2 m s^{-1} all day except briefly between 1500 and 1830 hours, when a westerly wind with a speed of 5.5 m s^{-1} sprang up. The night had been relatively cool and still, with only a small northerly breeze around midnight which reached 4 m s^{-1}. This stopped completely by 0700 hours.

An early morning transect along the central valley of Salmon Brook from the upstream extreme to Station C10 (see Figure 26a) is shown in Figure 26b. Cold water, which had formed around the perimeter of the side arm during the night, fell under gravity from the surface to about 7 m. A transverse transect taken across SB20 (Figure 26c) shows the symmetric plunging of the boundary water. Drogue tracking indicated a drawing of central water into the side arm along the surface consistent with the temperature observations that the night cooling had formed cold boundary water, which downwelled around the perimeter of the side arm. The inflow occurred over 50–70% of the depth of the underflow (7 m), comparable to the laboratory results of Harashima and Watanabe (1986).

In the early afternoon, after strong heating all day, the stratification changed to that shown in Figures 26d and 26e, indicating strong boundary heating around the perimeter with baroclinic pressure gradients of the temperature field forcing the water away from the perimeter into the main basin. However, Monismith and Imberger (1988) observed that during this time the surface water was still slowly moving into the side arm, holding the warm water near the perimeter.

By late afternoon, the surface water motion had reversed and a strong, buoyant surface overflow (velocity approximately 0.10 m s^{-1}) confined to the top 1 m was observed to be moving the water out into the main basin. By 2200 hours, the whole of the surface water in the top 2 m had been replaced by this mechanism.

The westerly wind that sprang up at 1520 hours (see Figures 26f, 26g, and 26h) immediately moved the thin surface buoyant plume onto the east bank; the influence of the wind was progressively stronger towards the mouth of the brook. It is interesting, in passing, to observe the general shape of the upwelling pattern in Salmon Brook as one moves

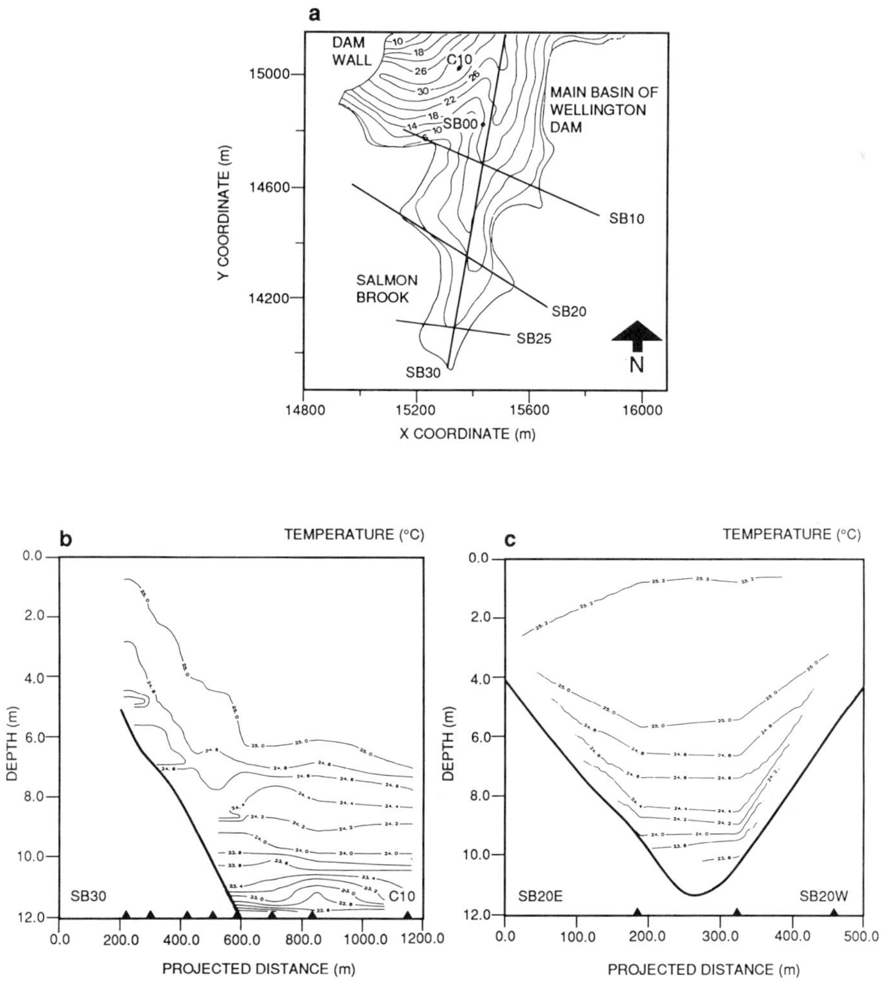

FIG. 26. An example of differential heating and cooling, illustrated with data collected in Salmon Brook on 25 February 1985. (a) Bottom bathymetry of the site showing site and transect positions. (b) Longitudinal transect taken early morning, starting at 0849 hours. (c) Transect across Salmon Brook at Station SB20, early morning, commencing at 0909 hours. (d) Longitudinal transect illustrating midday heating, taken at 1333 hours. (e) Cross-sectional transect at Station SB20, showing thermal heating near boundaries, data taken at 1340 hours. (f) Transect taken at SB25 at 1639 hours, illustrating the influence of a very weak westerly wind. (g) Transect taken at SB20 at 1536 hours, the wind at this station was somewhat stronger. (h) Transect taken at SB00, commencing at 1639 hours. This section was almost completely exposed to the wind, which now had reached a value of 4 m s^{-1}. (i) The net heat flux variation over the five days of the experiment as computed by the rate of change of heat of the total water column at Station SB30.

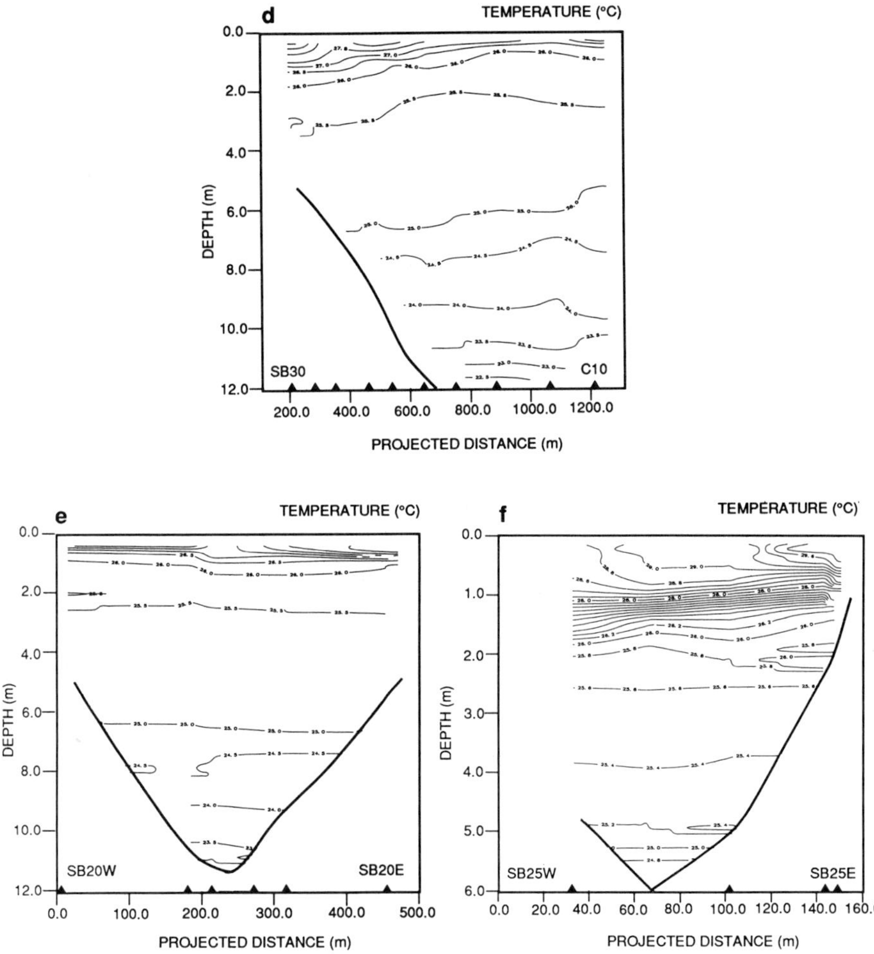

FIG. 26. *(contd.)*

from the sheltered SB25 station to the more exposed SB00 (Figure 26h) station, and to compare this with the discussion in Section VI. At SB25, the Wedderburn number was around 5–10, as indicated by the very weak surface upwelling and the almost horizontal isotherms immediately below. At SB20, the whole diurnal thermocline was tilted, surfacing at the west end of the side arm, indicating that the Wedderburn number was around 1. Lastly, at SB00, the Wedderburn number was calculated as 0.2 and the isotherms indicate a very similar pattern to that shown in Figure 19h, where the Wedderburn number was 0.4.

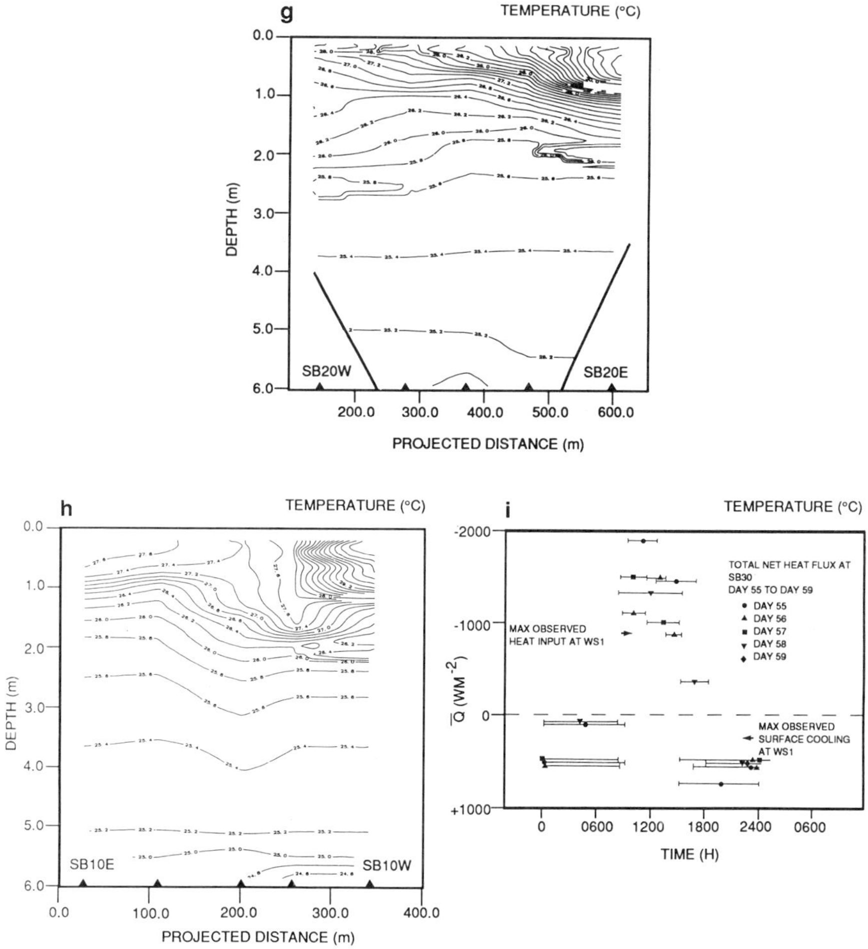

FIG. 26. (*contd.*)

The consequences for the heat budget of this cyclic flow pattern are illustrated in Figure 26i. The data show that the water column at Station SB30 (the head of the side arm) responded as if twice the surface heat load had been applied in both the heating and cooling phases. This is further strong evidence that the inertia of the water leads to a large phase lag between the thermal forcing and the flow response.

Flow patterns associated with the cooling phase are driven by the greater temperature drop around the boundary; it is yet to be established whether a variable heat flux also contributed to the excess temperature.

On the heating phase, both the reduced depth and the reduced latent heat loss (lower wind speed in the sheltered areas) around the perimeter lead to the greater temperatures around the boundaries. Steady convection in a long, shallow cavity, an approximation of the side arm, has received a great deal of attention for two-dimensional configurations (Hart, 1972; Cormack *et al.*, 1974; Imberger, 1974; Bejan and Tien, 1978; Bejan and Rossie, 1981; Bejan *et al.*, 1981; Inaba *et al.*, 1981; Simpkins and Dudderar, 1981; Hart, 1983a, 1983b; Drummond and Korpela, 1987) and in three-dimensions by Imberger (1976) and Scott and Imberger (1988). These solutions rely on a viscous buoyancy balance at first order and all have stable stratification that changes from one with vertical isotherms at low Rayleigh numbers to one with horizontal isotherms and boundary jets along the top and bottom boundaries at very large Rayleigh numbers, such as those that operate in a side arm of a reservoir. These solutions are of limited value, however, as they require an imposed steady horizontal density gradient, with zero heat flux at the upper and lower boundaries. Surface heat fluxes can be incorporated into the solution technique (Cormack *et al.*, 1975; Scott and Imberger, 1988) but unless vertical velocities are allowed at first order, these heat fluxes can only be incorporated at second order, making the solution techniques less applicable.

Most importantly, in these diurnally cycled flows the inertia of the water leads to a velocity field almost 90° out of phase with the thermal forcing. In the data shown in Figure 26, the velocities did not reverse until about 1500 hours, seven hours after the net heat flux had changed from a loss to a gain (Monismith and Imberger, 1988). The explanation of this observation lies in the following scaling arguments. A thermal gradient is instantaneously imposed on a long cavity of fluid. This gradient would induce a flow which would be governed by the balance

$$\rho_0 \frac{\partial u}{\partial t} = -\frac{\partial p}{\partial x}, \tag{7.1}$$

$$\frac{\partial p}{\partial z} = -g\rho, \tag{7.2}$$

leading to a time scale for spin-up;

$$T_s = \frac{u}{Kgh}, \tag{7.3}$$

where u is a typical velocity of the final flow, K is the density gradient

$$\frac{1}{\rho_0}\frac{\partial \rho}{\partial x},$$

and h is the depth of the flow. Inserting typical values derived from Figure 26d leads to a time scale of two hours, which is reasonable given the gradual build-up of the gradient.

The unsteady cavity problem has also been investigated by Patterson and Imberger (1980) and Patterson (1984). The latter paper used internal radiative heating to drive the flow rather than heating and cooling at the vertical end boundaries. The critical parameter to emerge from the long box problem is $\mathrm{Gr}\,A^4$, where

$$\mathrm{Gr} = \frac{g\alpha\,\Delta\theta h^3}{\nu^2}, \tag{7.4}$$

$$A = \frac{h}{L}, \tag{7.5}$$

α is the coefficient of thermal expansion, and $\Delta\theta$ is the temperature differential. Once again, the applicability of these theories is marginal due to the cyclic nature of the heat sources and the very different boundary conditions. However, evaluation of the parameter $P = \mathrm{Gr}\,A^4$ leads to a value well in excess of 1 for a typical side arm, indicating that the flow in side arms must be dominated strongly by inertia even if an effective diffusivity ν of $10^{-4}\,\mathrm{m^2\,s^{-1}}$ is used, as would be indicated by the turbulence measurements in these flows (Imberger, 1988).

The integral solutions advanced by Sturm (1981) and Jain (1982) seek to explain the laboratory findings of Brocard and Harleman (1980). All three deal with side arms with a strong surface heat loss, sustaining an unstable upper water column and a longitudinal circulation driven by a longitudinal temperature gradient. These investigations were motivated by cooling pond configurations where there is a net surface heat loss at all times and a steady state solution can be contemplated. Harashima and Watanabe (1986) presented further laboratory verification that the solutions found by Sturm (1981) and Jain (1982) meaningfully describe these flows. However, since the surface temperature in a cooling pond is generally above the normal equilibrium temperature for the prevailing meteorological fluxes, a longitudinal surface temperature gradient always develops. By comparison (Figure 26b) no such gradient develops during the cooling phase, which is better described by a source of negative buoyancy distributed around the perimeter of the lake. By contrast, the heating cycle may be modeled by a line source of buoyancy in combination with a variable heat flux along the length of the side arm.

Imberger (1985) pointed to a potentially important convection problem in lakes: differential absorption. Light and short wave radiation penetrate the water column to considerable depth; the exact depth is dependent on the concentration of particulate matter and the color of the water (Kirk, 1983). Where the water is transparent, short wave energy is distributed to considerable depth; conversely, in areas of high turbidity, radiation will penetrate to a much lesser depth and more heat will be absorbed near the surface of the water column. This causes the surface temperature in the turbid area to rise, and the water immediately beneath, shaded by the turbidity, will become colder than in neighboring areas. Large temperature changes may therefore be expected to build up over successive days of heating where the turbidity distribution is uneven.

Data collected recently in the upper reaches of the Canning Reservoir are shown in Figures 27a and 27b, for the transect from C45 to C240 (see Figure 27c). The isotherms show that a strong surface warm water lens had formed where the water was turbid near Station CA100. Underneath this lens, the water was considerably colder than the water at the same depth at Station CA45. At the time of these measurements, a weak variation in salinity existed at the upstream end of the reservoir, enough to be used as a tracer, but at a low enough concentration not to significantly affect the density field. The corresponding salinity structures are shown in Figure 27b. An intrusion had formed at a depth of 3 m and was propagating downstream and upstream. At the surface, the hot water lens was spreading out, as evidenced by the spread of the isohalines, although the speed was considerably slower than in the subsurface intrusion. These flows are driven by the weak density gradients induced by differential absorption. The flow may be termed a thermal siphon, and will in many cases be a major contributor to a horizontal redistribution of water in areas of patchy turbidity, an important factor in the kinetics of phytoplankton communities.

Two theoretical models have been developed from the results shown in Figures 27a and 27b. Patterson (1984) used the long cavity with a longitudinally variable absorption as a model. He assumed a longitudinal dependence of the radiation flux

$$Q(x, z, t) = 2Q_0 h(x - \tfrac{L}{2}), \tag{7.6}$$

where h is the depth of the cavity, L is the length of the cavity, and Q_0 is a constant, the value of which was used to fix the intensity of the radiative heating.

The flow was assumed to be laminar so that the thermal energy

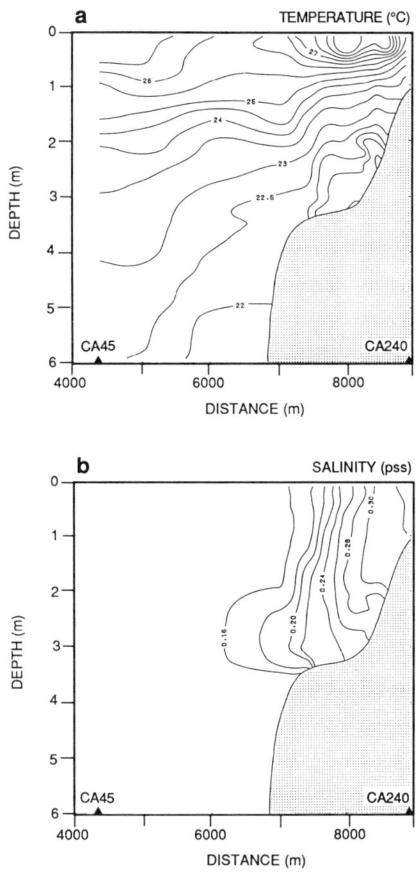

FIG. 27. (a) Isotherms (°C) along a transect from Station CA45 (distance = 4377 m) in the main Canning River tributary to Station CA24D (distance = 9147 m). Data was collected between 1704 hours and 1813 hours, 15 December 1983. For station locations, see Figure 7b. (From Imberger (1985).) (b) Isohalines (pss) for the same data set as in (a). (From Imberger (1985).) (c) Map of Canning Reservoir showing measurement stations.

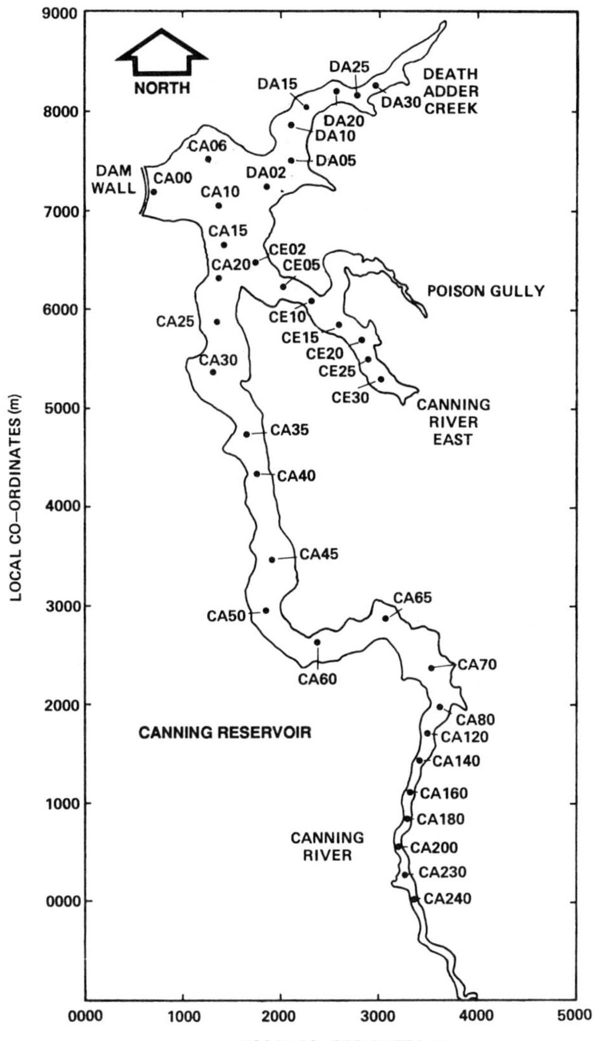

FIG. 27. (*contd.*)

equation became

$$\frac{\partial \theta}{\partial t} + u\frac{\partial \theta}{\partial x} + w\frac{\partial \theta}{\partial z} = \kappa\,\nabla^2\theta + Q(x, z, t). \tag{7.7}$$

A scaling analysis revealed that the flow development, for the initial value problem, depended critically on the magnitude of the parameter

$$P = \mathrm{Gr}\,A^4, \tag{7.8}$$

where

$$\mathrm{Gr} = \frac{g\alpha Q_0 h^4 L^3}{\nu^3} \tag{7.9}$$

is the Grashof number for the problem and A is defined by (7.5). Eight regimes of flow were defined. There are two extremes: $P > \mathrm{Pr}^4$ (where $\mathrm{Pr} = \nu/\kappa$, the Prandtl number) is the regime where inertia and convection dominate the final stages of flow. Second, $P < \mathrm{Pr}^{-2}A^4$ gives a thermal field described by a simple balance between conduction and radiative heating.

Introducing typical values for the problem illustrated in Figures 27a and 27b leads to the first regime (assuming $\mathrm{Pr} \sim 1$) as P is much greater than one. For this regime Patterson (1984) observed a single gyre centered at the transition between heating and cooling. The eddy was flat and supported horizontal isotherms.

Trevisan and Bejan (1986) modeled the turbidity cloud as the perfect reflector. This is the extreme in differential absorption and, once again, a simple gyre was generated at the boundary between the absorbing and reflecting areas although the gyre computed by these authors was not dissimilar to that found by Patterson (1984). Neither model describes the real situation; the absorption is differential in horizontal and vertical directions and the patches are usually three-dimensional.

The original problem that motivated research into convective motions was in cold lakes and concentrated on the thermal bar problems (Rodgers, 1971; Spain *et al.*, 1976; Mortimer, 1987). This phenomenon occurs when a lake, initially at a temperature less than 4°C, begins to warm. Similar to the observations described above, this warming occurs more rapidly around the perimeter due to the shallow depth and the warmer river inflows. In this sense, it is identical to what is believed to be a good model for the data shown in Figures 26d and 26e. The extra complication is that the equation of state is nonlinear at these low temperatures, with a maximum density at 4°C.

Once the boundary fluid has a temperature greater than 4°C, it is lighter and propagates along the surface. As it meets the colder (<4°C) water offshore, mixing produces some 4°C water that sinks (Marmoush *et al.*, 1984) and produces a complicated frontal mixing pattern, distinct from the normal buoyant surface overflow which has a well-defined frontal roller (Luketina and Imberger, 1987). Elliott and Elliott (1970) and Ivey and Hamblin (1988) have carried out laboratory experiments in long cavities and Ivey and Hamblin (1988) in particular present graphic data for the formation of a strong region of convergence, where the 4°C water plunged in a plumelike sheet to the bottom of the tank. By analogy with the long box problem, Ivey and Hamblin (1988) suggest the thermal bar (the region of strong convergence) has a width of $O(h\ \mathrm{Ra}^{-1/4})$, where

$$\mathrm{Ra} = \frac{g\alpha(\Delta\theta)^2 h^3}{\nu\kappa}, \tag{7.10}$$

and

$$\rho = \rho_0[1 - \alpha(\theta - \theta_0)^2] \tag{7.11}$$

describes the equation of state in the range $0°\mathrm{C} \leq \theta \leq 8°\mathrm{C}$, $\theta_0 = 4°\mathrm{C}$.

Bennett (1971) included rotation to show that a long shore shear was generated by the thermal bar density gradients. Further, a transverse gyre motion supporting the convergence reinforced the thermal bar. This analysis showed that the time scales for the establishment and propagation into the center of the lake is of the order of months, distinct from the behavior in temperate lakes where, as shown above, it is a matter of hours.

In summary, we have illustrated a new class of convective flows driven by what has been called differential heating and cooling. The main features of this flow are that it is three-dimensional and unsteady with the inertia of the water introducing a large phase lag between the thermal forcing and the resulting motions. The impact of these motions should not be underestimated. In the smaller lakes discussed above, these convective transports will communicate boundary water to the center of the lake within hours and in large lakes, where rotation is important, Ookubo *et al.* (1984) have shown that the boundary thermal inputs lead to basin scale gyres, which again ensure a greatly enhanced horizontal transport.

VIII. Outflow

The outflow from a stratified fluid is influenced by the buoyancy force, which inhibits differential vertical motion, causing tilting of the isopycnals. When an outlet is suddenly opened, the pressure wave instantaneously sets up a radial flow pattern towards the outlet. However, the radial convergence near the sink quickly distorts the isopycnals, leading to uneven vertical displacements. Buoyancy initiates a set of internal waves that propagate out, adjusting the isopycnals back to a horizontal neutral position, and the initial radial flow collapses into a jet-like structure, flowing horizontally toward the outlet. Such a flow field is called selective withdrawal, as most of the flow comes from the level of the sink.

Selective withdrawal has wide application as a management technique for the supply of water with desired water quality properties. The technique was used intuitively to scour out poor quality water from the bottom of the reservoir, long before any theory was developed. It was not until the early 1960s that Raphael (1962), Koh (1966b), and Wunderlich and Elder (1969) developed the foundations of the selective withdrawal process.

The effect of selective withdrawal on the stratification of the reservoir has received considerably less attention. The thermal structure shown in Figure 6, measured in Wellington Reservoir in Western Australia, is an example. Prior to 1976, the lake had a very strong thermocline from November to June, which caused severe oxygen depletion in the hypolimnion. In August 1976, the scour policy was first used (see Imberger and Hebbert (1980)) to rid the reservoir of highly saline water from the first inflows. The procedure weakened the density structure, lowering the value of the Lake number so that the thermal structure was completely changed for the rest of the year. This change influenced the mixing, which affected the vertical flux of oxygen, and thus the reservoir's water quality. While the Wellington Reservoir is an extreme example because of the influence of salinity on density structure, it displays the principle well.

Selective withdrawal may thus be used to optimize the water quality in the reservoir and so obviate the need for mechanical devices such as pumps or bubblers. Research into selective withdrawal has been intensive, and there is a large body of literature. This discussion is divided into the following categories:

(1) line sink in a two-layered system,
(2) point sink in a two-layered flow,
(3) line sink in a linear stratification,
(4) point sink in a linear stratification,
(5) selective withdrawal in a rotating stratified fluid.

An empirical procedure for determining withdrawal characteristics from a nonlinear stratification is described in Fischer *et al.* (1979); few fundamental results exist.

1. *Line Sink in a Two-Layered System*

Consider first the simple case of an infinitely deep fluid of density $(\rho + \Delta\rho)$ on top of which is a thin, finite layer of water with a density of ρ. The case where there is no surface tension between the fluids, the viscosity is small and a line sink is located a depth H below the interface was first investigated by Craya (1949). The two layer case is mathematically identical to that of a single fluid with $g' = (\Delta\rho/\rho)g$ replacing g (Imberger, 1980). Since the fluid is infinitely deep and the pressure is constant along the free surface, the Bernoulli equation can be applied along the free surface. Craya (1949) postulated a solution with a sharp cusp extending from the surface to distance h above the sink. This is distinct from the flow where a small stagnation mount forms over the sink (Imberger, 1980). The flow at infinity is at rest and, assuming the flow in the vicinity of the sink is radial and unaffected by the presence of the cusp, the Bernoulli equation reads

$$H = h + \left(\frac{q}{2\pi h}\right)^2 \frac{1}{2g'}, \tag{8.1}$$

where q is the two-dimensional volume flux at the sink. Craya (1949) solved this equation for the Froude number F:

$$F^2 = \frac{q^2}{g'H^3}, \tag{8.2}$$

such that

$$F^2 = 8\pi^2\varepsilon^2(1 - \varepsilon), \tag{8.3}$$

where ε is the ratio h/H. He postulated that the critical Froude number (the Froude number at which the upper surface is drawn down into the cusp) was given by

$$\frac{dF^2}{d\varepsilon} = 0. \tag{8.4}$$

This leads to the solution $\varepsilon = \frac{2}{3}$ and

$$F_c = 3.42. \tag{8.5}$$

Gariel (1949) performed experiments with a line sink at the bottom of a flat bed tank. Modifying his results for the full q as above yields a critical Froude number

$$F_c = 3.0 \tag{8.6}$$

Tuck and Vanden Broeck (1984) used conformal mapping and a series solution technique to determine a cusp solution for the above semi-infinite fluid domain problem. They obtained a value (for the full q)

$$F_c = 3.55. \tag{8.7}$$

Hocking (1985) treated the case of a sink above a sloping bottom with an arbitrary angle. In the limit of a line sink embedded at the apex of the sloping bottom and the slope going to zero, he found the critical value

$$F_c = 2.88, \tag{8.8}$$

which is smaller but of similar order to the Bernoulli solution (8.6).

Hocking (1988) gave experimental evidence that the critical value was 1.2, much smaller than that predicted by (8.5) or (8.7). Yih (1980), however, speculated upon the existence of solutions with a cusp to the problem with a linearized (constant velocity) free surface condition and hence infinite Froude number. Such solutions are impossible in a fluid of infinite depth, since the velocity at all points in the fluid must approach zero at large distances from the sink. This condition would mean that the velocity everywhere on the free surface was zero. If solutions exist in the finite depth infinite Froude number problem, then solutions must also exist for finite, but large, Froude numbers. This would mean that for this geometry, there would be more than one solution with a cusp. Yih (1980) suggested that the critical withdrawal layer would then be the lowest value of the Froude number for which a cusp solution existed. Collings (1985) and Hocking (1988) found solutions, the latter over a complete range of sink to bottom depth ratios, into the range where the sink rises above the level of the free surface at infinity (negative ratios).

Vanden Broeck and Keller (1987) found not only the infinite Froude number solutions but, using a variation on the series method, solutions of the full range of sink to bottom depth ratios between zero and one, for a wide range of finite Froude numbers. They found that as the Froude number was decreased from infinity to a given ratio, cusp-like solutions were obtained at all values until some critical Froude number was reached, or the Froude number reached 2. They speculated that where

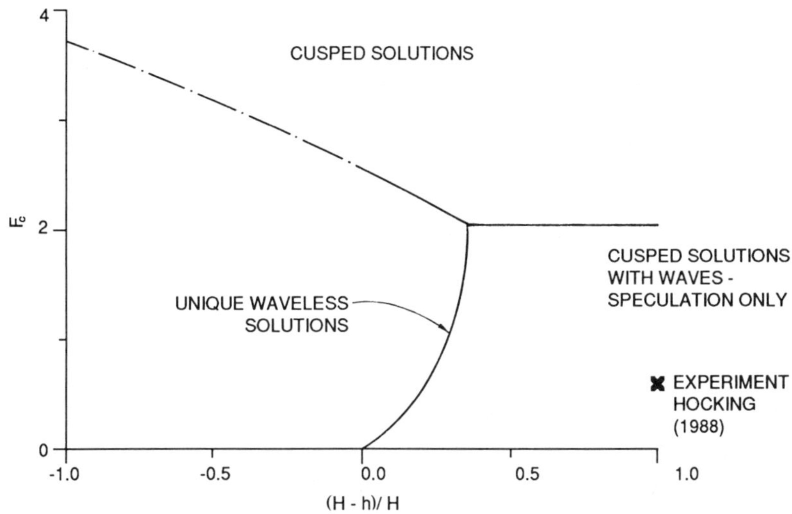

FIG. 28. Solid line: theoretical results from Vanden Broeck and Keller (1987). Dot–dash line: theoretical results with a sink (Hocking, 1988). The Froude number F_c in this diagram is based on the depth of water and not the sink depth.

solutions existed down to a Froude number of 2, solutions with waves on the free surface could also be found for a Froude number less than 2, although they did not calculate them. They also found the branch of solutions on which no waves existed and which allowed solutions into the range of Froude numbers much less than 2. This branch contains solutions in which the fluid is of infinite depth. These results have been summarized in the phase diagram in Figure 28.

2. *Point Sink in a Two-Layered Fluid*

Here, the Bernoulli approach for the semi-infinite domain becomes (using the same notation)

$$H = h + \frac{Q^2}{(4\pi h^2)^2} \frac{1}{2g'} \tag{8.9}$$

where Q is the three-dimensional volume flux. Rearranging and finding the maximum of this expression leads to the critical Froude number

$$F_c = 5.09, \tag{8.10}$$

where

$$F = \frac{Q}{(gH^5)^{1/2}}. \tag{8.11}$$

This result was first derived by Craya (1949). More recently Wood (1978) generalized (8.10) to flows in a sector of a sphere and Bryant and Wood (1976) treated the flow of multiple layers in a sector. Forbes and Hocking (1988) derived mathematical solutions for the stagnation flow solutions with no cusps up to a Froude number of 6.4. Experimental verification was carried out by Lawrence and Imberger (1979) who found a value of $F_c = 4.9$, which was confirmed by Wood (1978). Harleman *et al.* (1959) had found a value of 3.4, whereas the experiments of Jirka and Katavola (1979) yielded a value of 3.15.

Blake and Ivey (1986a,b) also looked at the case where viscosity is the controlling force and not inertia to derive the result

$$h_c = 2.1\left(\frac{\nu Q}{g'}\right)^{1/4}, \tag{8.12}$$

where h_c is the critical draw-down depth. The transition between (8.12) and (8.10) occurs when the parameter (Ivey and Blake, 1985)

$$S = \left(\frac{Q^3 g'}{\nu^5}\right)^{1/20} = 2 \rightarrow 5. \tag{8.13}$$

In both the line and point sinks a diffuse interface will lead to lower critical Froude numbers (Jirka and Katavola, 1979). This may be explained by comparing the flow with the upwelling flow in a high Wedderburn number regime; the three-layer solution of Monismith (1986) showed a distinct divergence of the interface at the upwind end. Such a mode 2 response is identical to the interface spreading near the sink.

For discharges greater than the critical value given by (8.10) or (8.12), both layers flow toward the sink, such that the interface makes a well-defined angle at the sink and has a critical point at which the flow in both layers is controlled. Wood (1978) derived an expression $\lambda = Q_E/Q_H$ as a function of the Froude number

$$F_3 = \frac{Q}{2\sqrt{2g'H^5}},$$

where Q_E is the flow from the epilimnion and Q_H is the flow from the

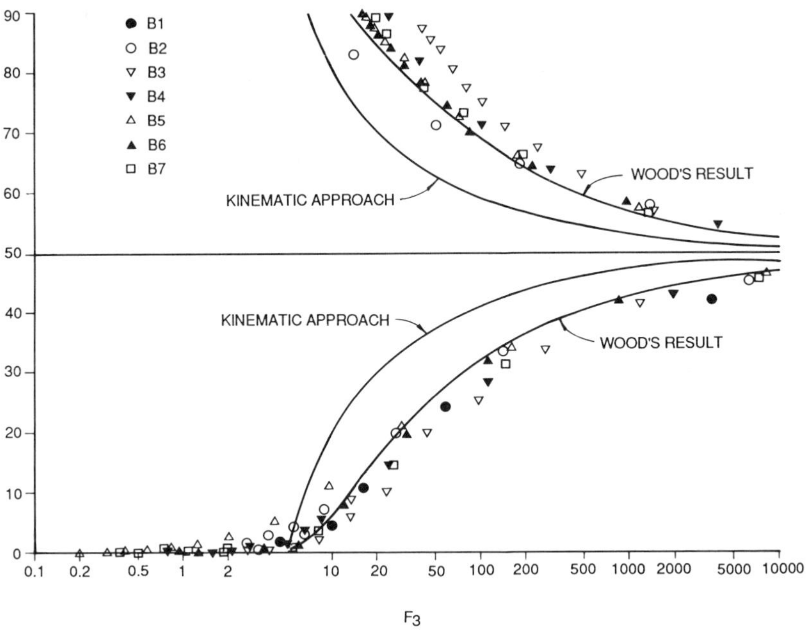

FIG. 29. The ratio λ of epilimnion to hypolimnion discharge as a function of the Froude number for flow towards a point sink. (After Lawrence and Imberger (1979).)

hypolimnion and $g' = (\rho_2 - \rho_1)g/\rho_2$. He also presented laboratory findings that verify the concept of the critical point and the flow ratio once the value of the Froude number was adjusted to compensate for the non-radial flow at the critical section. Lawrence and Imberger (1979) carried out further experimental verification of the above theory. Their results and Woods' are shown in Figure 29. Murota and Michioku (1986a) generalized these results to a three-layer fluid.

3. *Line Sink in a Linear Stratification*

The horizontal duct is the simplest configuration in which to consider the initial value problem for a linearly stratified fluid. Initiation of the sink discharge leads to an instantaneous potential flow solution

$$\psi = \left(\frac{1}{4} - \frac{z}{2}\right) - \sum_{m=1}^{\infty} \frac{1}{2m\pi} e^{-2m\pi x} \sin 2m\pi z, \tag{8.14}$$

where ψ is the streamfunction nondimensionalized with respect to q, and

z and x are the vertical and horizontal coordinates nondimensionalized with respect to H. However, as soon as the fluid particles begin to move, they experience the buoyancy forces introduced by the stratification. During these initial stages, inertia and viscous forces will be negligible and the flow represents a balance between the unsteady vorticity term and the baroclinic generation of vorticity as expressed by the streamfunction equation:

$$\nabla^2 \psi_{tt} + \psi_{xx} = 0, \tag{8.15}$$

where $\psi = \psi/q$, $z = z/H$, $x = x/H$, $t = tN$, N is the buoyancy frequency of the stratification, and ∇^2 is the Laplacian operator.

Imberger (1970) used transform techniques to verify that close to the sink, the solution to the duct problem was identical to the solution in the unbounded fluid domain solved earlier by Koh (1966a). These authors were able to show that for large time the horizontal velocity is

$$u(x, z, t) \sim \frac{\cos\theta}{\pi r}\left[J_0(t) + \frac{1-\beta^2}{\beta}\int_0^t J_0(\omega)\sin\beta(t-\omega)\,d\omega\right], \tag{8.16}$$

where $\beta = \sin\theta$, $\theta = \arctan(z/x)$, $r = (x^2 + z^2)^{1/2}$, and J_0 is the Bessel function of zeroth order. In the limit $t \to \infty$ (x and z fixed), this solution reduces to

$$u(x, z, t) \sim \delta(z), \tag{8.17}$$

which represents a singular jet flow at the level of the sink. The internal waves responsible for changing the initial uniform flow in the duct into the line jet must move at finite speed and hence steady state can never be achieved in the model problem described by (8.15).

It was not until the work described in Kao *et al.* (1974) and Pao and Kao (1974) that the role of the internal waves was fully understood.

These authors simplified (8.15) by omitting the x derivatives in the Laplacian operator and showed that the double Fourier series describing both the horizontal and vertical wave number space could be simplified to a single sum of columnar disturbances.

$$\psi = \left(\frac{1}{4} - \frac{z}{2}\right) + \sum_{n=1}^{\infty} \frac{2}{n\pi}\sin 2n\pi z\, H(n\pi x + t). \tag{8.18}$$

McEwan and Baines (1974) independently defined these columnar disturbances, but named them shear waves, in connection with a different problem. They derived, by a modified method of steepest descent, the behavior of the fluid in the vicinity of wave fronts. Their analysis showed that instead of remaining compact, as suggested by the step function in (8.18), the shear waves are analogous to expansion waves in compressible flow, and a wave travelling with a speed $c_n = HN/n\pi$ has a frontal width $W_n = H(Nt)^{1/3}/n$.

Physically, the initiation of the sink discharge thus leads to a potential flow which, in turn, may be decomposed into a complete set of progressive internal waves, containing the whole spectrum of vertical and horizontal wave numbers. Once established this initial wave field propagates away from the sink leaving a steady sink flow in its wake. As Mahony and Pritchard (1977) point out the sole purpose of these waves is to adjust the isopycnals so that they can return to a horizontal position at steady state. McEwan and Baines (1974) proved that it is possible to sum over the horizontal wave number space leaving an infinite series of shear waves.

It may now be shown (Clarke, pesonal communication) that the long wave limit of (8.15) is the Airy Equation (Cohen, 1979) and so shear waves may be thought of as solutions to the Airy Equation. Clarke has also recently shown that the inclusion of the nonlinear convective acceleration term leads to the Korteweg–de Vries equation (Benney, 1966; Djordjevic and Redekopp, 1978) so that the nonlinear aspects of shear waves can be thought of as analogous to an undular bore (Smyth and Holloway, 1988).

The nonlinear aspects of a shear wave generation, distinct from the propagating properties, were investigated experimentally by Monismith *et al.* (1988). As the inertia of the flow increased, the observed shear wave amplitude decreased relative to their theoretical values. When the outflow was repeatedly pulsed, the strength of the average flow towards the sink indicated a withdrawal layer thicker than that which would have been found given either the average flow rate or the maximum instantaneous flow rate.

Each shear wave has a vertical wave length $\lambda_n = 2H/n$ and a wave celerity $C_n = \lambda_n N/2\pi$, so that the approach to the δ function velocity distribution given by (8.17) is brought about by the passage of successive shear waves, each with a progressively smaller vertical wave length. Steady state is achieved either by the dissipation of these waves or by advection sweeping the waves back out through the sink.

First, consider the action of viscosity, which will dissipate the nth wave in a time of $O(\lambda_n^2/\nu)$. However, in this time the wave has moved the distance $\lambda_n^3 N/2\pi\nu$ so that the non-dimensional thickness δ/x becomes

$$\frac{\delta}{x} = \frac{\lambda_n}{2x} = O(\mathrm{Gr}_x^{-1/6}), \tag{8.19}$$

where $\mathrm{Gr}_x = N^2x^4/\nu^2$. Thus, a viscosity dominated layer has an $x^{1/3}$ upstream growth. This layer was first derived theoretically and verified experimentally by Koh (1966a). The velocity profile is bell shaped with a weak return flow immediately above and below the main layer.

Second, as a shear adds to its predecessor, the central velocity growth is of order $2\pi x/Nt$. Advection will dominate once the induced velocity $O(\nu/\lambda_n)$ is equal to the nth wave celerity $O(\lambda_n N/2\pi)$. Successive waves of order greater than n will no longer be able to propagate upstream. This condition leads to a layer thickness

$$\frac{\delta}{x} = O(F_x^{1/2}), \tag{8.20}$$

where $F_x = q/2Nx^2$ and q is the total flow from both sides. Hence, in the inertial case, the layer is of constant thickness converging only in the potential flow region near the sink.

Combining (8.19) and (8.20) shows that the withdrawal layer changes from one dominated by viscous forces to an inertially dominated layer at a distance x_c given by:

$$\frac{x_c}{L} = R_L^{3/2}, \tag{8.21}$$

where $R_L = F_L\,\mathrm{Gr}_L^{1/3}$ and L is a reference length replacing x. The length x_c was first recognized by Imberger (1972), and Silvester (1979) confirmed its importance with a series of laboratory experiments.

The viscous flow given by (8.19) will no longer form the correct upstream flow when the experimental tank is shorter than x_c. Debler (1959) described experiments in a short tank with an adjoining large upstream reservoir. In such circumstances, ρU^2 is constant upstream, and the motion is governed by the Euler equation together with the constraint $D\rho/Dt = 0$. As demonstrated by Dubreil-Jacotin (1934), this leads to the modified streamfunction equation

$$\nabla^2\psi' = \frac{gz}{\rho_0}\frac{d\rho}{d\psi'} + \frac{d\tilde{H}}{d\psi'}, \tag{8.22}$$

where $\psi' = (\rho/\rho_0)^{1/2}\psi$ and $\tilde{H}$ is the total upstream head. The first term on the right side of (8.22) represents the vorticity generated by the buoyancy forces, while $d\tilde{H}/d\psi'$ is the vorticity inherited from the upstream boundary condition. Yih (1965) showed that if $\rho U^2 = \text{constant}$ and the density varies linearly with z, then the right side of (8.22) is constant and a solution may be found by separation of variables for $F = q/2NH^2 > 1/\pi$. This solution consists of uniform flow upstream necking down to a withdrawal layer near the sink. Above and below this layer are weakly rotating eddies, which extend to infinity at $F = 1/\pi$, thus violating the upstream boundary conditions. This result was also derived by Trustrum (1964). Kao (1970) constructed a numerical solution for $F < 1/\pi$, along the lines of the method used by Huber (1960). Kao (1970) showed that the layer thickness was given by (8.20), with a numerical constant $\sqrt{\pi}$ instead of the order sign. This value agrees with the experimental results of Debler (1959).

In all experimental and prototype applications of the theory, some type of rear wall exists that blocks the horizontal velocities. Imberger and Fandry (1975) investigated the influence of such blocking by solving the flow in a vertical duct. They demonstrated that internal waves were indeed initiated but the vertical duct prevented their separation into vertical shear waves. The internal wave group merely moved vertically through a uniform flow, leaving behind the flow into the sink. This model showed that the horizontal velocity varied linearly from $x = 0$ to $x = L/2$.

Imberger *et al.* (1976) solved the flow in a finite tank with a free surface. The vertical constraints reintroduced shear waves that traversed the tank, reflecting at the vertical boundaries. Their solution highlighted a number of points:

(a) The horizontal jet discharge varied linearly from the sink to the end wall.

(b) For the parameter $R_L < 1$, the withdrawal layer thickness was given by

$$\frac{\delta}{L} = 5.5\,\mathrm{Gr}_L^{-1/6}\,\mathrm{Pr}^{-1/6}, \tag{8.23}$$

with 64% of the layer lying above the sink center line and where Pr is the Prandtl number.

(c) For $R_L > 1$, the flow was completely inertial and the withdrawal layer thickness was given by

$$\frac{\delta}{L} = 4.0 F_L^{1/2}. \tag{8.24}$$

(d) The withdrawal layer flow may be assumed to be steady throughout the tank when it is steady at the end of the tank. This led to a set-up time $t = O(N^{-1}\,\mathrm{Gr}_L^{1/6})$ for $R_L < 1$ and $t = O(N^{-1}F_L^{-1/2})$ for $R_L > 1$.

These conclusions, including the numerical factors in (8.23) and (8.24), have been verified experimentally by Silvester (1979). Further, data given by Bohan and Grace (1973) for the inertial flow regime may be interpreted via the above formula, (8.24) yielding a numerical factor of 3.4.

In the above, it was implicitly assumed that the Prandtl number was of order unity. In many applications, especially where the water is stratified with salt, this is not the case. Imberger *et al.* (1976) showed that the flow structure was much more complicated for fluids with a large Prandtl or Schmidt number. Basically, two complications arise. First, there are two critical horizontal length scales x_c, one where the viscous buoyancy diffusive layer changes to a nondiffusive layer and one where the influence of inertia becomes important. Second, the layer of thickness order $(\mathrm{Gr}^{-1/6})$ is no longer stable but undergoes a secondary and tertiary collapse. This leads to a hierarchy of layer types depending on the relative magnitude of R_L compared to the Prandtl (Schmidt) number Pr. One of these layers has a thickness $\delta/L = O(\mathrm{Gr}^{-1/6}R_L^{1/5})$, experimentally found by Mahony and Pritchard (1977), and one that is basically the nondiffusive layer documented by Gelhar and Mascolo (1966) and Walesh and Monkmeyer (1973). Silvester (1979) has experimentally verified the existence of the flow hierarchy, but there remains a difficulty in the regime $\mathrm{Pr}^{-5/6} < R_L < \mathrm{Pr}^{-1/3}$, where the potential energy of the flow is zero over certain parts of the layer (see Imberger and Fischer (1970)) and where Mahony and Pritchard (1977) have observed hysterisis effects.

4. *Point Sink in a Linear Stratification*

The axisymmetric withdrawal layer was first investigated by Koh (1966a) in the limit of very small discharge so that the flow was governed by a viscous buoyancy balance. Lawrence and Imberger (1979) demonstrated that this steady flow was established by the passage of a series of shear waves originating from the sink at the initiation of the outflow. However, in the axisymmetric case, these waves are cylindrical with amplitudes depending inversely on the radius of propagation. These authors postulated, since the layer thickness was independent of the discharge, that the thickness should be the same as in the two-dimensional case. This is borne out by the data from Koh (1966a), who found

$$\frac{\delta}{L} = 5.8\,\mathrm{Gr}^{-1/6}\,\mathrm{Pr}^{-1/6} \tag{8.25}$$

for the case where inertia was negligible. Spigel and Farrant (1984) investigated the inertia–buoyancy limit and verified that the layer thickness was given by the scale presented by Imberger (1980):

$$\delta = C_1\left(\frac{Q}{N}\right)^{1/3}, \tag{8.26}$$

where $C_1 = 1.58$ (Bohan and Grace, 1973), 1.32 (Lawrence and Imberger, 1979), 1.6 (Spigel and Farrant, 1984), and 1.32 (Ivey and Blake, 1985). Ivey and Blake (1985) also showed there was a layer thickness

$$\delta = 4.2\left(\frac{\nu Q}{N^2}\right)^{1/5} \tag{8.27}$$

that described selective withdrawal when inertia was small but species convection was important.

The transition between these various layers is more complicated than in the two-dimensional case. Ivey and Blake (1985) found a transition parameter

$$S = \left(\frac{Q^2 N}{\nu^3}\right)^{1/15}, \tag{8.28}$$

but this parameter is now independent of distance from the sink so that the inertia-free convective regime thickness (8.27) is also no longer dependent on distance from the sink. A double criterion thus arises depending on whether the convective layer can exist at all. These are

A:	$S > 3$,	$S > (\mathrm{Gr}\,\mathrm{Pr}^{-2})^{1/30}$	Inertial layer	(8.26);
B:	$S > 3$,	$S < (\mathrm{Gr}\,\mathrm{Pr}^{-2})^{1/30}$	Diffusive layer	(8.25);
C:	$S < 3$,	$S > (\mathrm{Gr}\,\mathrm{Pr}^{-2})^{1/18}$	Convective layer	(8.27);
D:	$S < 3$,	$S < (\mathrm{Gr}\,\mathrm{Pr}^{-2})^{1/18}$	Diffusive layer	(8.25);

The difference here is that in Cases C and D, the inertial layer is thinner than the convective layer and thus cannot exist. The first criterion on S is a measure of the relative size of the inertial and convective layers, while the second criterion on S differentiates the inertia and diffusive layers in Cases A and B and the convective and diffusive layers in Cases C and D.

5. *Selective Withdrawal in a Rotating Stratified Fluid*

A homogeneous rotating fluid behaves analogously to a stratified rotating fluid (Veronis, 1967), so it is not surprising that selective withdrawal has its counterpart in a rotating fluid. Long (1956) obtained a solution for steady flow of a rotating inviscid fluid into a point sink located at the axis of the rotation to an infinitely tall cylinder. Using the full nonlinear equations (Yih, 1965), he found necking near the sink with an annular eddy around the axisymmetric layer. This eddy elongated to infinity as the Rossby number

$$R_0 = \frac{Q}{2\pi b^3 f} \tag{8.29}$$

(where Q is the discharge, b the radius of the cylinder, and f the Coriolis parameter) approached 0.261. Shih and Pao (1971) carried out experiments and found that the withdrawal layer formed with a radius

$$\delta_\tau = 0.7\left(\frac{Q}{2\pi f}\right)^{1/3}, \tag{8.30}$$

whenever $R_0 < 0.261$. Motivated by these experiments, Pao and Shih (1973) constructed a slip line solution and found excellent agreement with the prediction from (8.30). The viscous dominated solution was also derived by Pao and Kao (1969) and they found in this case that the layer thickness was given by

$$\delta_\tau = 4.1\left(\frac{\nu L}{f}\right)^{1/3}, \tag{8.31}$$

the numerical fctor being obtained both theoretically and experimentally. The transition between the two does not appear to have been studied, but by analogy with a purely stratified case, the transition should depend on the parameter

$$R_f = \frac{Q}{\nu L}. \tag{8.32}$$

For $R_f > 1$, (8.30) applies; for $R_f < 1$, (8.31) is relevant.

The above formulae predict a vertical withdrawal layer whereas in the stratified case, the withdrawal layer is horizontal. Investigations are underway (McDonald, personal communication) in an effort to reveal the transition of the flow as the parameter f/N is varied from 0 to infinity.

Monismith and Maxworthy (1988) performed experiments on the establishment of a selective withdrawal layer in a rectangular channel under conditions of strong stratification and weak rotation ($f/N = 0.8 \rightarrow 0.25$). Their observations show that the initiation of the outflow resulted in a Kelvin shear wave propagating cyclonically around the perimeter of the tank, leaving in its wake an anti-cyclonic withdrawal layer. The thickness of the withdrawal layer was, however, not influenced by the rotation and was given by (8.26).

In summary, a great deal is known about selective withdrawal and most situations can be predicted. There are three exceptions, however. First, arbitrary stratification leads to premature withdrawal because of the higher mode response, but no predictive procedure exists. Second, the influence of sills or contractions has received only preliminary treatment (Kao, 1976) and much work remains to be done. In particular, transitions that occur through a contraction where the withdrawal layer changes from a subcritical condition to supercritical and back to subcritical with an associated hydraulic jump would appear to be extremely important to mixing in lakes and the ocean. Third, the transition flow from a fully rotating fluid to a fully stratified fluid has not been studied in detail.

IX. Inflow

Reservoirs and lakes are supplied from rivers; as the river water meets the relatively stagnant reservoir water, it usually encounters water of a slightly different temperature, salinity or turbidity. These variables lead to an inflow that is either lighter or heavier than the surface water of the lake. When the inflowing water is very much heavier (Hebbert *et al.*, 1979; Ford *et al.*, 1980) the river water enters the reservoir, and plunges to the bottom of the lake. Alternatively, examples of where the inflow water was not sufficiently heavy to take it all the way to the bottom are given by Howard (1953), Wunderlich and Elder (1969), Ford (1978), Ford and Johnson (1981), Fischer and Smith (1983), and Imberger (1985). In all these cases, the inflowing water plunged beneath the lake water at the entrance then flowed down the drowned river valley, entraining lake water as it moved downstream. At some depth, the total entrainment made the underflow neutrally buoyant relative to the immediately adjacent water, and the underflow became an intrusion

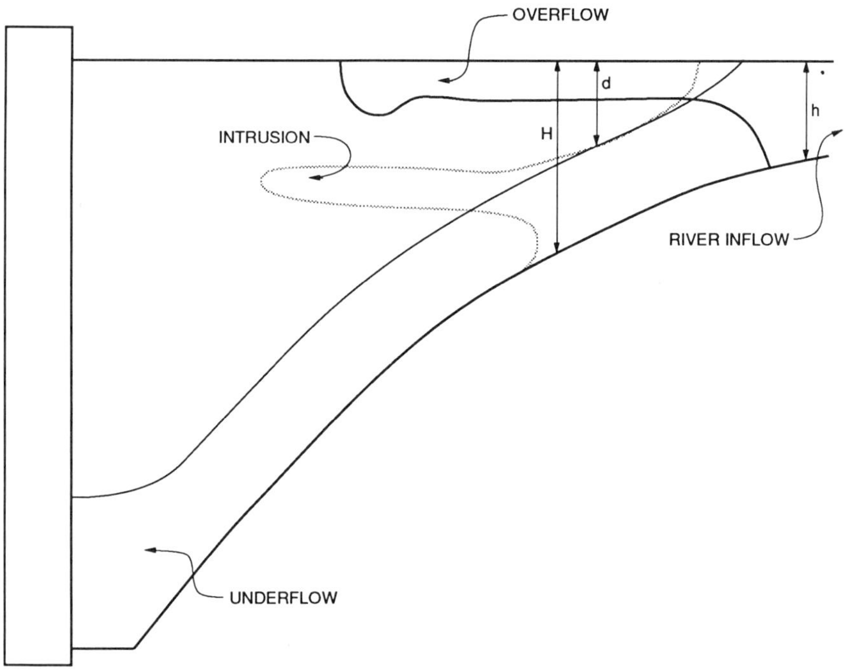

FIG. 30. Schematic of inflow possibilities: overflow, plunge-point, intrusion, and underflow.

(Figure 30). Last, there is the case where the inflowing water is lighter than the resident surface water and the river overflows the lake as a buoyant surface flow (Figure 30). The dynamics of such an overflow are similar to the situation where a source of fresh water enters an estuary or coastal environment (Chen, 1980; Ookubo and Muramota, 1981; Chao, 1988; Luketina and Imberger, 1987, 1988).

The surface buoyant jet occurs only rarely in a lake in tropical and temperate regions but is normal in high latitude cold lakes, where such inflows induce thermal bars (see Section VII). The dynamics of these flows is a subject in itself and has been extensively treated by Chen (1980), Sargent and Jirka (1982), Fischer and Smith (1983), O'Donnell and Garvine (1983), Chu and Jirka (1985), Arita *et al.* (1986), and Luketina and Imberger (1987, 1988), with an excellent review of the frontal region being given by Simpson (1982). The reader is referred to these articles for further information on the overflow case.

Consider the case where the density of the inflowing water is somewhat heavier than the resident surface water in the lake and the river water will plunge down the river bed until it reaches neutral depth. The propagation down the inclining plane of the leading head of the gravity current has been investigated by Stacey and Bowen (1988) and Wilkinson and Wood (1972). However, in most cases, the inflowing discharge is sufficiently steady to allow the underflowing river to be analyzed as a steady or quasi-steady gradually varying flow.

In this case, a balance is struck in the underflow such that bottom friction, interfacial entrainment and inertia retard the flow, and the downslope component of gravity accelerates it. This momentum balance is influenced by the slope and flatness of the river bed and the rotation of the earth. A discussion of the latter influence is omitted here.

As mentioned above, most authors have assumed the underflow discharge to be steady, and if it is further assumed that the underflow is confined laterally by the drowned river valley, a gradually varying flow analysis yields satisfactory results (Ellison and Turner, 1959; Savage and Brimberg, 1975; Hebbert *et al.*, 1979; Akiyama and Stefan, 1984). The main conclusion reached in these studies (see Fischer *et al.* (1979)) is that the flow quickly adjusts, so that as it propagates downstream along a particular bed slope, the Froude number of the underflow becomes a constant equal to the normal flow Froude number F_n, where the Froude number is given by

$$F = \frac{u}{(g'd)^{1/2}} = F_n, \tag{9.1}$$

where u is the velocity of the downflow, g' is the modified acceleration due to gravity between the lake and underflowing water, and d is the hydraulic depth of the underflow.

For the case where the density difference is derived from particulate matter (turbidity current), the fallout or accumulation of particles must be accounted for. However, provided the rate of change of density due to deposition or erosion is slow compared to the time it takes to travel a number of depths of the underflow, it is possible to retain the gradually varying flow assumption, so that (9.1) is still valid even if only locally (Akiyama and Stefan, 1985). Fischer *et al.* (1979) solved the vertically integrated momentum and mass equation to derive an expression for the normal Froude number:

$$F_n^2 = \frac{\dfrac{TE}{P} + \left(\dfrac{(h - h_c)T^2E}{aP} + \dfrac{T}{P}\tan\phi\right)\left(\dfrac{a}{3T^2}\dfrac{dT}{dh} - 1\right)}{C_D\left(\dfrac{a}{3T^2}\dfrac{dT}{dh} - 1\right) + E\left(\dfrac{2a}{3TP}\dfrac{dT}{dh} - 1\right)}, \tag{9.2}$$

where T is the top width of the underflow, E is the entrainment coefficient, P is the perimeter of the underflow, h is the underflow depth, h_c is the depth of the underflow centroid below the interface, a is the area of the underflow, ϕ the valley slope, and C_D is the channel bottom drag coefficient.

The entrainment coefficient may be determined by applying a model similar to the one used for the surface layer (see Section IV), but as discussed in Fischer *et al.* (1979), the coefficients of the model may be different due to the very different velocity profiles within the underflowing current compared to the velocity distribution in the wind driven surface layer. The entrainment at the interface is energized by the shear there and by the turbulence imported from that generated at the river bed. For mild slopes ($\phi = 10^{-3}$), Hebbert *et al.* (1979) have shown that shear production was negligible and that E had a value of 1.9×10^{-4}.

The location of the plunge point can now be found by neglecting the local entrainment. This is done by substituting the entrance flow and density difference into (9.1) and equating this to the normal Froude number evaluated from (9.2). This will yield the first estimate of the plunging depth h_1, which may be improved by accounting for the non-hydrostatic pressure distribution near the plunge point (Wilkinson, 1972; Hebbert *et al.*, 1979; Jain, 1982; Akiyama and Stefan, 1984, 1985, 1987). This leads to a relationship

$$\frac{d}{H} = f(F_n), \tag{9.3}$$

which implies an energy loss at the plunge point.

The three equations (9.1 to 9.3) can now be solved together for d and H with only the assumption of no entrainment at the plunge point. Comparison with field data is excellent (Hebbert *et al.*, 1979).

When the slope becomes steeper (F_n larger) or the inflowing momentum is increased, the energy loss implied by (9.3) can no longer be neglected and the associated entrainment at the plunge point will invalidate the above analysis and the theory must be modified as suggested by Britter and Simpson (1978) (see also Luketina and Imberger (1987) and Jirka and Arita (1987)). The analysis for the inclusion for such entrainment has been carried out by Akiyama and Stefan (1987), but no data exist to fix the amount of entrainment except for the buoyant overflow documented in Luketina and Imberger (1987, 1988). There are some numerical simulations available (Farrell and Stefan, 1986; Buchak and Edinger 1984). However, these all use closure schemes so that the estimates of entrainment would be overestimated (Franke *et al.*, 1987).

For a divergent river valley with a relatively flat bottom, the flow may separate from the side walls. Johnson *et al.* (1987a,b) observed six different flow regimes depending on the entrance aspect ratio, the Froude number, and the angle of divergence. These are schematically reproduced in Figure 31. In general, the higher the Froude number, the closer the flow is to a simple, entering jet. Once separated from the side wall, the inflow behaves as a free jet and Johnson *et al.* (1988) found that the entrainment was much higher for such separated flows and could be estimated from simple jet analysis (Fischer *et al.*, 1979; Rodi, 1982).

The case of inflow into a lake with a flat bed morphology is similar to the large divergence case and is characterized by a central momentum jet followed by a radially spreading buoyant plume, originating from a virtual origin. The virtual origin occurs at a distance from the inflow of order $M^{3/4}B^{-1/2}$ (Fischer *et al.*, 1979; Luketina and Imberger, 1987), where M is the entrance momentum flux and B is the corresponding buoyancy flux. Luketina and Imberger (1987, 1988) presented an analysis and a comparison with field data of a buoyant overflow, whereas Hauenstein and Dracos (1984) presented a similar analysis and a comparison with laboratory data for the case when the inflow is denser

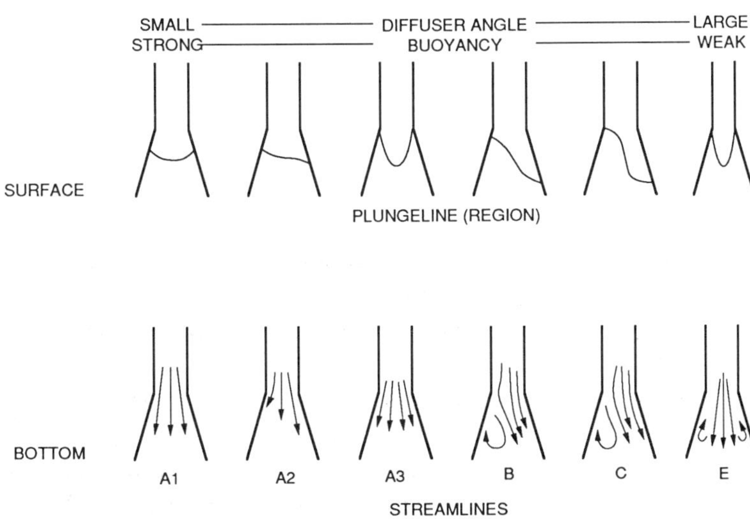

FIG. 31. Principal regimes of negatively buoyant flow into horizontal diverging channel. (After Johnson *et al.* (1987a).)

than the lake water. Recently, Tsihrintzis and Alavian (1987) developed a further integral model similar to that of Hauenstein and Dracos (1984) for three-dimensional flow down an inclined plane.

Once the inflow reaches a depth of neutral buoyancy, the inflowing water leaves the river valley and intrudes horizontally into a stratified lake. Imberger (1985) presents field data from the Wellington Reservoir that illustrate this lift-off process. The dynamics of the intrusion appear to have only been thoroughly investigated for two-dimensional flow. In this case, an intrusion entering a linearly stratified ambient water body will have a length and width as summarized by Imberger *et al.* (1976) with the propagation length being given by

$$l = \begin{cases} 0.44\, LR^{1/2}t', & t' \leq R, \qquad (9.4) \\ 0.57\, LR^{2/3}t'^{5/6}, & R < t' < \mathrm{Pr}^{2/3}, \qquad (9.5) \\ CLR^{3/4}t'^{3/4}, & \mathrm{Pr}^{2/3} < t' < R^{-1}, \qquad (9.6) \end{cases}$$

where the coefficients in (9.4) to (9.6) were determined from experimental data by Manins (1976) and Maxworthy (1980), but where the constant C in (9.6) is still unknown. In these formulations, the parameters are defined as

$$R = F\,\mathrm{Gr}^{1/3}, \tag{9.7}$$

$$F = \frac{q}{NL^2}, \tag{9.8}$$

$$\mathrm{Gr} = \frac{N^2L^4}{\nu^2}, \tag{9.9}$$

$$t' = \frac{t}{N^{-1}\,\mathrm{Gr}^{1/6}R}, \tag{9.10}$$

and l is the length of the intrusion at a time t after initiation of the flow.

As already discussed for intrusions formed by differential deepening, (9.4) to (9.6) are not valid when the intrusion approaches the rear wall (Fischer *et al.*, 1979).

Darden *et al.* (1975) showed that the accompanying velocity structure due to such an intrusion consists of multiple reversing jets above and below the entering intrusion, and Maxworthy (1980) demonstrated that simple intrusions can degenerate into solitons if the volume flux to the intrusion is terminated.

The second main case in a lake is where the downflow encounters a sudden change of density at a density interface. If the water below the density interface is heavier than that of the downflowing water, an intrusion will form along the interface. If the interface is very thin compared to intrusion thickness, the intrusion behaves as two separate gravity currents, the depth of the intrusion relative to the interface adjusting so that the speeds match (Holyer and Huppert, 1980; Britter and Simpson, 1981). Faust (1981) gives the speed of propagation V as

$$V = \left\{ \frac{g\Delta\rho}{2} \frac{(\rho_1/\rho_3)(d_1^2 - h_1^2) + (d_2^2 - h_2^2)}{\rho_1 d_1[2(d_1/h_1) - 1] + \rho_2 d_2[2(d_2/h_2) - 1]} \right\}^{1/2}, \tag{9.11}$$

where d_1 is the depth of the upper layer, d_2 is the depth of the lower layer, ρ_1 is the density of the upper layer, ρ_2 is the density of the lower layer, h_1 is the depth of the upper fluid above the intrusion, h_2 is the depth of the lower layer below the intrusion, and $\Delta\rho = \rho_2 - \rho_1$. Faust (1981) generalized the same analysis and showed that a similar expression could be derived for an intrusion into a three-layer fluid. However, in this case, he found that internal waves generated at the interfaces could lead to a slow-down of the intrusion.

The above intrusion dynamics have only considered laminar flow. However, in lakes, the downflow is strongly turbulent (Imberger, 1985) and it is important to assess the collapse of this turbulence as the intrusion enters the lake. Lin and Pao (1979) and later Hopfinger (1987) have reviewed this subject and concluded that the collapse takes place in times of $O(N^{-1})$ and that the vertical motions decay most quickly, leaving, in the absence of additional shear, two-dimensional turbulence with a vorticity vector orientated predominantly vertically. The boundary mixing experiments of Browand *et al.* (1987) suggest that the decay is extremely rapid and that we would not expect to find turbulence within the intrusion at a horizontal distance from the lift-off point much greater than the thickness of the intrusion. This has been confirmed by Roberts and Matthews (1987) who injected a horizontally neutral buoyant momentum source from a round nozzle into a linearly stratified fluid and found that the mean shape of the jets spread horizontally at a distance of $M^{1/4}N^{-1/2}$, where M is the momentum flux of the jet and N is the buoyancy frequency of the ambient water. Although these authors did not measure the turbulent flux directly, they inferred that the largest turbulent eddies became anisotropic at the same distance.

Now in the case of an underflow, the Froude number based on the density difference between the underflow and the neighboring ambient water becomes infinity since g' tends to zero at the point of neutral buoyancy. However, the inertia of the underflow would carry the fluid through the lift-off into the intrusion, and a finite sized intrusion based on an inertia buoyancy balance would form with a thickness (Manins, 1976)

$$\delta = 4\left(\frac{q}{N}\right)^{1/2} \tag{9.12}$$

for the two-dimensional flow case. Assuming that the collapse starts at a distance order $(MN^{-2})^{1/3}$, which is the equivalent result for the two-dimensional flow, the turbulence would commence the decay at a distance $O(QN^{-1/2})$ or one intrusional layer thickness into the lake. Given these two pieces of evidence, one would expect not to find turbulence in an intrusion. However, recently Imberger (1987) has shown, by carrying out temperature microstructure measurements, that the shear from the intrusions together with background shear from internal wave activity can sustain a low level of turbulence. In a further analysis of the same data, Imberger (1988) has shown that the turbulence was at the point of decay, with

$$\mathrm{Fr}_\gamma = \left(\frac{\varepsilon}{\nu N^2}\right)^{1/2} = 15^{1/2} = 3.87. \tag{9.13}$$

Thus, even though the gradient Richardson number associated with intrusion was of the order of 11, the turbulence remained viable, although at the point of fossilization (Rohr *et al.*, 1988).

In summary, we have seen that the inflow dynamics may be conveniently partitioned into three separate problems: the underflow, the plunge point, and the intrusion dynamics. The inflow of the heavier water mass into a drowned river valley appears to be adequately described by existing gradually varying flow theories, although the entrainment over steep slopes where shear production is the dominant source of energy for entrainment has not been verified in the field. Equally, under these conditions, we may expect severe entrainment at the plunge point, which so far defies parameterization. The interaction of the inflow with a resident internal wave field (Fischer and Smith, 1983) or with a surface mixed layer deepening (Imberger, 1985) can introduce enhanced mixing and must be treated on a case by case basis for each lake.

The most important problem remaining concerns the mixing associated with inflows, both directly in the intrusion and also, as discussed in the next section, in the lake as a whole.

X. Mixing below the Surface Layer

Mixing below the surface layer is generally patchy and sporadic, with individual events occupying no more than tens of cubic meters and remaining at full intensity for no more than a few minutes (Imberger, 1985). The two central questions remain: What is the efficiency of energy conversion at turbulent sites and what is the patch statistical distribution (Imberger and Hamblin, 1982)?

No direct measurements of the local value of the vertical diffusion coefficient K_z exist, but many investigators have taken advantage of the closed nature of the lake basin and obtained estimates of the vertical diffusion coefficient averaged over the whole basin (Kullenburg *et al.*, 1974; Jassby and Powell, 1975; Imboden and Emerson, 1977; Imboden *et al.*, 1977; Robarts and Ward, 1978; Weiss *et al.*, 1979; Lewis and Perkin, 1982; Imboden *et al.*, 1983; Imboden and Joller, 1984; Sanderson *et al.*, 1986; Colman and Armstrong, 1987; Lake Biwa Research Institute, 1987; Wuest, 1987; Wuest *et al.*, 1988). The data confirmed (Imberger and Hamblin, 1982) that the lake-wide averaged vertical exchange coefficient fluctuates between $2 \times 10^{-6}\,m^2\,s^{-1}$ (near molecular) during periods of very strong stratification and weak winds (Wuest, 1987) to values as high as $10^{-4}\,m^2\,s^{-1}$ during periods of very weak stratification in tropical lakes (Lewis and Perkin, 1982). It has also been shown (Ward, 1977; Straskraba, 1980; Wuest, 1987) that the vertical diffusivity depends on the buoyancy frequency, the area of the lake and the depth of the lake.

Wuest (1987) attempted a correlation of the rate of mixing with the Wedderburn number (2.5) and the buoyancy frequency, writing the vertical diffusion coefficient as

$$K_z = \frac{a}{N^{2\alpha}}. \tag{10.1}$$

He found most data sets required a value of α in the range of $0.3 < \alpha < 0.9$. This may be compared with oceanic variations, summarized by Gargett and Holloway (1984), where $0.25 < \alpha < 0.5$. The dependence on the Wedderburn number is shown in Figure 32, reproduced from Wuest (1987), and indicates that if W ranges from between 200 and 0, the vertical diffusivities range from around $10^{-6}\,m^2\,s^{-1}$ to around $10^{-2}\,m^2\,s^{-1}$. In computing the value of the Wedderburn number W, Wuest (1987) used the depth of the epilimnion, but does not give details of the definition used. However, judging from

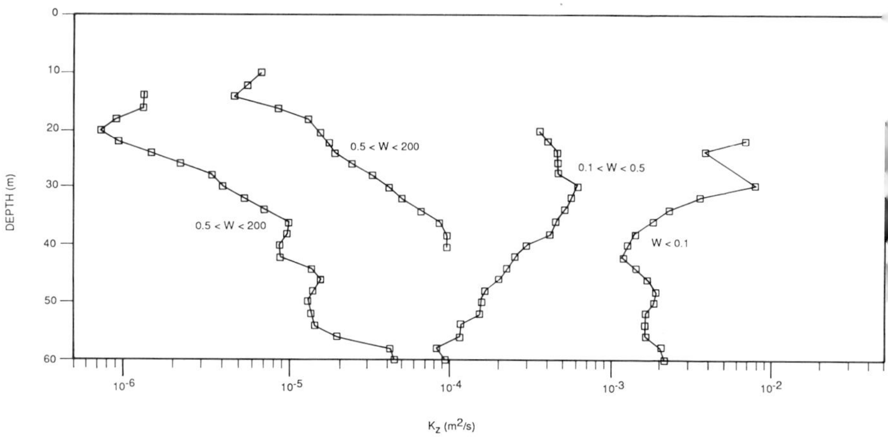

FIG. 32. A comparison of K_z in Sempachersee for four different periods. The value of K_z was calculated using traces and are typical of the three regimes discussed in the text: $0.5 < W < 200$, $0.1 < W < 0.5$, and $W < 0.1$. (After Wuest (1987).)

his profile shapes, this would have been close to the depth of the seasonal thermocline, which would make W close to L_N. His W is really an indicator of the stability of the seasonal thermocline and not the surface layer.

We saw in Section II that in the majority of cases a lake is adequately approximated by a three-layer stratification, so there are two numbers, W and L_N, which determine its response to a wind force. Small W and large L_N lead to a strong surface layer response, and small L_N causes the main thermocline to tilt with the associated stronger mixing in the hypolimnion. Three examples provide an illustration. In Figure 33, data from the Canning Reservoir show a case where $W = 0.1$ and $L_N = 288$. The dissipation levels are close to zero everywhere, except in the top two meters, where a very weakly stratified surface layer resided. During the time of this profile, nearly 200 microstructure casts were taken over a period of 10 days and only one profile showed the slightest sign of turbulence in the hypolimnion. By contrast, Figure 17f depicts a microstructure profile at $W = 2.1$ and $L_N = 0.24$, whereas in Figure 24c the conditions were such that $W = 4.5$ and $L_N = 0.44$; in both cases the whole profile had become turbulent with dissipation values in the hypolimnion reaching $10^{-5}\ \mathrm{m^2\,s^{-3}}$. These data are presently being analyzed more fully, but this preliminary analysis, together with the evidence from Wuest (1987), strongly suggests that the response of the

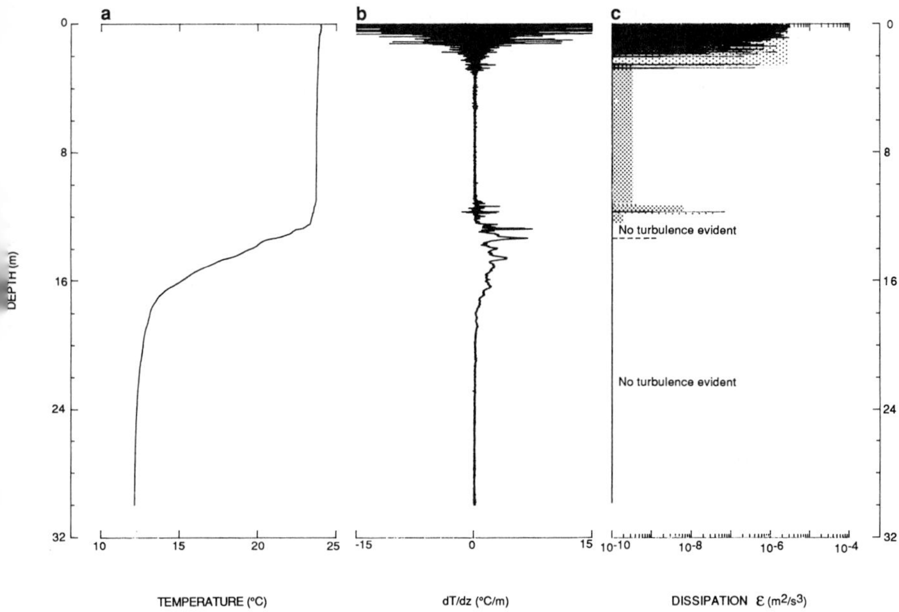

FIG. 33. Microstructure taken in the Canning Reservoir during a period of large Lake number $L_N = 288$. (a) Temperature. (b) Temperature gradient. (c) Dissipation as calculated from the Wigner–Ville technique (sticks) and the Batchelor spectra fitting technique (bar chart). Note that there is no turbulence below the top 2 m.

seasonal thermocline, which determines the deep mixing in a lake, is in general determined by the value of the Lake number L_N.

The variability within the hypolimnion must depend on some empirical relationship similar to that given by (10.1). Weinstock (1984) has determined the following theoretical formulation:

$$K_z = \alpha \frac{\varepsilon}{N^2 + k_0^2 u^2}, \tag{10.2}$$

where ε is the dissipation of turbulent kinetic energy, N is the buoyancy frequency, k_0 is the wave number of the energy bearing eddies, u is the root mean square velocity of the motion in the turbulent patch, and α is a constant between 0.2 and 0.8 (Imberger and Hamblin, 1982). It is clear, however, that the dependence on N is too strong in (10.2) and is not supported by the empirical measurements of the average diffusion coefficient (10.1). However, it would be interesting to investigate whether the averaging process could decrease the dependency from N^{-2} to N^{-1}.

The following overall scenario is appealing: as the wind, inflow, or artificial mixing device (e.g., an aerator) start up, they induce seiching of the thermocline (mode 1 response in a three-layer system). This seiching provides a background shear within which internal waves generated by differential deepening and by leakage through the base of the surface layer increase the frequency and intensity of turbulent patches. This leads to an increase in the buoyancy flux and turbulent kinetic energy dissipation. This is a much simpler picture than in the ocean where the background shear arises from a host of motions.

The Lake number can be extended easily to account for these three major disturbances to read

$$L_N = \frac{S_t g(1 - z_T/z_m)}{\rho_0 u_*^2 A_0^{3/2}(1 - z_g/z_m) + \rho_0 g^{2/3} Q_0^{2/3} z_m^{1/3} A_0 + \rho U Q_i A_0^{1/2}(z_g - z_i)/z_m} \tag{10.3}$$

where Q_0 is the air flow rate of the aerator, Q_i is the river inflow rate at the point where the intrusion forms, U is the inflow velocity of the intrusion, and z_i is the height of the intruding intrusion. All other terms are defined in Sections II and XII. The additional terms are the additional disturbing moments offered by the bubble flow $\rho_0 g^{2/3} Q_0^{2/3} z_m^{1/3} A_0$ (Section XII) and the river momentum flux $\rho_0 U Q_i$ acting at a moment of $(z_g - z_i)$ (Section XI). Whenever L_N is small, the seasonal thermocline may be expected to respond more actively, thus leading to higher transport rates.

The origins of the mechanisms leading to internal mixing are numerous, consisting of a range of instabilities, wave–wave, wave–shear, Kelvin–Helmholtz billows, Holmboe instabilities, critical layer absorption, double diffusion, bores, and hydraulic jumps. These mechanisms have recently been thoroughly reviewed and the reader is referred to Hopfinger (1987) and Thorpe (1987). Mixing in the metalimnion is also reviewed by Gregg (1987). For a review of double diffusion, the reader is referred to Turner (1985).

The nature of the turbulence in a lake can be determined from the turbulent Froude versus Reynolds number diagrams (Gibson, 1980, 1982, 1986, 1987a,b, 1988; Imberger and Boashash, 1986; Imberger, 1988). The Froude number is defined

$$\mathrm{Fr}_t = \left(\frac{\varepsilon^2}{g'^3 L_c}\right)^{1/6}, \tag{10.4}$$

and the Reynolds number is given by

$$\mathrm{Re} = \left(\frac{\varepsilon L_c^4}{\nu^3}\right)^{1/3}, \tag{10.5}$$

where ε is the dissipation of the turbulent kinetic energy, g' is the acceleration due to gravity derived from the density anomaly within the turbulence, and L_c is the displacement scale (Imberger and Boashash, 1986). Turbulence close to the line

$$\mathrm{Fr}_\gamma = \left(\frac{\varepsilon L_c}{\nu g'}\right)^{1/2} = 3.87 \tag{10.6}$$

is at the point of fossilization (Itsweire *et al.*, 1986; Imberger, 1988), whereas active turbulence lies in the region $\mathrm{Fr}_t > 1$ and $\mathrm{Re} > 1$. In Figure 34, we reproduce data derived from turbulence measurements from 10 different physical mechanisms operating in lakes. The mixed layer turbulence is active; turbulence in the hypolimnion is at the point of decay. It is necessary, and possible, to design a field experiment aimed at correlating the location of the data points in the Froude versus Reynolds number phase diagrams with the values of W and L_N. Given that the mixing efficiency (the flux Richardson number) can be correlated to the location of the data in Figure 34, this would lead to an operational model for the prediction of vertical transport in a lake (Imberger, 1988).

So far we have discussed only mixing internal to the hypolimnion. There is evidence in the ocean (Ivey, 1987; Garrett and Gilbert, 1988; Gregg and Sanford, 1988) that the energy input due to tidal motion exceeds that appearing in the water column at depth, suggesting that dissipation in the benthic boundary layer is a major sink for this energy. Lueck and Osborn (1985) presented evidence that close to the bottom in a submarine canyon, dissipation values increased considerably. Ivey and Boyce (1982) showed that a benthic boundary layer existed at the bottom of Lake Erie.

In Figure 35, we show the development of turbulence in the water column of the Harding Reservoir in response to a surface wind stress on 27 November 1988. Early in the morning the meteorological conditions were warm and calm. At 1000 hours a wind with a speed ranging between 2 and 3 m s^{-1} commenced and continued until 1400 hours, leading to a Wedderburn number of 0.05 and a Lake number of 2.2. The sequence of profiles (all extending from the bottom to the lake surface) shows the establishment of turbulence both with the hypolimnion and in the benthic

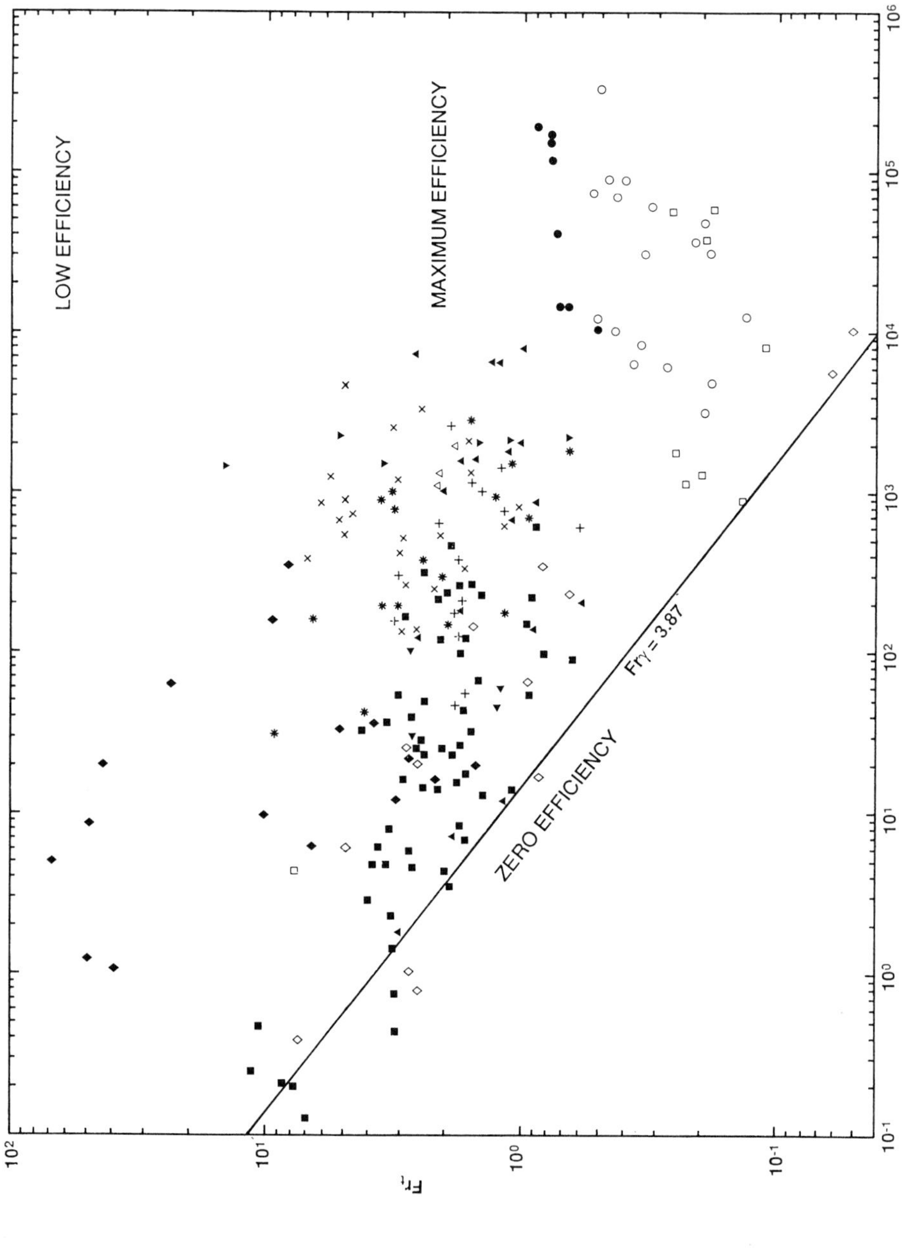
LOW EFFICIENCY
MAXIMUM EFFICIENCY
ZERO EFFICIENCY
Frγ = 3.87
Fr_i
10^2
10^1
10^0
10^{-1}
10^{-1}
10^0
10^1
10^2
10^3
10^4
10^5
10^6

boundary layer with dissipation values as high as $6 \times 10^{-6}\, m^2\, s^{-3}$ in the bottom 1 m of the profile (Figure 35f). This data set is presently being analyzed but was included as it offers the first conclusive evidence of the development of turbulence in the benthic boundary layer of a lake.

It is tempting to carry out a budget of the turbulent kinetic energy. For instance, the data shown in Figure 33 indicate dissipation in the surface layer (depth of 2 m) that almost exactly balances the energy input $C_N^3 u_*^3$. However, by comparison, integration of the total dissipation per m^2 of the lake's surface for the profile shown in Figures 17 and 24 greatly exceeds that introduced at the surface. Sufficient measurements are required to account for the very strong variability (see Section VI) of wind stress over the lake surface.

Boundary mixing can be sustained by two major mechanisms. First, internal wave seiching will cause a velocity shear at the boundary and set up a turbulent boundary layer over the rough lake bottom (see Imberger and Hamblin (1982) for a review). Numerous experiments (Ivey and Corcos, 1982; Thorpe, 1982; Phillips *et al.*, 1986; Browand *et al.*, 1987; Ivey, 1987) have modeled this mechanism with a grid stirring the water adjacent to a vertical or sloping wall in a stratified tank. These experiments all produce a turbulent boundary layer close to the grid, which then initiates intrusions that flow horizontally into the tank. The mean motion induced by these intrusions is such that the density structure in front of the intrusion changes in a way which mimics an enhanced vertical diffusion adjustment. Ivey (1987) has summarized this work and inferred an ocean averaged diapycnal eddy diffusivity of

FIG. 34. The Froude versus Reynolds number plot containing the data from microstructure segments with well-defined spectral components protruding above a step gradient form. All dissipations used in this plot were estimated by fitting a Batchelor spectrum to the spectra computed for the whole segment.

- ● Bubble plume thermal PT
- ○ Thermal falling in a weak ambient stratification
- □ Thermal impinging on a sharp interface
- ▲ Penetrative convection
- ▼ Surface buoyant plume roller region
- ▼ Surface buoyant plume wake
- × Well mixed surface region
- + Strong stratification, strong shear
- ■ Strong stratification, weak unknown shear
- ◆ Strong stratification, no known shear
- $*$ Gravitational underflow
- △ Penetrative convection thermals impinging on an interface
- ◇ Intrusion flows.

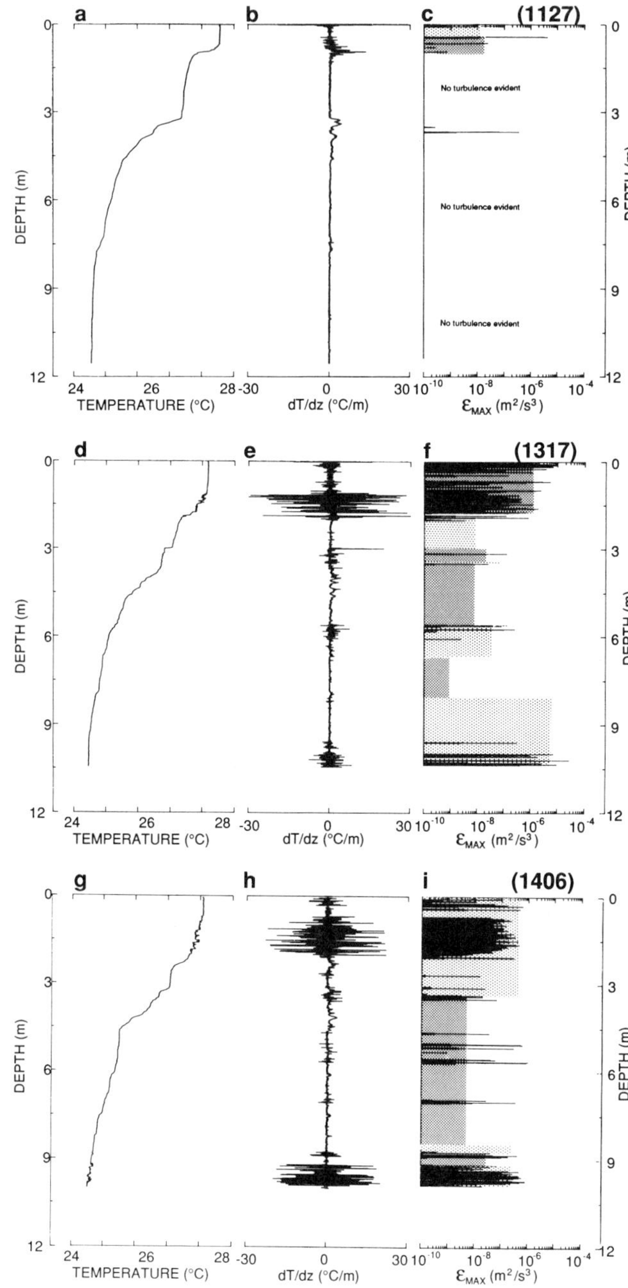

FIG. 35. Examples of the development of turbulence in the Harding Reservoir at an extremely low Lake number $L_N = 2.2$. All profiles extend from 0.2 m above the rocky bottom to the surface. (a) Temperature profile at 1127 hours. (b) Temperature gradient profile at 1127 hours. (c) Dissipation. Sticks: Wigner–Ville estimate; bar chart: Batchelor spectra fitting estimate. (d) As (a) at 1317. (e) As (b) at 1317. (f) As (c) at 1317. (g) As (a) at 1406. (h) As (b) at 1406. (i) As (c) at 1406.

$3 \times 10^{-5}\ m^2\ s^{-1}$ for the interior of the ocean. This result was derived from the formula

$$K_z = \frac{Kl}{\theta L}, \tag{10.7}$$

where K is the diffusion coefficient in the benthic boundary layer, l is the turbulent benthic boundary layer thickness, θ is the bottom slope, and L is the basin scale.

The other mechanism is breaking internal waves (Cacchione and Wunsch, 1974; Murota and Hirata, 1979; Murota *et al.*, 1980; Thorpe, 1987; Garrett and Gilbert, 1988; Ivey and Nokes, 1989; Wallace and Wilkinson, 1988). Internal waves shoal and break as they propagate over a sloping bottom, especially if the bottom slope is at the same angle as the rays of the internal wave field (a critical slope). Ivey and Nokes (1989) found the turbulence in the benthic boundary layer resulting from the breaking waves to be sustained by shear leading to Richardson numbers as low as 0.01. For a critical slope, the steady state boundary layer thickness resulting from the wave breaking was given by

$$l = 5\zeta, \tag{10.8}$$

where ζ is the internal wave amplitude.

The mixing efficiency within this active boundary layer measured as the flux-Richardson number (buoyancy flux divided by production) reached a maximum of 20%. The diffusion coefficient K_z can now be estimated from (10.7) and the estimate of the diffusion coefficient in the benthic boundary layer (Ivey and Nokes, 1989):

$$K = 0.09\ \omega\zeta^2, \tag{10.9}$$

where ζ is the wave amplitude and ω the wave frequency. Thus, given (10.7)–(10.9), a recipe for estimations of the vertical diffusion coefficient due to boundary mixing is now available. However, application still requires a knowledge of the internal wave spectrum in a lake as a function of L_N, in order to determine what percentage of the bottom is at critical slope. This is an urgent priority. Further, given the importance of the first mode seiche, it is not clear that boundary mixing plays the same important role in lakes as it appears to do in the ocean.

In summary, a great deal of progress has been made in the area of hypolimnetic mixing. Recent measurements have conclusively shown the turbulence level in the interior of the lake increased as L_N decreases and also that the effective basin average diffusion coefficient correlates to $N^{-\alpha}$ where α is approximately 1. The role of the benthic boundary layer and the turbulence due to shear and internal wave breaking has been quantified in laboratory experiments and these results are now ready to be tested in field experiments with the data set from Harding Reservoir. This should determine the functional relationship between the proportion of transport in the interior compared with that in the benthic boundary layer as a function of the Lake number L_N.

XI. Modeling

The previous sections have given an overview of the seasonal behavior of lakes and reservoirs, and discussed in detail the major processes that determine their density structure. The complex interactions of these processes which produce the overall thermal characteristics are best studied with computer simulations. The need for these detailed descriptions has been motivated by lake management requirements, which has, in the last few decades, led to the development of a variety of modeling techniques for stratified lakes and reservoirs.

Much of this development has occurred under the assumption of one-dimensionality, where vertical motions are inhibited and transverse and longitudinal variations are quickly evened out. Even with this great simplification, it is difficult to model the interaction of a number of complex processes to predict a density structure, and a number of models of varying levels of complexity and success have been produced. There has also been, to a lesser extent, some development of two- and three-dimensional stratification models; the increasing complexity and computational requirements have severely limited this development. The question of circulation models is a separate issue and is not addressed in this review.

The early one-dimensional models of stratification concerned the solution of the one-dimensional heat transport equation. Perhaps the first of these models was the Dake and Harleman (1969) formulation, in which molecular diffusivity was assumed to be the vertical transport mechanism that redistributed the heat input by the surface heat fluxes. In a preliminary version of later developments, this model generated a surface mixed layer by redistributing density instabilities produced by surface cooling. The true diffusivity models followed: Orlob and Selna

(1970), Huber *et al.* (1972), Sundaram and Rehm (1973), and Markofsky and Harleman (1973) all developed models that simulated the vertical transport of heat in the epilimnion by means of a diffusion-like process, characterized by an eddy diffusivity E_z. The definition of E_z was the major difference in these models. For example, Huber *et al.* (1972) assumed a constant value of E_z, calibrated from a data sequence. Orlob and Selna (1970) used a constant value of E_z in the epilimnion, with a value exponentially decaying with depth in the hypolimnion, also calibrated against data. These models were poor representatives of the actual processes. Sundaram and Rehm (1973) incorporated some of the physics into the representation by defining an E_z that depended on the stratification and the wind, through an overall Richardson number. Similar formulations for E_z have been suggested by, for example, Thomas (1975) and Witten and Thomas (1976). Henderson-Sellers (1976) has further developed these ideas for deriving E_z for thermocline models based on boundary layer theory and a mixing length turbulence closure scheme.

Although the early models based on diffusivities were moderately successful and, being relatively insensitive to the value of E_z, were simple to calibrate, Harleman (1982) showed that this was the result of the dominance of other processes for the vertical transport of heat, and the insensitivity may not occur in all cases. Further, in all cases, calibration was required and transfer of the model from one lake or reservoir to another required recalibration. Even within one lake, conditions outside the range of calibration could not be modeled with confidence. More importantly, the description of the turbulent transport processes by a single bulk eddy transport coefficient meant that the effects of the individual processes, such as turbulence generated by the wind, convective mixing, internal waves, boundary mixing, and basin scale seiching for example, were integrated into a single parameterization. The coefficient was particular to both the lake and the combination of processes acting at the time, and could have no real physical interpretation. Because of its integrated, non-physical nature, there was no independent means of verifying the parameter value or dependence. Finally, the application of these diffusivities to other lake properties, for example salinity or dissolved oxygen, was not necessarily correct. Fischer *et al.* (1979) pointed out that the distribution of different properties may arise from different combinations of physical processes, and the same diffusivity may therefore not be appropriate.

Recent developments for the incorporation of specific processes, for example Henderson-Sellers (1984a, 1985), have minimized some of these difficulties, but if an understanding of the lake behavior in terms of its short time scale dynamics is required, models based entirely on diffusivity formulations will be inadequate.

This conclusion led to a number of investigators taking a different approach to modeling the epilimnion region of the reservoir or lake. The early developments were based on the Kraus and Turner (1967) model of thermocline formation in the ocean. As detailed in Section IV, this provided a deepening rate based on a balance between the turbulent kinetic energy (TKE) produced at the surface by the stirring of the surface wind and the energy required at the thermocline to entrain water from below the thermocline. The first lake models to incorporate a parameterization of this process were those of Stefan and Ford (1975) and Harleman and Hurley-Octavio (1977). Tucker and Green (1977) utilized a version of the Kraus–Turner model for the epilimnion with a diffusive model with a diffusivity that depended on wind speed and stratification in the hypolimnion. Tucker and Green also included a dependence on fetch length of the TKE made available for mixing.

The model development by Harleman and Hurley-Octavio (1977) was expanded on by Hurley-Octavio *et al.* (1977) and Bloss and Harleman (1979). The latter incorporated an effective efficiency of the transfer of TKE from the surface to the base of the mixed layer, based on a Richardson number formula. This incorporated the storage effects and dissipation. This model formed the base for the standard MIT model.

These models all solved for the vertical heat transport on a fixed discretization of the vertical direction; the lake was effectively split into a number of horizontal layers of uniform property and fixed thickness. Vertical transport was then achieved by transport across these boundaries by mixing from the mixed layer model, diffusion from the eddy diffusion model in the hypolimnion, or vertical advection as the result of inflow and outflow. This latter process involved the calculation of vertical velocities, and consequently, potential problems of numerical diffusion arose.

A different approach was taken by Imberger *et al.* (1978) in the first version of the model DYRESM. Here, the layers were Lagrangian in character, of variable thickness, and able to move vertically. Thus, inflow and outflow simply changed the thickness of the layers affected, with those above moved vertically to accommodate the change in storage. The TKE resulting from penetrative convection was formally incorporated in this version. Later versions of DYRESM also included the effects of interfacial shear, parameterization of the TKE storage, an eddy diffusivity for the hypolimnion based on the dissipation of the TKE and the stratification, and the travel time of the inflows (Spigel and Imberger, 1980; Imberger and Patterson, 1981; Imberger, 1982; Spigel *et al.*, 1986).

The clear advantage of these process-based models was the ability to characterize the processes independently of the reservoir model. Thus, although the mixed layer models, for example, contain a number of parameters, these may be identified as efficiencies of the various processes and their values derived from field or laboratory experiments. This independence means that the resulting model may be transferred from lake to lake without recalibration; a failure of the model to predict observed data is not necessarily an indication of a faulty parameterization, but rather an indication of the existence of processes not parameterized. Further, process-based models enable the user to determine the detailed response of the lake to external forcing.

A third class of one-dimensional model also exists, based on a full solution of the one-dimensional momentum and energy equations incorporating a turbulence closure scheme of a level higher than the eddy diffusivities discussed above. A number of k–ε models of this kind have been developed (for example, Spalding and Svensson (1977), Svensson (1978), and Sahlberg (1983)). The computational requirements of this approach for long term simulations, however, are prohibitive.

The application of one-dimensional models is restricted to those lakes for which the assumptions hold. Water Resources Engineers, Inc. (1969) proposed an internal Froude number relating the average throughflow velocity u from inflows and outflows and the stratification

$$\mathrm{Fr} = \frac{u}{(g'H)^{1/2}}, \tag{11.1}$$

where g' was based on the density difference between inflow and surface water, and H was the total depth. For $\mathrm{Fr} < 1/\pi$, they concluded that the reservoir was sufficiently stratified for the one-dimensional assumption to hold. Bloss and Harleman (1979) pointed out that the surface fluxes, particularly the effects of the surface wind, were not represented in this criterion and suggested that a limitation on wind speed U_w be used in addition. This was

$$U_w < 2 \times 10^3 \left(d \frac{H}{L} \frac{\Delta\rho}{\rho_0} \right)^{1/2}, \tag{11.2}$$

where d was the mixed layer depth, L the length of the lake, H the total depth, and $\Delta\rho$ and ρ_0 the density jump at the thermocline and the mean hypolimnion density respectively. If this inequality was not satisfied, it was assumed that the deviation of the thermocline from the horizontal was sufficient to prevent the one-dimensional assumption from holding. The numerical parameter was based on earlier empirical data relating surface set up to wind speed.

This criterion was similar in character to that established by the analysis of Spigel and Imberger (1980), based on rigorous dynamical grounds. Here, as discussed in detail in Section IV, the stratification and wind forcing were related through the Wedderburn number W (Thompson and Imberger, 1980) in a series of criteria which categorized the response of the lake to external wind forcing. These criteria, together with modified inflow and outflow criteria and a parameter that characterized the importance of the earth's rotation, were discussed in Patterson *et al.* (1984). These may be summarized as follows.

The deviation of the mixed layer was characterized by the value of W (defined in Section IV). For $W > 10$, the deviation was negligible, and the mixed layer deepening was characterized by the one-dimensional effects of surface stirring. For $3 < W < 10$, increasing evidence of shear production was likely, however the one-dimensional model, with a shear production algorithm was still valid. For $W < 3$, the interface was sufficiently close to the surface at the upwind end to be affected by surface stirring, and the lake became mixed through an upwelling front moving downwind. These criteria were based on the first mode wave behavior in a rectangular basin containing a fluid with two densities. Even with this simplification, the criteria proved a useful classification of lakes (Patterson *et al.*, 1984).

The inflow criterion was effectively the original Water Resources Engineers form above (11.1). The approximate nature of the calculation of Fr meant that there was little significance in the factor $1/\pi$, and $\mathrm{Fr} < 1$ was taken to indicate one-dimensional behavior. Similarly, an outflow Froude number was formed as

$$F_o = \frac{Q}{H^2(g'H)^{1/2}}, \tag{11.3}$$

where Q was the discharge, which again reflected the relative importance of inertial and gravitational adjustments. For $F_o < 1$, the one-dimensional structure was not disturbed. A criterion based on comparing the Rossby radius of deformation with the lake scale was used to characterize the effects of the earth's rotation.

The difficulty with the Wedderburn number and Froude number calculations is that they are based on simple geometries and simple two-layer density structures. To calculate the value of W appropriate to a real lake involved solving the eigenvalue problem for the internal wave field in the actual stratification and utilizing this to form an equivalent two layer structure; depending on the stratification ambiguity could arise. Likewise, the Froude number criteria were based on very simple geometries, and took no account of, for example, the ratio of inflow to storage.

The Lake number L_N introduced above minimizes these difficulties of application. The form discussed in Section II has been shown to be equivalent, in the two layer rectangular case, to W, and its interpretation is clear. For $L_N \gg 1$, the stratification is sufficiently strong to minimize any disturbance of the metalimnion from the surface wind, and the parameterizations of Section IV are appropriate. Thus in an equivalent way to W, the value of L_N characterizes the assumption of the overall one-dimensionality with respect to surface wind rather than just the surface layer. Further, by redefining the maximum angle of deviation, the higher mode responses may also be characterized.

The values of L_N calculated for the Canning Reservoir, based on actual values of u_* (rather than the fixed value used in Section II) and the field profiles of temperature and salinity, are shown in Figure 36a. This shows the same trend as Figure 8: L_N values of order 50 for the summer period when the stratification is strong and L_N values of order 0–2 in winter when the stratification is weak and the winds high. Clearly the one-dimensional assumption is appropriate in summer, but there are periods during winter when L_N is close to 0, and the assumption may be invalid. Examination of those times, however, reveals that the stratification is extremely weak, and the error made by a vertical mixing model compared to a model based on upwelling is small; both produce a homogeneous reservoir over short time scales. On the other hand, if L_N were small in times of moderate to strong stratification, or W is small but L_N is large, upwelling will occur at the upwind end from the hypolimnion in the former case and the surface layer in the latter. In both cases, the one-dimensional assumption does not hold. In these cases, a two- or three-dimensional circulation model may be more appropriate.

The form of L_N for river inflow may also be easily determined.

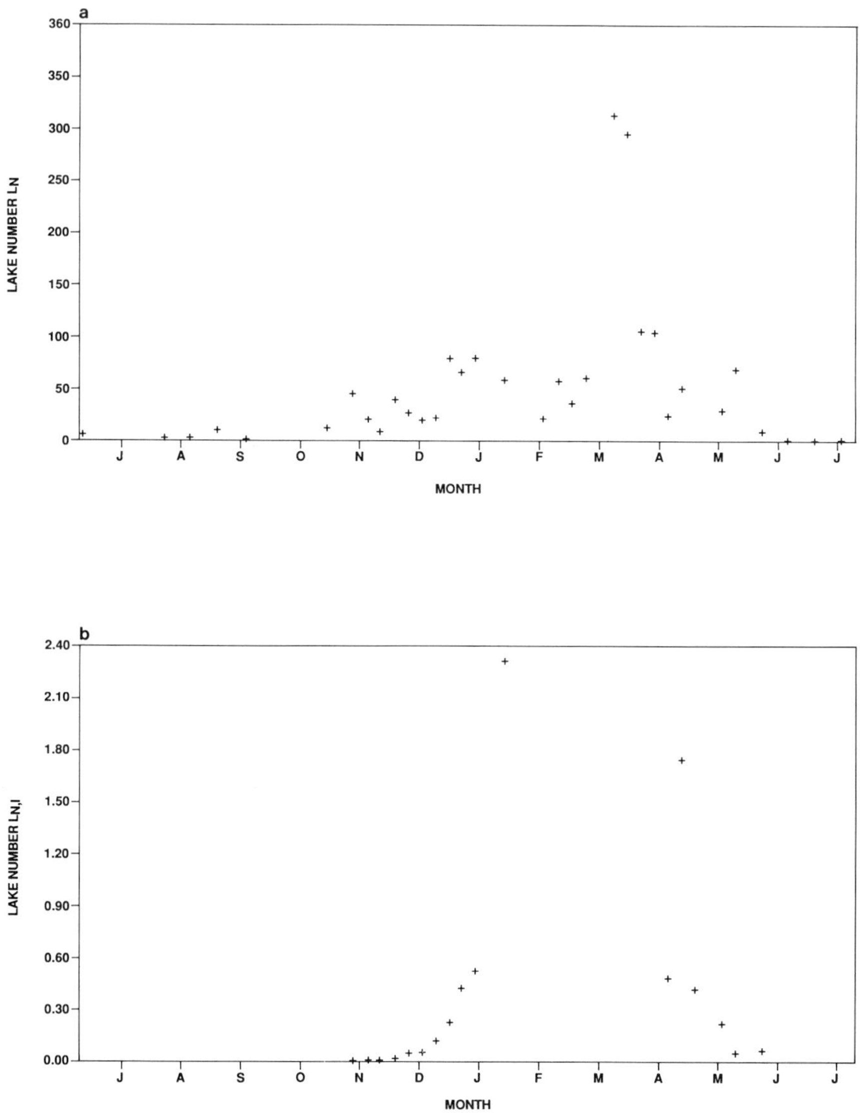

FIG. 36. (a) The Lake number L_N for Canning Reservoir for those days on which field data are available, based on the actual values of wind speed for the period 11 June 1986 to 7 September 1987. (b) The Lake number $L_{N,I}$ for Canning Reservoir, based on the inflows for the same period as Figure 36a.

Following Section X, the Lake number based on inflow can be defined by

$$L_{N,I} = \frac{gS_t(1 - z_T/z_m)}{\rho_i U Q_i A_0^{1/2}(z_g - z_i)/z_m}, \tag{11.4}$$

where the length scale for the reservoir is taken as $A_0^{1/2}$. Thus, large values of $L_{N,I}$ correspond to small interface deviations.

The values of $L_{N,I}$ for the Canning Reservoir on those days on which profile data are available are shown in Figure 36b. These are based on the conservative assumption that $z_i = 0$, that is, that the river underflows. Clearly, $L_{N,I}$ increases to relatively large values in the summer when the stratification, and therefore, S_t are large, and the inflows are small. $L_{N,I}$ becomes small in periods of very weak stratification, and as in the previous case, the weakness of the one-dimensional assumption causes little damage.

Using L_N and $L_{N,I}$ then gives an indication of the appropriateness of the one-dimensional assumption for surface and inflow driven deformations of the density structure. The extension of L_N to outflows and deformations resulting from the earth's rotations are not obvious and have not been pursued here.

Given that the indicators for a lake or reservoir suggest that a one-dimensional parameterization is appropriate and that a process-based mixing model is preferred to a diffusivity model, the remaining question is which of the available models should be chosen. As indicated above, most of the development has been in the context of the MIT and DYRESM models. In addition, the U.S. Army Corps of Engineers model CE-QUAL-R1 is available. The temperature prediction component of this model is, in its present form (U.S. Army Corps of Engineers, 1982), very similar to DYRESM. In the following, the structure and performance of DYRESM only will be discussed in detail.

As noted above, DYRESM is based on a Lagrangian representation of the lake, with each horizontal layer of uniform property but variable thickness and location. Thus, each layer expands or contracts as inflow and outflow affect its volume, and those above move up or down. In this way all vertical advection of mass is accounted for by layer movement, and problems of numerical diffusion are not present. Further, conservation of mass is simpler to achieve without the necessity of computing vertical velocities.

The model contains five basic process descriptions: surface fluxes of heat, mass, and momentum; mixed layer dynamics; mixing below the surface layer; inflow; and outflow. Each of these is based on the parameterizations given in the earlier sections and will be described only briefly. The evolution of the model is described in the literature (Imberger *et al.*, 1978; Imberger and Patterson, 1981; Spigel and Imberger, 1980; Imberger, 1982; Patterson *et al.*, 1984; Hocking *et al.*, 1988). Various extensions to the model are given in Patterson *et al.* (1985) (dissolved oxygen), Patterson and Hamblin (1988) (ice cover), and Ivey and Patterson (1984) (bottom current induced mixing and effect of rotation on surface layer dynamics).

1. *Surface Heat, Mass, and Momentum Fluxes*

The fluxes of heat, mass, and momentum at the surface are described in detail in Section III. Following the comments in that section with respect to the self-regulation of the thermal budget, the bulk formulae equations (3.1)–(3.3) are appropriate for modeling that reports on a daily time scale, and frequently utilizes meteorological data from a single site, averaged over a full day. The evaporative heat flux is calculated from (3.3), using a bulk value of C_W of 2.6×10^{-3}, within the range of values suggested. This corresponds to the formulation arising from the Lake Hefner data (Tennessee Valley Authority, 1972) and is similar to the form used by Orlob and Selna (1970). Sensible heat transfer (3.2) uses $C_H = C_W$ and is related by the Bowen ratio (Henderson-Sellers, 1986). The value of C_D for momentum transfer (3.1) is wind speed dependent, in the form (Donelan, 1982; Ivey and Patterson, 1984)

$$C_D = \begin{cases} 1.124 \times 10^{-3}, & U < 4 \text{ m s}^{-1}, \\ (0.96 + 0.041\, U) \times 10^{-3}, & U > 4 \text{ m s}^{-1}. \end{cases} \tag{11.5}$$

The remaining heat transfers are treated in the usual way. Incoming short wave radiation is absorbed by the water column, after reflection at the surface. The reflection is characterized by the albedo, which is measured for the particular site. The albedo A_s is a function of sun angle, surface roughness, water color, etc., but is usually taken, for averaged data, as a constant. Typically, A_s is of order 0.04, corresponding to 4% reflection. The absorption of short wave radiation is characterized by Beer's Law

$$Q(z) = Q_0 \sum_i \beta_i e^{-\eta_i z}, \tag{11.6}$$

where the sum is taken over a number of wavelength bands,

$$\sum_i \beta_i = 1, \tag{11.7}$$

and the η_i are the attenuation coefficients for each band. Although two and higher band relationships have been used, for example for the absorption of short wave radiation in ice (Patterson and Hamblin, 1988), in general a single band formulation is sufficient for use with daily data.

Incoming long wave radiation is either measured or predicted from formulae such as the Swinbank equation (Swinbank, 1963), modified to include the effect of cloud (Henderson-Sellers, 1986). Likewise, the long wave emission from the surface may be measured or calculated from the Stefan–Boltzmann Law

$$Q_{\mathrm{L}} = \varepsilon \sigma T^4, \tag{11.8}$$

where ε is the surface emissivity, σ the Stefan–Boltzmann constant, and T the surface temperature in °K. The emissivity is fixed at 0.975.

These heat transfers are implemented on the layer structure of the model, with all fluxes except short wave affecting the surface layer only. The time step of the heating and surface mixing calculations is also set here; the time is limited to that in which the surface layer temperature changes by 3°C, up to a maximum of 12 hours. An additional limitation is placed on the time step in terms of momentum transfer; the change in the mean mixed layer velocity from the previous value is also limited.

2. *Mixed Layer Dynamics*

The mixed layer dynamics is modeled by an algorithm that represents a simplified form of (4.18)–(4.22). Specifically, since the use of daily data precludes rapid (within a diurnal time scale) changes, which prompted the Spigel *et al.* (1986) formulation, the assumption that the turbulence in the mixed layer adjusts rapidly to changing external inputs is made. This is equivalent to putting (4.19) equal to zero. Equation (4.18) then collapses onto the form used in DYRESM,

$$\left[\frac{C_{\mathrm{T}}}{2} q_*^2 + \frac{\Delta\rho g h}{2\rho_0} + \frac{g\delta^2}{24\rho_0}\frac{d(\Delta\rho)}{dh} + \frac{g\Delta\rho\delta}{12\rho_0}\frac{d\delta}{dh}\right]\frac{dh}{dt}$$
$$= \frac{C_{\mathrm{K}}}{2}(w_*^3 + \eta^3 u_*^3) + \frac{C_{\mathrm{s}}}{2}\left[U_{\mathrm{s}}^2 + \frac{U_{\mathrm{s}}^2}{6}\frac{d\delta}{dh} + \frac{U_{\mathrm{s}}\delta}{3}\frac{d(U_{\mathrm{s}})}{dh}\right]\frac{dh}{dt}, \tag{11.9}$$

where

$$q_*^3 = w_*^3 + \eta^3 u_*^3, \tag{11.10}$$

and the following equivalences apply:

$$\eta \equiv C_N, \tag{11.11}$$

$$E_s^{3/2} \equiv \frac{1}{C_F + C_E} q_*^3, \tag{11.12}$$

$$C_K \equiv \frac{C_F}{C_F + C_E}, \tag{11.13}$$

$$C_T \equiv (C_F + C_E)^{-2/3}. \tag{11.14}$$

The values used are $\eta = 1.23$, $C_s = 0.2$, $C_T = 0.51$, and $C_K = 0.125$. These yield values of C_F and C_E similar to those given in Section IV. Equation (4.22) is used to calculate the shear velocity.

3. *Mixing below the Surface Layer*

As discussed in Section X, mixing below the surface mixed layer is characterized by patchy, sporadic, individual events that span regions of only a few meters and last for only a few minutes. These events may arise from, for example wave–wave interaction, double diffusion, boundary mixing, or Kelvin–Helmholtz billows. It is not possible, in the context of a model such as DYRESM, to model individually each of these processes on the appropriate time and length scales. Rather, the turbulent processes are modeled by an eddy diffusivity. This approach has been commonly followed with a wide range of field determinations of K_z (see Section X).

DYRESM utilizes the formulation given by (10.2), as described by Imberger (1982). Here, the energy inputs from the surface wind and the inflow are assumed to be uniformly distributed over the surface layer and the thermocline region. The wave characteristics are determined by matching the K_z with extreme values. Thus, for the case where wind power input exceeds river inflow, K_z from (10.2) is matched with $0.067(H - z_T)u_*$, assuming that $u \sim u_*$. This yields $k_0 \sim 12.4\, A_s/[V'(H - z_0)]$, where V' is the volume over which the energy is dissipated, the lower boundary of which is at z_0. Similarly, for the inflow dominated case, $k_0 \sim 2\pi/d_0$, where d_0 is the depth of underflow, and $u \sim 0.1\, Q/d_0^2 \tan \alpha'$, where Q is the flowrate and α' the stream angle.

The parameter α in (10.2) is given the value 0.5 (Imberger and Hamblin, 1982).

4. *Inflow*

The inflow of the rivers into the reservoir is modeled in three separate processes, as described in Section IX: determination of the plunge point, underflow, and intrusion. Following Section IX, the plunge point determination is parameterized by (9.1) and (9.3), the underflow by (9.2), and the intrusions by (9.4), (9.5), and (9.7)–(9.11). The implementation of (9.4) and (9.5) is described in Imberger and Patterson (1981); briefly the balance is changed from inertia-buoyancy at $R \sim 1$ (R defined by (9.7)). The cases of steep slopes, unconfined, or broad river valleys are not modeled by DYRESM. The impact of a downflow on a sudden density change is treated by the method of calculation of N^2, as for the withdrawal case below.

The downflow is modeled by tracking the day's inflow down the slope and locating its position and flowing depth. Thus, the underflow is made up of a number of parcels of inflows placed along the slope, moving down on each day until their level of neutral buoyancy is reached. Insertion following (9.4) or (9.5) generates the intrusion thickness from conservation of volume. The insertion lengths calculated from (9.4) and (9.5) together with the downflow characteristics are essentially two-dimensional parameterizations of the inflow; these are later utilized in a quasi two-dimensional version (see below).

5. *Outflow*

In the context of the assumptions of both one- and two-dimensional forms of DYRESM, the effects of the earth's rotation are not considered in the description of the process of withdrawal. Similarly, the behavior in two-layer fluids is not explicitly modeled, with the behavior relevant to sudden changes described in Section VIII being modeled in the use of density gradients averaged across the interface. Although this procedure is not strictly correct, the error is small. The averaging procedure is described fully in Imberger and Patterson (1981). The outflow algorithm is therefore based on (8.23) to (8.26), describing point and line sink withdrawal from a continuously stratified fluid. Briefly, these equations are implemented as follows. The inertial withdrawal layer thickness δ

corresponding to the point source in a stratified fluid is calculated according to (8.26), with a value of C_1 of 2, as discussed below. If $\delta > W$, where W is the lake width, the layer has spread the width of the lake, and for all but this immediate spreading region, the layer behaves as though it is generated by a line sink across the full width of the lake (Lawrence and Imberger, 1979).

For the point sink case, the inertial limit is described by (8.26), and the viscous limit by (8.25). The transition flow is modeled, with the distinction between the two cases being determined by the value of the parameter S (8.28), following Cases A and B. The species convection cases described by Ivey and Blake (1985) are not covered by the algorithm.

If the line sink model is invoked by the condition referred to above, the layer thickness is described by (8.23) and (8.24), depending on the value of the parameter R_L, as described in Section VIII. The value of C_1 for the inertial point sink case is chosen so that the inertial layer thickness of the point sink case will equal the inertial layer thickness arising from the line sink, at the discharge at which the transition from point to line sink occurs. Although slightly larger than the experimental determinations of C_1, the error is small and is accepted to ensure a smooth transition from one type to the other.

The region from which the fluid is drawn is determined by the algorithm developed by Hocking *et al.* (1988). Here, a velocity profile is assumed, following Spigel and Farrant (1984), to be

$$u = \begin{cases} \frac{1}{2}u_0\left(1 - \dfrac{x}{L}\right)\left[1 + \cos\left(\pi \dfrac{z - z_s}{\delta_{1/2}}\right)\right], & 0 \leqslant \dfrac{z - z_s}{\delta_{1/2}} < 1, \\ 0, & \dfrac{z - z_s}{\delta_{1/2}} \geq 1, \end{cases} \tag{11.15}$$

which is a cosine profile about the level of the sink with a linear decay away from the sink to the far end of the lake, where L is the lake length, $\delta_{1/2}$ the layer half thickness, u_0 the centerline velocity, and z_s the height of the sink.

Invoking conservation of volume enables computation of the vertical velocities, and the streamlines were shown by Hocking *et al.* (1988) to have the form

$$x(z) = L - \Psi\left[1 + \frac{(z - z_s)}{\delta_{1/2}} + \frac{1}{\pi}\sin\left(\pi \frac{(z - z_s)}{\delta_{1/2}}\right)\right]^{-1}, \tag{11.16}$$

where Ψ is a constant. Hocking *et al.* integrated along the streamlines from the sink to obtain the envelope from which the withdrawal in a given time step would be taken. Although closed form solutions were not available, approximate forms for the curves for both large and small withdrawal were found. The results were verified by comparison with the Spigel and Farrant (1984) experimental results.

The calculation of these withdrawal envelopes gave an accurate representation of the quantity of fluid taken from each of the model layers. This allowed for a proper representation of the horizontal transport in the withdrawal layer, noting that the envelope extended above the calculated withdrawal layer near the sink; fluid from outside the withdrawal layer translated vertically into the withdrawal layer before moving horizontally into the sink.

These process descriptions are linked together in the framework of Lagrangian layers introduced above. Thus, mixing in the mixed layer algorithm takes place by the amalgamation of layers with redistribution of the layer properties across the amalgamated layer according to conservation of mass and energy. The thickness of the new layer will equal the sum of the thicknesses of the individual layers; the lost resolution is acceptable since there is no variability within the amalgamated layer. Before a process affects that layer, for example heat fluxes at the surface, the layer is checked against a maximum thickness criterion and split if necessary. In other words, only the spatial resolution required for each process is retained. In high gradient regions, the resolution is correspondingly fine. The temporal resolution is set by the limitations on heat and mass fluxes described above; again the resolution is fine only in periods of relatively rapid change. The result is resolution of length scales down to a few centimeters and time scales down to 15 minutes, but only when necessary. This means a computationally efficient and accurate code.

The quality of a simulation by any model depends largely on the quality of the meteorological, inflow, and withdrawal data. As an example of the application of DYRESM with relatively good data, a simulation of the Canning Reservoir, described above, for the period 11 June 1986 to 7 September 1987, (450 days, covering a full seasonal cycle) has been carried out. The result, presented as isotherm-depth histories, is shown in Figure 37. This result should be directly compared with Figure 2. Clearly, the simulation has almost exactly followed the data, particu-

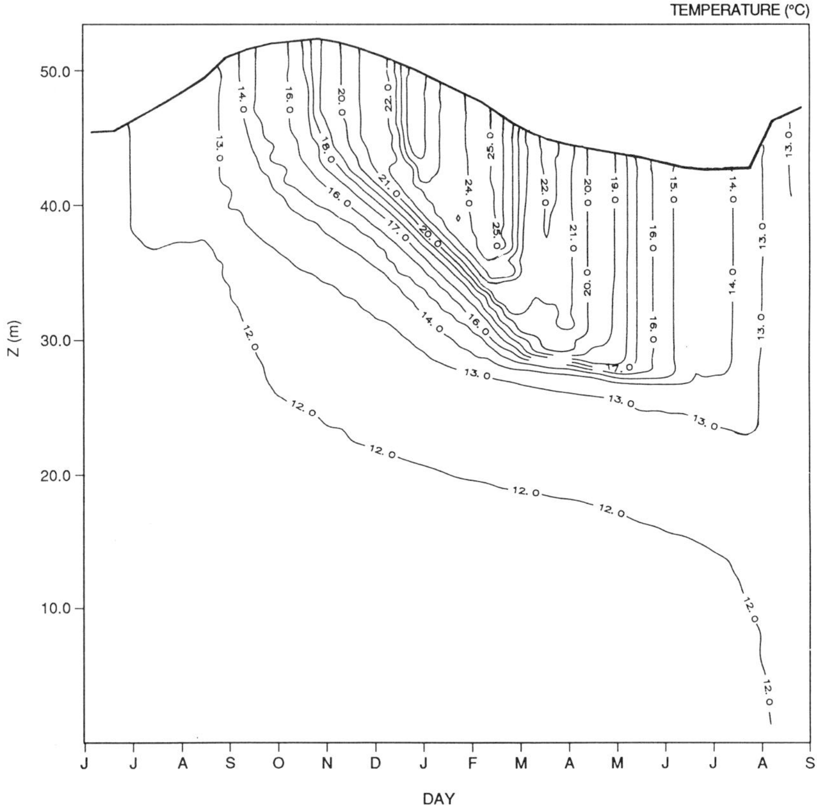

FIG. 37. Isotherm-depth histories from the simulation of Canning Reservoir by DYRESM for the period 11 June 1986 to 7 September 1987. This figure should be compared with Figure 2, which shows the data for the corresponding period.

larly with respect to formation of the stratification through summer and the subsequent mixed layer deepening in mid-May. There are two points of mismatch. First, the shape of the 12°C isotherm in the deeper part of the reservoir differs from the observed data. The error is minor, however, as the temperature gradients in this region are extremely weak and small deviations shift the isotherm large distances. Second, the surface temperature early in the summer period is simulated as being slightly higher than the data indicate. The error is within 1°C, however.

As a second example of the use of the model for longer term simulations, an eight-year data set for the Wellington Reservoir was assembled. Simulations of part of this period have been presented

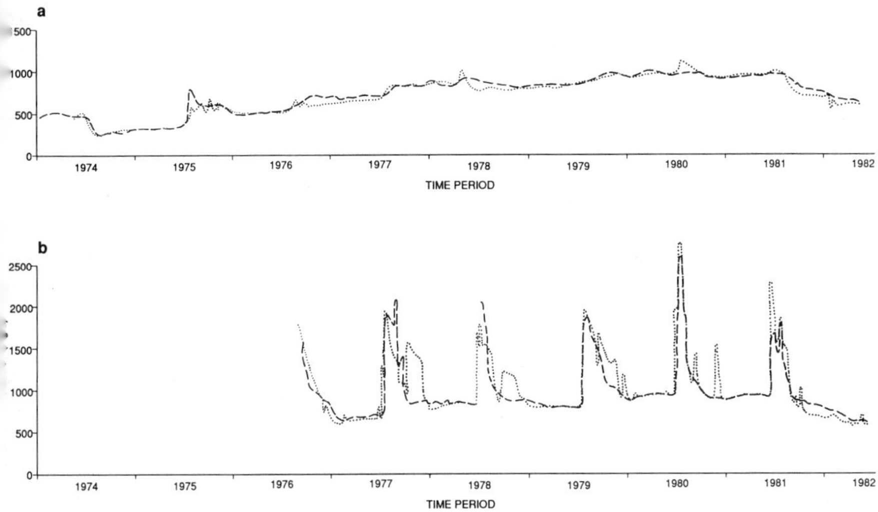

FIG. 38. The offtake salinities at Wellington Reservoir, measured (dashed line) and predicted by DYRESM (dotted line), over the extended period 1974 to 1982: (a) mid-level offtake and (b) scour offtake.

elsewhere (Imberger and Patterson, 1981); here, comparisons of the predicted offtake properties from the two operating offtakes with the measured values are shown (Figure 38). Clearly, there are some deviations from the measured values, but over the long term the agreement is remarkably good.

These applications show the strength of the one-dimensional version of DYRESM, both for short term (one year) and longer term (eight years) simulations. In both cases however, only the one-dimensional variations in the stratifying species may be modeled, consistent with the assumption of one-dimensionality. In many cases, however, the parameters of interest may not influence the density, and even if the density field is one-dimensional, these parameters may exhibit considerable horizontal variability. Further, the motions in reservoirs and lakes are in general not one-dimensional; for example, the wind accelerates the surface waters and induces a return flow as well as tilting the surface, perhaps leading to upwelling and internal and surface waves; the density gradients established by these and other mechanisms such as unequal heat capture may drive horizontal motions; inflows to a reservoir are usually of a different density to the surface water and underflow for some distance before being inserted horizontally. All of these two- and three-dimensional effects are averaged out in a one-dimensional representation and the details are lost.

Rodi (1987) reviewed the available models for a broad scope of lake and reservoir problems. He concluded that in three dimensions the difficulty of adequately resolving the important time and particularly length scales limited the quantitative application of these models. In two dimensions the situation is somewhat better, although models with stratification are of limited application. These take a vertical surface along the river valley of the lake, and simulate variations and motions in the longitudinal and vertical directions. The best known of these models is the LARM model of Edinger and Buchak (1983) (Gordon, 1981; Kim *et al.* 1983) and its more recent version GLVHT (Buchak and Edinger, 1984). This model solves the laterally integrated equations of motion on a finite difference grid, subject to a number of assumptions. Turbulence closure was with a mixing length model with a Richardson number correction. Again, however, resolution remains a major difficulty.

A different approach was taken by Jokela and Patterson (1985) with a quasi two-dimensional version of DYRESM. The formulation considered the development of horizontal structure only from inflow intrusions along the previously defined DYRESM horizontal layers. To identify the horizontal gradients, each horizontal layer was divided into a number of Lagrangian parcels that moved horizontally within the layer, changing length as the layer thickness changed, retaining the Lagrangian character of the model. The total inflow for the day was distributed across a number of horizontal layers in the usual way, with each layer's apportionment forming a new parcel for that layer. The length of the parcel was determined from the calculation of the intrusion distance, and all previous parcels in the layer were forced forward, contracting in length as the layer thickness increased.

This preliminary model had no provision for adjustment of horizontal gradients, and was applicable only for situations in which the one-dimensional assumption held. However, the simulation of an inflow event showed good agreement with the measured data.

An extended version of this model, with attention paid to other processes besides inflow, has been completed by Hocking and Patterson (1988). This model builds on the original Jokela and Patterson model, including a gravitational adjustment. The vertical mixing processes from the original one-dimensional model are retained, but are applied on a parcel by parcel basis.

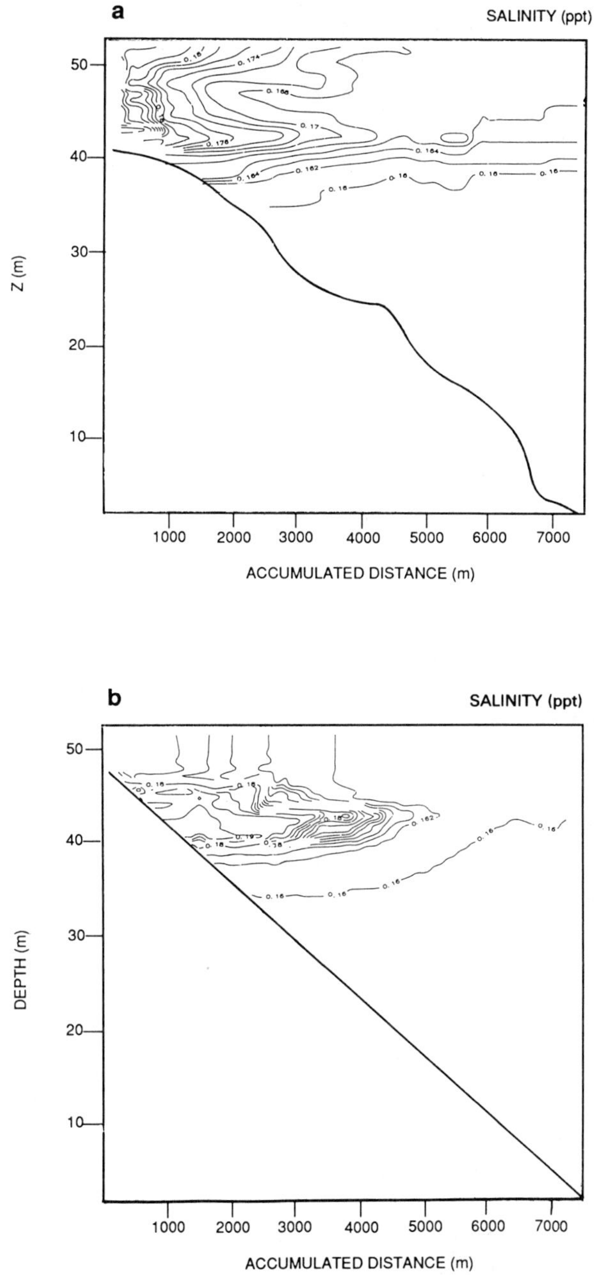

FIG. 39. (a) The isohalines resulting from salinity measurements at a sequence of stations in Canning Reservoir on 27 October 1986. (b) The isohalines drawn from the simulation of Canning Reservoir by two-dimensional DYRESM, after a short (30-day) simulation.

A simulation of an inflow event in the Canning Reservoir is compared with the field data in Figure 39, following a 30-day simulation. The intrusion is characterized by the isohalines shown, salinity being effectively a tracer in the Canning Reservoir. The simulated intrusion is a good representation of the field data in terms of the character of the intrusion and the actual salinity values. The depth and progress of the intrusion into the reservoir are accurately predicted, and the shape is qualitatively the same. The thickness is slightly overpredicted. The tilting of the 0.16 ppt isohaline in the deeper part of the reservoir is also well modeled.

The advantages of this model over those which solve some form of the equations of motion on a fixed grid are obvious. The horizontal resolution of the model is determined by the horizontal process acting and is only fine where required. This means that the algorithm is computationally economical and the dispersion problems associated with fixed grids are dispensed with. Further, the tracking of tracers through the reservoir is made particularly simple by the Lagrangian nature of the representation.

XII. Reservoir Destratification by Bubble Aerators

The development of modeling techniques for lakes and reservoirs from detailed process descriptions resulted from an increasing water resource management requirement. For example, it may be desired to operate a reservoir to optimize the quality of the withdrawn water, to minimize evaporation, to prevent long-term build up of some dissolved substance such as salt, or to prevent algae growth. To design operational strategies to satisfy such diverse requirements, a process-based model of the kind described above is essential. Only then, within the limitations of the assumptions of the model, can the strategies be tested with the processes properly represented.

As a case in point, reservoir destratification systems are frequently installed to prevent the build up of stratification, or dismantle an existing stratification in a reservoir, usually to improve water quality. In summer the metalimnion is a barrier between the actively mixed epilimnion and the hypolimnion. The hypolimnion may become deoxygenated, causing water quality problems. Removing the stratification artificially or directly aerating the hypolimnion have been practiced for several decades to combat this problem. The use of these management techniques in reservoirs apparently dates from 1954 (Cooley and Harris, 1954), although applications of bubble curtains as artificial breakwaters existed in 1907 (see Bulson (1967) and Wilkinson (1979)). A number of reviews have outlined the means available for destratification (Tolland, 1977; Henderson-Sellers, 1982), including hydraulic pumping systems, mechanical mixers, and air bubbler systems.

For reservoirs, the most common destratification device is an air bubbler system. Compressed air is pumped to the bottom of the reservoir and released; the buoyant plume rises to the surface, carrying with it water from the hypolimnion, which is then ejected from the plume where the density of the air–water mixture is approximately equal to the ambient density. Once ejected, the fluid is locally heavy, and mixing occurs, reducing the stratification. The release of the air occurs either as one or more distinct plumes, or as a curtain, depending on the design. The success of this procedure depends on the bubble plume being buoyant enough to carry the hypolimnion water into the epilimnion before discharge. The efficiency therefore depends on both the stratification and the air discharge rate.

A typical bubble plume operation was recently documented in the Harding Reservoir and is shown in Figure 40(a) and (b). In both examples the wind was negligible, but during the acquisition of the data shown in Figure 40(a) (see p. 448) the water surface layer was only very weakly stratified. As evidenced by the shape of the isotherms the bubbler was sufficiently active ($P_N = 0.3$, see (12.1)) to completely penetrate the whole water column and yet the stratification was sufficient to return the cold surface water over the bubbler back to a depth of 6 m; the falling plume being nearly 30 m wide.

By contrast the data illustrated in Figure 40(b) was collected during a windless hot afternoon when the stratification was much stronger at the surface. Again the isotherms display the impact of the bubbler; the cold water rising in the plume could not completely penetrate the surface stratification ($P_N = 800$ for the surface stratification) and the stronger stratification led to a much thinner plume.

In spite of their relatively long period of use, the design of bubbler systems falls far short of ideal, with much of the existing design based on empirical rules determined for a particular water body. For example, Davis (1980) gives a design procedure based on the stability of the water body and the energy flux through the surface. Based on a calculation of the stability (2.3), the design procedure assumes an efficiency of 5% to determine the energy input required by the destratifying device. Similarly, determination of the length of diffuser pipe, the diffuser hole diameters, and the diffuser separation are empirically based. The energy requirement is based on the strongest stratification and no wind effects, and is therefore overestimated. The calculations are sensitive to the value of S_t, and the resulting design depends strongly on the initial assumptions made. Applications using these or similar procedures are described by Brim and Beard (1980), Brown *et al.* (1982), and Brady *et al.* (1983). The latter demonstrated the sensitivity of the Davis procedure to the assumed initial profile. Burns (1977) describes some hydraulic model tests intended to assist in the design procedure.

Even though the design procedures are inadequate for lake and reservoir destratification systems, the bulk of research into bubble plumes has been done in the context of a homogeneous environment (Kobus, 1968; Wilkinson, 1979; Milgram, 1983; Tacke *et al.*, 1985; Sun and Faeth, 1986; Cheung and Epstein, 1987). These papers describe various aspects of the plume characteristics, including determinations of the velocity distribution, entrainment coefficient, and bubble slip velocity. In general these are based on an integral model of the plume dynamics, with the exception of the Sun and Faeth (1986) paper, which deals with the turbulent properties of the plume.

Ditmars and Cederwall (1974), Milgram (1983), and Cheung and Epstein (1987) presented integral models based on a homogeneous ambient fluid. The conclusion drawn by these papers has been that, first, it is not appropriate to treat the plume as a single phase flow unless the airflow rate is very small, and second, that it is possible to treat the plume as a simple plume with the buoyancy term modified by the presence of the bubbles, provided that the slip velocity of the bubbles is incorporated.

Stratified surrounding fluid was introduced in the models of McDougall (1978) and Hussain and Narang (1984). The McDougall paper deals with experiments in a strongly linearly stratified environment and shows that the behavior of the plume is somewhat more complex than in the unstratified case. Briefly, the plume rises, carrying relatively heavy water and entraining ambient fluid, to the point where the mixture density is approximately equal to the ambient, where the fluid detrains from the plume. At this point, the detrained fluid has lost the additional buoyancy of the bubbles and is locally heavy. It therefore plunges to a new level before spreading horizontally. The bubble plume continues to rise, creating a new buoyant plume. In a strongly stratified fluid, this process may occur a number of times, resulting in multiple intrusions, the spacing of which depends on the stratification and the buoyancy flux. McDougall proposes two integral models for the plume; the first based on the simple plume model modified by the buoyancy flux input by the bubbles and the bubble slip velocity; the second based on a double plume structure, with an interior plume in which all the bubbles are contained, and surrounded by an annular outer plume containing only liquid.

Hussain and Narang (1984) developed a similar double plume structure model for application in weakly stratified fluids. The condition placed on the stratification implicitly meant that multiple intrusions were not possible.

In both stratified and unstratified experimental and model studies little attention had been placed on the behavior under widely varying stratification conditions and airflow rates. The paper by Asaeda and Imberger (1988) attempted to resolve this question by conducting experiments in both two layer and linearly stratified fluids of varying degrees and varying airflow rates.

Asaeda and Imberger (1988) confirmed experimentally that the efficiency of the destratification depended on both the stratification and the airflow rate. For the linear stratification, they showed that the behavior of a single plume, or, more accurately, the conversion of the energy input by the plume into mixing, depended on the parameter

$$P_N = \frac{N^3 H^4}{Q_0 g}, \tag{12.1}$$

where N is the buoyancy frequency, H the depth of the aerator, and Q_0 the airflow rate. In particular, they showed that the conversion efficiency reached a maximum for a value of P_N at approximately 10^3. The efficiency was calculated as

$$\eta = \frac{g\,\Delta S_t}{\rho_0 Q_0 H g\,\Delta t}, \tag{12.2}$$

representing the ratio of the change in potential energy $g\,\Delta S_t$ in the stratification to the energy input by the plume in time Δt (Tolland, 1977). The maximum efficiency observed was approximately 0.12. The maximum was also related to the character of the intrusions. For small values of P_N (relative to 10^3), the plume travelled to the surface before detraining, similar to the homogeneous experiments of Milgram (1983) and others; for large values, multiple, unsteady intrusions formed. With P_N near 10^3, the plume detrained at one or more locations, with steady subsurface intrusions forming.

For a two-layer fluid, similar results were obtained. For a given stratification, low airflow rates produced a plume which detrained at or below the level of the interface, with the result that very little exchange between upper and lower layers occurred. At high rates, the detrainment occurred at the surface and plunged back to the interface. The critical parameter here was

$$P_\Delta = \frac{g'^{3/2} h_l^{3/2} H}{g Q_0} \left(\frac{h_u}{H}\right)^3 \tag{12.3}$$

For $P_\Delta < 30$, the plume broke through the interface and mixed with the upper layer. For $P_\Delta > 30$, the stratification was strong enough to prevent the entrained fluid from penetrating the interface, and detrainment occurred at or below the interface. In the first case, mixing occurred from above, with a deepening of the interface; in the second, mixing was weak, and characterized by a thickening of the interface.

These results were consistent with others; the McDougall (1978) experimental results were of moderate P_N, with steady subsurface intrusions. The two layer results were similar to those obtained by Graham (1980). While insufficient details are given by Graham to calculate P_Δ, it is clear that in the early part of the experiments, the "diffusive" profiles are characteristic of $P_\Delta > 30$; the density jump at the interface is sufficiently strong to force detrainment below the interface. At later times, the stratification has been reduced to allow the effective $P_\Delta < 30$, and the plume penetrates fully, resulting in a sharper interface as mixing from above deepens the upper layer. The experiments of Kranenburg (1979) may also be put in the context of P_Δ. The series of profiles shown in Kranenburg indicate, after the initial stages, a rapidly deepening interface, consistent with a calculated value of P_Δ of 0.1, much less than the suggested value of 30 for penetration of the interface. This is consistent with the qualitative description given by Kranenburg of an intrusion plunging from the surface to the interface. The experiments of Dortch and Holland (1980) were directed at comparisons with hydraulic methods and insufficient details were given for a comparison to be made.

These conclusions have been incorporated into the reservoir model described above. Although integral models of the plume itself exist, none have been coupled to a far field model, other than that of Kranenburg (1979). Kranenburg developed a three component model; the plume, the intrusion, and the far field two-layer stratification, in a cylindrical geometry. Good comparisons with the experimental results were obtained, but the model was not intended for reservoir applications.

The development of the coupled bubble plume and reservoir model is fully described in Patterson and Imberger (1989); briefly, the single plume model of McDougall (1978) was discretized using the layer structure determined by the reservoir model, with the plume entraining from each layer as it passes. As the plume rises, the effective buoyancy anomaly decreases as entrainment lowers the plume density and the ambient density decreases. At some level, the vertical velocity of the entrained fluid becomes zero, the fluid is ejected and mixes with the underlying reservoir water. It is assumed that the resulting horizontal gravitational adjustment is sufficiently rapid for the one-dimensional adjustment to remain in force. The bubble plume continues, starting a new entrainment cycle.

The coupled model was verified against a one-year data series from Myponga Reservoir in South Australia in which an aerator had been operated over the summer at a depth of 14 m below the surface (Patterson and Imberger, 1989). The model in general reproduced the resulting thermal structure, and it was concluded that it adequately represented the coupling of the bubble plume dynamics with the processes acting in the reservoir.

This enabled the use of the coupled model to determine various aspects of plume behavior. Of particular interest was the maximum efficiency airflow rate found in the Asaeda and Imberger (1988) experiments for a simple geometry and simple stratifications. The question of the applicability of this to real geometries and stratifications is fundamental to the optimum design of destratification systems.

One measure of performance of a destratification system is the efficiency η, defined above. Asaeda and Imberger (1988) showed that the efficiency depended on the parameters P_N and P_Δ for the two specific stratifications described. For the more general case the stability of the system is characterized by S_t and the airflow rate by Q_0. Similar to the cases for wind stirring and inflow, it is possible to form a version of the Lake number appropriate to the bubble plume. Again, following Section II and taking moments about the center of volume yields $S_t g\beta$, where β is the angle of deviation of an isopycnal. If the disturbing force is taken as the bubble plume, characterized by a friction velocity w_*, the moment of the force scales with $\rho_0 w_*^2 H^2$, since the diameter of the plume scales with H. The moment is therefore $\rho_0 w_*^2 H^2 L$, where L is the length from the plume to the center of volume. The work done by the plume is characterized by $\rho_0 Q_0 Hg\ \Delta t$ (Dortch and Holland, 1980); this must be equivalent to the work done by the disturbing force. Thus,

$$\rho_0 w_*^2 H^2 w_*\ \Delta t \sim \rho_0 Q_0 Hg\ \Delta t \tag{12.4}$$

or

$$w_* \sim \left(\frac{Q_0 g}{H}\right)^{1/3}. \tag{12.5}$$

In the same way as in Section X, a generalized number may be defined:

$$L_{\mathrm{N,B}} = \frac{S_t g \beta_{\max}}{\rho_0 (Q_0 g/H)^{2/3} H^2 L}, \tag{12.6}$$

where $\beta_{\max}$ is the maximum deviation. For the two-layer case $\beta_{\max} \sim (z_{\mathrm{m}} - z_{\mathrm{T}})/L$; for the linearly stratified case $\beta_{\max} \sim H/L$. In general, if $L_{\mathrm{N,B}}$ is large, the deviation is small compared with $\beta_{\max}$ and the density structure will not be disturbed by the bubble plume; if $L_{\mathrm{N,B}}$ is small, the deviation is large. In the first case, the airflow rate is insufficient to penetrate the stratification and cause mixing. The second case corresponds to an airflow rate well in excess of that required to penetrate the stratification. If $L_{\mathrm{N,B}}$ is large the efficiency will clearly be small since very little mixing in a relatively strongly stratified environment will occur; if $L_{\mathrm{N,B}}$ is small, although mixing is occurring, the disturbance is more than is required, and the efficiency is again low. The manifestation of the small $L_{\mathrm{N,B}}$ case is a plume detraining at the surface, which would also have detrained at the surface with a much lower airflow rate. This qualitative argument indicates that a maximum value of efficiency should occur at intermediate values of $L_{\mathrm{N,B}}$ and that, following the experiments of Asaeda and Imberger (1988), the achievable efficiency should be of the order of 0.12.

The parameters derived by Asaeda and Imberger (1988) are directly related to $L_{\mathrm{N,B}}$. It is straightforward to show, using the definition of $\beta_{\max}$ above, that, for the linear stratification,

$$L_{\mathrm{N,B}} = \frac{A_0}{12L^2} P_N^{2/3}, \tag{12.7}$$

where A_0 is the surface area of the experimental tank, and for the two layer case,

$$L_{\mathrm{N,B}} = \frac{1}{2}\frac{A_0}{L^2} P_{\Delta}^{2/3}. \tag{12.8}$$

For the linear stratification experiments, the value of $P_N \sim 10^3$ gives $L_{\mathrm{N,B}} \sim 8$. For the two-layer experiments that yielded a transition of character at $P_{\Delta} \sim 30$, this puts transition at $L_{\mathrm{N,B}} \sim 5$. The definition of $L_{\mathrm{N,B}}$ then unifies the two parameters defined by the experiments of Asaeda and Imberger, and suggests that for other stratifications, $L_{\mathrm{N,B}} \sim$ 5–8 will maximize efficiency of the bubble plume-induced mixing.

The numerical experiments with reservoir data by Patterson and Imberger (1989) described above yielded single plume efficiency curves which indicated a maximum efficiency of order 0.15, consistent with the Asaeda and Imberger experiments. Patterson and Imberger introduced the parameter P, where

$$P = \frac{(S_t g/M)^{3/2}}{(gQ_0/H)} \tag{12.9}$$

and M is the total mass of the reservoir, derived from the ratio of energy stored in the stratification to the energy input by the plume.

The simulations in Patterson and Imberger (1989) showed that a maximum efficiency occurred at $P \sim 1$, as shown in Figure 41a for Myponga Reservoir. Over the simulation period, a wide range of stratification is present, and by varying Q_0, a wide range of P values may be obtained. Figure 41 is constructed from the results of several simulations, with the efficiency plotted as a function of the P calculated for a particular stratification and airflow without regard to the temporal order of occurrence.

Myponga Reservoir is a small reservoir of storage $26.8 \times 10^6\ \mathrm{m}^3$ and surface area $2.8 \times 10^6\ \mathrm{m}^2$. The same calculation has been carried out for the Harding Reservoir in the northwest of Western Australia. The Harding Reservoir is substantially larger (storage $63.8 \times 10^6\ \mathrm{m}^3$ and surface area $14.1 \times 10^6\ \mathrm{m}^2$) and is exposed to quite different meteorological forces, being located at latitude 20°S (compared with Myponga Reservoir at 35°S). The calculation of P, however, yields a maximum efficiency again at $P \sim 1$, with a peak efficiency of 0.15 (Figure 41b). This parameter therefore unifies the simulation data.

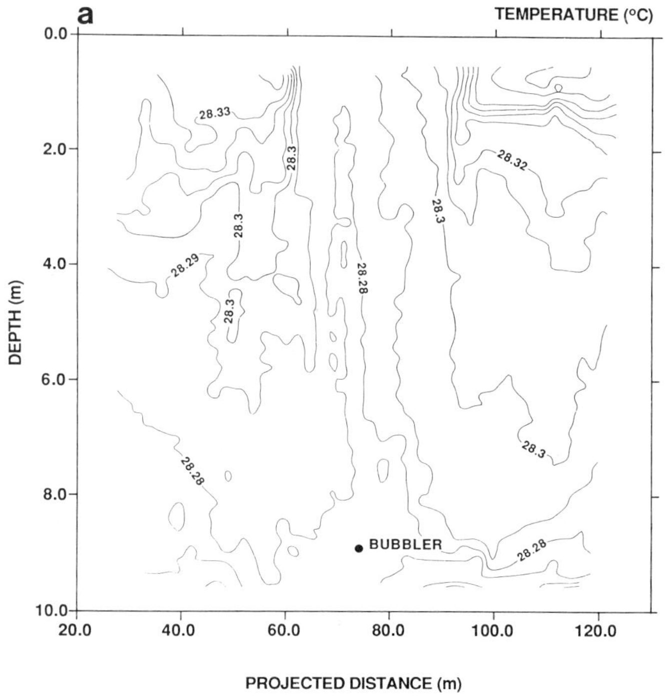

FIG. 40. Transverse transect across the center of a 400 m long bubble diffuser (1 mm diameter holes spaced at 4 m) located at a depth of 9 m. Data was obtained by yo-yoing a temperature probe through the water column at intervals of approximately 8 meters from the surface to within 1 m of the bottom. Gas flow rate was $40\,\mathrm{l\,s^{-1}}$. (a) Morning with weak stratification (April 1989) and (b) Afternoon with strong stratification (April 1989).

The relationship between P and $L_{\mathrm{N,B}}$ is given by

$$L_{\mathrm{N,B}} = \frac{M}{\rho_0 H^2 L} P^{2/3} \beta_{\max}. \tag{12.10}$$

For the reservoir, the appropriate $\beta_{\max}$ is $(H - z_{\mathrm{T}})/L$, and the relationship between $L_{\mathrm{N,B}}$ and P is not direct, as both M and z_{T} may change daily. The calculation of $L_{\mathrm{N,B}}$ would require new simulations; these have not been performed.

The performance of a single plume is characterized by the value of P and therefore indirectly by $L_{\mathrm{N,B}}$; for optimum performance, a single plume should be operated such that $P \sim 1$. As the stratification changes, the airflow rate should change. Further, the degree of destratification obtained by a plume is given by $\Delta(gS_t)$, which will generally be much less

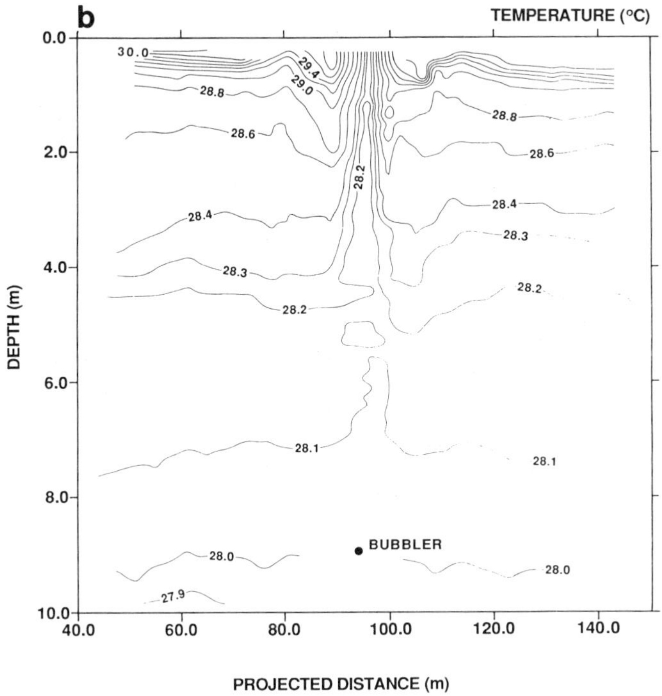

FIG. 40. (*contd.*)

than gS_t. Either more plumes are required or the time over which the destratification occurs must be extended. An estimate for the number of plumes required is given by $n \sim gS_t/\Delta(gS_t)$. Patterson and Imberger (1989), however, show that the efficiency of multiple plumes is higher than a single plume carrying the same total airflow. The optimum operational strategy then is to operate as many separate plumes as possible, each with $P \sim 1$. This of couse involves altering the airflow rate at perhaps daily intervals, depending on the stratification, a level of control which may not be possible. In reality, both the number of plumes and the airflow rate are usually fixed, or variable only in the simplest sense, such as stopping and starting the aerator.

The effect of this nonvariable operational strategy was demonstrated by Patterson and Imberger (1989); for Myponga Reservoir, the overall efficiency achieved was only 0.025 for a $50\,\mathrm{l\,s^{-1}}$ airflow, the result of many times when the stratification was weak and the P value extremely low. This result was strongly dependent on the number of plumes and

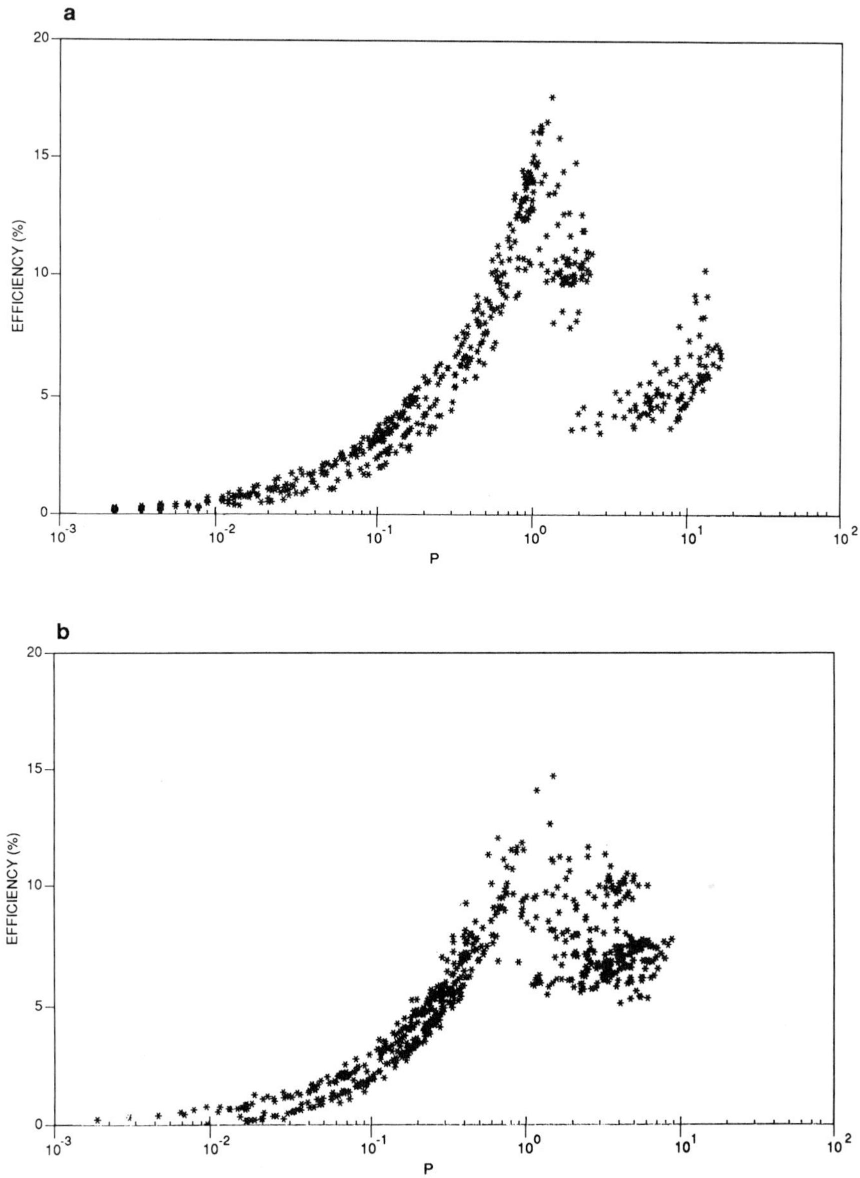

FIG. 41. (a) The efficiency of a single bubble plume as a function of the parameter P for Myponga Reservoir, showing a clear maximum at $P \sim 1$. (b) The efficiency for a single bubble plume as a function of P for Harding Reservoir. The maximum is again at $P \sim 1$.

was achieved with 10 separate plumes. A relatively simple control strategy, in which the aerator was activated only on those days when the temperature difference between aerator and surface exceeded 1°C, doubled the total efficiency to 0.052. This was achieved with 100 separate plumes. The interaction between the optimum number of plumes, the total airflow rate, and the operational strategy is a complex one, though it is clear that further substantial gains in efficiency could be made if control over the airflow rate is introduced.

XIII. Summary

Advances in physical limnology in the last 10 years, both in the more traditional area of internal wave seiching and in the newer area of small-scale motions and mixing, has created a very large discipline. We have highlighted this progress in this review, and now wish to draw together those areas which we feel require more work.

(a) *Seasonal behavior.* Lakes in the tropics may be compared with those in temperate regions if the stability and forcing are matched. This may be done by evaluating two nondimensional numbers: the Wedderburn number W, which describes the behavior of the surface layer, and the Lake number L_N, which describes the response of the entire lake. This hypothesis was formed from a review of earlier work on the intercomparison of lakes; a detailed intercomparison study of a large range of lakes in different locations should now be conducted. On the basis of this it may be found that further nondimensional numbers are required to specify higher mode responses.

(b) *Surface fluxes.* A great deal of progress has been made, mainly by oceanographers, in determining the exchange of momentum, heat, and water vapor at the air–water interface. Provided the atmospheric internal boundary layer is well established and the air is either not too stable or unstable, there are adequate procedures for calculating surface fluxes. However, few lakes studies have data of sufficient quality to enable the use of the stability correction. This, together with the well-documented fact that surface roughness has a strong influence, requires that meteorological stations be located in the internal boundary layer and that both stability and roughness be incorporated through a model for the fetch duration and radiation of waves. Last, the wind is spatially highly variable over the surface of the lake. This leads to energy input at a basin-scale length that causes deep hypolimnetic mixing and contributes to the strong, horizontal transport processes in the lake, a process so far completely ignored and one requiring a measure of the wind variability over the lake.

(c) *The surface layer.* The one-dimensional, vertical energy-based mixed layer models appear to describe the surface layer well if local wind and flux data are used. What is now required is a generalization of the one-dimensional model incorporating the Wedderburn number W, which governs the behavior of this layer, to account for horizontal advection of momentum and mass. This should include gravitation-induced advection, which becomes dominant at the cessation of a wind.

(d) *Upwelling.* When the Wedderburn number becomes low, the mixed layer upwells at the upwind end. This upwelling occurs at all values of the Wedderburn number; it is more severe at low values, but is still present at values as high as 10 or 20. Recent models of upwelling have indicated an entrainment process with the same deepening rate as the rate usually attributed to surface-introduced turbulence. Once again, these models need to be generalized to allow for two-dimensional effects and the subsequent horizontal advection spawned at the cessation of wind.

(e) *Differential deepening.* This new mechanism describes the uneven deepening process brought about by the spatial variability of wind stress. The high degree of variability of the wind over a lake causes extreme deepening in exposed areas and almost no deepening in sheltered areas, embayments, and side arms. This observation affects two areas. First, a lake-averaged surface layer model must somehow account for this variability, since at the cessation of each wind period, gravitational motions even out the isopycnals and the net amount of deepening will depend on the exact distribution which, in turn, is determined by the variability of the wind. Second, data from the Wellington Reservoir have shown that the strong undulations introduced at the base of the mixed layer by differential deepening lead to the introduction of energy from the wind through the surface layer directly into the hypolimnion at a basin scale. This is a much more effective way of getting energy into the deep part of a lake than with a simple uniform upwelling mechanism because not only are the first and second modes of the basin excited but differential deepening energizes higher order internal wave modes. The relationship between upwelling, basin seiching, and differential deepening as a function of Wedderburn and Lake numbers must be fully investigated and the consequences for internal mixing assessed.

(f) *Differential heating and cooling.* This subject has received a great deal of attention through the study of the thermal bar in cold lakes but, because it is the strongest single process for horizontal advection in lakes, it should receive a great deal more. An assessment of the influence of frequency of cycling (diurnal) on embayments of various sizes must be carried out to understand the influence of inertia of the water column. Only then can we predict the horizontal transport in and out of sidearms. This is important for both ecology and engineering management, since recreation in reservoirs expose the lake to pollution. Differential absorption, the mechanism by which phytoplankton blooms are torn apart by self-generated convective motions, is a new field. Much work needs to be done in the interaction of convection and growth dynamics of various species of plankton.

(g) *Outflow.* Selective withdrawal has received a great deal of attention over the last 20 years and most fundamental flow configurations can be predicted with reasonable accuracy. The outstanding problems center around selective withdrawal in a variable stratification, and the influence of rotation on the initial value problem and the final degeneration into eddies. Last, the interaction of selective withdrawal and topographic constraints (sills and contractions) is ready for analysis. This is especially important since it has been shown that shear waves, which set up the selective withdrawal layer, are solutions to the Korteweg de Vries equation, allowing the full interaction to be explored while retaining some of the nonlinearity of the fluid motion.

(h) *Inflows.* An inflow plunges underneath the reservoir water, flows down the drowned river valley until it reaches a point of neutral buoyancy, then intrudes into the lake as a horizontal layer. The case where the entrainment of the underflow is dominated by shear production does not seem to have been investigated. Further, the entrainment properties of the plunge point need urgent attention. The generation of turbulence or the maintenance of turbulence in the intrusion layer is important because intrusions can interact with the background internal wave field to sustain a definite level of turbulent transport. Last, lake-specific problems such as the interaction of the surface layer with an inflowing stream and the interaction of the heaving (associated with internal wave basin scale motions) and the inflow require generalization.

(i) *Mixing below the surface layer.* The central problem here remains. However, with the introduction of the Lake number, the way to proceed is clearer. The distribution of turbulent patches and mixing efficiency within each of these patches must be identified as the Wedderburn and Lake numbers vary. Since, in a typically stratified lake, the volume of fluid actively involved in turbulent motions at any time is at a maximum of 1%, this is a most urgent and useful avenue to pursue. Once an algorithm has been achieved that allows for a statistical description of the patch distribution and identification of the mechanism underlying each patch (and thus a description of the mixing efficiency), we will be able to model deep mixing in a lake. A strong effort should be mounted to investigate the formation of internal patches and the maintenance of the turbulent benthic boundary layer.

(j) *Modeling.* Modeling has proceeded along two avenues: One-dimensional parametric modeling and full eddy simulation or closure scheme models. Neither approach is appropriate: two- and three-dimensional motions are of central importance in a lake, and the scales of motion range from the Kolmogorov scale to the basin scale. There does, however, appear to be a spectral gap between the internal waves at the basin scale and the motions within a turbulent patch. Further, as mentioned above, the volume occupied by turbulent motions in a lake at any one time is likely to be less than 1%. A two-dimensional model is needed that accounts for the mean motions generally described by basin seiching, superimposed on which is a model that determines the statistical distribution of turbulence within the lake, in the benthic boundary layer and surface layer. Last, a further model must be superimposed on these two that adjusts the density field according to certain mixing efficiencies. Such a closure scheme has a real chance of success in even quite large lakes, with present computing power. Many variations are of course possible, ranging from a simple parameterization of the patch distribution and the patch efficiency to full turbulent simulations within each patch once they have been identified.

(k) *Reservoir destratification by bubble aeration.* It is unwise to try to destratify a lake completely by introducing mechanical energy. It is much more energy efficient to weaken the structure through the bubble flow and allow the wind to complete the mixing. This requires an examination of the bubble flow and the natural reservoir dynamics. This area of research seems to be in its infancy, as does the equivalent research where propellors are used rather than bubble plumes. Given that the field of reservoir destratification is an enormous industry, large cost savings are no doubt possible.

Acknowledgements

Sections I to X were written by Jörg Imberger and Sections XI and XII were written by John Patterson. We would like to thank Greg Ivey and Graeme Hocking for their comments on the manuscript. Carolyn Oldham prepared the figures and performed the computations, and Colleen Henry-Hall edited the manuscript and prepared it for publication. This assistance is gratefully acknowledged. This work was supported by the Centre for Environmental Fluid Dynamics, the Australian Research Council, the Australian Water Research Advisory Council, and the West Australian Water Authority.

References

Akiyama, J., and Stefan, H. G. (1984). Plunging flow into a reservoir: Theory. *J. Hydr. Eng.* **110**, 484–499.

Akiyama, J., and Stefan, H. G. (1985). Turbidity current with erosion and deposition. *J. Hydr. Eng.* **111**, 1473–1496.

Akiyama, J., and Stefan, H. G. (1987). Gravity currents in lakes, reservoirs and coastal regions: Two-layer stratified flow analysis. Proj. Rept. No. 253, University of Minnesota.

Anderson, R. J., and Smith, S. D. (1981). Evaporation coefficient for the sea surface from eddy flux measurement. *J. Geophys. Res.* **86**, 449–456.

Andre, J. E., and Lecarrere, P. (1985). Mean and turbulent structures of the oceanic surface layer as determined from one-dimensional third-order simulations. *J. Phys. Oceanogr.* **15**, 121–132.

Arai, J. (1964). Some relations between the thermal property of lake and its fetch size. *Geogr. Rev. of Japan* **37**, 131–137.

Arita, M., Jirka, G. H., and Tamai, N. (1986). Classification and mixing of two-dimensional buoyant surface discharges. *J. Hydr. Res.* **24**, 333–345.

Asaeda, T., and Imberger, J. (1988). Structures of bubble plumes in stratified environments. Environmental Dynamics Rept. ED-88-250, Centre for Water Research, University of Western Australia.

Atkinson, J. F. (1988). Interfacial mixing in stratified flows. *J. Hydr. Res.* **26**, 27–31.

Atkinson, J. F., and Harleman, D. R. F. (1983). A wind-mixer layer model for solar ponds. *Solar Energy* **31**, 243–259.

Bean, B. R., Emmanuel, C. B., Gilmer, R. O., and McGavin, R. E. (1975). The spatial and temporal variations of the turbulent fluxes of heat, momentum and water vapor over Lake Ontario. *J. Phys. Oceanogr.* **5**, 532–540.

Bejan, A., Al-Homoud, A. A., and Imberger, J. (1981). Experimental study of high-Rayleigh number convection in a horizontal cavity with different end temperatures. *J. Fluid Mech.* **109**, 283–299.

Bejan, A., and Rossie, A. N. (1981). Natural convection in a horizontal duct connecting two fluid reservoirs. *J. Heat Transfer* **103**, 108–113.

Bejan, A., and Tien, C. L. (1978). Laminar natural convection heat transfer in a horizontal cavity with different end temperatures. *J. Heat Transfer* **100**, 641–647.

Bella, D. A. (1970). Simulating the effect of sinking and vertical mixing on algal population dynamics. *J. Water Pollution Control Fed.* **42**(5), 140–152.

Bennett, J. R. (1971). Thermally driven lake currents during the spring and fall transition periods. *In* "Proc. 14th Conf. Great Lakes Res.," Ann Arbor p. 535. *Intl. Assoc. for Great Lakes Res.*

Benney, D. J. (1966). Long nonlinear waves in fluid flows. *J. Math. Phys. Sci.* **45**, 52–63.

Blake, S., and Ivey, G. N. (1986a). Density and viscosity gradients in zoned magma chambers, and their influence on withdrawal dynamics. *J. Volcanology and Geothermal Res.* **30**, 201–230.

Blake, S., and Ivey, G. N. (1986). Magma-mixing and the dynamics of withdrawal from stratified reservoirs. *J. Volcanology and Geothermal Res.* **27**, 153–178.

Blanc, T. V. (1983). A practical approach to flux measurements of long duration in the marine atmospheric surface layer. *J. Clim. Appl. Met.* **22**, 1093–1110.

Blanc, T. V. (1985). Variation of bulk-derived surface flux, stability, and roughness results due to the use of different transfer coefficient schemes. *J. Phys. Oceanogr.* **15**, 650–669.

Blanc, T. V. (1987). Accuracy of bulk-method-determined flux, stability and sea surface roughness. *J. Geophys. Res.* **92**, 3867–3876.

Blanton, J. O. (1973). Vertical entrainment into the epilimnia of stratified lakes. *Limnol. Oceanogr.* **18**, 697–704.

Bloss, S., and Harleman, D. R. F. (1979). Effect of wind-induced mixing on the seasonal thermocline in lakes and reservoirs. Rept. No. 249, Massachusetts Institute of Technology.

Bohan, J. P., and Grace, J. L. (1973). Selective withdrawal from man-made lakes. Tech. Rept. No. H-73-4, Waterways Experiment Station, Vicksburg, Mississippi.

Boyce, F. M., Robertson, D. G., and Ivey, G. N. (1983). Summer thermal structure of Lake Ontario off Toronto: Cooling the big city. *Atmos. Ocean.* **21**, 397–417.

Brady, J. A., Davis, J. M., and Douglas, E. W. (1983). First filling of Kielder Reservoir and future operational considerations. *J. Inst. Water Eng. Sci.* **37**, 295–312.

Brim, W. D., and Beard, J. D. (1980). Reservoir management by pneumatic induced circulation. Paper No. 5-16. In "Proc. Symp. on Surface Water Impoundments," (H. G. Stefan, ed.), ASCE, pp. 877–885. Minneapolis, Minnesota.

Britter, R. E., and Simpson, J. E. (1978). Experiments on the dynamics of a gravity current head. *J. Fluid Mech.* **88**, 223–240.

Britter, R. E., and Simpson, J. E. (1981). A note on the structure of the head of an intrusive gravity current. *J. Fluid Mech.* **112**, 459–466.

Brocard, D. N., and Harleman, D. R. F. (1980). Two-layer model for shallow horizontal convective circulation. *J. Fluid Mech.* **100**, 129–146.

Browand, F. K., Guyomar, D., and Yoon, S. C. (1987). The behavior of a turbulent front in a stratified fluid: Experiments with an oscillating grid. *J. Geophys. Res.* **92**, 5329–5341.

Brown, I. K., Wooley, D. A., and Jory, A. G. (1982). Artificial destratification of Lake Morris to improve water quality. In "Proc. Hydr. and Water Resources Symp.," Institution of Engineers, Australia, Melbourne, Victoria.

Brubaker, J. M. (1987). Similarity structure in the convective boundary layer of a lake. *Nature* **330**, 742–745.

Bryant, P. J., and Wood, I. R. (1976). Selective withdrawal from a layered fluid. *J. Fluid Mech.* **77**, 581–591.

Buchak, E. M., and Edinger, J. E. (1984). Generalized, longitudinal–vertical hydrodynamics and transport: Development, programming and applications. Doc. No. 84-18-R, Waterways Experiment Station, Vicksburg, Mississippi.

Bulson, P. S. (1967). Theory and design of bubble breakwaters. *Dock and Harbour Auth.* **48**(560), 47–54.

Bunn, S. E., and Edward, D. H. D. (1984). Seasonal meromixis in three hypersaline lakes on Rottnest Island, Western Australia. *Aust. J. Mar. Freshw. Res.* **35**, 261–265.

Burns, F. L. (1977). Localized destratification of large reservoirs to control discharge temperatures. *Prog. Wat. Tech.* **9**, 53–63.

Busch, N. E. (1977). Fluxes in the surface boundary layer over the sea. In "Modeling and Prediction of the Upper Layers of the Ocean" (E. B. Kraus ed.), pp. 72–91, Pergamon Press, New York.

Businger, J. A. (1955). On the structure of the atmospheric surface layer. *J. Meteor.* **12**, 553–561.

Businger, J. A. (1973). Turbulent transfer in the atmospheric surface layer. In "Workshop on Micrometeorology" (D. A. Hauger ed.), pp. 67–100. Science Press, Ephrata, Pennsylvania.

Cacchione, D., and Wunsch, C. (1974). Experimental study of internal waves over a slope. *J. Fluid Mech.* **66**, 223–239.

Caldwell, D. R. (1983). Oceanic turbulence: Big bangs or continuous creation? *J. Geophys. Res.* **88**, 7543–7550.

Carmack, E. C. (1986). Circulation and mixing in ice-covered waters. In "The Geophysics of Sea Ice" (N. Untersteiner ed.), pp. 641–712. Plenum Press, New York.

Carslaw, H. S., and Jaeger, J. C. (1978). *Conduction of Heat in Solids,* Third Edition. Oxford University Press.

Carson, D. J., and Richards, P. J. R. (1978). Modeling surface turbulent fluxes in stable conditions. *Boundary Layer Met.* **14**, 67–81.

Cavaleri, L., and Zecchetto, S. (1987). Reynolds stresses under wind waves. *J. Geophys. Res.* **92**, 3894–3904.

Chao, S. Y. (1988). River-forced estuarine plumes. *J. Phys. Oceanogr.* **18**, 72–88.

Charnock, H. (1955). Wind stress on a water surface. *Quart. J. Roy. Meteor. Soc.* **82**, 639–640.

Chen, J. C. (1980). Studies on gravitational spreading currents. Rept. No. KH-R-40, California Institute of Technology.

Cheung, F. B., and Epstein, M. (1987). Two-phase gas bubble-liquid boundary layer flow along vertical and inclined surfaces. *Nuclear Eng. and Design.* **99**, 93–100.

Christodoulou, G. G. (1986). Interfacial mixing in stratified flows. *J. Hydr. Res.* **24**, 77–92.

Chu, V. H., and Baddour, R. E. (1984). Turbulent gravity-stratified shear flows. *J. Fluid Mech.* **138**, 353–378.

Chu, V. H., and Jirka, G. H. (1985). Surface buoyant jets and plumes. In "Encyclopedia of Fluid Mechanics" (N. Cheremisinoff ed.), pp. 1053–1084. Gulf Publishing, Houston, Texas.

Clarke, R. H., Dyer, A. J., Brook, R. R., Reid, D. G., and Troup, A. J. (1971). The Wangara experiment: Boundary layer data. Tech. Paper No. 19, CSIRO Div. Met. Phys., Melbourne, Victoria.

Claussen, M. (1987). The flow in a turbulent boundary layer upstream of a change in surface roughness. *Boundary Layer Met.* **40**, 31–86.

Cohen, A. (1979). Existence and regularity for solutions of the Korteweg–de Vries equation. *Arch. Rat. Mech. Anal.* **71**, 143–175.

Collings, I. L. (1985). Two large Froude numbers cusped free-surface flows due to a submerged line source or sink. Manuscript. Deakin University, Victoria.

Colman, J. A., and Armstrong, D. E. (1987). Radon-222 determined vertical eddy diffusivity in the benthic boundary layer of ice-covered lakes. *Limnol. Oceanogr.* **32**, 577–591.

Cooley, P., and Harris, S. L. (1954). The prevention of stratification in reservoirs. *J. Inst. Water Eng.* **8**, 517–537.

Cormack, D. E., Leal, L. G., and Imberger, J. (1974). Natural convection in a shallow cavity with differentially heated end walls. Part 1: Asymptotic theory. *J. Fluid Mech.* **65**, 209–230.

Cormack, D. E., Stone, G. P., and Leal, L. G. (1975). The effect of upper surface conditions on convection in a shallow cavity with differentially heated end walls. *Intl. J. Heat Mass Transfer* **18**, 635–648.

Craya, A. (1949). Recherches theoriques sur l'ecoulement de couches superposees de fluides de densites differentes. *La Houille Blanche* Jan.–Feb., 44–55.

Csanady, G. T. (1972). Response of large stratified lakes to wind. *J. Phys. Oceanogr.* **2**, 3–13.

Csanady, G. T. (1982). "Circulation in the Coastal Ocean." D. Reidel Publishing, Boston.

Dake, J. M. K., and Harleman, D. R. F. (1969). Thermal stratification in lakes: Analytical and laboratory studies. *J. Water Resources Res.* **5**, 484–495.

Darbyshire, J., and Colclough, M. (1972). Measurement of thermal conductivities in a freshwater lake. *Geofisica Pura i Applicata* **93**, 151–158.

Darden, R. B., Imberger, J., and Fischer, H. (1975). Jet discharge into a stratified reservoir. *J. Hydr. Div. ASCE* **101**, 1211–1220.

Davis, J. M. (1980). Destratification of reservoirs—A design approach for perforated-pipe compressed-air systems. *Water Services* **84**, 497–504.

de Szoeke, R. A. (1980). On the effects of horizontal variability of wind stress on the dynamics of the ocean mixed layer. *J. Phys. Oceanogr.* **10**, 1439–1454.

de Szoeke, R. A., and Rhines, P. B. (1976). Asymptotic regimes in mixed-layer deepening. *J. Marine Res.* **34**, 111–116.

Deardorff, J. W. (1968). Dependence of air–sea transfer coefficients on bulk stability. *J. Geophys. Res.* **73**, 2549–2557.

Deardorff, J. W. (1970). Convective velocity and temperature scales for the unstable planetary boundary layer. *J. Atmos. Sci.* **27**, 1211–1213.

Deardorff, J. W. (1974). Three-dimensional numerical study of the height and mean structure of a heated planetary boundary layer. *Boundary Layer Met.* **7**, 81–106.

Debler, W. R. (1959). Stratified flow into a line sink. *J. Eng. Mech. Div.* **85**, 51–65.

Denton, R. A., and Wood, I. R. (1981). Penetrative convection at low Peclet number. *J. Fluid Mech.* **113**, 1–21.

Dickey, T. D., and Simpson, J. J. (1983). The influence of optical water type on the diurnal response of the upper ocean. *Tellus* **35B**, 142–154.

Dillon, T. M., Richman, J. G., Hansen, C. G., and Pearson, M. D. (1981). Near surface turbulence measurements in a lake. *Nature* **290**, 390–392.

Ditmars, J. D., and Cederwall, K. (1974). Analysis of air-bubble plumes. In "Proc. Coastal Eng. Conf.," Copenhagen. Ch. 128, pp. 2209–2226.

Djordjevic, V. D., and Redekopp, L. G. (1978). The fission and disintegration of internal solitary waves moving over two-dimensional topography. *J. Phys. Oceanogr.* **8**, 1016–1024.

Dobson, F. W., Hasse, L., and Davis, R. (1980). "Air–Sea Interaction Instruments and Methods." Plenum Press, New York.

Donelan, M. A. (1980). Similarity theory applied to the forecasting of wave heights, periods and directions. In "Proc. Canadian Coastal Conf.," pp. 47–61. Burlington, Ontario.

Donelan, M. A. (1982). The dependence of the aerodynamic drag coefficient on wave parameters. In "First Intl. Conf. on Meteorology and Air/Sea Interaction of the Coastal Zone" (preprint volume), pp. 381–387. The Hague, American Met. Soc., Boston.

Donelan, M. A., and Peirson, W. J. (1987). Radar scattering and equilibrium ranges in wind-generated waves with application to scatterometry. *J. Geophys. Res.* **92**, 4971–5029.

Dortch, M. S., and Holland, J. P. (1980). Methods of total lake destratification. Paper No. 5-19. In "Proc. Symp. on Surface Water Impoundments," ASCE, pp. 913–922. Minneapolis, Minnesota.

Drazin, P. G., and Reid, W. H. (1981). "Hydrodynamic stability." Cambridge University Press, New York.

Drummond, J. E., and Korpela, S. A. (1987). Natural convection in a shallow cavity. *J. Fluid Mech.* **182**, 543–564.

Dubreil-Jacotin, M. L. (1934). Sur les theorems dexistence relatifs aux ondes permanentes periodiques a deux dimensions dan les liquids heterogenes. *J. de Mathematiques* **3**, 217–291.

Dyer, A. J. (1974). A review of flux-profile relationships. *Boundary Layer Met.* **7**, 363–372.

E, X., and Hopfinger, E. J. (1986). On mixing across an interface in stably stratified fluid. *J. Fluid Mech.* **166**, 227–244.

Edinger, J. E., and Buchak, E. M. (1983). Developments in LARM2: A longitudinal-vertical, time-varying hydrodynamic reservoir model. Tech. Rept. E-83-1, Waterways Experiment Station, Vicksburg, Mississippi.

Elder, J. W. (1959). The dispersion of marked fluid in turbulent shear flow. *J. Fluid Mech.* **5**, 544–560.

Elliott, G. H., and Elliott, J. A. (1970). Laboratory studies on the thermal bar. In "Proc. 13th Conf. Great Lakes Res.," pp. 413–418. *Intl. Assoc. Great Lakes Res.* Ann Arbor.

Ellison, T. H., and Turner, J. S. (1959). Mixing of a dense fluid in a turbulent pipe flow. Part 1: Overall description of the flow. *J. Fluid Mech.* **33**, 514–528.

Emmanuel, C. B. (1975). Drag and bulk aerodynamic coefficients over shallow water. *Boundary Layer Met.* **8**, 465–474.

Farmer, D. M. (1978). Observations of long nonlinear waves in a lake. *J. Phys. Oceanogr.* **8**, 63–73.

Farmer, D. M., and Carmack, E. (1982). Wind mixing and restratification in a lake near the temperature of maximum density. *J. Phys. Oceanogr.* **11**, 1516–1533.

Farrell, G. J., and Stefan, H. G. (1986). Buoyancy induced plunging flow into reservoirs and coastal regions. Proj. Rept. No. 241, St. Anthony Falls Hydraulic Lab., University of Minnesota.

Faust, K. M. (1981). Modelldarstellung von Warmeinselstromungen durch Konvektionsstrahlen. Paper No. SFB 80/ET/201, Universitat Karlsruhe.

Findikakis, A. N., and Street, R. L. (1982a). Finite element simulation of stratified turbulent flows. *J. Hydr. Div. ASCE* **108**, 904–920.

Findikakis, A. N., and Street, R. L. (1982b). Mathematical description of turbulent flows. *J. Hydr. Div. ASCE* **108**, 887–903.

Findikakis, A. N., and Street, R. L. (1983). Approximate numerical model for stratified flow. *J. Eng. Mech.* **109**, 950–969.

Fischer, H. B., and Smith, R. D. (1983). Observations of transport to surface waters from a plunging inflow to Lake Mead. *Limnol. Oceanogr.* **28**, 258–272.

Fischer, H. B., List, E. J., Koh, R. C. Y., Imberger, J., and Brooks, N. H. (1979). "Mixing in Inland and Coastal Waters." Academic Press, New York.

Forbes, L. K., and Hocking, G. C. (1988). Flow caused by a point sink in a fluid having a free surface. *J. Aust. Math. Soc. Ser. B* (in press).

Ford, D. (1978). Unpublished data. Environ. Lab., Waterways Experiment Station, Vicksburg, Mississippi.

Ford, D. E., and Johnson, M. C. (1981). Field observations of density currents in impoundments. Surface Water Impoundments, H. G. Stefan, ed. ASCE. Minneapolis, Minnesota. pp. 1239–1248.

Ford, D. E., Johnson, M. C., and Monismith, S. G. (1980). Density inflows to Degray Lake, Arkansas. In "Second Intl. Symp. on Stratified Flows," pp. 977–987. Trondheim.

Franke, R., Leschziner, M. A., and Rodi, W. (1987). Numerical simulation of wind-driven turbulent flow in stratified water bodies. In "Proc. 3rd Intl. Symp. on Stratified Flows," (E. J. List, ed.) ASCE Pasadena, California.

Friehe, C. A., and Schmitt, K. F. (1976). Parameterization of air–sea interface fluxes of sensible heat and moisture by the bulk aerodynamic formulas. *J. Phys. Oceanogr.* **6**, 801–809.

Fritts, D. C. (1984). Gravity wave saturation in the middle atmosphere: A review of theory and observations. *Rev. Geophys. Space Phys.* **22**, 275–308.

Gargett, A. E., and Holloway, G. (1984). Dissipation and diffusion by internal waves breaking. *J. Marine Res.* **42**, 15–27.

Gariel, P. (1949). Recherches experimentales sur l'ecoulement de couches superposees de fluides de densites differentes. *La Houille Blanche* Jan.–Feb., 44–55.

Garratt, J. R. (1977). Review of drag coefficients over oceans and continents. *Monthly Weather Review* **105**, 915–929.

Garratt, J. R. (1987). The stably stratified internal boundary layer for steady and diurnally varying offshore flow. *Boundary Layer Met.* **38**, 369–394.

Garratt, J. R., and Ryan, B. F. (1988). The structure of the stably stratified internal boundary layer in offshore flow over the sea. *Boundary Layer Met.* **47**, 17–40.

Garrett, C., and Gilbert, D. (1988). Estimates of vertical mixing by internal waves reflected off a sloping bottom. In "Small-Scale Turbulence and Mixing in the Ocean: Proc. 19th Intl. Leige Coll. on Ocean Hydrodyn." (J. J. Nihoul and B. M. Jamart eds.), pp. 405–424. Elsevier, New York.

Garrett, C., and Munk, W. (1979). Internal waves in the ocean. *Ann. Rev. Fluid Mech.* **11**, 339–369.

Garwood, R. W. (1977). An oceanic mixed layer model capable of simulating cyclic states. *J. Phys. Oceanogr.* **7**, 455–468.

Geernaert, G. L., and Katsaros, K. B. (1986). Incorporation of stratification effects on the oceanic roughness length in the derivation of the neutral drag coefficient. *J. Phys. Oceanogr.* **16**, 1580–1584.

Geernaert, G. L., Katsaros, K. B., and Richter, K. (1986). Variation of the drag coefficient and its dependence on sea state. *J. Geophys. Res.* **91**, 7667–7679.

Geernaert, G. L., Larsen, S. E., and Hansen, F. (1987). Measurements of the wind stress, heat flux and turbulence intensity during storm conditions over the North Sea. *J. Geophys. Res.* **92**, 13127–13139.

Gelhar, L. W., and Mascolo, D. M. (1966). Non-diffusive characteristics of slow viscous stratified flow towards a line sink. Massachusetts Institute of Technology Hydrodynamics Lab. Rept. No. 88, pp. 1–39.

Gibson, C. H. (1980). Fossil temperature, salinity and vorticity turbulence in the ocean. In "Marine Turbulence" (J. C. J. Nihoul ed.), pp. 221–257. Elsevier, New York.

Gibson, C. H. (1982). On the scaling of vertical temperature gradient spectra. *J. Geophys. Res.* **87**, 8031–8038.

Gibson, C. H. (1986). Internal waves, fossil turbulence and composite ocean microstructure spectra. *J. Fluid Mech.* **168**, 89–117.

Gibson, C. H. (1987a). Fossil turbulence and intermittency in sampling oceanic mixing processes. *J. Geophys. Res.* **92**, 5383–5404.

Gibson, C. H. (1987b). Oceanic turbulence: Big bangs and continuous creation. *J. Physiochem. Hydrodyn.* **8**, 1–22.
Gibson, C. H. (1988). Evidence and consequences of fossil turbulence in the ocean. In "Small-Scale Turbulence and Mixing in the Ocean, Proc. 19th Intl. Leige Coll. on Ocean Hydrodyn." (J. J. Nihoul and B. M. Jamart, eds.), pp. 319–334. Elsevier, New York.
Gill, A. E. (1966). The boundary-layer regime for convection in a rectangular cavity. *J. Fluid Mech.* **26**, 515–536.
Gill, A. E., and Clarke, A. J. (1974). Wind induced upwelling, coastal currents and sea level changes. *Deep-Sea Res.* **21**, 325–345.
Gordon, J. A. (1981). LARM two-dimensional model: An evaluation. *J. Environ. Eng. Div.* **107**, 877–886.
Graf, W. H., Merzi, N., and Perrinjaquet, C. (1984). Aerodynamic drag measured at a nearshore platform on Lake Geneva. *Arch. Met. Geophys. Bioklim.* **A33**, 151–173.
Graham, D. S. (1980). Destratification characteristics and energy efficiency of air-plume mixing of stratified fluids. Paper No. 5–20. In "Proc. Symp. on Surface Water Impoundments," ASCE pp. 923–932. Minneapolis, Minnesota.
Gregg, M. C. (1987). Diapycnal mixing in the thermocline: A review. *J. Geophys. Res.* **92**, 5249–5286.
Gregg, M. C., and Briscoe, M. G. (1979). Internal waves, finestructure, microstructure and mixing in the ocean. *Rev. Geophys. Space Phys.* **17**, 1524–1548.
Gregg, M. C., and Sanford, T. B. (1988). The dependence of turbulent dissipation on stratification in a diffusively stable thermocline. *J. Geophys. Res.* **93**, 12381–12392.
Hannoun, I. A., and List, E. J. (1988). Turbulent mixing at a shear-free density interface. *J. Fluid Mech.* **189**, 211–234.
Hannoun, I. A., Fernando, H. J. S., and List, E. J. (1988). Turbulence structure near a sharp density interface. *J. Fluid Mech.* **189**, 189–209.
Harashima, A., and Watanabe, M. (1986). Laboratory experiments on the steady gravitational circulation excited by cooling of the water surface. *J. Geophys. Res.* **91**, 13056–13064.
Harleman, D. R. F. (1982). Hydrothermal analysis of lakes and reservoirs. *J. Hydr. Div. ASCE* **108**, 302–325.
Harleman, D. R. F., and Hurley-Octavio, K. A. (1977). Heat transport mechanisms in lakes and reservoirs. In "Proc. 17th Cong. Intl. Assoc. Hydr. Res.," Baden Baden.
Harleman, D. R. F., Morgan, R. L., and Purple, R. A. (1959). Selective withdrawal from a vertically stratified fluid. In "Proc. 8th Congr. Intl. Assoc. for Hydr. Res.," Montreal, pp. 10-C-1–10-C-16.
Hart, J. E. (1972). Stability of thin non-rotating Hadley circulations. *J. Atmos. Sci.* **29**, 687–697.
Hart, J. E. (1983a). Low Prandtl number convection between differentially heated end walls. *Intl. J. Heat Mass Transfer* **26**, 1069–1074.
Hart, J. E. (1983b). A note on the stability of low-Prandl number Hadley circulations. *J. Fluid Mech.* **132**, 271–281.
Hauenstein, W., and Dracos, T. H. (1984). Investigation of plunging density currents generated by inflows in lakes. *J. Hydr. Res.* **22**, 157–179.
Heaps, N. S. (1984). Vertical structure of current in homogeneous and stratified waters. In "Hydrodynamics of Lakes" (K. Hutter ed.), pp. 153–207. Springer-Verlag, New York.
Heaps, N. S., and Ramsbottom, A. E. (1966). Wind effects on water in a narrow two-layered lake. *Phil. Trans. Roy. Soc. London Ser. A* **259**, 391–430.

Heathershaw, A. D., and Martin, J. (1987). The theory and tuning of the Met 0.20 mixed layer model. *Admirality Research Establishment* Accn. No. 75636.

Hebbert, R., Imberger, J., Loh, I., and Patterson, J. (1979). Coilie River underflow into the Wellington Reservoir. *J. Hydr. Div. ASCE* **105**, 533–545.

Henderson-Sellers, B. (1976). Role of eddy diffusivity in thermocline formation. *Proc. ASCE* **102**, 517–531.

Henderson-Sellers, B. (1982). Reservoir destratification techniques and models. In ASCE "Proc. Applying Res. to Hydraulic Practice Conf.," pp. 697–706. Mississippi.

Henderson-Sellers, B. (1984a). Development and application of "U.S.E.D.": A hydroclimate lake stratification model. *Ecol. Modelling* **21**, 233–246.

Henderson-Sellers, B. (1984b). "Engineering Limnology." Pitman, Marshfield, Massachusetts.

Henderson-Sellers, B. (1985). New formulation of eddy diffusion thermocline models. *Appl. Math. Modelling* **9**, 441–446.

Henderson-Sellers, B. (1986). Calculating the surface energy balance for lake and reservoir modeling: A review. *Rev. Geophys.* **24**, 625–649.

Hicks, B. B. (1972). Some evaluations of drag and bulk transfer coefficients over water bodies of different sizes. *Boundary Layer Met.* **3**, 201–213.

Hicks, B. B. (1975). A procedure for the formulation of bulk transfer coefficients over water. *Boundary Layer Met.* **8**, 315–324.

Hicks, B. B. (1975). Wind profile relationships from the 'Wangara' experiment. *Quart. J. Roy. Met. Soc.* **102**, 535–551.

Hicks, B. B., Drinklow, R. L., and Grauze, G. (1974). Drag and bulk transfer coefficients associated with a shallow water surface. *Boundary Layer Met.* **6**, 287–297.

Hocking, G. C. (1985). Cusp-like free-surface flows due to a submerged source or sink in the presence of a flat or sloping bottom. *J. Aust. Math. Soc. Ser. B.* **26**, 470–486.

Hocking, G. C. (1988). Critical withdrawal from a two-layer fluid through a line sink. Environmental Dynamics Rept. No. ED-88-240, Centre for Water Research, University of Western Australia.

Hocking, G. C., and Patterson, J. C. (1988). Quasi two-dimensional modelling of reservoirs. Environmental Dynamics Rept. ED-88-283, Centre for Water Research, University of Western Australia.

Hocking, G. C., Sherman, B. S., and Patterson, J. C. (1988). Algorithm for selective withdrawal from stratified reservoir. *J. Hydr. Div. ASCE* **114**, 707–719.

Holloway, G. (1986). Eddies, waves, circulation and mixing: Statistical geofluid mechanics. *Ann. Rev. Fluid Mech.* **18**, 91–147.

Holyer, J. Y., and Huppert, H. E. (1980). Gravity currents entering a two-layer fluid. *J. Fluid Mech.* **100**, 739–767.

Hopfinger, E. J. (1987). Turbulence in stratified fluids: A review. *J. Geophys. Res.* **92**, 5287–5303.

Hopfinger, E. J., and Toly, J. A. (1976). Spatially decaying turbulence and its relation to mixing across density interfaces. *J. Fluid Mech.* **78**, 155–175.

Horn, W., Mortimer, C. H., and Schwab, D. J. (1986). Wind-induced internal seiches in Lake Zurich observed and modeled. *Limnol. Oceanogr.* **31**, 1232–1254.

Howard, C. S. (1953). Density currents in Lake Mead. In "Proc. Intl. Hydr. Convention," ASCE. pp. 355–368. Minnesota.

Hsu, S. A. (1974). A dynamic roughness equation and its application to wind stress determination at the air–sea interface. *J. Phys. Oceanogr.* **4**, 116–120.

Hsu, S. A. (1986). A mechanism for the increase of wind stress (drag) coefficient with wind speed over water surfaces: A parametric model. *J. Phys. Oceanogr.* **16**, 144–150.

Huber, D. G. (1960). Irrotational motion of two fluid strata towards a line sink. *J. Eng. Mech. Div.* **86**, 71–86.

Huber, W. C., Harleman, D. R. F., and Ryan, P. J. (1972). Temperature prediction in stratified reservoirs. *J. Hydr. Div. ASCE* **98**, 645–666.

Hurley-Octavio, K. A., Jirka, G. H., and Harleman, D. R. F. (1977). Vertical heat transport mechanisms in lakes and reservoirs. Tech. Rept. 227, Massachusetts Institute of Technology.

Hussain, N. A., and Narang, B. S. (1984). Simplified analysis of air-bubble plumes in moderately stratified environments. *J. Heat Transfer* **106**, 543–551.

Hutchinson, G. E. (1957). "A Treatise on Limnology. I. Geography, Physics and Chemistry." John Wiley & Sons, New York.

Hutter, K. (1984). "Hydrodynamics of Lakes." Springer-Verlag, New York.

Hutter, K. (1986). Hydrodynamic modeling of lakes. In "Encyclopedia of Fluid Mechanics," pp. 897–998. Gulf Publishing, Houston, Texas.

Idso, S. B. (1973). On the concept of lake stability. *Limnol. Oceanogr.* **18**, 681–683.

Idso, S. B., and Cole, G. A. (1973). Studies on a Kentucky Knobs Lake: V. Some aspects of the vertical transport of heat in the hypolimnion. *J. Ecol.* **61**, 413–420.

Imberger, J. (1970). Selective withdrawal from a stratified reservoir. Ph.D. thesis, University of California, Berkeley.

Imberger, J. (1972). Two-dimensional sink flow of a stratified fluid contained in a duct. *J. Fluid Mech.* **53**, 329–349.

Imberger, J. (1974). Natural convection in a shallow cavity with differentially heated end walls. Part 3: Experimental results. *J. Fluid Mech.* **65**, 247–260.

Imberger, J. (1976). Dynamics of a longitudinally stratified estuary. In "Proc. 15th Coastal Eng. Conf.," pp. 3108–3115. ASCE Hawaii.

Imberger, J. (1980). Selective withdrawal: A review. In "Proc. 2nd Intl. Symp. on Stratified Flows," Int. Ass. Hydraulic Research (IAHR) Trondheim.

Imberger, J. (1982). Reservoir dynamics modeling. In "Prediction in Water Quality" (E. M. O'Loughlin and P. Cullen eds.), pp. 223–248. Australian Academy of Science, Canberra.

Imberger, J. (1985). The diurnal mixed layer. *Limnol. Oceanogr.* **30**, 737–770.

Imberger, J. (1987). Hydrodynamics of lakes. In "Proc. 12th Convention Australian Water and Wastewater Assoc.," pp. 401–423. AWWA. Adelaide, South Australia.

Imberger, J. (1988). On the nature of turbulence in a stratified water body. Environmental Dynamics Rept. No. ED-87-216, Centre for Water Research, University of Western Australia.

Imberger, J., and Boashash, B. (1986). Application of the Wigner–Ville distribution to temperature gradient microstructure: A new technique to study small-scale variations. *J. Phys. Oceanogr.* **16**, 1997–2012.

Imberger, J., and Fandry, C. (1975). Withdrawal of a stratified fluid from a vertical two-dimensional duct. *J. Fluid Mech.* **70**, 321–332.

Imberger, J., and Fischer, H. B. (1970). Selective withdrawal from a stratified reservoir. Water Pollution Control Res. Series 15040 EJZ 12/70, E.P.A. Water Quality Office.

Imberger, J., and Hamblin, P. F. (1982). Dynamics of lakes, reservoirs and cooling ponds. *Ann. Rev. Fluid Mech.* **14**, 153–187.

Imberger, J., and Hebbert, R. H. B. (1980). Management of water quality in reservoirs. Tech. Paper No. 49, Res. Proj. No. 74/70, Aust. Water Res. Council.

Imberger, J., and Monismith, S. G. (1986). A model for deepening due to upwelling when $W > 1$. *J. Fluid Mech.* **171**, 432–439 (Appendix to Monismith (1986)).

Imberger, J., and Parker, G. (1985). Mixed layer dynamics in a lake exposed to a spatially variable wind field. *Limnol. Oceanogr.* **39**, 473–488.

Imberger, J., and Patterson, J. C. (1981). A dynamic reservoir simulation model—DYRESM. In "Transport Models for Inland and Coastal Waters" (H. B. Fischer ed.), pp. 310–361. Academic Press, New York.

Imberger, J., and Spigel, R. H. (1987). Circulation and mixing in Lake Rotongaio and Lake Okaro under conditions of light to moderate winds: Preliminary results. *N.Z. J. Marine and Freshwater Res.* **21**, 515–519.

Imberger, J., Thompson, R. O. R. Y., and Fandry, C. (1976). Selective withdrawal from a finite rectangular tank. *J. Fluid Mech.* **78**, 489–512.

Imberger, J., Patterson, J. C., Hebbert, R., and Loh, I. (1978). Dynamics of reservoirs of medium size. *J. Hydr. Div. ASCE* **104**, 725–743.

Imboden, D. M., and Emerson, S. (1977). Natural radon and phosphorus as limnologic tracers: Horizontal and vertical eddy diffusion in Greifensee. *Limnol. Oceanogr.* **23**, 77–90.

Imboden, D. M., and Joller, T. (1984). Turbulent mixing in the hypolimnion of Baldeggersee (Switzerland) traced by natural radon-222. *Limnol. Oceanogr.* **29**, 831–844.

Imboden, D. M., Weiss, R. F., Craig, H., Michel, R. L., and Goldman, C. R. (1977). Lake Tahoe geochemical study. 1. Lake chemistry and tritium mixing study. *Limnol. Oceanogr.* **22**, 1039–1051.

Imboden, D. M., Lemmin, U., Joller, T., and Schurter, M. (1983). Mixing mechanisms in lakes: Mechanisms and ecological relevance. *Schweiz. Z. Hydrol.* **45**, 11–44.

Inaba, H., Seki, N., Fukusako, S., and Kanyama, K. (1981). Natural convective heat transfer in a shallow rectangular cavity with different end temperatures. *Numer. Heat Transfer* **4**, 459–468.

Itsweire, E. C., Helland, K. N., and van Atta, C. W. (1986). The evolution of grid generated turbulence in a stably stratified fluid. *J. Fluid Mech.* **162**, 299–338.

Ivetic, M., Iwasa, Y., and Inoue, K. (1986). Large eddy simulation of a shear driven flow in a test reservoir. In "Proc. 2nd Intl. Conf. Hydrosoft 86," (M. Radojkovic, C. Maksimovic, and C. A. Brebbia, eds.) Southampton. pp. 209–216, Springer Verlag.

Ivey, G. N. (1987). The role of boundary mixing in the deep ocean. *J. Geophys. Res.* **92**, 11873–11878.

Ivey, G. N., and Blake, S. (1985). Axisymmetric withdrawal and inflow in a density-stratified container. *J. Fluid Mech.* **161**, 115–137.

Ivey, G. N., and Boyce, F. M. (1982). Entrainment by bottom currents in Lake Erie. *Limnol. Oceanogr.* **27**, 1029–1038.

Ivey, G. N., and Corcos, G. M. (1982). Boundary mixing in a stratified fluid. *J. Fluid Mech.* **121**, 1–26.

Ivey, G. N., and Hamblin, P. F. (1989). Convection near the temperature of maximum density for high Rayleigh number, low aspect ratio, rectangular cavities. *J. Heat Transfer* **111**, 100–105.

Ivey, G. N., and Nokes, R. I. (1989). Vertical mixing due to the breaking of initial waves on sloping boundaries. *J. Fluid Mech.* **204**, 479–500.

Ivey, G. N., and Patterson, J. C. (1984). A model of the vertical mixing in Lake Erie in summer. *Limnol. Oceanogr.* **29**, 553–563.

Jain, S. C. (1982). Buoyancy-driven circulation in free-surface channels. *J. Fluid Mech.* **122**, 1–12.

Janowitz, G. S. (1986). A surface density and wind-driven model of the thermocline. *J. Geophys. Res.* **91**, 5111–5118.

Jassby, A., and Powell, T. (1975). Vertical patterns of eddy diffusion during stratification in Castle Lake, California. *Limnol. Oceanogr.* **20**, 530–543.

Jirka, G. H., and Arita, M. (1987). Density currents or density wedges: Boundary-layer influence and control methods. *J. Fluid Mech.* **177**, 187–206.

Jirka, G. H., and Katavola, D. S. (1979). Supercritical withdrawal from two-layered fluid systems. Part 2: Three-dimensional flow into round intake. *J. Hydr. Res.* **17**, 53–62.

Johnson, T. R., Ellis, C. R., Farrell, G. J., and Stefan, H. G. (1987a). Negatively buoyant flow in a diverging channel. I: Flow regimes. *J. Hydr. Eng.* **113**, 716–730.

Johnson, T. R., Ellis, C. R., Farrell, G. J., and Stefan, H. G. (1987b). Negatively buoyant flow in a diverging channel. II: 3-D flow field descriptions. *J. Hydr. Eng.* **113**, 731–742.

Johnson, T. R., Ellis, C. R., and Stefan, H. G. (1988). Negatively buoyant flow in a diverging channel: IV: Entrainment and dilution. *J. Hydr. Eng.* **115**, 437–456.

Jokela, J. B., and Patterson, J. C. (1985). Quasi-two-dimensional modelling of reservoir inflow. In "Proc. 21st IAHR Intl. Congr.," Vol. 2, pp. 317–322. Melbourne, Institution of Engineers, Aust. Natl. Conference Publication No. 85/13.

Kamial, J. C., Wyngaard, J. C., Haugen, D. A., Cote, O. R., and Izumi, Y. (1976). Turbulence structure in the convective boundary layer. *J. Atmos. Sci.* **33**, 2152–2169.

Kantha, L. H., Phillips, O. M., and Azad, R. S. (1977). On turbulent entrainment at a stable density interface. *J. Fluid Mech.* **79**, 753–768.

Kao, T. W. (1970). Free streamline theory for inviscid stratified flow into a line sink. *Phys. Fluids* **13**, 558–564.

Kao, T. W. (1976). Selective withdrawal criteria of stratified fluids. *J. Hydr. Div. ASCE* **102**, 717–729.

Kao, T. W., Pao. H. P., and Wei, S. N. (1974). Dynamics of establishment of selective withdrawal of a stratified fluid from a line sink. Part 2: Experiment. *J. Fluid Mech.* **65**, 689–710.

Kao, T. W., Park, C., and Pao, H. P. (1978). Inflows, density currents and fronts. *Phys. Fluids* **21**, 1912–1922.

Kato, H., and Phillips, O. M. (1969). On the penetration of a turbulent layer into stratified fluid. *J. Fluid Mech.* **37**, 643–655.

Keijman, J. Q. (1974). The estimation of the energy balance of a lake from simple weather data. *Boundary Layer Met.* **7**, 399–407.

Keller, W. C., Plant, W. J., and Weissman, D. E. (1985). The dependence of X band microwave sea return on atmospheric stability and sea state. *J. Geophys. Res.* **90**, 1019–1029.

Keulegan, G. G., and Brame, V. (1960). Fourteenth progress report on model laws for density currents mixing effect of wind induced waves. NBS Project No. 0603-11-06431, U.S. Dept. of Commerce.

Khomskis, W. R. (1969). "Dynamics and Thermics of Small Lakes" Vilnyus: Mintis (in Russian).

Khomskis, W. R., and Filatova, T. N. (1972). Principles of typology of stratified lakes in relation to vertical exchange. *Verhandlungen der Internationalen Vereinigung fur Theoretische und Angewandte Limnologie* **18**, 528–536.

Kielmann, J., and Simons, T. J. (1984). Some aspects of baroclinic circulation models. In "Hydrodynamics of Lakes" (K. Hutter ed.), pp. 235–285. Springer-Verlag, New York.

Kim, B. R., Higgins, J. M., and Bruggink, D. J. (1983). Reservoir circulation patterns and water quality. *J. Environ. Eng.* **109**, 1284–1294.

Kirk, J. T. O. (1983). "Light and Photosynthesis in Aquatic Ecosystems." Cambridge University Press, Cambridge.

Kitaigorodskii, S. A. (1968). On the calculation of the aerodynamic roughness of the sea surface. *Izv. Atmos. Oceanic Phys.* **4**, 498–502.

Kitaigorodskii, S. A., and Zaslavskii, M. M. (1974). A dynamical analysis of the drag conditions at the sea surface. *Boundary Layer Met.* **6**, 53–61.

Kitaigorodskii, S. A., Kutznetsov, O. A., and Panin, G. N. (1973). Coefficients of drag, sensible heat and evaporation in the atmosphere over the surface of the sea. *Izv. Atmos. Oceanic Phys.* **9**, 1135–1141.

Kitaigorodskii, S. A., Donelan, M. A., Lumley, J. A., and Terray, E. A. (1983). Wave-turbulence interactions in the upper ocean. Part II: Statistical characteristics of wave and turbulent components of the random velocity field in the marine surface layer. *J. Phys. Oceanogr.* **13**(11), 1988–1999.

Kobus, H. E. (1968). Analysis of the flow produced by air bubble systems. In "Proc. 11th Coastal Eng. Conf.," pp. 1016–1031. ASCE London.

Koh, R. C. Y. (1966a). Unsteady stratified flow into a sink. *J. Hydr. Res.* **4**(2), 21–34.

Koh, R. C. Y. (1966b). Viscous stratified flow towards a sink. *J. Fluid Mech.* **24**, 555–575.

Kondo, J. (1975). Air-sea bulk transfer coefficients in diabatic conditions. *Boundary Layer Met.* **9**, 91–112.

Kranenburg, C. (1979). Destratification of lakes using bubble columns. *J. Hydr. Div. ASCE* **105**, 333–349.

Kranenburg, C. (1984). Wind-induced entrainment in a stably stratified fluid. *J. Fluid Mech.* **145**, 253–273.

Kranenburg, C. (1985). Mixed-layer deepening in lakes after wind set-up. *J. Hydr. Div. ASCE.* **111**, 1279–1297.

Kraus, E. B. (1977). "Modeling and Prediction of the Upper Layers of the Ocean." Pergamon Press, New York.

Kraus, E. B., and Turner, J. S. (1967). A one-dimensional model of the seasonal thermocline: II. The general theory and its consequences. *Tellus* **14**, 98–106.

Kullenberg, G., Murthy, C. R., and Westerberg, H. (1974). Vertical mixing characteristics in the thermocline and hypolimnion regions of Lake Ontario. (IFYGL). In "Proc. 17th Conf. Great Lakes Res.," Vol. 1, pp. 425–434. *Intl. Assoc. Great Lake Res.* Ann Arbor.

Lake Biwa Research Institute (1987). Fundamental studies on the vertical transport of materials in Lake Biwa. Rept. No. 86-A05.

Langmuir, I. (1938). Surface motion of water induced by wind. *Science* **87**, 119–123.

Large, W. G., and Pond, S. (1981). Open ocean momentum flux measurements in moderate to strong winds. *J. Phys. Oceanogr.* **11**, 324–336.

Lawrence, G. A., and Imberger, J. (1979). Selective withdrawal through a point sink in a continuously stratified fluid with a pycnocline. Environmental Dynamics Rept. No. ED-79-002, Centre for Water Research, University of Western Australia.

Lawrence, G. A., Lasheras, J. C., and Browand, F. K. (1987). Shear instabilities in stratified flow. *Preprint from* "3rd Intl. Symp. on Stratified Flows," Vol. 1. Pasadena, California.

Leibovich, S. (1977a). Convective instability of stably stratified water in the ocean. *J. Fluid Mech.* **82**, 561–585.

Leibovich, S. (1977b). On the evolution of the system of wind drift currents and Langmuir circulations in the ocean. Part 1: Theory and the averaged current. *J. Fluid Mech.* **79**, 715–743.

Leibovich, S. (1980). On wave-current interaction theories of Langmuir circulation. *J. Fluid Mech.* **99**, 715–724.

Leibovich, S. (1983). The form and dynamics of Langmuir circulations. *Ann. Rev. Fluid Mech.* **15**, 391–427.
Leibovich, S., and Paolucci, S. (1980). The Langmuir circulation instability as a mixing mechanism in the upper ocean. *J. Phys. Oceanogr.* **10**, 186–207.
Leibovich, S., and Radhakrishnan, K. (1977). On the evolution of the system of wind drift currents and Langmuir circulations in the ocean. Part 2: Structure of the Langmuir vortices. *J. Fluid Mech.* **80**, 481–507.
Lerman, A. (1971). Time to chemical steady-state in lakes and ocean. *Adv. in Chem. Ser.* **106**, 30–76.
Lerman, A., and Stiller, M. (1969). Vertical eddy diffusion in Lake Tiberias. *Verhandlungen der Internationalen Vereinigung fur Theoretische und Angewandte Limnologie* **17**, 323–333.
Lewis, W. M. (1973). The thermal regime of Lake Lanao (Philipines) and its theoretical implication for tropical lakes. *Limnol. Oceanogr.* **18**, 200–217.
Lewis, E. L., and Perkin, R. G. (1982). Seasonal mixing processes in an Arctic fjord system. *J. Phys. Oceanogr.* **12**, 74–83.
Lighthill, M. J. (1969). Dynamic response of the Indian Ocean to the southwest monsoon. *Phil. Trans. R. Soc. London Ser. A.* **265**, 45–92.
Lin, J. T., and Pao, Y. H. (1979). Wakes in stratified fluids. *Ann. Rev. Fluid Mech.* **11**, 317–338.
Linden, P. F. (1975). The deepening of a mixed layer in a stratified fluid. *J. Fluid Mech.* **71**, 385–405.
Linden, P. F., and Simpson, J. E. (1986). Gravity-driven flows in a turbulent fluid. *J. Fluid Mech.* **172**, 481–497.
Lo, A. K., and McBean, G. A. (1978). On the relative errors in methods of flux calculations. *J. Appl. Met.* **17**, 1704–1711.
Loffler, H. (1968). Tropical high-mountain lakes. Their distribution, ecology and zoogeographical importance. *Colloquium Geographicum* **9**, 57–76.
Long, R. R. (1956). Sources and sinks at the axis of a rotating liquid. *Quart. J. Mech. and Appl. Math.* **9**, 385–393.
Lueck, R. G., and Osborn, T. R. (1985). Turbulence measurements in a submarine canyon. *Continental Shelf Res.* **4**, 681–698.
Luketina, D. A., and Imberger, J. (1987). Characteristics of a surface buoyant jet. *J. Geophys. Res.* **92**, 5435–5447.
Luketina, D. A., and Imberger, J. (1988). Turbulence and entrainment in a buoyant surface plume. *J. Geophys. Res.* (in press).
MacIntyre, S., and Melack, J. M. (1982). Meromixis in an equatorial African soda lake. *Limnol. Oceanogr.* **27**, 595–609.
Mahony, J. J., and Pritchard, W. G. (1977). Withdrawal from a reservoir of stratified fluid. Rept. No. 79, Fluid Mech. Res. Institute, University of Essex.
Mahrt, L., and Lenschow, D. H. (1976). Growth dynamics of the convectively mixed layer. *J. Atmos. Sci.* **33**, 41–51.
Manins, P. C. (1976). Intrusion into a stratified fluid. *J. Fluid Mech.* **74**, 547–560.
Markofsky, M., and Harleman, D. R. F. (1973). Prediction of water quality in stratified reservoirs. *J. Hydr. Div. ASCE* **99**, 729–745.
Marmoush, Y. R., Smith, A. A., and Hamblin, P. F. (1984). Pilot experiments on thermal bar in lock exchange flow. *J. Energy Eng.* **110**, 215–227.
Marti, V. D. E., and Imboden, D. M. (1986). Thermische energieflusse an der wasseroberflache: beispiel sempachersee. *Schweiz. Z. Hydrol.* **48**, 196–229.
Maxworthy, T. (1972). Experimental and theoretical studies of horizontal jets in a

stratified fluid. In "IAHR Intl. Symp. on Stratified Flows," Novosibirsk. IAHR. pp. 611–619.

Maxworthy, T. (1980). On the formation of nonlinear internal waves from the gravitational collapse of mixed regions in two and three dimensions. *J. Fluid Mech.* **96**, 47–72.

Maxworthy, T., and Monismith, S. G. (1988). Differential mixing in a stratified fluid. *J. Fluid Mech.* **189**, 571–598.

McBean, G. A., and Paterson, R. D. (1975). Variations of the turbulent fluxes of momentum, heat and moisture over Lake Ontario. *J. Phys. Oceanogr.* **5**, 523–531.

McDougall, T. J. (1978). Bubble plumes in stratified environments. *J. Fluid Mech.* **85**, 655–672.

McEwan, A. D., and Baines, P. G. (1974). Shear fronts and an experimental stratified shear flow. *J. Fluid Mech.* **63**, 257–272.

Mellor, G. L., and Durbin, P. A. (1975). The structure and dynamics of the ocean surface mixed layer. *J. Phys. Oceanogr.* **5**, 718–728.

Mellor, G. L., and Yamada, T. (1982). Development of a turbulence closure model for geophysical fluid problems. *Rev. Geophys. Space Phys.* **29**, 851–875.

Melville, W. K. (1977). Wind stress and roughness length over breaking waves. *J. Phys. Oceanogr.* **7**, 702–710.

Milgram, J. H. (1983). Mean flow in round bubble plumes. *J. Fluid Mech.* **133**, 345–376.

Miyake, M., Donelan, M., McBean, G., Paulson, C., Badgley, F., and Leavitt, E. (1970). Comparison of turbulent fluxes over water determined by profile and eddy correlation techniques. *Quart. J. Roy. Met. Soc.* **96**, 132–137.

Monin, A. S., and Obukhov, A. M. (1954). Basic laws of turbulent mixing in the atmosphere near the ground. *Jr. Akad. Nauk SSSR Geofiz. Inst.* **24**(151), 163–187.

Monismith, S. G. (1985). Wind-forced motions in stratified lakes and their effect on mixed-layer shear. *Limnol. Oceanogr.* **30**, 771–783.

Monismith, S. G. (1986). An experimental study of the upwelling response of stratified reservoirs to surface shear stress. *J. Fluid Mech.* **171**, 407–439.

Monismith, S. G. (1987). The modal response of reservoirs to wind stress. *J. Hydr. Div. ASCE* **113**, 1290–1326.

Monismith, S. G., and Imberger, J. (1988). Horizontal convection in the sidearm of a small reservoir. Environmental Dynamics Rept. ED-86-161, Centre for Water Research, University of Western Australia.

Monismith, S. G., and Maxworthy, T. (1988). Selective withdrawal and spin-up of a rotating, stratified fluid. *J. Fluid Mech.* **199**, 377–401.

Monismith, S. G., Imberger, J., and Billi, G. (1988). Shear-waves and unsteady selective withdrawal. *J. Hydr. Div. ASCE* **114**, 1134–1152.

Mortimer, C. H. (1974). Lake hydrodynamics. *I.V.L.* **20**, 124.

Mortimer, C. H. (1987). Fifty years of physical investigations and related limnological studies on Lake Erie, 1928–1977. *J. Great Lakes Res.* **13**, 407–435.

Mulhearn, P. J. (1981). On the formation of a stably stratified internal boundary-layer by advection of warm air over a cooler sea. *Boundary Layer Met.* **21**, 247–254.

Muller, P., and Garwood, R. W. (1988). Mixed layer dynamics: Progress and new directions. *EOS* **69**, 2–4, 10–12.

Muller, P., and Henderson, D. (1987). Dynamics of the oceanic surface mixed layer. In "Proc. Aha Huliko'a Hawaiian Winter Workshop," Hawaii Institute of Geophysics Special Publication, Honolulu.

Muller, P., and Pujalet, R. (1984). Internal gravity waves and small-scale turbulence. In "Proc. Aha Huliko'a Hawaiian Winter Workshop," Hawaii Institute of Geophysics Special Publication, Honolulu.

Munk, W., and Anderson, E. R. (1948). Notes on a theory of the thermocline. *J. Marine Res.* **7**, 276–295.

Murota, A., and Hirata, T. (1979). Breaking of internal gravity wave on slope. *Tech. Repts. of Osaka University* **29**(1487), 255–264.

Murota, A., and Michioku, K. (1986a). Analysis on selective withdrawal from three-layered stratified systems and its practical application. *J. Hydrosci. Hydr. Eng.* **4**, 31–50.

Murota, A., and Michioku, K. (1986b). Characteristics of internal fluctuations in thermally stratified fields induced by combination of mechanical and thermal stirring. *Proc. JSCE* **375**(II-6), 171–179.

Murota, A., Hirata, T., and Michioku, K. (1980). Breaking of internal gravity wave on slope: II. *Tech. Repts. of Osaka University* **30**(1543), 255–261.

Murota, A., Michioku, K., and Kuchida, M. (1988). Numerical analysis on vertical mixing phenomena induced by composite stirring. *Proc. JSCE* II-9, 67–76.

Myrup, L. O., Powell, T. M., Godden, D. A., and Goldman, C. R. (1979). Climatological estimate of the average monthly energy and water budgets of Lake Tahoe, California–Nevada. *Water Resources Res.* **15**, 1499–1508.

Narimousa, A., and Fernando, H. J. S. (1987). On the sheared-density interface of an entraining stratified fluid. *J. Fluid Mech.* **174**, 1–22.

Narimousa, S., Long, R. R., and Kitaigorodskii, S. A. (1986). Entrainment due to turbulent shear flow at the interface of a stably stratified fluid. *Tellus* **38A**, 76–87.

Newman, F. C. (1976). Temperature steps in Lake Kivu: A bottom heated saline lake. *J. Phys. Oceanogr.* **6**, 157–163.

Niiler, P. P. (1975). Deepening of the wind-mixed layer. *J. Marine Res.* **33**, 405–422.

Niiler, P. P., and Kraus, E. B. (1977). One-dimensional models of the upper ocean. In "Modeling and Prediction of the Upper Layers of the Ocean" (E. B. Kraus ed.), pp. 143–172, Pergamon Press, New York.

Nishida, S., and Yoshida, S. (1984). Stability of a two-layer shear flow. *Theoretical and App. Mech.* **32**, 35–45.

Nokes, R. I. (1988). On the entrainment rate across a density interface. *J. Fluid Mech.* **188**, 185–204.

O'Donnell, J., and Garvine, R. W. (1983). A time dependent, two-layer frontal model of buoyant plume dynamics. *Tellus* **35A**, 73–80.

Ookubo, K., and Muramoto, Y. (1981). Coriolis' effects on density currents—Experiments on the rotating exchange flow. *Bulletin of the Disaster Prevention Research Institute* **24B-2**, 339–365.

Ookubo, K., Muramoto, Y., Oonishi, Y., and Kumagai, M. (1984). Laboratory experiments on thermally induced currents in Lake Biwa. *Bulletin of the Disaster Prevention Research Institute* **34/2**(304), 19–54.

Orlob, G. T., and Selna, L. G. (1970). Temperature variations in deep reservoirs. *J. Hydr. Div. ASCE* **96**, 391–410.

Panofsky, H. A. (1983). Determination of stress from wind and temperature measurements. *Quart. J. Roy. Meteor. Soc.* **89**, 85–94.

Pao, H. P., and Kao, T. W. (1969). Sources and sinks at the axis of a viscous rotating fluid. *Phys. Fluids* **12**, 1536–1546.

Pao, H. P., and Kao, T. W. (1974). Dynamics of establishment of selective withdrawal of a stratified fluid from a line sink. Part I: Theory. *J. Fluid Mech.* **65**, 657–688.

Pao, H. P., and Shih, T. H. (1973). Selective withdrawal and blocking wave in rotating fluids. *J. Fluid Mech.* **57**, 459–480.

Parker, G. J., and Imberger, J. (1986). Differential mixed-layer deepening in lakes and reservoirs. In "Limnology in Australia" (P. de Deckker and W. D. Williams eds.), pp. 63–92. W. Junk Publishers, Melbourne, Victoria.

Patalas, K. (1960). Mieszanie wody jako czynnik okreslajacy intensywnose krazenia materii w roznych morfologicznie jeziorach okolic Wegorzewa. *Roczniki Nauk Rolniczych.* **B77**, 223–242.

Patalas, K. (1961). Wind und morphologiebedingte wasserwegunstypen als bestimmender Faktor fur die Intersitat des Stoffkreislaufes in nordpolnischen Seen. *Verhandlungen der Internationalen Vereinigung fur Theoretische und Angewandte Limnologie.* **14**, 59–64.

Patterson, J. C. (1984). Unsteady natural convection in a cavity with internal heating and cooling. *J. Fluid Mech.* **140**, 135–151.

Patterson, J. C., and Hamblin, P. F. (1988). Thermal simulation of a lake with winter ice cover. *Limnol. Oceanogr.* **33**, 323–338.

Patterson, J., and Imberger, J. (1980). Unsteady natural convection in a rectangular cavity. *J. Fluid Mech.* **100**, 65–86.

Patterson, J. C., and Imberger, J. (1989). Simulation of bubble plume destratification systems in reservoirs. *Aquatic Sciences* **51**, 3–18.

Patterson, J. C., Hamblin, P. F., and Imberger, J. (1984). Classification and dynamic simulation of the vertical density structure of lakes. *Limnol. Oceanogr.* **29**, 845–861.

Patterson, J. C., Allanson, B. R., and Ivey, G. N. (1985). A dissolved oxygen budget model for Lake Erie in summer. *Freshwater Biology* **15**, 683–694.

Paulson, C. A. (1970). The mathematical representation of wind speed and temperature profiles in the unstable atmospheric surface layer. *J. Appl. Met.* **9**, 857–861.

Peterson, E. W. (1969). Modification of mean flow and turbulent energy by a change in surface roughness under conditions of neutral stability. *Quart. J. Roy. Met. Soc.* **95**, 561–575.

Phillips, O. M., Shyu, J. H., and Salmun, H. (1986). An experiment on boundary mixing: Mean circulation and transport rates. *J. Fluid Mech.* **173**, 473–499.

Pollard, R. T. (1977). Observations and theories of Langmuir circulations and their role in near suface mixing. In "A Voyage of Discovery: George Deacon 70th Anniversary Volume" (M. Angel ed.), pp. 235–251. Pergamon Press, New York.

Pollard, R. T., Rhines, P. B., and Thompson, R. O. R. Y. (1973). The deepening of the wind mixed layer. *Geophys. Fluid Dyn.* **3**, 381–404.

Pond, S., Fissel, D. B., and Paulson, C. A. (1974). A note on bulk aerodynamic coefficients for sensible heat and moisture fluxes. *Boundary Layer Met.* **6**, 333–339.

Price, J. F., Mooers, C. N. K., and Van Leer, J. C. (1978). Observations and simulation of storm-induced mixed-layer deepening. *J. Phys. Oceanogr.* **8**, 582–599.

Raphael, J. M. (1962). Prediction of temperature in rivers and reservoirs. *J. Power Div.* **88**, 157–182.

Rayner, K. N. (1981). Diurnal energetics of a reservoir surface layer. Environmental Dynamics Rept. ED-80-005, Centre for Water Research, University of Western Australia.

Robarts, R. D., and Ward, P. R. B. (1978). Vertical diffusion and nutrients transport in a tropical lake (Lake McIlwaine, Rhodesia). *Hydrobiologia* **59**, 213–221.

Robarts, R. D., Ashton, P. J., Thornton, J. A., Taussig, H. J., and Sephton, L. M. (1982). Overturn in a hypertonic, warm, monomictic impoundment (Hartbeespoort Dam, South Africa). *Hydrobiologia* **97**, 209–224.

Roberts, P. J. W., and Matthews, P. R. (1987). Behavior of low buoyancy jets in a linearly stratified fluid. *J. Hydr. Res.* **25**, 503–519.

Rodgers, G. K. (1971). Field investigation of the thermal bar in Lake Ontario: Precision temperature measurements. In "Proc. 14th Conf. Great Lakes Res.," pp. 618–624. *Intl. Assoc. Great Lakes Res.* Ann Arbor.

Rodi, W. (1982). "HMT, Vol. 6: Turbulent buoyant jets and plumes." Pergamon Press, Sydney.

Rodi, W. (1987). Examples of calculation methods for flow and mixing in stratified fluids. *J. Geophys. Res.* **92**(C5), 5305–5328.

Rohr, J. J., Itsweire, E. C., Helland, K. N., and Van Atta, C. W. (1988). An investigation of the growth of turbulences in a uniform-mean-shear flow. *J. Fluid Mech.* **187**, 1–33.

Sadhuram, Y., Vethamony, P., Suryanarayana, A., Swamy, G. N., and Sastry, J. S. (1988). Heat energy exchange over a large water body under stable atmospheric conditions. *Boundary Layer Met.* **44**, 171–180.

Sahlberg, J. (1983). A hydrodynamical model for calculating the vertical temperature profile in lakes during cooling. *Nordic Hydrology* 239–254.

Sanderson, B., Perry, K., and Pederson, T. (1986). Vertical diffusion in meromictic Powell Lake, British Columbia. *J. Geophys. Res.* **91**, 7647–7655.

Sargent, R. E., and Jirka, G. H. (1982). A comparative study of density currents and density wedges (with application to intermediate field dynamics of ocean thermal energy conversion plants). Contract No. ET-78-S-02-4683, U.S. Dept. of Energy.

Savage, S. B., and Brimberg, J. (1975). Analysis of plunging phenomena in water resources. *J. Hydr. Res.* **13**, 187–204.

Schertzer, W. M., Saylor, J. H., Boyce, F. M., Robertson, D. G., and Rosa, F. (1987). Seasonal thermal cycle of Lake Erie. *J. Great Lakes Res.* **13**, 468–486.

Scott, C. F., and Imberger, J. (1988). Three-dimensional estuary circulation and classification. In "Proc. Intl. Conf. Phys. of Shallow Estuaries and Bays, (R. Cheng, ed.) Springer-Verlag. Pacific Grove, California. Lecture Notes on Coastal and Estuarine Studies.

Scott, J. T., Myer, G. E., Stewart, R., and Walther, E. G. (1969). On the mechanism of Langmuir circulations and their role in epilimnion mixing. *Limnol. Oceanogr.* **14**, 493–503.

Shay, T. J., and Gregg, M. C. (1984). Turbulence in an oceanic convective mixed layer. *Nature* **310**, 282–285.

Shay, T. J., and Gregg, M. C. (1986). Convectively driven turbulent mixing in the upper ocean. *J. Phys. Oceanogr.* **16**, 1777–1798.

Shay, T. J., and Imberger, J. (1988). Mixing in a strongly stratified low-energy lake. Environmental Dynamics Rept. ED-88-255, Centre for Water Research, University of Western Australia.

Sherman, F. S., Imberger, J., and Corcos, G. M. (1978). Turbulence and mixing in stably stratified waters. *Ann. Rev. Fluid Mech.* **10**, 267–288.

Shih, H. H., and Pao, H. S. (1971). Selective withdrawal in rotating fluids. *J. Fluid Mech.* **49**, 509–527.

Shih, T. H., Lumley, J. L., and Janicka, J. (1987). Second-order modeling of a variable-density mixing layer. *J. Fluid Mech.* **180**, 93–116.

Silvester, R. (1979). An experimental study of end wall effects on selective withdrawal of linearly stratified liquid from a reservoir. Ph.D. thesis, University of Western Australia.

Simpkins, P. G., and Dudderar, J. D. (1981). Convection in rectangular cavities with differentially heated end walls. *J. Fluid Mech.* **110**, 433–456.

Simpson, J. E. (1982). Gravity currents in the laboratory, atmosphere, and ocean. *Ann. Rev. Fluid Mech.* **14**, 213–234.

Smith, S. D. (1980). Wind stress and heat flux over the ocean in gale force winds. *J. Phys. Oceanogr.* **5**, 709–726.

Smith, S. D., and Anderson, R. J. (1984). Spectra of humidity, temperature and wind over the sea at Sable Island, Nova Scotia. *J. Geophys. Res.* **89**, 2029–2040.

Smith, S. D., and Banke, E. G. (1975). Variation of the sea surface drag coefficient with wind speed. *Quart. J. Roy. Met. Soc.* **101**, 665–673.

Smith, J., Pinkel, R., and Weller, R. A. (1987). Velocity structure in the mixed layer during MILDEX. *J. Phys. Oceanogr.* **17**, 425–439.

Smyth, N. F., and Holloway, P. E. (1988). Hydraulic jump and undular bore formation on a shelf break. *J. Phys. Oceanogr.* **18**, 947–962.

Sorbjan, Z. (1986). On similarity in the atmospheric boundary layer. *Boundary Layer Met.* **34**, 377–397.

Spain, J. D., Wernert, G. M., and Hubbard, D. W. (1976). The structure of the spring thermal bar in Lake Superior. *J. Great Lakes Res.* **2**, 296–306.

Spalding, D. B., and Svensson, U. (1977). The development and erosion of the thermocline. In "Proc. Seminar Intl. Centre for Heat and Mass Transfer," Hemisphere Publ. Corp. Washington D.C. Vol. 1, pp. 113–122.

Spigel, R. H. (1978). Wind mixing in lakes. Ph.D. thesis, University of California, Berkeley.

Spigel, R. H. (1980). Coupling of internal wave motion with entrainment at the density interface of a two-layer lake. *J. Phys. Oceanogr.* **10**, 144–155.

Spigel, R. H., and Farrant, B. (1984). Selective withdrawal through a point sink and pycnocline formation in a linearly stratified flow. *J. Hydr. Res.* **22**, 35–51.

Spigel, R. H., and Imberger, J. (1980). The classification of mixed layer dynamics in lakes of small to medium size. *J. Phys. Oceanogr.* **19**, 1104–1121.

Spigel, R. H., Imberger, J., and Rayner, K. N. (1986). Modeling the diurnal mixed layer. *Limnol. Oceanogr.* **31**, 533–556.

Stacey, M. W., and Bowen, A. J. (1988). The vertical structure of density and turbidity currents: Theory and observations. *J. Geophys. Res.* **93**, 3528–3542.

Stefan, H. G., and Ford, D. E. (1975). Temperature dynamics in dimictic lakes. *J. Hydr. Div. ASCE* **101**, 97–114.

Stefan, H. G., Cardoni, J. J., Schiebe, F. R., and Cooper, C. M. (1983). Model of light penetration in a turbid lake. *Water Resources Res.* **19**, 109–120.

Steinhorn, I., and Assaf, G. (1980). The physical structure of the Dead Sea water column: 1975–1977. In "Hypersaline Brines and Evaporitic Environments" (A. Nissenbaum ed.), pp. 145–153. Elsevier, Amsterdam.

Steinhorn, I., and Gat, J. R. (1983). The Dead Sea. *Scientific American.* **249**(4), 102–109.

Stewart, R. W. (1974). The air–sea momentum exchange. *Boundary Layer Met.* **6**, 151–167.

Stocker, T., and Hutter, K. (1986). One-dimensional models for topographic Rossby waves in elongated basins on the f-plane. *J. Fluid Mech.* **170**, 435–459.

Straskraba, M. (1980). The effects of physical variables on freshwater production: Analyses based on models. In "The Functioning of Freshwater Ecosystems" (E. D. Le Cren and R. H. McConnell eds.), pp. 13–84. Cambridge University Press, England.

Strub, P. T. (1983). The response of a small lake to atmospheric forcing during fall cooling. Ph.D. thesis, University of California.

Strub, P. T., and Powell, T. M. (1986). Wind-driven surface transport in stratified closed basins: Direct versus residual circulations. *J. Geophys. Res.* **91**, 8497–8508.

Strub, P. T., and Powell, T. M. (1987). The exchange coefficients for latent and sensible heat flux over lakes: Dependence upon atmospheric stability. *Boundary Layer Met.* **40**, 349–361.

Stull, R. (1976). The energetics of entrainment across a density interface. *J. Atmos. Sci.* **33**, 1260–1267.

Sturm, T. W. (1981). Laminar gravitational convection of heat in dead-end channels. *J. Fluid Mech.* **110**, 97–113.

Sun, T. Y., and Faeth, G. M. (1986). Structure of turbulent bubbly jets—1. Methods and centerline properties. *Intl. J. Multiphase Flow* **12**, 99–114.

Sundaram, T. R., and Rehm, R. G. (1973). The seasonal thermal structure of deep temperate lakes. *Tellus* **25**, 157–168.

Svensson, U. (1978). A mathematical model of the seasonal thermocline. Rept. No. 1002, Dept. of Water Resources Eng., University of Lund, Sweden.

Sverdrup, H. U. (1945). "Oceanography for Meteorologists." George Allen and Unwin, London.

Swinbank, W. C. (1963). Long-wave radiation from clear skies. *Quart. J. Roy. Met. Soc.* **89**, 339–348.

Tacke, K. H., Schubert, H. G., Weber, D. J., and Schwerdtfeger, K. (1985). Characteristics of round vertical gas bubble jets. *Metallurgical Trans. B.* **16**, 263–275.

Taylor, G. I. (1954). The dispersion of matter in turbulent flow through a pipe. *Proc. Roy. Soc. London Ser. A.* **223**, 446–468; (1960). *Sci. Pap.* **2**, 466–488.

Taylor, P. A. (1970). A model of airflow above changes in surface heat flux, temperature and roughness for neutral and unstable conditions. *Boundary Layer Met.* **1**, 474–497.

Taylor, M., and Aquise, E. (1984). A climatological energy budget of Lake Titicaca (Peru/Bolivia). *I.V.L.* **22**, 1246–1251.

Tennekes, H., and Lumley, J. L. (1972). "A First Course in Turbulence," MIT Press, Cambridge, Massachusetts.

Tennessee Valley Authority (1972). Heat and Mass Transfer Between a Water Surface and the Atmosphere. Lab. Rept. No. 14, Norris, Tennessee.

Thomas, J. H. (1975). A theory of steady wind-driven currents in shallow water with variable eddy viscosity. *J. Phys. Oceanogr.* **5**, 136–142.

Thompson, R. O. R. Y., and Imberger, J. (1980). Response of a numerical model of a stratified lake to wind stress. In "Proc. 2nd IAHR Symp. on Stratified Flows, Trondheim." (T. Carstens and T. McClimans eds.), pp. 562–570. IAHR.

Thorpe, S. A. (1969). Experiments on stability of stratified shear flow. *Radio Sci.* **4**, 1327–1331.

Thorpe, S. A. (1973). Turbulence in stably stratified fluids: A review of laboratory experiments. *Boundary Layer Met.* **5**, 95–119.

Thorpe, S. A. (1974). Near-resonant forcing in a shallow two-layer fluid: A model for the internal surge in Loch Ness? *J. Fluid Mech.* **63**, 509–527.

Thorpe, S. A. (1978). The near-surface ocean mixing layer in stable heating conditions. *J. Geophys. Res.* **8**, 2875–2885.

Thorpe, S. A. (1982). On the layers produced by rapidly oscillating a vertical grid in a uniformly stratified fluid. *J. Fluid Mech.* **124**, 391–409.

Thorpe, S. A. (1987). Transitional phenomena and the development of turbulence in stratified fluids: A review. *J. Geophys. Res.* **92**, 5231–5248.

Thorpe, S. A., and Hall, A. J. (1982). Observations of the thermal structure of Langmuir circulation. *J. Fluid Mech.* **114**, 237–250.

Tolland, H. G. (1977). Destratification/aeration in reservoirs. Tech. Rept. TR 50. Water Research Centre, England.

Trevisan, O. V., and Bejan, A. (1986). Convection driven by the non-uniform absorption of thermal radiation at the free surface of a stagnant pool. *Numerical Heat Transfer* **10**, 483–506.

Trustum, K. (1964). Rotating and stratified flow. *J. Fluid Mech.* **19**, 415–432.

Tsihrintzis, V. A., and Alavian, V. (1987). A model for predicting the properties of three-dimensional negatively buoyant starting plumes on a sloping surface. In "Proc. 3rd Intl. Symp. on Stratified Flows," California Institute of Technology, Pasadena. IAHR.

Tuck, E. O., and Vanden Broeck, J. M. (1984). A cusp-like free-surface flow due to a submerged source or sink. *J. Aust. Math. Soc. Ser. B* **25**, 443–450.

Tucker, W. A., and Green, A. W. (1977). A time-dependent model of the lake averaged, vertical temperature distribution of lakes. *Limnol. Oceanogr.* **22**, 687–699.

Turner, J. S. (1985). Multicomponent convection. *Ann. Rev. Fluid Mech.* **17**, 11–44.

Tzur, Y. (1973). One-dimensional diffusion equations for the vertical transport in an oscillating stratified lake of varying cross-section. *Tellus* **25**, 266–271.

UNESCO (1981). Tenth report of the joint panel on oceanographic tables and standards. Sidney, B.C., Canada.

U.S. Army Coastal Engineering Research Center (1977). "Shore Protection Manual," Vol. 1, 3.35–3.37. U.S. Govt. Printing Office, Washington.

U.S. Army Corps of Engineers (1982). CE-QUAL-R1: A numerical one-dimensional model of reservoir water quality. Instr. Rept. No. E-82-1, Waterways Experiment Station.

Vanden Broeck, J. M., and Keller, J. B. (1987). Free surface flow due to a sink. *J. Fluid Mech.* **175**, 109–117.

Venkatram, A. (1977). A model of internal boundary-layer development. *Boundary Layer Met.* **11**, 419–437.

Venkatram, A. (1986). An examination of methods to estimate the height of the coastal internal boundary layer. *Boundary Layer Met.* **36**, 149–156.

Ventz, D. (1973). Die Einzugsgebietsgrosse, ein Geofaktor fur den Trophiczustand stehender Gewasser. *Fortschnitte der Wasserchemie und ihrer Grenzgebiete* **14**, 105–118.

Veronis, G. (1967). Analogous behavior of homogeneous, rotating fluids and stratified, nonrotating fluids. *Tellus* **19**, 620–634.

Walesh, S. G., and Monkmeyer, P. L. (1973). Bottom withdrawal of viscous stratified fluid. *J. Hydr. Div. ASCE* **99**, 1401–1418.

Wallace, B. C., and Wilkinson, D. L. (1988). Run-up of internal waves on a gentle slope in a two-layered system. *J. Fluid Mech.* **191**, 419–442.

Ward, P. R. B. (1977). Diffusion in lake hypolimnia. In "Proc. 17th Congr. Intl. Assoc. Hydr. Res.," Vol. 2, pp. 103–110. Baden-Baden. IAHR.

Water Resources Engineers, Inc. (1969). Mathematical models for prediction of thermal energy changes in impoundments. U.S. E.P.A. Water Pollution Control Res. Series, 16130 EXT, Contract 14-22-422. U.S. Govt. Printing Office, Washington, D.C.

Wedderburn, E. M. (1906). The temperature of the freshwater lochs of Scotland with special reference to Loch Ness. *Trans. Roy. Soc. Edinburgh* **45**, 407–490.

Wedderburn, E. M. (1911). The temperature seiche: Observations in the Madusee. *Trans. Roy. Soc. Edinburgh* **47**, 619–642.

Wedderburn, E. M. (1912). Temperature observations in Loch Earn with a further contribution to the hydrodynamical theory of temperature seiches. *Trans. Roy. Soc. Edinburgh* **48**, 629–695.

Weinstock, J. (1984). Effect of gravity waves on turbulence decay in stratified fluids. *J. Fluid Mech.* **140**, 11–26.

Weiss, W., Lehn, H., Munnich, K. O., and Fischer, K. H. (1979). On the deep-water turnover of Lake Constance. *Arch. Hydrobiol.* **86**, 405–422.

Weller, R. A., Dean, J. P., Marra, J., Price, J. F., Francis, E. A., and Boardman, D. C.

(1985). Three dimensional flow in the upper ocean. *Science* **227**, 1552–1556.
Wetzel, R. G. (1975). "Limnology." W. B. Saunders, Philadelphia, Pennsylvania.
Wieringa, J. (1974). Comparison of three methods for determining strong wind stress over Lake Flevo. *Boundary Layer Met.* **7**, 3–19.
Wilkinson, D. L. (1972). Dynamics of contained oil slicks. *J. Hydr. Div. ASCE* **98**, 1013–1030.
Wilkinson, D. L. (1979). Two-dimensional bubble plumes. *J. Hydr. Div. ASCE* **105**, 139–154.
Wilkinson, D. L., and Wood, I. R. (1972). Some observations on the motion of the head of a density current. *J. Hydr. Res.* **10**, 305–324.
Willis, G. E., and Deardorff, J. W. (1974). A laboratory model of the unstable planetary boundary layer. *J. Atmos. Sci.* **31**, 1297–1307.
Witten, A. J., and Thomas, J. H. (1976). Steady-wind driven currents in a large lake with depth-dependent eddy viscosity. *J. Phys. Oceanogr.* **6**, 85–92.
Wood, I. R. (1978). Selective withdrawal from two-layer fluid. *J. Hydr. Div. ASCE* **104**, 1647–1659.
Wright, J. C. (1961). The limnology of Canyon Ferry Reservoir. IV. The estimation of primary production from physical limnological data. *Limnol. Oceanogr.* **6**, 330–337.
Wu, J. (1973). Wind-induced turbulent entrainment across a stable density interface. *J. Fluid Mech.* **61**, 275–287.
Wu, J. (1980). Wind-stress coefficients over sea surface near neutral conditions–A revisit. *J. Phys. Oceanogr.* **10**, 727–740.
Wu, J. (1985). Parameterization of wind-stress coefficients over water surfaces. *J. Geophys. Res.* **90**, 9069–9072.
Wucknitz, J. (1976). Determination of turbulent fluxes of momentum and sensible heat from fluctuation measurements and the structure of wind field over waves above the tropical Atlantic during ATEX. "*Meteor*" *Forschungsergab* **11**, 25–50.
Wuest, A. (1987). Ursprung und grosse von mischungsprozessen im hypolimnion naturlicher seen. Rept. No. 8350, Eidgenossischen Technischen Hoschschule, Zurich.
Wuest, A., Imboden, D. M., and Schurter, M. (1988). Origin and size of hypolimnic mixing in Urnersee, the southern basin of Vierwaldstattersee (Lake Lucerne). *Schweiz. Z. Hydrol.* **50**.
Wunderlich, W. D., and Elder, R. A. (1969). Selective withdrawal from density stratified reservoirs. Rept. No. 21, Water Resources Res. Advance.
Yih, C. S. (1965). "Dynamics of non-homogeneous fluids." Macmillan, New York.
Yih, C. S. (1980). "Stratified Flows." Academic Press, New York.
Yoshimura, S. (1936). A contribution to the knowledge of deep water temperature of Japanese lakes. *Japanese J. Astro. Geophys.* **13**, 61–120.
Zeman, O., and Tennekes, H. (1977). Parameterization of the turbulent energy budget at the top of the daytime atmospheric boundary layer. *J. Atmos. Sci.* **34**, 111–123.

Author Index

Numbers in italics refer to the pages on which the complete references are listed.

I

J

Subject Index

O

P

R